Progress in Nonlinear Differential Equations and Their Applications
Volume 50

Evolution Equations, Semigroups and Functional Analysis

In Memory of Brunello Terreni

Alfredo Lorenzi
Bernhard Ruf
Editors

Springer Basel AG

Editors' address:

Dipartimento di Matematica
Università degli Studi di Milano
20133 Milano
Italy

e-mail: lorenzi@mat.unimi.it
e-mail: ruf@mat.unimi.it

2000 Mathematics Subject Classification 34G, 35G, 35L, 49J, 65N

A CIP catalogue record for this book is available from the Library of Congress, Washington D.C., USA

Deutsche Bibliothek Cataloging-in-Publication Data

Evolution Equations, Semigroups and Functional Analysis : in memory of Brunello Terreni / Alfredo Lorenzi ; Bernhard Ruf ed. – Basel ; Boston ; Berlin : Birkhäuser, 2002
(Progress in nonlinear differential equations and their applications ; Vol. 50)
ISBN 978-3-0348-9480-7 ISBN 978-3-0348-8221-7 (eBook)
DOI 10.1007/978-3-0348-8221-7

Originally published by Birkhäuser Verlag Basel in 2002
Printed on acid-free paper produced of chlorine-free pulp. TCF ∞

ISBN 978-3-0348-9480-7

9 8 7 6 5 4 3 2 1 www.birkhauser-science.com

Contents

Preface

This volume is dedicated to the memory of Brunello Terreni who passed away very prematurely on March 4, 2000.

Brunello was our friend and colleague. We continue to appreciate his contribution to Mathematics and his dedication to his students. We will miss his human warmth and sense of humor.

On September 27 and 28, 2000, many of Brunello's friends and colleagues met in Milano, in a conference in his honor. In this volume we collected the contributions of many of the speakers in that meeting, and included also articles of other friends and collaborators of Brunello.

We would like to thank the Rettore of the Università di Milano, the Dipartimento di Matematica dell'Università di Milano and the Dipartimento di Scienze dei Materiali dell'Università di Milano-Bicocca who made this conference and the publication of this book possible.

We also thank all authors who, by contributing an article to this collection, helped to establish a lasting testament to the work and ideas of Brunello Terreni.

Milano, February 2002

Bernhard Ruf and Alfredo Lorenzi

Brunello Terreni †

Brunello Terreni (1953–2000)

Brunello Terreni was born in Pisa on January 25, 1953. He graduated in 1975 and in 1981 he became an assistant professor at the Department of Mathematics of the University of Pisa (and, strangely, for many years, Brunello was the only native from Pisa in the Department).

In 1988 he became an associate professor of mathematical analysis at the University of Milan and in 1994 a full professor. He spent a short period at the University of L'Aquila and then returned to Milan.

In the meantime, his struggle against illness had already started, during which he showed remarkable courage; he underwent numerous surgical operations both in Italy and in America. Unfortunately, none of these operations proved to be conclusive and while physically he started to deteriorate, he remained fully alert. He is survived by his wife Raimonda and his daughters Ester, Maria Pia and Noemi.

Those who knew him remember his great warmth, his enthusiasm, his capacity for self-examination, and his strong sense of humour. He had a very keen didactic sensitivity and he always endeavoured to convey to his students the feeling for an "alive" mathematics, but at the same time rigorous one. He considered the degree program in Material Science, where he was teaching, a privileged meeting point between himself as a mathematician and his colleagues, both physicists and chemists. It is in that course that his commitment to teaching had found space for an initiative – the mathematical laboratory – which he valued very much. The idea was to involve the students, who were divided into small groups, in solving mathematical problems in an active and constructive way. Under the guidance of young supervisors, students were encouraged to overcome their fear of mathematics which was connected to their previous experiences.

His relationship with his students, who perceived his dedication, had always been intense and rewarding. In his last year he sorely missed his contact with them. Brunello was also the initiator, with two other colleagues, of the "Lezioni Leonardesche", a series of extremely high level mathematical talks which the Department of Mathematics of the University of Milan has been organizing in the past few years in collaboration with the Polytechnic.

Brunello's 33 scientific papers can be subdivided into three main areas: the first being linear evolution equations of parabolic type in Banach spaces; the second being the theory of sums of linear operators; the third being linear-quadratic control for parabolic problems. In the first area, the aim is to generalize and unify all types of known results concerning the existence and uniqueness of solutions for this type of equations.

In collaboration with Paolo Acquistapace, a general hypothesis is formulated which allows to describe a framework for much of the existing literature. Equations of parabolic type verify in a natural way this hypothesis. Necessary and sufficient conditions for the existence of continuously differentiable solutions are furnished together with Hölder regularity. The Acquistapace-Terreni hypothesis has been adopted by many authors as the basis for further developments of the theory.

In the second area of research, the equation $(A + B)u = f$ is studied, where A and B are closed linear operators on a Banach space and suitable hypotheses on the spectra of A and B are introduced. As a result, a considerable generalization can be obtained of the theory of sums of operators which was developed in the seventies by Da Prato and Grisvard.

The third area analyses control problems on the boundary for linear parabolic equations with time-dependent coefficients and a quadratic cost functional. The problem is posed in an abstract way in Hilbert spaces; the main aim is to represent the optimal control in a feedback way and to solve the operator-valued Riccati equation which is associated to the problem. The theory obtained generalizes and sharpens all known results in the case of coefficients which are independent of time.

Brunello had also been very active in the organization of Erasmus projects; he and Bernhard Ruf coordinated a project with included nine universities in four different countries with over forty participating students.

A final aspect of Brunello's work was his commitment to making mathematics accessible to a wider public: he had collaborated with Franco Conti in Pisa ten years ago at the early stages of the exhibition "Oltre il compasso" and with Maria Dedò in Milan in the setting up of the exhibition on symmetries.

Sandro Levi

Remembering Brunello

In 1988 I was a member of our departmental committee that decided to give the position of Associate Professor in Mathematical Analysis to Brunello Terreni, one of the winners of a national competition.

The choice of a young expert in Semigroup Theory seemed natural as it enlarged our competence to such an important research sector in Mathematical Analysis.

At that time, my own knowledge of Semigroup Theory was rather limited. However, I was beginning to turn from Inverse Problems in Potential Theory and Identification Problems for specific model parabolic equations to Identification Problems for general integro-differential parabolic equations.

Although I was still suspicious about Abstract Differential Equations – they seemed such a rough way of attacking a problem! – I was forced, while dealing with complicated inverse problems, to find out some general procedure that could get rid off a lot of cumbersome explicit computations. Moreover, an attempt to make use of general Functional Analysis had been criticized by a referee who had proposed to come back to an explicit example, abandoning the general treatment.

Actually I had already solved, with the aid of Eugenio Sinestrari from the University of Rome *La Sapienza*, my first identification problem consisting in recovering a time dependent kernel, i.e. a *memory kernel*, in an abstract integro-differential first-order equation.

By a lucky coincidence, Brunello and Eugenio were part of the *same* research group headed by Giuseppe Da Prato from the *Scuola Normale Superiore* in Pisa. That group was analyzing, at that time, also direct problems for the diffusion of heat in rigid bodies with memory, i.e. the same problems I myself was interested in!

The arrival in Milan of Brunello was the beginning of an interesting and friendly exchange of experiences. My knowledge of Semigroup Theory came mainly from a few books and research articles. But what I needed to know was exactly what books usually do not explain.

Brunello turned out to be the right person at the right time. He was naturally inclined to listen to new problems – e.g. how to formulate integro-differential identification problems in an appropriate operator setting – as well as to talk about informal topics such as "why it is not possible to use Dunford integrals to construct solutions to operator second-order equations". I remember that for some time I was working on this topic, obtaining partial results. Never I heard from him a sentence such as "This is nonsense", "I have no time". He was always ready to listen to people asking for an opinion or advice from him. He supplied technical information and new ideas to face a problem in a friendly fashion. I vividly remember that we had no contrast in our discussions.

Two more questions I proposed to him:

1) how to translate an evolution problem in a cylindrical space domain into a suitable operator form and how to solve it;
2) why not to develop a theory for partial abstract equations.

Regarding the first question, I was deeply impressed by his skill in obtaining representations for semigroup generated, e.g., by sums of linear (closed) operators. Even now I feel that these kinds of investigations are basic for identification problems related to operator differential or integro-differential equations where we have to recover an unknown parameter depending on time.

I must confess that our distant discussions are fruitful even now: the old ideas finally take their own mathematical forms.

Another important aspect to be stressed was Brunello's curiosity about everything in mathematical research, his open-minded way of behaving himself, his generosity and his being independent and, at the same time, on good terms with everyone.

I cannot help remembering that the cycle of International Scientific Lessons called *Lezioni Leonardesche* was proposed by Brunello and connected by himself with Leonardo's name – an Italian genius known all over the world and related to Milan, where he lived and worked for some years.

Without Brunello's enthusiasm I think that such an important achievement would have not been carried out!

The transmission of culture and new ideas was a permanent necessity in his way of living and thinking. And this can be witnessed not only by Marco Fuhrmann, one of his P.H.D. students – now an associate professor at the Polytechnic of Milan –, but also by any of his degree students. I remember how he paid attention to motivate Mathematical Analysis even to first-year students. His major efforts were devoted to supplying material to deepen the comprehension and to using a language actually allowing his students to take possession of the concepts he was outlining. In this attempt he understood very well the power of modern P.C.'s and he explored this way with his students in Material Sciences.

Furthermore, I have still in mind how he helped me to propagandize a course of mine on Inverse Problems that was quite outside the tradition of our Department. I stress that actions of this kind are quite uncommon in our academic world.

Finally, his courage in facing his terrible and long illness makes Brunello an example for many of us.

My last words are devoted to a proposal: to set up a Scientific International Meeting, to be held biennially here in Milan and to be named after Brunello Terreni.

Alfredo Lorenzi

Progress in Nonlinear Differential Equations
and Their Applications, Vol. 50, 1–25

Γ-Convergence for Infinite Dimensional Optimal Control Problems

Paolo Acquistapace and Ariela Briani

Dedicated to our unfortunate friend Brunello

1. Introduction

Among the general goals in the various theories of variational convergence, an important one consists in singling out a notion of Γ-limit problem $(\mathcal{P})$ for a sequence of minimum problems $(\mathcal{P}_n)$ of the form

$$(\mathcal{P}_n) \qquad \min\{F_n(y,u) : (y,u) \in Y \times \mathcal{U}\}.$$

Loosely speaking, a good definition of Γ-limit problem should guarantee the following property:

"*If (y_n, u_n) is an optimal pair for problem $(\mathcal{P}_n)$, or simply a minimizing sequence, and if $(y_n, u_n) \to (y, u)$ in $Y \times \mathcal{U}$, then (y, u) is an optimal pair for the Γ-limit problem $(\mathcal{P})$.*"

The next Theorem indicates which are the relevant features that the Γ-limit problem should possess in order to satisfy this property (the theorem is proved in [7], Proposition 2.1).

Theorem 1.1. *Let Y and $\mathcal{U}$ be topological spaces and let $F_n : Y \times \mathcal{U} \to \overline{\mathbb{R}}$ be a sequence of functions. Let (y_n, u_n) be a minimum point for F_n, or simply a pair such that*

$$\lim_{n\to\infty} F_n(y_n, u_n) = \lim_{n\to\infty}\left[\inf_{Y\times\mathcal{U}} F_n\right].$$

Assume further that $(y_n, u_n) \to (y, u)$ in $Y \times \mathcal{U}$, and there exists

$$F(y,u) = \Gamma(\mathbb{N}, Y^-, \mathcal{U}^-)\lim_{n\to\infty} F_n(y,u). \tag{1.1}$$

Then (y, u) is a minimum point for F on $Y \times \mathcal{U}$, and

$$\lim_{n\to\infty}\left[\inf_{Y\times\mathcal{U}} F_n\right] = \min_{Y\times\mathcal{U}} F.$$

The definition of the Γ-limit (1.1) is given in Section 4 (Definition 4.1). The above theorem motivates the following definition of Γ-limit problem.

Definition 1.2. *When* (1.1) *is satisfied, we say that the problem*

$$(\mathcal{P}) \qquad \min\{F(y,u) : (y,u) \in Y \times \mathcal{U}\},$$

is the Γ-limit *of problems* $(\mathcal{P}_n)$.

See [7], [8] for the explicit calculation of Γ-limits in various situations.

In this paper we consider the following family of optimal control problems: for fixed (small) $\varepsilon > 0$, we look for

$$\min_{u \in L^2(0,T;V)} J_\varepsilon(u), \tag{1.2}$$

where

$$J_\varepsilon(u) = \int_0^T \left(\|u(s,\cdot)\|_V^2 + \|y_\varepsilon(s,\cdot)\|_H^2\right) ds; \tag{1.3}$$

here V is the space $H_{00}^{1/2}(\Gamma_0, \mathbb{C}^N) \times H^{-1/2}(\Gamma_1, \mathbb{C}^N)$, and for each control $u \in L^2(0,T;V)$ the state function y_ε is the solution of the second order regular parabolic initial-boundary value problem

$$\begin{cases} \frac{\partial y}{\partial t}(t,x) - A(x,D)y(t,x) = 0 & \text{in } [0,T] \times \overline{\Omega} \\ \vartheta_\varepsilon(x)\frac{\partial y}{\partial \nu_A}(t,x) + (1-\vartheta_\varepsilon(x))y(t,x) = u(t,x) & \text{in } [0,T] \times \partial\Omega \\ y(0,x) = y_0(x) & \text{in } \overline{\Omega}, \end{cases} \tag{1.4}$$

where the scalar function ϑ_ε is defined as follows:

$$\vartheta_\varepsilon(x) = \begin{cases} 1-\varepsilon & \text{if } x \in \Gamma_1 \\ 1-\varepsilon - \frac{1-2\varepsilon}{\varepsilon}\, d(x,\Gamma_1) & \text{if } x \in \Gamma_0 \text{ and } d(x,\Gamma_1) < \varepsilon \\ \varepsilon & \text{if } x \in \Gamma_0 \text{ and } d(x,\Gamma_1) \geq \varepsilon. \end{cases} \tag{1.5}$$

Here Γ_0 and Γ_1 are disjoint smooth submanifolds of $\partial\Omega$, such that $\overline{\Gamma_0} \cup \overline{\Gamma_1} = \partial\Omega$.

Our main result is that the Γ-limit of this family as $\varepsilon \to 0^+$ is the following optimal control problem: look for

$$\min_{u \in L^2(0,T;U)} J(u) \tag{1.6}$$

where

$$J(u) = \int_0^T \left(\|u(s,\cdot)\|_U^2 + \|y(s,\cdot)\|_H^2\right) ds; \tag{1.7}$$

the space U is $H^{1/2}(\Gamma_0, \mathbb{C}^N) \times [H_{00}^{1/2}(\Gamma_1, \mathbb{C}^N)]^*$, and for each control $u = (u_0, u_1) \in L^2(0,T;U)$ the state function y is the solution of the second order parabolic mixed initial-boundary value problem

$$\begin{cases} \frac{\partial y}{\partial t}(t,x) - A(x,D)y(t,x) = 0 & \text{in } [0,T] \times \Omega \\ y(t,x) = u_0(t,x) & \text{in } [0,T] \times \Gamma_0 \\ \frac{\partial y}{\partial \nu_A}(t,x) = u_1(t,x) & \text{in } [0,T] \times \Gamma_1 \\ y(0,x) = y_0(x) & \text{in } \Omega. \end{cases} \tag{1.8}$$

We note that the space of controls for problem (1.6) is larger than that of problems (1.2), and in the framework of Γ-convergence we are forced to choose the smaller one: thus we can only approximate the solutions of (1.6) which belong to the smaller space, with a little loss of generality (compare with Remark 2.7 in Section 2).

A special case of the above situation occurs when $\Gamma_1 = \emptyset$, so that the approximating boundary condition has the form

$$\varepsilon \frac{\partial y}{\partial \nu_A}(t,x) + y(t,x) = u(t,x) \qquad \text{in } [0,T] \times \partial\Omega;$$

in this case the Γ-limit problem contains the Dirichlet boundary condition $y = u$ in $[0,T] \times \partial\Omega$, and the natural space of controls is just $L^2(0,T; L^2(\partial\Omega, \mathbb{C}^N))$ (see [6] for details).

This kind of result is interesting also from the numerical point of view. Indeed, in many programs that calculate numerically the solution of partial differential equations, the Neumann and Dirichlet conditions are treated by considering a unique boundary conditions of type

$$a\,\frac{\partial y}{\partial \nu_A} + by = u \qquad \text{in } [0,T] \times \partial\Omega,$$

where the biggest constant appears near the condition one wants to consider (an example is the program [18]).

Let us shortly describe the content of the following sections. In order to apply the general result given in [8], we need to find a representation formula for the solutions of equations (1.4) and (1.8). Therefore, in Section 2 we analyze the mixed problem and we prove that the solution of (1.8) can be written as

$$y(t,\cdot) = e^{tA} y_0(\cdot) - \int_0^t A e^{(t-s)A} G u(s,\cdot)\, ds, \tag{1.9}$$

where $A{:}D(A) \to L^2(\Omega, \mathbb{C}^N)$ and $G{:}U \to D([-A]^\vartheta)$ (for $0 \le \vartheta < 1/4$) are suitable operators. Next, in Section 3 we prove the same kind of formula for the solution of (1.4), i.e.

$$y_\varepsilon(t,\cdot) = e^{tA_\varepsilon} y_0(\cdot) + \int_0^t A_\varepsilon e^{(t-s)A_\varepsilon} G_\varepsilon u(s,\cdot)\, ds, \tag{1.10}$$

where now A_ε: $D(A_\varepsilon) \to L^2(\Omega, \mathbb{C}^N)$ and $G_\varepsilon{:}V \to D([-A_\varepsilon]^\vartheta)$ (for $0 \le \vartheta < 1/2$). In Section 4 we prove our Γ-convergence result.

Finally we set some notations.

If $T > 0$ and X is a Banach space, we will use the standard spaces $L^p(0,T;X)$, $1 \le p \le \infty$, and $C^0([0,T];X)$, with their usual norms. As a rule, the space X will be a Sobolev space $H^m(A)$ of functions defined on some subset $A \subseteq \mathbb{R}^N$, with values in $\mathbb{R}^N$ or $\mathbb{C}^N$. We will simply write $H^m(A)$ instead of $H^m(A;\mathbb{R}^N)$ or $H^m(A;\mathbb{C}^N)$; here the number $m > 0$ may be integer or not. In particular, we denote by $H_0^m(A)$ the closure of the space $C_0^\infty(A)$ in $H^m(A)$, where $C_0^\infty(A)$ is the set of infinitely differentiable functions with compact support contained in A; the set A may be an open set of $\mathbb{R}^N$, or a submanifold of the boundary of a smooth open set $\Omega \subset \mathbb{R}^N$.

In the latter case, we also define $H_{00}^{1/2}(A)$ as the Hilbert space of functions belonging to $H^{1/2}(A)$ whose trivial extension to $\partial\Omega$ is an element of $H^{1/2}(\partial\Omega)$. We recall that $H_{00}^{1/2}(A)$ is a proper, closed subspace of $H^{1/2}(A)$.

We denote by $(\cdot,\cdot)$ and $|\cdot|$ the scalar product and the norm in $\mathbb{C}^N$, whereas $\langle\cdot,\cdot\rangle_X$ denotes the duality pairing between a space X and its dual space X^*. In our estimates we will write C for a generic constant possibly varying from line to line.

2. The mixed problem

We start with describing in a more precise way the optimal control problem we are going to approximate.

Let Ω be an open set of $\mathbb{R}^N$ with $\partial\Omega = \Gamma_0 \cup \Gamma_1$. For fixed $T > 0$, we consider the state space $L^2(0,T;H)$ and the control space $L^2(0,T;U)$, where $H := L^2(\Omega)$ and $U := H^{1/2}(\Gamma_0) \times [H_{00}^{1/2}(\Gamma_1)]^*$. We look for

$$\min_{u\in L^2(0,T;U)} J(u) \tag{2.1}$$

where

$$J(u) = \int_0^T \left(\|u(s,\cdot)\|_U^2 + \|y(s,\cdot)\|_H^2\right) ds \tag{2.2}$$

and for each control $u = (u_0, u_1) \in L^2(0,T;U)$ the state function y is the solution of the parabolic mixed initial-boundary value problem

$$\begin{cases} \frac{\partial y}{\partial t}(t,x) - A(x,D)y(t,x) = 0 & \text{in } [0,T]\times\Omega \\ y(t,x) = u_0(t,x) & \text{in } [0,T]\times\Gamma_0 \\ \frac{\partial y}{\partial \nu_A}(t,x) = u_1(t,x) & \text{in } [0,T]\times\Gamma_1 \\ y(0,x) = y_0(x) & \text{in } \Omega. \end{cases} \tag{2.3}$$

Here

$$A(x,D) = \sum_{i,j=1}^n D_i(a_{ij}(x)\cdot D_j)\ , \qquad \frac{\partial}{\partial\nu_A} = \sum_{i,j=1}^n \nu_i(x)a_{ij}(x)\cdot D_j\ ,$$

and $\nu(x)$ is the unit outward normal vector at $x \in \partial\Omega$; Γ_0 and Γ_1 are suitable subsets of $\partial\Omega$. We interpret this problem in the variational sense: namely, we set $H^1_{\Gamma_0}(\Omega) = \{u \in H^1(\Omega) : u = 0 \text{ on } \Gamma_0\}$, and introduce the bilinear form

$$a(u,v) = \sum_{i,j=1}^n \int_\Omega (a_{ij}(x)\cdot D_j u, D_i v)\, dx, \qquad u,v \in H^1_{\Gamma_0}(\Omega); \tag{2.4}$$

finally we fix a function $g_0 \in L^2(0,T;H^1(\Omega))$ such that $g_0|_{[0,T]\times\Gamma_0} = u_0$. Then, following [14], the variational version of problem (1.8) consists in looking for a function y such that $v := y - g_0 \in L^2(0,T;H^1_{\Gamma_0}(\Omega))$ and

$$\int_0^T [-\langle v(t), \varphi'(t)\rangle_{L^2(\Omega)} + a(v(t),\varphi(t))]\, dt = \langle y_0, \varphi(0)\rangle_{L^2(\Omega)} + F(\varphi) \tag{2.5}$$

for all test functions φ such that

$$\varphi \in L^2(0,T;H^1_{\Gamma_0}(\Omega)) \cap H^1(0,T;L^2(\Omega)), \qquad \varphi(T) = 0, \tag{2.6}$$

where the functional F is defined, for all functions φ of the above type, by

$$F(\varphi) = \int_0^T \langle g_0(t), \varphi'(t)\rangle_{L^2(\Omega)}\, dt - a(g_0,\varphi) + \langle u_1, \varphi|_{\Gamma_1}\rangle_{H^{1/2}_{00}(\Gamma_1)}. \tag{2.7}$$

Of course, any solution of system (1.8) solves equation (2.5) too, and any sufficiently smooth solution of equation (2.5) is also a solution of system (1.8).

We list now our basic assumptions.

(H0) Ω is a bounded connected open set of $\mathbb{R}^N$ with boundary $\partial\Omega$ of class $C^{1,1}$.
(H1) $y_0 \in L^2(\Omega)$.
(H2) $a_{ij} \in L^\infty(\Omega;\mathbb{C}^{N^2})$, with $M = \sum_{i,j=1}^n |a_{ij}(\cdot)|_{L^\infty(\Omega;\mathbb{C}^{N^2})}$, and there exists $\nu > 0$ such that

$$Re \sum_{i,j=1}^n (a_{ij}(x)\cdot \eta_j, \eta_i) \geq \nu \sum_{i=1}^n |\eta_i|^2$$

for all $\eta_1, \ldots, \eta_n \in \mathbb{C}^N$, and for a.a. $x \in \Omega$.
(H3) Γ_0 and Γ_1 are $(n-1)$-manifolds of class C^2, such that $\Gamma_0 \subset \partial\Omega$, $\Gamma_1 = \partial\Omega \setminus \overline{\Gamma_0}$ and $\overline{\Gamma_0} \cap \overline{\Gamma_1} \neq \emptyset$.

For the solution of system (1.8) in the variational form (2.5) we have the following existence and uniqueness result.

Theorem 2.1. *We assume* (H0), (H1), (H2), (H3). *If* $u = (u_0, u_1) \in L^2(0,T;U)$, *then equation* (2.5) *has a unique solution* y *in the class* $L^2(0,T;H^1(\Omega))$; *moreover* $y \in C^0([0,T];L^2(\Omega)) \cap H^{1/2}(0,T;L^2(\Omega))$ *and*

$$\begin{aligned}\|y\|_{C^0([0,T];L^2(\Omega))} &+ \|y\|_{L^2(0,T;H^1(\Omega))} + \|y\|_{H^{1/2}(0,T;L^2(\Omega))}\\ &\leq C\Big[\|y_0\|_{L^2(\Omega)} + \|u_0\|_{L^2(0,T;H^{1/2}(\Gamma_0))} + \|u_1\|_{L^2(0,T;[H^{1/2}_{00}(\Gamma_1)]^*)}\Big].\end{aligned}$$

Proof. For each $\lambda > 0$ the bilinear form $a(u,v) + \lambda\langle u, v\rangle_{L^2(\Omega)}$ is coercive on the space $H^1(\Omega)$; hence the same holds for the space $H^1_{\Gamma_0}(\Omega)$. Thus the result follows by Theorems 1.1 and 2.2, Chapter IV, in [14]. □

In order to obtain a representation for the solution of equation (2.5), we define the following operators A and G:

$$\begin{cases} D(A) := \{v \in H^1(\Omega) : & A(\cdot, D)v \in L^2(\Omega), \\ & v|_{\Gamma_0} = 0,\ \frac{\partial v}{\partial \nu_A}|_{\Gamma_1} = 0\} \\ Av := A(\cdot, D)v, \end{cases} \tag{2.8}$$

$$v := G(u_0, u_1) \qquad \Longleftrightarrow \qquad \begin{cases} A(\cdot, D)v &= 0 & \text{in } \Omega \\ v &= u_0 & \text{on } \Gamma_0 \\ \frac{\partial v}{\partial \nu_A} &= u_1 & \text{on } \Gamma_1. \end{cases} \tag{2.9}$$

We prove now that A is the generator of an analytic semigroup in $L^2(\Omega)$.

Proposition 2.2. *Under assumptions* (H0), (H2), (H3), *let the operator A be defined by* (2.8). *Then A is densely defined and generates an analytic semigroup in $H = L^2(\Omega)$. Moreover for any $\delta > 0$ we have the estimate*

$$\|R(\lambda, A)f\|_H \le \frac{C(\delta)}{|\lambda|} \|f\|_H \qquad \forall \lambda \in S_{\vartheta_0}, \tag{2.10}$$

where $S_{\vartheta_0} := \{\lambda \in \mathbb{C} : |\arg(\lambda)| \le \vartheta_0\}$, and $\vartheta_0 = \pi - \arctan \delta$.

Proof. Obviously, $C_0^\infty(\Omega) \subset D(A)$ so that $D(A)$ is dense in H. Next, we show that the resolvent set $\rho(A)$ contains the positive real half-line: indeed, by the coerciveness in $H^1(\Omega)$ of the form $a(u, \varphi) + \lambda\langle u, \varphi\rangle_H$ for all $\lambda > 0$, we get that for all $f \in H$ the problem

$$\begin{cases} u \in H^1_{\Gamma_0}(\Omega) \\ a(u, \varphi) + \lambda\langle u, \varphi\rangle_H = \langle f, \varphi\rangle_H \qquad \forall \varphi \in H^1_{\Gamma_0}(\Omega) \end{cases} \tag{2.11}$$

is uniquely solvable. Choosing $\varphi \in H_0^1(\Omega)$ we easily find $A(\cdot, D)u = f - \lambda u \in H$, i.e. $u \in D(A)$ and $\lambda u - Au = f$. This shows in particular that the solution u of (2.11) solves the problem

$$\begin{cases} \lambda u - A(\cdot, D)u &= f & \text{in } \Omega \\ u &= 0 & \text{on } \Gamma_0 \\ \frac{\partial u}{\partial \nu_A} &= 0 & \text{on } \Gamma_1. \end{cases} \tag{2.12}$$

Thus we only need to prove (2.10). Fix $\delta > 0$ and take $\lambda \in S_{\vartheta_0}$, with $\vartheta_0 = \pi - \arctan \delta$. Let $v \in D(A)$ and set $f = \lambda v - Av$. Multiplying this equation by v in H and integrating by parts we get

$$\int_\Omega \lambda |v|^2 \, dx + \int_\Omega \sum_{i,j=1}^n (a_{ij}(x) \cdot D_j v, D_i v) \, dx = \int_\Omega (f, v) \, dx. \tag{2.13}$$

By taking the real part and recalling hypothesis (H2) we obtain

$$Re\lambda \int_\Omega |v|^2 \, dx + \nu \int_\Omega |Dv|^2 \, dx \le \|f\|_H \|v\|_H \, ;$$

by taking the imaginary part we get

$$|Im\lambda| \int_\Omega |v|^2\, dx \le \|f\|_H\, \|v\|_H + M \int_\Omega |Dv|^2\, dx.$$

Now if $|Re\lambda| \le \delta |Im\lambda|$, then

$$\begin{aligned} |\lambda| \textstyle\int_\Omega |v|^2\, dx &\le \sqrt{1+\delta^2}\, |Im\lambda| \textstyle\int_\Omega |v|^2\, dx \\ &\le \sqrt{1+\delta^2}\, (1+\tfrac{M}{\nu})\, \|f\|_H \|v\|_H\,, \end{aligned}$$

whereas if $Re\lambda > 0$ and $\delta |Im\lambda| < Re\lambda$ we have

$$\begin{aligned} |\lambda| \textstyle\int_\Omega |v|^2\, dx &\le \sqrt{1+\tfrac{1}{\delta^2}}\, Re\lambda \textstyle\int_\Omega |v|^2\, dx \\ &\le \sqrt{1+\tfrac{1}{\delta^2}}\, \|f\|_H \|v\|_H\,. \end{aligned}$$

These two estimates show that for any $\delta > 0$ we have

$$|\lambda| \|v\|_H \le \sqrt{1+\delta^2} \left[\left(1+\frac{M}{\nu}\right) + \frac{1}{\delta} \right] \|f\|_H \quad \forall \lambda \in S_{\vartheta_0}\,. \tag{2.14}$$

Note that the smaller is δ the larger is the sector S_{ϑ_0} but also the larger is the constant in the estimate.

As $\rho(A) \supseteq]0,\infty[$, by estimate (2.14) and standard arguments we deduce that $\rho(A)$ also contains the right half-plane; hence Proposition 2.1.11 in [16] implies that A is sectorial. Therefore A is the infinitesimal generator of an analytic semigroup; moreover, by (2.14) we immediately deduce (2.10). This completes the proof. □

The adjoint operator of A is defined as follows:

$$\begin{cases} D(A^*) := \{ v \in H^1(\Omega) : \ \overline{A(\cdot, D)} v \in L^2(\Omega), \\ \qquad\qquad\qquad\qquad v|_{\Gamma_0} = 0,\ \frac{\partial v}{\partial \nu_{A^*}}|_{\Gamma_1} = 0 \} \\ A^* v := \overline{A(\cdot, D)} v, \end{cases} \tag{2.15}$$

where $\overline{A(x,D)} v := \sum_{i,j=1}^n D_i [\overline{a_{ij}(x)^t} \cdot D_j v]$ and $\overline{a_{ij}(x)^t}$ is the matrix whose elements are the conjugates of the elements of the transposed $a_{ij}(x)^t$ of $a_{ij}(x)$. Consequently it is clear that the following statement holds:

Proposition 2.3. *Under assumptions* (H0), (H2), (H3), *let the operator A^* be given by* (2.15). *Then A^* is densely defined and generates an analytic semigroup in $H = L^2(\Omega)$. Moreover for any $\delta > 0$ we have the estimate*

$$\|R(\lambda, A^*) f\|_H \le \frac{C(\delta)}{|\lambda|} \|f\|_H \qquad \forall \lambda \in S_{\vartheta_0}. \tag{2.16}$$

□

Concerning the operator G, we have the following result:

Proposition 2.4. *Let* (H0), (H1), (H2), (H3) *be fulfilled. If A is defined in (2.8) then the operator G, given by* (2.9), *is well defined and continuous from $U = H^{1/2}(\Gamma_0) \times [H_{00}^{1/2}(\Gamma_1)]^*$ into $D([-A]^\vartheta)$ for $0 \le \vartheta < \frac{1}{4}$.*

Proof. Let $(u_0, u_1) \in U$ and fix a lifting $g_0 \in H^1(\Omega)$ of the datum u_0, with $\|g_0\|_{H^1(\Omega)} \le C\|u_0\|_{H^{1/2}(\Gamma_0)}$. Then the problem

$$\begin{cases} u - g_0 \in H^1_{\Gamma_0}(\Omega) \\ a(u,\varphi) = \langle u_1, \varphi\rangle_{H^{1/2}_{00}(\Gamma_1)} \qquad \forall \varphi \in H^1_{\Gamma_0}(\Omega) \end{cases}$$

has a unique solution $u \in H^1(\Omega)$. Hence $u = G(u_0, u_1)$ by definition of G, and, in particular, we have $G(u_0, u_1) \in H^{2\vartheta}(\Omega)$ for all $\vartheta \in [0, \frac{1}{2}]$.

On the other hand, by Theorem 3.1 of [5] we know that $D([-A]^{\vartheta})$ coincides with $H^{2\vartheta}(\Omega)$ if and only if $\vartheta \in [0, \frac{1}{4}[$. This completes the proof. □

We are finally ready to prove the representation formula for the solution of system (1.8) in the variational form (2.5).

Theorem 2.5. *Under assumptions* (H0), (H1), (H2), (H3), *let* $u = (u_0, u_1)$ *be an element of* $L^2(0,T;U)$, *where* $U = H^{1/2}(\Gamma_0) \times [H^{1/2}_{00}(\Gamma_1)]^*$. *Then the solution of* (2.5) *is given for each* $t \in [0,T]$ *by*

$$y(t,\cdot) = e^{tA}y_0(\cdot) - \int_0^t Ae^{(t-s)A}Gu(s,\cdot)\, ds. \tag{2.17}$$

Proof. By Theorem 2.1 we know that equation (2.5) has a unique solution $y \in C^0([0,T];H) \cap H^{1/2}(0,T;H) \cap L^2(0,T;H^1(\Omega))$, where $H = L^2(\Omega)$. Set now

$$v(t) = e^{tA}y_0 - \int_0^t Ae^{(t-s)A}Gu(s)\, ds, \qquad t \in [0,T]; \tag{2.18}$$

by the standard properties of analytic semigroups and by Proposition 2.4 it is easily seen that $v \in C^0([0,T];H) \cap L^2(0,T;H^{2\vartheta}(\Omega))$ for any $\vartheta \in [0, \frac{1}{4}[$. We have to show that $y \equiv v$.

We suppose first that

$$\begin{cases} y_0 \in H^2(\Omega), \quad u = (u_0, u_1) \in C^\infty([0,T],U), \\ y_0 - Gu(0) \in D(A). \end{cases} \tag{2.19}$$

Under assumption (2.19) we can integrate by parts in (2.18):

$$\begin{aligned} v(t) &= e^{tA}y_0 + \left[e^{(t-s)A}Gu(s)\right]_0^t - \textstyle\int_0^t e^{(t-s)A}Gu'(s)\, ds = \\ &= e^{tA}y_0 + Gu(t) - e^{tA}Gu(0) - \textstyle\int_0^t e^{(t-s)A}Gu'(s)\, ds. \end{aligned} \tag{2.20}$$

Hence we can compute

$$v'(t) = Ae^{tA}[y_0 - Gu(0)] - \int_0^t Ae^{(t-s)A}Gu'(s)\, ds \qquad \forall t \in [0,T]; \tag{2.21}$$

thus we see that

$$v(t) - Gu(t) \in D(A) \qquad \forall t \in [0,T], \tag{2.22}$$

and in addition, by (2.8) and (2.9),

$$v'(t) = A[v(t) - Gu(t)] = A(\cdot, D)v(t,\cdot) \qquad \forall t \in [0,T]. \tag{2.23}$$

In particular, $v' \in C^0([0,T];H)$. Moreover, $v(0) = y_0$ and, by (2.22), $v(t)$ behaves at $\partial\Omega$ just like $Gu(t)$, i.e.

$$v(t)|_{\Gamma_0} = u_0(t) \in H^{1/2}(\Gamma_0), \qquad \frac{\partial v(t)}{\partial \nu_A}|_{\Gamma_1} = u_1(t) \in [H_{00}^{1/2}(\Gamma_1)]^*.$$

This shows that v solves problem (1.8). In particular, v solves equation (2.5) too; by uniqueness, this implies $v \equiv y$, provided (2.19) holds.

Consider now the general case, i.e. $y_0 \in H$ and $u = (u_0, u_1) \in L^2(0,T;U)$; let y be the solution of equation (2.5) and let v be the function (2.18). There exist sequences $\{y_{0,n}\}$ and $\{u_n\}$, satisfying assumption (2.19) for all $n \in \mathbb{N}$, such that $y_{0,n} \to y_0$ in H and $u_n \to u$ in $L^2(0,T;U)$. Denoting by y_n the corresponding solution of (2.5) and by v_n the corresponding function (2.18), we have by the above argument $v_n \equiv y_n$ for all $n \in \mathbb{N}$. But since the $C^0([0,T];H)$-norm of both y_n and v_n depends continuosly on the H-norm of $y_{0,n}$ and on the $L^2(0,T;U)$-norm of g_n, as $n \to \infty$ we immediately deduce that $v \equiv y$. This proves the result. □

After the above preparations, we see that the control problem (1.6) fits in the abstract setting described in [12]: thus, by the results of [12] (see also Theorem 3.14 in [1], and Theorem 8.2 in [3]) we can characterize the optimal control through a feedback formula involving the Riccati operator. The synthesis of the optimal control problem is summarized in the following statement.

Theorem 2.6. *Let* (H0), (H1), (H2), (H3) *be fulfilled. Then:*

(i) *There exists a unique optimal pair* $(\hat{u}, \hat{y}) \in L^2(0,T;U) \times L^2(0,T;H))$ *for problem* (1.6), *where* $H = L^2(\Omega)$ *and* $U = H^{1/2}(\Gamma_0) \times [H_{00}^{1/2}(\Gamma_1)]^*$.

(ii) *The Riccati equation in integral form, i.e.*

$$P(t) = \int_t^T e^{(s-t)A^*} \left[I - P(s)AGG^*A^*P(s)\right] e^{(s-t)A}\, ds,$$

has a unique solution $P \in C^1([0,T[;\mathcal{L}(H)) \cap C^0([0,T];\mathcal{L}(H))$.
Moreover, $P(t) = P(t)^* \geq 0$ *and* $P(t) \in D([-A^*]^{1-\vartheta})$ *for each* $\vartheta \in]0,1]$, *with* $\|[-A^*]^{1-\vartheta}P(t)\|_{\mathcal{L}(H)} \leq C(T-t)^{-(1-\vartheta)}$ *for all* $t \in [0,T[$; *in addition it holds*

$$J(\hat{u}) = \langle P(0)y_0, y_0\rangle_H \; .$$

(iii) *We have the feedback formula for* $\hat{u}$:

$$\hat{u}(t,\cdot) = G^*A^*P(t)\hat{y}(t,\cdot), \qquad t \in [0,T[.$$

(iv) *The optimal trajectory* $\hat{y}$ *is expressed by* $\hat{y}(t,\cdot) = \Phi(t,0)y_0(\cdot)$, *where* $\Phi(t,s)$ *is defined by the integral equation*

$$\Phi(t,s) = e^{(t-s)A} - \int_s^t Ae^{(t-r)A}GG^*A^*P(r)\Phi(r,s)\, dr, \quad t \in [s,T].$$

The expressions $P(s)AG$ *and* $G^*A^*P(s)$ *are shorter forms relative to the well-defined operator* $\left[-[(-A^*)^{1-\vartheta}P(s)]^*(-A)^{\vartheta}G\right]$ *and to its adjoint, with fixed* $\vartheta \in]0, \frac{1}{4}[$. □

Remark 2.7. As noted in the Introduction, in order to obtain our Γ-convergence result we will need to restrict somewhat the control space U, replacing $H^{1/2}(\Gamma_0) \times [H_{00}^{1/2}(\Gamma_1)]^*$ by its closed subspace $V = H_{00}^{1/2}(\Gamma_0) \times H^{-1/2}(\Gamma_1)$. Of course, all results proved in this section still hold if we use the restricted control space.

3. The approximating problems

Our goal is to approach problem (1.6) by a family of more regular problems which we now describe.

Fix $\varepsilon > 0$. In the same open set Ω as in the preceding section, consider again the state space $L^2(0,T;H)$, with $H = L^2(\Omega)$, and take as control space $L^2(0,T;V)$, with

$$V = \{u \in \mathcal{D}'(\partial\Omega) : u|_{\Gamma_0} \in H_{00}^{1/2}(\Gamma_0),\ u|_{\Gamma_1} \in H^{-1/2}(\Gamma_1)\}, \tag{3.1}$$

endowed with its natural norm

$$\|u\|_V = \|u|_{\Gamma_0}\|_{H_{00}^{1/2}(\Gamma_0)} + \|u|_{\Gamma_1}\|_{H^{-1/2}(\Gamma_1)} \,.$$

We remind that a distribution $u \in \mathcal{D}'(\partial\Omega)$ belongs to V if the restrictions of the functional u to the subspaces $\mathcal{D}(\Gamma_0)$ and $\mathcal{D}(\Gamma_1)$ verify respectively

$$\begin{aligned} |\langle u,\varphi\rangle| &\le C_0\|\varphi\|_{H^{1/2}(\Gamma_1)} \quad \forall \varphi \in \mathcal{D}(\Gamma_1), \\ |\langle u,\varphi\rangle| &\le C_1\|\varphi\|_{[H_{00}^{1/2}(\Gamma_0)]^*} \quad \forall \varphi \in \mathcal{D}(\Gamma_0); \end{aligned} \tag{3.2}$$

by density, the above equalities are true for all $\varphi \in H^{1/2}(\Gamma_1)$ and for all $\varphi \in [H_{00}^{1/2}(\Gamma_0)]^*$ respectively.

Lemma 3.1. *The space V, defined in* (3.1), *is isomorphic to $H_{00}^{1/2}(\Gamma_0) \times H^{-1/2}(\Gamma_1)$ by the map $j(u) = (u|_{\Gamma_0}, u|_{\Gamma_1})$.*

Proof. We just verify that the map j is onto, the other properties being quite easy. Fix $(u_0,u_1) \in H_{00}^{1/2}(\Gamma_0) \times H^{-1/2}(\Gamma_1)$, and set $u = U + V$, where U is the trivial extension of u_0 to $\partial\Omega$ and V is the element of $H^{-1/2}(\partial\Omega)$ defined by $V(\varphi) = u_1(\varphi)$ for all $\varphi \in H^{1/2}(\partial\Omega) \subset H^{1/2}(\Gamma_1)$. Then it is straightforward to check that $u|_{\Gamma_0} = u_0$ and $u|_{\Gamma_1} = u_1$, i.e. $j(u) = (u_0,u_1)$. □

We recall here our approximating control problem: in view of Lemma 3.1, we set

$$V = H_{00}^{1/2}(\Gamma_0) \times H^{-1/2}(\Gamma_1), \tag{3.3}$$

and we look for

$$\min_{u \in L^2(0,T;V)} J_\varepsilon(u), \tag{3.4}$$

where

$$J_\varepsilon(u) = \int_0^T \left(\|u(s,\cdot)\|_V^2 + \|y_\varepsilon(s,\cdot)\|_H^2\right) ds \tag{3.5}$$

and for each control $u \in L^2(0,T;V)$ the state function y_ε is the solution of the regular parabolic initial-boundary value problem

$$\begin{cases} \frac{\partial y}{\partial t}(t,x) - A(x,D)y(t,x) &= 0 & \text{in } [0,T]\times\overline{\Omega} \\ \vartheta_\varepsilon(x)\frac{\partial y}{\partial \nu_A}(t,x) + (1-\vartheta_\varepsilon(x))y(t,x) &= u(t,x) & \text{in } [0,T]\times\partial\Omega \\ y(0,x) &= y_0(x) & \text{in } \overline{\Omega}. \end{cases} \tag{3.6}$$

Here the scalar function ϑ_ε is defined as follows:

$$\vartheta_\varepsilon(x) = \begin{cases} 1-\varepsilon & \text{if } x\in\Gamma_1 \\ 1-\varepsilon-\frac{1-2\varepsilon}{\varepsilon}\,d(x,\Gamma_1) & \text{if } x\in\Gamma_0 \text{ and } d(x,\Gamma_1)<\varepsilon \\ \varepsilon & \text{if } x\in\Gamma_0 \text{ and } d(x,\Gamma_1)\geq\varepsilon. \end{cases} \tag{3.7}$$

The classical results on regular elliptic and parabolic problems guarantee that, at least for smooth data, a unique solution of the state equation (1.4) exists; however the standard estimates (see e.g. [4]) depend on ε, so that the solution might be unbounded in certain spaces with respect to ε as $\varepsilon \to 0$.

On the other hand, problem (3.6) has also a variational formulation: for fixed $u\in L^2(0,T;V)$, problem (3.6) consists in looking for a function $y\in L^2(0,T;H^1(\Omega))$ such that

$$\begin{aligned} \int_0^T \Big[-\langle y(t),\varphi'(t)\rangle_{L^2(\Omega)} &+ a(y(t),\varphi(t)) + \\ &+ \langle \tfrac{1-\vartheta_\varepsilon}{\vartheta_\varepsilon}\, y(t),\varphi(t)\rangle_{H^{1/2}(\partial\Omega)}\Big]\,dt = \\ &= \langle y_0,\varphi(0)\rangle_{L^2(\Omega)} + \int_0^T \langle u(t),\varphi(t)\rangle_{H^{1/2}(\partial\Omega)}\,dt \end{aligned} \tag{3.8}$$

for all test functions φ such that

$$\varphi \in L^2(0,T;H^1(\Omega)) \cap H^1(0,T;L^2(\Omega)), \qquad \varphi(T)=0. \tag{3.9}$$

Our first task is to give an existence and uniqueness result for the variational problem (3.8), with an estimate not depending on ε.

Theorem 3.2. *We assume* (H0), (H1), (H2), (H3). *If $u \in L^2(0,T;V)$, with V defined in* (3.3), *then equation* (3.8) *has a unique solution y_ε in the class*

$$C^0([0,T];L^2(\Omega)) \cap L^2(0,T;H^1(\Omega)) \cap H^{1/2}(0,T;L^2(\Omega)).$$

Moreover there exists $C \geq 0$, independent of ε, such that

$$\begin{aligned} &\|y_\varepsilon\|_{C^0([0,T];L^2(\Omega))} + \|y_\varepsilon\|_{L^2(0,T;H^1(\Omega))} + \|y_\varepsilon\|_{H^{1/2}(0,T;L^2(\Omega))} \\ &\leq C\Big[\|y_0\|_{L^2(\Omega)} + \|u\|_{L^2(0,T;H^{1/2}_{00}(\Gamma_0))} + \|u\|_{L^2(0,T;H^{-1/2}(\Gamma_1))}\Big]. \end{aligned}$$

Proof. The bilinear form $a(y,v) + \langle \frac{1-\vartheta_\varepsilon}{\vartheta_\varepsilon}\, y, v\rangle_{H^{1/2}(\partial\Omega)}$ is clearly weakly coercive on $H^1(\Omega)$; in fact it is even coercive, since the norm $\|u\|_{L^2(\partial\Omega)} + \|Du\|_{L^2(\Omega)}$ is equivalent to the usual norm of $H^1(\Omega)$, as shown e.g. by Lemma 4.4 in [2]. Hence the result follows by Theorems 1.1 and 2.2, Chapter IV in [14]. □

As we did for the case of the mixed problem, we want to find now a representation formula for the solution y_ε of problem (3.6). We start with defining the operators

$$\begin{cases} D(A_\varepsilon) := \left\{ v \in H^2(\Omega) : \ \vartheta_\varepsilon \frac{\partial v}{\partial \nu_A} + (1-\vartheta_\varepsilon)v = 0 \text{ on } \partial\Omega \right\} \\ A_\varepsilon v := A(\cdot, D)v, \end{cases} \tag{3.10}$$

and

$$w := G_\varepsilon h \qquad \Longleftrightarrow \qquad \begin{cases} A(\cdot, D)w = 0 & \text{in } \Omega \\ \vartheta_\varepsilon \frac{\partial w}{\partial \nu_A} + (1-\vartheta_\varepsilon)w = h & \text{on } \partial\Omega. \end{cases} \tag{3.11}$$

As (3.10) is a regular elliptic operator, it is clear that A_ε is the infinitesimal generator of an analytic semigroup; we show now that this property holds "uniformly" with respect to ε.

Proposition 3.3. *Let* (H0), (H1), (H2) *be fulfilled. Then the operator A_ε defined by* (3.10) *is densely defined and generates an analytic semigroup in $L^2(\Omega)$. Moreover for any $\delta > 0$ we have the estimate*

$$\|R(\lambda, A_\varepsilon)f\|_{L^2(\Omega)} \le \frac{C(\delta)}{|\lambda|} \|f\|_{L^2(\Omega)} \qquad \forall \lambda \in S_{\vartheta_0}, \tag{3.12}$$

where $S_{\vartheta_0} := \{\lambda \in \mathbb{C} : |\arg(\lambda)| \le \vartheta_0\}$, and $\vartheta_0 = \pi - \arctan\delta$. In particular, the constant $C(\delta)$ does not depend on ε.

Proof. We just need to prove (3.12).

Let $f \in L^2(\Omega)$ and set $v_\varepsilon = R(\lambda, A_\varepsilon)f$: then v_ε solves

$$\begin{cases} \lambda v_\varepsilon - A(\cdot, D)v_\varepsilon = f & \text{in } \Omega \\ \vartheta_\varepsilon \frac{\partial v_\varepsilon}{\partial \nu_A} + (1-\vartheta_\varepsilon)v_\varepsilon = 0 & \text{on } \partial\Omega. \end{cases} \tag{3.13}$$

Let us multiply by v_ε, integrate over Ω and use the boundary condition: we get

$$\begin{aligned} &\lambda \int_\Omega |v_\varepsilon|^2\, dx + a(v_\varepsilon, v_\varepsilon) \\ &\quad + \int_{\Gamma_0} \tfrac{\vartheta_\varepsilon}{1-\vartheta_\varepsilon} \left|\tfrac{\partial v_\varepsilon}{\partial \nu_A}\right|^2 d\sigma + \int_{\Gamma_1} \tfrac{1-\vartheta_\varepsilon}{\vartheta_\varepsilon} |v_\varepsilon|^2\, d\sigma = \int_\Omega |(f, v_\varepsilon)|\, dx. \end{aligned} \tag{3.14}$$

This estimate plays the role of (2.13) in the proof of Proposition 2.2: thus, just repeating the argument used in that proof, we get for any $\delta > 0$

$$\|v_\varepsilon\|_{L^2(\Omega)} \le \frac{C(\delta)}{|\lambda|} \|f\|_{L^2(\Omega)} \qquad \forall \lambda \in S_{\vartheta_0},$$

where $\vartheta_0 = \pi - \arctan\delta$. This proves the result. □

Remark 3.4. By (3.14) it follows in particular the useful estimate, which holds true for all $\lambda \in S_{\vartheta_0}$:

$$\|v_\varepsilon\|^2_{H^1(\Omega)} + \int_{\Gamma_0} \frac{\vartheta_\varepsilon}{1-\vartheta_\varepsilon} \left|\frac{\partial v_\varepsilon}{\partial \nu_A}\right|^2 d\sigma + \int_{\Gamma_1} \frac{1-\vartheta_\varepsilon}{\vartheta_\varepsilon} |v_\varepsilon|^2\, d\sigma \le \frac{C(\delta)}{|\lambda|} \|f\|^2_{L^2(\Omega)} .$$

The adjoint operator of A_ε is defined by:

$$\begin{cases} D(A_\varepsilon^*) := \left\{ v \in H^2(\Omega) : \ \vartheta_\varepsilon \frac{\partial v}{\partial \nu_{A^*}} + (1-\vartheta_\varepsilon) v = 0 \text{ on } \partial\Omega \right\} \\ A_\varepsilon^* v := \overline{A(\cdot, D)} v. \end{cases} \tag{3.15}$$

Hence it is clear that the following result, parallel to that of Proposition 3.3, holds:

Proposition 3.5. *Let* (H0), (H1), (H2) *be fulfilled and fix* $\varepsilon > 0$. *Then the operator* A_ε^* *given by* (3.15) *is densely defined and generates an analytic semigroup in* $L^2(\Omega)$. *Moreover for any* $\delta > 0$ *we have the estimate*

$$\|R(\lambda, A_\varepsilon^*) f\|_{L^2(\Omega)} \le \frac{C(\delta)}{|\lambda|} \|f\|_{L^2(\Omega)} \qquad \forall \lambda \in S_{\vartheta_0}\ , \tag{3.16}$$

where $S_{\vartheta_0} := \{\lambda \in \mathbb{C} : |\arg(\lambda)| \le \vartheta_0\}$, *and* $\vartheta_0 = \pi - \arctan\delta$. *In particular, the constant* $C(\delta)$ *does not depend on* ε. □

Concerning the operator G_ε, we have:

Proposition 3.6. *Let* (H0), (H1), (H2), (H3) *be fulfilled; fix* $\varepsilon > 0$ *and let* A_ε *be defined by* (3.10). *The operator* G_ε , *given by* (3.11), *is well defined from the space* V, *defined by* (3.3), *into the space* $D([-A_\varepsilon]^\vartheta)$ *for each* $\vartheta \in]0, \frac{1}{2}[$. *In addition, there exists* $C > 0$, *independent of* ε, *such that*

$$\begin{aligned} C(\varepsilon)^{-1} \|[-A_\varepsilon]^\vartheta G_\varepsilon h\|^2_{L^2(\Omega)} &\le \|D G_\varepsilon h\|^2_{L^2(\Omega)} \\ &\le C \left[\|h|_{\Gamma_0}\|^2_{H^{1/2}_{00}(\Gamma_0)} + \|h|_{\Gamma_1}\|^2_{H^{-1/2}(\Gamma_1)} \right] \qquad \forall h \in V. \end{aligned}$$

Proof. Let first $w = G_\varepsilon h$, where $h \in C^\infty(\partial\Omega)$ and h vanishes in a neighbourhood of $\overline{\Gamma_0} \cap \overline{\Gamma_1}$; then w is smooth and solves problem (3.11). We repeat the argument used in the proof of Proposition 3.3. Multiply by $w(x)$ the equation $A(x, D)w = 0$ and integrate by parts: using Hypothesis (H2) and the boundary condition, we get

$$\begin{aligned} &\nu \int_\Omega |Dw|^2\, dx + \int_{\Gamma_0} \tfrac{\vartheta_\varepsilon}{1-\vartheta_\varepsilon} \left|\tfrac{\partial w}{\partial \nu_A}\right|^2 d\sigma + \int_{\Gamma_1} \tfrac{1-\vartheta_\varepsilon}{\vartheta_\varepsilon} |w|^2\, d\sigma \\ &\qquad \le \left| \int_{\Gamma_0} (h, \tfrac{\partial w}{\partial \nu_A}) \tfrac{1}{1-\vartheta_\varepsilon}\, d\sigma \right| + \left| \int_{\Gamma_1} (h, w) \tfrac{1}{\vartheta_\varepsilon}\, d\sigma \right|. \end{aligned} \tag{3.17}$$

Let us estimate separately the addenda of the last term in (3.17). We start with the second one, which is easier to handle. As $\vartheta_\varepsilon(x) \equiv 1 - \varepsilon$ on Γ_1 we have

$$\begin{aligned} \left| \int_{\Gamma_1} (h, w) \tfrac{1}{\vartheta_\varepsilon}\, d\sigma \right| &= \tfrac{1}{1-\varepsilon} \left| \langle h, w \rangle_{H^{1/2}(\Gamma_1)} \right| \\ &\le \tfrac{1}{1-\varepsilon} \|h\|_{H^{-1/2}(\Gamma_1)} \|w\|_{H^{1/2}(\Gamma_1)} \\ &\le \tfrac{1}{1-\varepsilon} \|h\|_{H^{-1/2}(\Gamma_1)} \|w\|_{H^{1/2}(\partial\Omega)} \\ &\le \tfrac{\nu}{2} \|h\|^2_{H^{-1/2}(\Gamma_1)} + C \|w\|^2_{H^1(\Omega)}. \end{aligned} \tag{3.18}$$

Concerning the first term on the last member of (3.17), set

$$\Gamma_\varepsilon = \{x \in \Gamma_0 : d(x, \Gamma_1) \le \varepsilon\}, \tag{3.19}$$

and denote by $\overline{h}$ the function which coincides with h on Γ_0 and vanishes on Γ_1: then $\overline{h} \in H^{1/2}(\partial\Omega)$ and we can write

$$\begin{aligned}
&\left|\int_{\Gamma_0}(h, \tfrac{\partial w}{\partial \nu_A})\ \tfrac{1}{1-\vartheta_\varepsilon}\ d\sigma\right| \\
&\le \left|\int_{\Gamma_0}(h, \tfrac{\partial w}{\partial \nu_A})\ \tfrac{1}{1-\varepsilon}\ d\sigma\right| + \int_{\Gamma_\varepsilon} |(h, \tfrac{\partial w}{\partial \nu_A})|\ (\tfrac{1}{1-\vartheta_\varepsilon} - \tfrac{1}{1-\varepsilon})\ d\sigma \\
&\le \left|\int_{\partial\Omega}(\overline{h}, \tfrac{\partial w}{\partial \nu_A})\ \tfrac{1}{1-\varepsilon}\ d\sigma\right| + \int_{\Gamma_\varepsilon} |(h, \tfrac{\partial w}{\partial \nu_A})|\ \tfrac{1}{1-\vartheta_\varepsilon}\ d\sigma \\
&= I + II.
\end{aligned} \tag{3.20}$$

On one hand, we get for all $\eta > 0$

$$\begin{aligned}
I &= \tfrac{1}{1-\varepsilon}\ \left|\int_{\partial\Omega}(\overline{h}, \tfrac{\partial w}{\partial \nu_A})\ d\sigma\right| \\
&\le C\ \|\tfrac{\partial w}{\partial \nu_A}\|_{H^{-1/2}(\partial\Omega)}\ \|\overline{h}\|_{H^{1/2}(\partial\Omega)} \\
&\le \eta\ \|w\|^2_{H^1(\Omega)} + C(\eta)\ \|h|_{\Gamma_0}\|^2_{H^{1/2}_{00}(\Gamma_0)};
\end{aligned} \tag{3.21}$$

on the other hand, we have

$$\begin{aligned}
II &= \int_{\Gamma_\varepsilon} |(h, \tfrac{\partial w}{\partial \nu_A})|\ \tfrac{1}{1-\vartheta_\varepsilon}\ d\sigma \\
&\le \int_{\Gamma_\varepsilon} \tfrac{1}{2}\ \tfrac{\vartheta_\varepsilon}{1-\vartheta_\varepsilon}\ |\tfrac{\partial w}{\partial \nu_A}|^2\ d\sigma + \int_{\Gamma_\varepsilon} \tfrac{2}{\vartheta_\varepsilon(1-\vartheta_\varepsilon)}\ |h|^2\ d\sigma,
\end{aligned} \tag{3.22}$$

and the second addendum in the last member of (3.22) is estimated by

$$\begin{aligned}
&\int_{\Gamma_\varepsilon} \tfrac{2}{\vartheta_\varepsilon(1-\vartheta_\varepsilon)}\ |h|^2\ d\sigma \le C \int_{\Gamma_\varepsilon} \tfrac{1}{\varepsilon}\ |h|^2\ d\sigma \\
&\quad \le C \int_{\Gamma_0} |h|^2\ \tfrac{1}{d(x,\Gamma_1)}\ d\sigma \le C\|h|_{\Gamma_0}\|^2_{H^{1/2}_{00}(\Gamma_0)}.
\end{aligned} \tag{3.23}$$

By (3.17), (3.18), (3.20), (3.21), (3.22) and (3.23), we deduce for a sufficiently small η:

$$\begin{aligned}
&\int_\Omega |Dw|^2\ dx + \int_{\Gamma_0} \tfrac{\vartheta_\varepsilon}{1-\vartheta_\varepsilon}\ |\tfrac{\partial w}{\partial \nu_A}|^2\ d\sigma + \int_{\Gamma_1} \tfrac{1-\vartheta_\varepsilon}{\vartheta_\varepsilon}\ |w|^2\ d\sigma \\
&\qquad \le C\left[\|h|_{\Gamma_0}\|^2_{H^{1/2}_{00}(\Gamma_0)} + \|h|_{\Gamma_1}\|^2_{H^{-1/2}(\Gamma_1)}\right].
\end{aligned} \tag{3.24}$$

Now the general case, i.e. the case $h \in V$, follows by a density argument, since we may approximate h by a sequence $\{h_n\} \subset C^\infty(\partial\Omega)$ of functions vanishing in a neighbourhood of $\overline{\Gamma_0} \cap \overline{\Gamma_1}$. Thus estimate (3.24) holds for $w = G_\varepsilon h$ too; this means that the map G_ε is bounded, uniformly with respect to ε, from $H^{1/2}_{00}(\Gamma_0) \times H^{-1/2}(\Gamma_1)$ to $H^1(\Omega)$.

We invoke finally a result due to Fujiwara (Theorem 2 in [10]), according to which we have $D([-A_\varepsilon]^\vartheta) = H^{2\vartheta}(\Omega)$ for all $\vartheta \in [0, 3/4[$ and

$$\|f\|_{D([-A_\varepsilon]^\vartheta)} \le C(\varepsilon)\|f\|_{H^{2\vartheta}(\Omega)} \qquad \forall f \in H^{2\vartheta}(\Omega), \tag{3.25}$$

with $C(\varepsilon)$ possibly depending on ε. This estimate, together with (3.24), implies our result. □

Remark 3.7. In the proof of Proposition 3.6 we have proved in particular the following useful estimate for $w = G_\varepsilon h$:

$$\begin{aligned} &\int_{\Gamma_0} \frac{\vartheta_\varepsilon}{1-\vartheta_\varepsilon} \left|\frac{\partial w}{\partial \nu_A}\right|^2 d\sigma + \int_{\Gamma_1} \frac{1-\vartheta_\varepsilon}{\vartheta_\varepsilon} |w|^2 \, d\sigma \\ &\quad \leq C \left[\|h|_{\Gamma_0}\|^2_{H^{1/2}_{00}(\Gamma_0)} + \|h|_{\Gamma_1}\|^2_{H^{-1/2}(\Gamma_1)} \right]. \end{aligned} \tag{3.26}$$

Remark 3.8. An estimate similar to (3.26) is valid for the solution y_ε of the parabolic problem (3.8), namely

$$\begin{aligned} &\int_0^T \int_{\Gamma_0} \frac{\vartheta_\varepsilon}{1-\vartheta_\varepsilon} \left|\frac{\partial y_\varepsilon}{\partial \nu_A}\right|^2 d\sigma dt + \int_0^T \int_{\Gamma_1} \frac{1-\vartheta_\varepsilon}{\vartheta_\varepsilon} |y|^2 \, d\sigma dt \\ &\quad \leq C \int_0^T \left[\|u_0(t,\cdot)\|^2_{H^{1/2}(\Gamma_0)} + \|u_1(t,\cdot)\|^2_{H^{-1/2}(\Gamma_1)} \right] dt. \end{aligned} \tag{3.27}$$

The proof requires the same argument used in the proof of Proposition 3.6, and we can omit it.

We are finally ready to prove the representation formula for the solution of problem (1.4).

Theorem 3.9. *Assume* (H0), (H1), (H2), (H3). *If* $u \in L^2(0,T;V)$, *with* V *defined by* (3.3), *then the solution* y_ε *of* (1.4) *is given for each* $t \in [0,T]$ *by*

$$y_\varepsilon(t,\cdot) = e^{tA_\varepsilon} y_0(\cdot) + \int_0^t A_\varepsilon e^{(t-s)A_\varepsilon} G_\varepsilon u(s,\cdot)\, ds. \tag{3.28}$$

Proof. As (1.4) is a regular parabolic initial-boundary value problem, the result follows by adapting the proof of Proposition 2.13 in [1]; otherwise, one may repeat the same argument used in proving Theorem 2.5 above. We omit the details. □

Finally we state the result on the synthesis of the optimal control problem (1.2), whose proof is parallel to that of Theorem 2.6 (compare again with [12], [1], and [3]).

Theorem 3.10. *Let* (H0), (H1), (H2), (H3) *be fulfilled. Then:*

(i) *There exists a unique optimal pair* $(\hat{u}_\varepsilon, \hat{y}_\varepsilon) \in L^2(0,T;V) \times L^2(0,T;H))$ *for problem* (1.6), *where* $H = L^2(\Omega)$ *and* V *is defined in* (3.3).

(ii) *The Riccati equation in integral form, i.e.*

$$P_\varepsilon(t) = \int_t^T e^{(s-t)A_\varepsilon^*} \left[I - P_\varepsilon(s) A_\varepsilon G_\varepsilon G_\varepsilon^* A_\varepsilon^* P_\varepsilon(s)\right] e^{(s-t)A_\varepsilon}\, ds,$$

has a unique solution $P_\varepsilon \in C^1([0,T[;\mathcal{L}(H)) \cap C^0([0,T];\mathcal{L}(H))$, *such that* $P_\varepsilon(t) = P_\varepsilon(t)^* \geq 0$ *and* $P_\varepsilon(t) \in D([-A^*]^{1-\vartheta})$ *for each* $\vartheta \in]0,1]$, *with* $\|[-A_\varepsilon^*]^{1-\vartheta} P_\varepsilon(t)\|_{\mathcal{L}(H)} \leq C(T-t)^{-(1-\vartheta)}$ *for all* $t \in [0,T[$; *in addition it holds*

$$J(\hat{u}_\varepsilon) = \langle P_\varepsilon(0) y_0, y_0 \rangle_H \, .$$

(iii) *We have the feedback formula for* $\hat{u}_\varepsilon$:

$$\hat{u}_\varepsilon(t,\cdot) = G_\varepsilon^* A_\varepsilon^* P_\varepsilon(t) \hat{y}_\varepsilon(t,\cdot), \qquad t \in [0,T[.$$

(iv) *The optimal trajectory $\hat{y}_\varepsilon$ is given by $\hat{y}_\varepsilon(t,\cdot) = \Phi_\varepsilon(t,0)y_0(\cdot)$, where $\Phi_\varepsilon(t,s)$ is defined by the integral equation*

$$\Phi_\varepsilon(t,s) = e^{(t-s)A_\varepsilon} - \int_s^t A_\varepsilon e^{(t-r)A_\varepsilon} G_\varepsilon G_\varepsilon^* A_\varepsilon^* P_\varepsilon(r)\Phi_\varepsilon(r,s)\, dr, \quad t \in [s,T].$$

The expressions $P_\varepsilon(s)A_\varepsilon G_\varepsilon$ and $G_\varepsilon^* A_\varepsilon^* P_\varepsilon(s)$ are shorter forms relative to the well-defined operator $\left[-[(-A_\varepsilon^*)^{1-\vartheta}P_\varepsilon(s)]^*(-A_\varepsilon)^\vartheta G_\varepsilon\right]$ and to its adjoint, with fixed $\vartheta \in]0,\frac{1}{2}[$. □

4. Γ-convergence

First of all, we observe that all Γ-convergence results in the literature deal with a discrete parameter n tending to $+\infty$. So, from now on we will consider a fixed subsequence of our approximating problems with parameter $\varepsilon = \varepsilon_n$ such that $\varepsilon_n \to 0^+$ for $n \to \infty$.

We remark that, by (1.5),

$$\lim_{n\to\infty} \vartheta_{\varepsilon_n}(x) = \begin{cases} 1 & \text{if } x \in \Gamma_1 \\ 0 & \text{if } x \in \Gamma_0. \end{cases} \tag{4.1}$$

In order to apply the abstract result of [8], we need to rewrite the optimal control problems in a different way. Therefore, we set

$$H = L^2(\Omega), \qquad V = H_{00}^{1/2}(\Gamma_0) \times H^{-1/2}(\Gamma_1), \tag{4.2}$$

and

$$Y = L^2(0,T;H), \qquad \mathcal{U} = L^2(0,T;V). \tag{4.3}$$

Define the operators $M_{\varepsilon_n}, M : Y \to Y$ and $B_{\varepsilon_n}, B : \mathcal{U} \to Y$ as follows:

$$\begin{cases} M_{\varepsilon_n}(y) := y - e^{tA_{\varepsilon_n}} y_0 \\ M(y) := y - e^{tA} y_0 \end{cases} \qquad \forall y \in Y, \tag{4.4}$$

$$\begin{cases} B_{\varepsilon_n}(u) := \int_0^t A_{\varepsilon_n} e^{(t-s)A_{\varepsilon_n}} G_{\varepsilon_n}(s)u(s)\, ds, \\ B(u) := \int_0^t A e^{(t-s)A} G(s)u(s)\, ds \end{cases} \qquad \forall u \in \mathcal{U}. \tag{4.5}$$

As we already noted, the space of controls for problems (1.2) is smaller than that of problem (1.6). See the remark in the Introduction and compare with Remark 2.7.

Due to Theorems 2.5 and 3.9 above, the state equations of the approximating problems and of the mixed problem can be written as

$$M_{\varepsilon_n}(y) = B_{\varepsilon_n}(u) \qquad \text{and} \qquad M(y) = B(u), \tag{4.6}$$

respectively. We also observe that the corresponding cost functionals J_{ε_n} and J, defined by (1.3) and (1.7), are in fact the same (they differed only in the choice of the space of controls, which are equal to V now). In order to stress their dependence on y, we relabel both $J_{\varepsilon_n}(u)$ and $J(u)$ as $J(y,u)$, i.e. we set

$$J(y,u) := \int_0^T \left(\|u(s,\cdot)\|_V^2 + \|y(s,\cdot)\|_H^2\right) ds. \tag{4.7}$$

Now our approximating problem (1.2) has the following equivalent formulation:

$$(\mathcal{P}_n) \qquad \min\left\{J(y,u) + \chi_{\{M_{\varepsilon_n}(y)=B_{\varepsilon_n}(u)\}} : (y,u) \in Y \times \mathcal{U}\right\},$$

where for each set A the function χ_A is given by

$$\chi_A(x) = \begin{cases} 0 & \text{if } x \in A \\ +\infty & \text{if } x \notin A. \end{cases}$$

Similarly, we rewrite the mixed problem (1.6) (with U replaced by V) as

$$(\mathcal{P}) \qquad \min\left\{J(y,u) + \chi_{\{M(y)=B(u)\}} : (y,u) \in Y \times \mathcal{U}\right\}.$$

Thus, setting

$$F_n(y,u) := J(y,u) + \chi_{\{M_{\varepsilon_n}(y)=B_{\varepsilon_n}(u)\}}(y,u), \tag{4.8}$$

the sequence of optimal control problems we consider is

$$(\mathcal{P}_n) \qquad \min\left\{F_n(y,u) : (y,u) \in Y \times \mathcal{U}\right\},$$

with Y and $\mathcal{U}$ given by (4.3).

Of course, a point $(y_n, u_n) \in Y \times \mathcal{U}$ is an optimal pair for problem $(\mathcal{P}_n)$ if

$$F_n(y_n,u_n) = \min_{Y \times \mathcal{U}} F_n(y,u).$$

We recall now the general definition of multiple Γ-limits.

Definition 4.1. *Let X and W be topological spaces and let $\{F_n\}_{n\in\mathbb{N}}$ be a sequence of functions from $X \times W$ to $\overline{\mathbb{R}}$; we denote by $Z(+)$ and $Z(-)$ the* sup *and* inf *operators respectively. For every $x \in X$ and $w \in W$ we set*

$$\Gamma(\mathbb{N}^\alpha, X^\beta, W^\gamma) \lim_{n\to\infty} F_n(x,w) = \underset{\{w_n\}\in S(w)}{Z(\gamma)} \; \underset{\{x_n\}\in S(x)}{Z(\beta)} \; \underset{k\in\mathbb{N}}{Z(-\alpha)} \; \underset{n\geq k}{Z(\alpha)} F_n(x_n,w_n)$$

where $\alpha, \beta, \gamma \in \{+,-\}$, and $S(x)$, $S(w)$ denote the set of all sequences $x_n \to x$ in X and $w_n \to w$ in W respectively. Note that when the Γ-limit does not depend on the signs + or − in one (or more) of its variables, the corresponding sign is customarily omitted.

Our aim is now to apply Proposition 2.1 in [7] (see also Theorem 1.1 in the Introduction). Therefore we take as F_n the functionals defined in (4.8). Our goal is to prove that the functional F corresponding to problem $(\mathcal{P})$, i.e.

$$F(y,u) = J(y,u) + \chi_{\{M(y)=B(u)\}}, \tag{4.9}$$

coincides precisely with the multiple Γ-limit (1.1):

$$F(y,u) = \Gamma(\mathbb{N}, Y^-, \mathcal{U}^-) \lim_{n\to\infty} F_n(y,u). \tag{4.10}$$

This will mean that in fact problem $(\mathcal{P})$ is the Γ-*limit problem* of the sequence of problems $\{(\mathcal{P}_n)\}$, in the sense given by Definition 1.2 in the Introduction.

In order to prove (4.10), we are going to apply an abstract result in [8]. We first recall the definition of G-convergence of operators.

Definition 4.2. *Let Z and W be topological spaces and for all $n \in \mathbb{N}$ let D_n be operators from Z to W. We say that the sequence $\{D_n\}$* G-converges *to the operator $D : Z \to W$ if*

$$\Gamma(\mathbb{N}, W, Z^-) \lim_{n\to\infty} \chi_{\{D_n(z)=w\}}(w, z) = \chi_{\{D(z)=w\}}(w, z) \qquad \forall w \in W,\ \forall z \in Z,$$

that is, if the following conditions are satisfied:

(i) *if $z_n \to z$ in Z, $w_n \to w$ in W and $D_n(z_n) = w_n$ for infinitely many $n \in \mathbb{N}$, then $D(z) = w$;*

(ii) *if $z \in Z$ and $w \in W$ are such that $D(z) = w$ ana $w_n \to w$ in W, then there exists $\{z_n\} \subseteq Z$ such that $z_n \to z$ in Z and $D_n(z_n) = w_n$ for all sufficiently large $n \in \mathbb{N}$.*

The following general result holds true (it is a special case of a result proved in [8], Proposition 3.3 and Section 3).

Theorem 4.3. *Let Z, W be topological spaces, let $G_n : W \times Z \to \mathbb{R}$ be a functional and let $D, D_n : Z \to Z$, $K, K_n : W \to Z$ be operators. Consider a sequence of control problems of the following form:*

$$\min_{Z\times W} \{G_n(z, w) + \chi_{\{D_n(z)=K_n(w)\}}\}.$$

Assume that:

(i) *the sequence $\{D_n\}$ G-converges to D;*

(ii) *if $w_n \to w$ in W, then $K_n(w_n) \to K(w)$ in W;*

(iii) *there exist a function $\Psi : W \to \mathbb{R}$, bounded on bounded sets of W, and a function $\omega : Z \times Z \to \mathbb{R}$, with $\lim_{v\to z} \omega(z, v) = 0$ for all $z \in Z$, such that*

$$G_n(z, w) \le G_n(v, w) + \Psi(w)\omega(z, v). \qquad \forall w \in W,\ \forall z, v \in Z,\ \forall n \in \mathbb{N}.$$

Then

$$\Gamma(\mathbb{N}, Z^-, W^-) \lim_{n\to\infty} [G_n + \chi_{\{D_n(z)=K_n(w)\}}](z.w) = [G + \chi_{\{D(z)=K(w)\}}](z, w),$$

where

$$G(z, w) = \Gamma(\mathbb{N}, W^-) \lim_{n\to\infty} G_n(z, w). \tag{4.11}$$

We will apply this theorem with $Z = Y$, $W = \mathcal{U}$ (defined in (4.3)), $G_n = G = J$ (given by (4.7)), and $D_n = M_{\varepsilon_n}$, $D = M$, $K_n = B_{\varepsilon_n}$, $K = B$ (see (4.4) and (4.5)). Thus what we have to do is to verify that assumptions **(i)**–**(iii)** of Theorem 4.3 are in fact satisfied.

We start with proving hypothesis **(i)**.

Lemma 4.4. *Let* (H0), (H1), (H2), (H3) *be fulfilled. If* M_{ε_n} *and* M *are defined by* (4.4), *then the sequence* $\{M_{\varepsilon_n}\}$ *G-converges to* M *as* $n \to \infty$.

Proof. Suppose we have shown that

$$e^{tA_{\varepsilon_n}} y_0 \to e^{tA} y_0 \quad \text{in } Y \qquad \forall y_0 \in H; \tag{4.12}$$

then we easily deduce the following properties. Firstly, if $\{y_n\}, \{v_n\} \subset Y$ are such that $y_n \to y$ in Y and $v_n = M_{\varepsilon_n}(y_n) = y_{\varepsilon_n} - e^{tA_{\varepsilon_n}} y_0$ for infinitely many $n \in \mathbb{N}$, then letting $n \to \infty$ we get $v = y - e^{tA} y_0 = M(y)$.

Secondly, if $v = M(y) = y - e^{tA} y_0$, and $\{v_n\} \subset Y$ is a sequence such that $v_n \to v$ in Y, then setting $y_n := v_n + e^{tA_{\varepsilon_n}} y_0$ we have $M_{\varepsilon_n}(y_n) = v_n$ for infinitely many n and $y_n \to y$ in Y.

Recalling Definition 4.2, these facts imply that M_{ε_n} G-converges to M; hence the proof of Lemma 4.4 is achieved provided we show (4.12).

Let us prove now (4.12). To begin with, fix $f \in L^2(\Omega)$ and $\lambda \in \rho(A_{\varepsilon_n}) \cap \rho(A)$. If we set

$$v_{\varepsilon_n} := R(\lambda, A_{\varepsilon_n}), \qquad v := R(\lambda, A), \tag{4.13}$$

then by (3.13) and (2.12) the function $v - v_{\varepsilon_n}$ solves the equation

$$\lambda(v - v_{\varepsilon_n}) - A(\cdot, D)(v - v_{\varepsilon_n}) = 0 \quad \text{in } \Omega, \tag{4.14}$$

whereas at the boundary v and v_{ε_n} verify

$$\begin{cases} \vartheta_{\varepsilon_n} \frac{\partial v_{\varepsilon_n}}{\partial \nu_A} + (1 - \vartheta_{\varepsilon_n}) v_{\varepsilon_n} = 0 & \text{in } \partial\Omega \\ v = 0 & \text{in } \Gamma_0 \\ \frac{\partial v}{\partial \nu_A} = 0 & \text{in } \Gamma_1. \end{cases} \tag{4.15}$$

Hence multiplying by $v - v_{\varepsilon_n}$ and integrating by parts we easily get

$$\begin{aligned} &|\lambda| \|v - v_{\varepsilon_n}\|^2_{L^2(\Omega)} + \nu \|D(v - v_{\varepsilon_n})\|^2_{L^2(\Omega)} \\ &\le \int_{\partial\Omega} (v - v_{\varepsilon_n}) \frac{\partial}{\partial \nu_A} (v - v_{\varepsilon_n}) d\sigma \\ &= -\int_{\Gamma_0} v_{\varepsilon_n} \frac{\partial}{\partial \nu_A} (v - v_{\varepsilon_n}) d\sigma - \int_{\Gamma_1} (v - v_{\varepsilon_n}) \frac{\partial v_{\varepsilon_n}}{\partial \nu_A} d\sigma \\ &= -\int_{\partial\Omega} v_{\varepsilon_n} \frac{\partial}{\partial \nu_A} (v - v_{\varepsilon_n}) d\sigma - \int_{\partial\Omega} (v - v_{\varepsilon_n}) \frac{\partial v_{\varepsilon_n}}{\partial \nu_A} d\sigma \\ &\quad - \int_{\partial\Omega} v_{\varepsilon_n} \frac{\partial v_{\varepsilon_n}}{\partial \nu_A} d\sigma = I + II + III. \end{aligned} \tag{4.16}$$

Let us estimate separately each term. We have

$$I \le c \|v_{\varepsilon_n}\|_{H^{1/2}(\partial\Omega)} \left\| \frac{\partial}{\partial \nu_A} (v - v_{\varepsilon_n}) \right\|_{H^{-1/2}(\partial\Omega)} \le C \|v_{\varepsilon_n}\|_{H^1(\Omega)} \|v - v_{\varepsilon_n}\|_{H^1(\Omega)} ,$$

and similarly

$$II \le C \left\| \frac{\partial v_{\varepsilon_n}}{\partial \nu_A} \right\|_{H^{-1/2}(\partial\Omega)} \|v - v_{\varepsilon_n}\|_{H^{1/2}(\partial\Omega)} \le C \|v_{\varepsilon_n}\|_{H^1(\Omega)} \|v - v_{\varepsilon_n}\|_{H^1(\Omega)} ;$$

hence by Remark 3.4 we obtain

$$I + II \leq \frac{\nu}{4}\|v - v_{\varepsilon_n}\|^2_{H^1(\Omega)} + \frac{C}{|\lambda|}\|f\|^2_{L^2(\Omega)} \,. \tag{4.17}$$

Concerning the third term, we get as before

$$III \leq C\|v_{\varepsilon_n}\|_{H^{1/2}(\partial\Omega)} \left\|\frac{\partial v_{\varepsilon_n}}{\partial \nu_A}\right\|_{H^{-1/2}(\partial\Omega)} \leq C\|v_{\varepsilon_n}\|^2_{H^1(\Omega)} \leq \frac{C}{|\lambda|}\|f\|^2_{L^2(\Omega)} \,. \tag{4.18}$$

Thus, by (4.16), (4.17) and (4.18) it follows that

$$|\lambda|\|v - v_{\varepsilon_n}\|^2_{L^2(\Omega)} + \nu\|D(v - v_{\varepsilon_n})\|^2_{L^2(\Omega)} \leq \frac{C}{|\lambda|}\|f\|^2_{L^2(\Omega)}. \tag{4.19}$$

By compactness there exist a subsequence, that we still call $v - v_{\varepsilon_n}$, and a function $w \in H^1(\Omega)$ such that

$$v - v_{\varepsilon_n} \rightharpoonup w \quad \text{in } H^1(\Omega) \quad \text{as } n \to \infty \tag{4.20}$$

and

$$v - v_{\varepsilon_n} \to w \quad \text{in } L^2(\Omega) \quad \text{as } n \to \infty. \tag{4.21}$$

Since the operator $A(\cdot, D)$ is closed on $L^2(\Omega)$ and $v - v_{\varepsilon_n}$ solves (4.14), we deduce that

$$\lambda w - A(\cdot, D)w = 0 \qquad \text{in } \Omega. \tag{4.22}$$

Now, using Remark 3.4, we can rewrite (4.19) as follows

$$\begin{aligned} &|\lambda|\|v - v_{\varepsilon_n}\|^2_{L^2(\Omega)} + \nu\|D(v - v_{\varepsilon_n})\|^2_{L^2(\Omega)} + \\ &\quad + \textstyle\int_{\Gamma_0} \frac{\vartheta_\varepsilon}{1-\vartheta_\varepsilon}\left|\frac{\partial v_\varepsilon}{\partial \nu_A}\right|^2 d\sigma + \int_{\Gamma_1} \frac{1-\vartheta_\varepsilon}{\vartheta_\varepsilon}|v_\varepsilon|^2 \, d\sigma \leq \frac{C}{|\lambda|}\|f\|^2_{L^2(\Omega)}. \end{aligned} \tag{4.23}$$

Using again compactness, passing possibly to another subsequence, we get the existence of two functions $\gamma \in L^2(\Gamma_0)$ and $\mu \in L^2(\Gamma_1)$ such that

$$z_{\varepsilon_n} := \sqrt{\tfrac{1-\vartheta_{\varepsilon_n}}{\vartheta_{\varepsilon_n}}}\, v_{\varepsilon_n} \rightharpoonup \gamma \text{ in } L^2(\Gamma_0), \qquad q_{\varepsilon_n} := \sqrt{\tfrac{\vartheta_{\varepsilon_n}}{(1-\vartheta_{\varepsilon_n})}}\, \tfrac{\partial v_{\varepsilon_n}}{\partial \nu_A} \rightharpoonup \mu \text{ in } L^2(\Gamma_1). \tag{4.24}$$

Now, since Ω is bounded, the $(N-1)$-dimensional measure of the set Γ_ε introduced in (3.19) is bounded by $C\varepsilon$. Consequently, by Remark 3.4 we get

$$\|v_{\varepsilon_n}\|_{L^1(\Gamma_{\varepsilon_n})} \leq \|1\|_{L^2(\Gamma_{\varepsilon_n})}\|v_{\varepsilon_n}\|_{L^2(\partial\Omega)} \leq \frac{C}{|\lambda|^{1/2}}\, \|f\|_{L^2(\Omega)}\, \sqrt{\varepsilon_n} \tag{4.25}$$

Thus, recalling (3.7), we deduce

$$\begin{aligned} \|v_{\varepsilon_n}\|_{L^1(\Gamma_0)} &= \|v_{\varepsilon_n}\|_{L^1(\Gamma_{\varepsilon_n})} + \|v_{\varepsilon_n}\|_{L^1(\Gamma_0\setminus\Gamma_{\varepsilon_n})} \\ &\leq C\sqrt{\varepsilon_n} + \left\|\tfrac{\vartheta_{\varepsilon_n}}{1-\vartheta_{\varepsilon_n}}\right\|_{L^2(\Gamma_0\setminus\Gamma_{\varepsilon_n})}\|z_{\varepsilon_n}\|_{L^2(\Gamma_0)} \\ &\leq C\sqrt{\varepsilon_n} + C\sqrt{\tfrac{\varepsilon_n}{1-\varepsilon_n}}\,, \end{aligned} \tag{4.26}$$

so that

$$v_{\varepsilon_n} \to 0 \qquad \text{in } L^1(\Gamma_0). \tag{4.27}$$

Similarly, as

$$\left\| \frac{\partial v_{\varepsilon_n}}{\partial \nu_A} \right\|_{L^1(\Gamma_1)} \leq \sqrt{\frac{\varepsilon_n}{1-\varepsilon_n}} \, \|q_{\varepsilon_n}\|_{L^2(\Gamma_1)} \leq C\sqrt{\frac{\varepsilon_n}{1-\varepsilon_n}}\,, \tag{4.28}$$

we get

$$\frac{\partial v_{\varepsilon_n}}{\partial \nu_A} \to 0 \qquad \text{in } L^1(\Gamma_1). \tag{4.29}$$

Comparing (4.20), (4.27) and (4.15), we see that

$$w = 0 \qquad \text{in } \Gamma_0. \tag{4.30}$$

Similarly, by the closedness of the operator $\frac{\partial}{\partial \nu_A}$ from $H^1(\Omega)$ into $H^{-1/2}(\partial\Omega)$, (4.20) implies that

$$\frac{\partial(v - v_{\varepsilon_n})}{\partial \nu_A} \rightharpoonup \frac{\partial w}{\partial \nu_A} \qquad \text{in } H^{-1/2}(\partial\Omega),$$

so that by (4.29) and (4.15) we obtain

$$\frac{\partial w}{\partial \nu_A} = 0 \qquad \text{in } \Gamma_1. \tag{4.31}$$

By (4.22), (4.30) and (4.31) we deduce that w solves the mixed problem (2.12) with homogeneous data, so that Theorem 2.1 allows us to conclude that $w = 0$.

As a result, by (4.21), ve get $v_{\varepsilon_n} \to v$ in $L^2(\Omega)$. Now an easy argument by contradiction proves that in fact the whole sequence $\{v_{\varepsilon_n}\}$ converges to v in $L^2(\Omega)$. Thus, recalling (4.13), we have proved that

$$R(\lambda, A_{\varepsilon_n})f \to R(\lambda, A)f \quad \text{in } L^2(\Omega) \qquad \forall f \in L^2(\Omega). \tag{4.32}$$

Let us show now that

$$e^{tA_{\varepsilon_n}} f \to e^{tA} f \quad \text{in } L^2(\Omega) \qquad \forall f \in L^2(\Omega),\ \forall t > 0. \tag{4.33}$$

We recall that, by the usual representation of analytic semigroups as Dunford integrals, we have

$$e^{tA_{\varepsilon_n}} f - e^{tA} f = \frac{1}{2\pi i} \int_\gamma e^{\lambda t} \left[R(\lambda, A_{\varepsilon_n}) - R(\lambda, A)\right] f \, d\lambda \qquad \forall f \in L^2(\Omega),$$

where γ is a path from $+\infty e^{-i\vartheta}$ to $+\infty e^{i\vartheta}$, with $0 < \vartheta < \vartheta_0$, contained in S_{ϑ_0} (see Proposition 2.2) and leaving 0 on its left side. Using this representation and a simple change of variable, as the integrand is a holomorphic function of λ, we have for all $f \in L^2(\Omega)$:

$$\begin{aligned} &\|e^{tA_{\varepsilon_n}} f - e^{tA} f\|_{L^2(\Omega)} \\ &= \left\| \tfrac{1}{2\pi i} \int_\gamma e^{\sigma} \left[R(\tfrac{\sigma}{t}, A_{\varepsilon_n}) - R(\tfrac{\sigma}{t}, A)\right] f \, \tfrac{d\sigma}{t} \right\|_{L^2(\Omega)} \\ &\leq C t^{-1} \int_\gamma e^{Re\ \sigma} \|[R(\tfrac{\sigma}{t}, A_{\varepsilon_n}) - R(\tfrac{\sigma}{t}, A)] f\|_{L^2(\Omega)} \, |d\sigma|. \end{aligned} \tag{4.34}$$

Now fix $\delta \in]1/2, 1[$: by (2.10) and (3.12) we can write

$$\|e^{tA_{\varepsilon_n}} f - e^{tA} f\|_{L^2(\Omega)}$$
$$\leq C(\delta)\|f\|^{\delta}_{L^2(\Omega)} t^{\delta-1} \int_\gamma e^{Re\ \sigma}\sigma^{-\delta}\|[R(\tfrac{\sigma}{t}, A_{\varepsilon_n}) - R(\tfrac{\sigma}{t}, A)]f\|^{1-\delta}_{L^2(\Omega)}\ |d\sigma|.$$

We observe that

$$\frac{e^{Re\ \sigma}}{\sigma^\delta}\left\|\left[R\left(\frac{\sigma}{t}, A_{\varepsilon_n}\right) - R\left(\frac{\sigma}{t}, A\right)\right] f\right\|^{1-\delta}_{L^2(\Omega)} \leq C\ \frac{e^{Re \sigma}}{\sigma} T^{1-\delta}\|f\|_{L^2(\Omega)}\ ,$$

and that the right-hand side has finite integral over γ; thus, recalling (4.32), by Lebesgue Theorem we get (4.33).

Moreover, since

$$\|e^{tA_{\varepsilon_n}} f - e^{tA} f\|_{L^2(\Omega)} \leq C(\delta)t^{\delta-1}\ \|f\|^{\delta}_{L^2(\Omega)} \left[\int_\gamma e^{Re\ \sigma}\frac{|d\sigma|}{|\sigma|}\right]^{1/2} T^{1-\delta} \tag{4.35}$$

and $\delta > 1/2$, applying again the Lebesgue Theorem we conclude that (4.12) holds true. This completes the proof of Lemma 4.4. □

We prove now that hypothesis **(ii)** in Theorem 4.3 is fulfilled too.

Lemma 4.5. *Assume* (H0), (H1), (H2), (H3), *let B_{ε_n} and B be defined by* (4.5) *and let Y and $\mathcal{U}$ be given by* (4.3). *If $u_{\varepsilon_n} \to u$ in $\mathcal{U}$, then $B_{\varepsilon_n}(u) \to B(u)$ in Y.*

Proof. Set $v_{\varepsilon_n} := B_{\varepsilon_n}(u_{\varepsilon_n})$ and write $u := (u_0, u_1)$. By Theorem 3.9, v_{ε_n} is the solution of

$$\begin{cases} \frac{\partial v_{\varepsilon_n}}{\partial t} - A(\cdot, D)v_{\varepsilon_n} = 0 & \text{in } [0,T] \times \overline{\Omega} \\ \vartheta_{\varepsilon_n}\frac{\partial v_{\varepsilon_n}}{\partial \nu_A} + (1-\vartheta_{\varepsilon_n})v_{\varepsilon_n} = u_{\varepsilon_n} & \text{in } [0,T] \times \partial\Omega \\ v_{\varepsilon_n}(0, \cdot) = 0 & \text{in } \overline{\Omega}. \end{cases} \tag{4.36}$$

Then, by Theorem 3.2 and Remark 3.8, v_{ε_n} satisfies

$$\begin{aligned} &\sup_{s\in[0,T]} \|v_{\varepsilon_n}(s,\cdot)\|^2_{L^2(\Omega)} + \int_0^T \|Dv_{\varepsilon_n}\|^2_{L^2(\Omega)}\ dt \\ &\quad + \int_0^T \int_{\Gamma_0} \frac{\vartheta_{\varepsilon_n}}{1-\vartheta_{\varepsilon_n}}\left|\frac{\partial v_{\varepsilon_n}}{\partial \nu_A}\right|^2 d\sigma\ dt + \int_0^T\int_{\Gamma_1} \frac{1-\vartheta_{\varepsilon_n}}{\vartheta_{\varepsilon_n}}|v_{\varepsilon_n}|^2\ d\sigma\ dt \\ &\leq C\int_0^T \left(\|u_0\|^2_{H^{1/2}_{00}(\Gamma_0)} + \|u_1\|^2_{H^{-1/2}(\Gamma_1)}\right)\ dt. \end{aligned} \tag{4.37}$$

By compactness, passing possibly to a subsequence still denoted by $\{v_{\varepsilon_n}\}$, there exists a function $v \in L^2(0,T; H^1(\Omega))$ such that

$$v_{\varepsilon_n} \rightharpoonup v \qquad \text{in } L^2(0,T; H^1(\Omega)). \tag{4.38}$$

Since this implies $v_{\varepsilon_n} \rightharpoonup v$ in $L^2(0,T; L^2(\partial\Omega))$, we also have

$$\vartheta_{\varepsilon_n}\frac{\partial v_{\varepsilon_n}}{\partial \nu_A} = u_{\varepsilon_n} - (1-\vartheta_{\varepsilon_n})v_{\varepsilon_n} \rightharpoonup u_0 - v \qquad \text{in } L^2(0,T; L^2(\Gamma_0)). \tag{4.39}$$

Now, using again compactness, (4.37) implies that there exists a function $w \in L^2(0,T;L^2(\Gamma_0))$ such that

$$\vartheta_{\varepsilon_n}\frac{\partial v_{\varepsilon_n}}{\partial \nu_A} = \sqrt{(1-\vartheta_{\varepsilon_n})\vartheta_{\varepsilon_n}}\sqrt{\frac{\vartheta_{\varepsilon_n}}{1-\vartheta_{\varepsilon_n}}}\frac{\partial v_{\varepsilon_n}}{\partial \nu_A} \rightharpoonup 0\cdot w = 0 \qquad \text{in } L^2(0,T;L^2(\Gamma_0)).$$

Then, by uniqueness, (4.39) implies

$$v = u_0 \qquad \text{on } \Gamma_0. \tag{4.40}$$

Moreover, since the operator $\frac{\partial}{\partial \nu_A}$ is closed from $H^1(\Omega)$ into $H^{-1/2}(\partial\Omega)$, by (4.38) we have

$$\frac{\partial v_{\varepsilon_n}}{\partial \nu_A} \rightharpoonup \frac{\partial v}{\partial \nu_A} \qquad \text{in } L^2(0,T;H^{-1/2}(\partial\Omega)). \tag{4.41}$$

Then, recalling that $\vartheta_{\varepsilon_n} = 1-\varepsilon_n$ on Γ_1 we obtain

$$\vartheta_{\varepsilon_n}\frac{\partial v_{\varepsilon_n}}{\partial \nu_A} \rightharpoonup \frac{\partial v}{\partial \nu_A} \qquad \text{in } L^2(0,T;[H^{1/2}_{00}(\Gamma_1)]^*),$$

so that, by (4.41), we get

$$\frac{\partial v}{\partial \nu_A} = u_1 \qquad \text{on } \Gamma_1. \tag{4.42}$$

Finally, letting $n \to \infty$ in problem (4.36), by (4.38), (4.40) and (4.42) we conclude that v solves the equation

$$\begin{cases} \frac{\partial v}{\partial t} - A(\cdot,D)v &= 0 & \text{in } [0,T]\times\overline{\Omega} \\ v &= u_0 & \text{in } [0,T]\times\Gamma_0 \\ \frac{\partial v}{\partial \nu_A} &= u_1 & \text{in } [0,T]\times\Gamma_1 \\ v(0,\cdot) &= 0 & \text{in } \overline{\Omega}. \end{cases} \tag{4.43}$$

Therefore, by uniqueness and by Theorem 2.5, we deduce $v = B(u)$; hence we have proved that $B_{\varepsilon_n}(u_{\varepsilon_n}) \rightharpoonup B(u)$ in $L^2(0,T;H^1(\Omega))$. By Rellich Theorem we also get $B_{\varepsilon_n}(u_{\varepsilon_n}) \to B(u)$ in Y. □

We are finally ready to prove our main result.

Theorem 4.6. *Assume* (H0), (H1), (H2), (H3). *In the spaces Y, $\mathcal{U}$ introduced in* (4.3), *let M_{ε_n}, M, B_{ε_n} and B be defined by* (4.4) *and* (4.5) *respectively, and let J be given by* (4.7). *Then the sequence of optimal control problems*

$$(\mathcal{P}_n) \qquad \min\left\{J(y,u) + \chi_{\{M_{\varepsilon_n}(y)=B_{\varepsilon_n}(u)\}}(y,u) : (y,u)\in Y\times\mathcal{U}\right\}$$

Γ-converges to the optimal control problem

$$(\mathcal{P}) \qquad \min\left\{J(y,u) + \chi_{\{M(y)=B(u)\}}(y,u) : (y,u)\in Y\times\mathcal{U}\right\}.$$

Proof. We apply Theorem 4.3. In Lemma 4.4 and in Lemma 4.5 we proved that, under our present assumptions, hypotheses **(i)** and **(ii)** in Theorem 4.3 hold true. On the other hand it is easy to see that the cost functional verifies **(iii)**. Thus by Theorem 4.3 and Theorem 1.1 we conclude that the sequence of problems $(\mathcal{P}_n)$ Γ-converges to the optimal control problem $(\mathcal{P})$. □

The above theorem holds for an arbitrary sequence $\varepsilon_n \to 0^+$; this allows us to say that problems $(\mathcal{P}_\varepsilon)$ Γ-converge to problem $(\mathcal{P})$ as $\varepsilon \to 0^+$.

Remark 4.7. For the sake of simplicity we considered a cost functional with a very simple form. One can also deal with more general cost functionals such as

$$\int_0^T (\|y(s,\cdot)\|_H^2 + \|u(s,\cdot)\|_V^2)\, ds + \langle P_T y(T), y(T)\rangle_{L^2(\Omega)}$$

with $P_T \in \mathcal{L}(L^2(\Omega))$, provided the operator P_T is regular enough in order to have existence of the optimal controls and to satisfy condition **(iii)** of Theorem 4.3 (see [1] and [2]). The cost functional might also depend explicitly on ε: in that case one has in addition to determine the Γ-limit in (4.11).

References

[1] P. Acquistapace, F. Flandoli, B. Terreni, *Initial boundary value problems and optimal control for nonautonomous parabolic systems*, SIAM J. Control Optim., **29** (1991) 89–118.

[2] P. Acquistapace, B. Terreni, *Infinite-horizon linear-quadratic regulator problems for nonautonomous parabolic systems with boundary control*, SIAM J. Control Optim. **34** (1996) 1–30.

[3] P. Acquistapace, B. Terreni, *Classical solutions of nonautonomous Riccati equations arising in parabolic boundary control problems*, Appl. Math. Optim. **39** (1999) 361–409.

[4] S. Agmon, A. Douglis, L. Niremberg, *Estimates near the boundary for solutions of elliptic partial differential equations satisfying general boundary conditions* II, Comm. Pure Appl. Math., **XVII** (1964) 35–92.

[5] J. Banasiak, *Domains of fractional powers of operators arising in mixed boundary value problems in non-smooth domains and applications*, Appl. Anal., **55** (1994) 79–89.

[6] A. Briani, *Hamilton-Jacobi-Bellman equations and* Γ*-convergence for optimal control problems*, Tesi di Dottorato, Dipartimento di Matematica, Università di Pisa (1999).

[7] G. Buttazzo, G. Dal Maso, Γ*-convergence and optimal control problems*, J. Optim. Theory Appl., **38** (1982) 385–407.

[8] G. Buttazzo, L. Freddi, *Optimal control problems with weakly converging input operators*, Discrete and Contin. Dynam. Systems, **1** (3) (1995) 401–420.

[9] G. Dal Maso, *An Introduction to* Γ*-convergence*, Birkhäuser, Boston 1993.

[10] C. Fujiwara, *Concrete characterization of the domains of fractionals powers of some elliptic differential operators of the second order*, Proc. Japan Acad. Ser. A Math. Sci., **43** (1967) 82–86.

[11] J.A. Goldstein, *Semigroups of linear operators and application*, Oxford University Press, Oxford 1985.

[12] I. Lasiecka, R. Triggiani, *Riccati differential equations with unbounded coefficients and non-smooth terminal condition: the case of analytic semigroups*, SIAM J. Math. Anal. **23** (1992) 449–481.

[13] X. Li, J. Yong, *Optimal control theory for infinite-dimensional systems*, Birkhäuser, Boston 1995.

[14] J.L. Lions, *Équations différentielles opérationnelles*, Springer Verlag, Berlin 1961.

[15] J.L. Lions, E. Magenes, *Problèmes aux limites non homogènes*, **I**, Dunod, Paris 1968.

[16] A. Lunardi, *Analytic semigroups and optimal regularity in parabolic problems*, Birkhäuser, Boston 1995.

[17] A. Pazy, *Semigroups of linear operators and applications to partial differential equations*, Springer Verlag, New York 1983.

[18] O. Pironneau, C. Prud'homme, *Free FEM, a language for the finite element method*, http://www.enseeiht.fr/travaux/CD9899/travaux/optmfn/newcod/manuals/freefem/freefem.htm.

[19] G. Savarè, *Regularity and perturbation results for mixed second order elliptic problems*, Commm. Partial Differential Equations **22** (1997) 869–899.

[20] G. Savarè, *Parabolic problems with mixed variable lateral conditions: an abstract approach*, J. Math. Pures Appl., **76** (1997) 321–351.

[21] H. Tanabe, *Equations of evolution*, Pitman, London 1979.

Paolo Acquistapace and Ariela Briani
Dipartimento di Matematica
Università di Pisa
via Filippo Buonarroti, 2
I-56127 Pisa
e-mail: `acquistp@dm.unipi.it`
e-mail: `briani@mail.dm.unipi.it`

Progress in Nonlinear Differential Equations
and Their Applications, Vol. 50, 27–37

A Degenerate Two-point Problem*

Viorel Barbu and Angelo Favini†

1. Introduction

This paper is concerned with the degenerate two point problem in a (complex) Hilbert space H, with inner product $\langle \cdot, \cdot \rangle$ and norm $|\cdot|$,

$$\left(\frac{d}{dt} + \epsilon\right)(M_0 y) = -l_0 y - B_0 u + f(t), \quad 0 < t < \tau, \tag{1.1}$$

$$\left(-\frac{d}{dt} + \epsilon\right)(M_1 u) = B_1 y - L_1 u + g(t), \quad 0 < t < \tau, \tag{1.2}$$

$$M_0 y(0) = M_0 y_0, \quad M_1 u(\tau) = M_1 u_\tau, \tag{1.3}$$

where ϵ is a non negative constant, $B_i \in \mathcal{L}(H)$, the space of all bounded linear operators from H into itself, L_i, M_i are closed linear operators from H into itself, $0 \in p(L_i)$, with domain $D(L_i) \subseteq D(M_i)$, $i = 0, 1$, $f, g \in L^2(0, \tau; H)$, $y_0 \in D(L_0)$, $u_\tau \in D(L_1)$ are given. No assumption is made on the invertibility of the operators M_i. We shall say that the pair (y, u) is a solution to (1.1)–(1.3) if $y(\cdot) \in L^2(0, \tau; D(L_0))$, $u(\cdot) \in L^2(0, \tau; D(L_1))$, $M_0 y(\cdot) \in H^1(0, \tau; H)$, $M_1 u \in H^1(0, \tau; H)$, the equations (1.1), (1.2) hold a.e. in $(0, \tau)$ and (1.3) is satisfied.

Problem (1.1)–(1.3) has had a large literature in the last decades when $M_0 = M_1 = I$, the identity operator in H. We refer, for example, to J.L. Lions [10], G. Da Prato [7], J.M. Cooper [5].

More recently, in order to study the regulator problem for linear degenerate control systems both in finite and infinite dimensional spaces, it was observed that a system like (1.1)–(1.3), with $\epsilon = 0$, describes the link between the state $y(\cdot)$ and the co-state $p(\cdot)$ for the degenerate equation

$$\frac{d}{dt}(My) = -Ly + Bu(t), \quad 0 < t < \tau, \tag{1.4}$$

with the initial condition

$$My(0) = My_0, \tag{1.5}$$

and the quadratic cost functional

$$J(u) = \int_0^\tau |Cy(t)|_Z^2 \, dt + \int_0^\tau \langle Nu(t), u(t) \rangle_U \, dt \tag{1.6}$$

* Work partially supported by the Italian M.I.U.R. and by University of Bologna Funds for selected research topics.

† The second author is a member of G.N.A.M.P.A. of the Italian Istituto di Alta Matematica (I.N.d.A.M.).

to be minimized, provided that $z = 0$ is a simple pole for $(z+T)^{-1}$ and $(z+S)^{-1}$, where $T = ML^{-1}$, $S = M^*L^{*-1}$.

Here U and Z are real Hilbert spaces (we always write $\langle \cdot, \cdot \rangle_W$ and $|\cdot|_W$ for the inner product and the norm in the Hilbert space W, respectively, and we shall omit the subscript where $W = H$, for brevity), L,M are closed linear operators in H, $0 \in \rho(L)$, $D(L) \subseteq D(M)$, $B \in \mathcal{L}(U,H)$, the space of bounded linear operators from U into H, $C \in \mathcal{L}(H,Z)$, $N = N^* \in \mathcal{L}(U) = \mathcal{L}(U,U)$ is a positive symmetric operator, the space of controls is all of $L^2(0,\tau;U)$, and $y_0 \in D(L)$. Indeed, it was shown (Barbu and Favini [1]) that under the above assumptions the optimal control problem to find $u^* \in L^2(0,\tau;U)$ such that

$$J(u^*) = \inf_u J(u) \tag{1.7}$$

has a unique solution.

Moreover, if the pair $(y,p) \in L^2(0,\tau;D(L)) \times L^2(0,\tau;D(L^*))$, with $My \in H^1(0,\tau;H)$, $p \in H^1(0,\tau;H)$, satisfies

$$\frac{d}{dt}(My) = -Ly - BN^*B^*p, \quad 0 < t < \tau, \tag{1.8}$$

$$-\frac{d}{dt}(M^*p) = -L^*p + C^*Cy, \quad 0 < t < \tau, \tag{1.9}$$

$$My(0) = My_0, \quad M^*p(\tau) = 0, \tag{1.10}$$

then $u^* = -N^{-1}B^*p$ is just the unique optimal control for (1.4)–(1.7).

Some results on the regulator problem for linear degenerate systems in finite-dimensional spaces were obtained by L. Pandolfi [11]. Different control problems, relative to $M = I$ but N singular are treated in S.L. Campbell [3].

The optimal control problem in the infinite interval $(0,\infty)$ with relative cost functional

$$\int_0^\infty \{\langle Cy(t), y(t)\rangle_{R^n} + \langle Nu(t), u(t)\rangle_{R^m}\}\, dt,$$

and C, N positive operators, was investigated by J.D. Cobb [4]. See also the monograph [6] by L. Dai.

In view of the polar singularity assumption, i.e. $z = 0$ a simple pole for $L(zL+M)^{-1}$, our results apply to Sobolev type linear equations, too. To this purpose, we recall that the optimal control for the Sobolev type equation

$$M\frac{dy}{dt} + Ly = Bu(t) + f(t), \quad 0 < t < \tau, \tag{1.11}$$

was discussed by G.A. Sviridyuk and A.A. Efremov [12] by means of a completely different approach. Since in (1.11) $y(\cdot)$ is assumed to have a derivative, different control spaces like $H_0^m(0,\tau;U)$, $m \geq 2$, and cost functional containing the derivatives of the controls $u(\cdot)$ must be required.

Our main results are found in Section 2. In Section 3 we outline some examples to clarify the range of application of our statements in Section 2.

2. The main results

In order to establish our results we need to recall some preliminary facts.

First of all, if $z = 0$ is a simple pole for the bounded linear operator $T_i = M_i L_i^{-1}$ from the complex Hilbert space H into itself, $i = 0, 1$, then the direct sum representation $H = N(T_i) \oplus R(T_i)$ holds, $R(T_i)$ being a closed subspace of H.

Let us denote by P_i the projection operator on $N(T_i)$ along $R(T_i)$.

Let $X = H \times H$ be endowed with the inner product

$$\langle (x, y), (u, v) \rangle_X = \langle x, u \rangle + \langle y, v \rangle, \quad x, y, u, v \in H.$$

The operator $P = (P_0, P_1)$ is a projection from X into itself and $\|P\|_{\mathcal{L}(X)} \leq (\|P_0\|^2_{\mathcal{L}(H)} + \|P_1\|^2_{\mathcal{L}(H)})^{1/2}$. In particular, if $\|P_i\|_{\mathcal{L}(H)} \leq 1$, then $\|P\|_{\mathcal{L}(X)} \leq \sqrt{2}$. Moreover, we want to reduce the boundary conditions (1.3) to the homogeneous ones. To this end it suffices to define $y(t) - y_0 = x(t)$, $u(t) - u_\tau = v(t)$, so that (1.1)–(1.3) become

$$\Big(\frac{d}{dt} + \epsilon\Big)(M_0 x) = -L_0 x - B_0 v + f(t) - L_0 y_0 - B_0 u_\tau - \epsilon M_0 y_0, \; 0 < t < \tau, \quad (2.1)$$

$$\Big(-\frac{d}{dt} + \epsilon\Big)(M_1 v) = B_1 x - L_1 v + g(t) + B_1 y_0 - L_1 u_\tau - \epsilon M_1 u_\tau, \; 0 < t < \tau, \quad (2.2)$$

$$M_0 x(0) = M_1 v(\tau) = 0. \quad (2.3)$$

We then introduce new operators as follows.

$$D(Q_0) = H_0^1(0, \tau; H) = \{y \in H^1(0, \tau; H); y(0) = 0\}, \quad Q_0 y = \frac{dy}{dt} (= y'), \quad (2.4)$$

$$D(Q_1) = H_\tau^1(0, \tau; H) = \{v \in H^1(0, \tau; H); v(\tau) = 0\}, \quad Q_1 v = -\frac{dv}{dt} (= -v'), \quad (2.5)$$

$$D(Q) = D(Q_0) \times D(Q_1), \quad Q(x, v) = (Q_0 x, Q_1 v), \quad (2.6)$$

$$D(M) = L^2(0, \tau; D(M_0) \times D(M_1)), \quad M(x, v)(t) = (M_0 x(t), M_1 v(t)), \quad (2.7)$$

$$D(R_i) = \{x \in L^2(0, \tau; H); x(t) \in D(L_i) \; a.e. \; t \in [0, \tau],$$
$$L_i x(\cdot) \in L^2(0, \tau; H)\}, \quad R_i x = L_i x(\cdot), \quad x \in D(R_i), \quad i = 0, 1, \quad (2.8)$$

$$D(R) = D(R_0) \times D(R_1), \quad R(x, v) = (R_0 x, R_1 v), \quad (2.9)$$

$$D(S) = L^2(0, \tau; H) \times L^2(0, \tau; H), \quad S(x, v)(t) = (B_0 v(t), -B_1 x(t)), \quad (2.10)$$

$$F(t) = (f(t) - (\epsilon M_0 + L_0) y_0 - B_0 u_\tau, g(t) - (\epsilon M_1 + L_1) u_\tau + B_1 y_0). \quad (2.11)$$

Then it is readily seen that with the hitherto adopted nomenclature, problem (2.1)–(2.3) is written under the form

$$((\epsilon + Q)M + R + S)w = F, \quad (2.12)$$

where $w = (x, v)$. Next step is to prove a lemma as follows.

Lemma 2.1. *Under the assumptions above, for all $G \in L^2(0,\tau;X)$, the equation*

$$(\epsilon + Q)M\phi + R\phi = G \tag{2.13}$$

has a unique solution for any $\epsilon > 0$. Moreover, $R\phi = PG + \tilde{T}^{-1}S_\epsilon(1-P)G$, where $\tilde{T}$ is the restriction of T to $R(T) = R(T_0) \times R(T_1)$ and $S_\epsilon \in \mathcal{L}(L^2(0,\tau;R(T)))$ tends to zero in norm as ϵ goes to ∞.

Proof. The change of variable $R\phi = \eta = (\eta_0, \eta_1)$ transforms (2.13) into the equivalent equation

$$(\epsilon + Q)T\eta + \eta = G. \tag{2.14}$$

Notice that $T\eta = (T_0\eta_0, T_1\eta_1)$. Again, just in virtue of the diagonal expression of the involved operators, we recognize that (2.14) in its turn reduces to the pair of equation

$$(\epsilon + Q_i)T_i\eta_i + \eta_i = G_i, \quad i = 0, 1. \tag{2.15}$$

In what follows, we will indicate P_i the projection operator in $L^2(0,\tau;H)$ induced by the previous one. If $\tilde{Q}_i$ denotes the operator

$$D(\tilde{Q}_0) = H_0^1(0,\tau;R(T_0)), \quad \tilde{Q}_0 x = x',$$
$$D(\tilde{Q}_1) = H_\tau^1(0,\tau;R(T_1)), \quad \tilde{Q}_1 v = -v',$$

and $\tilde{T}_i$ is the restriction of T_i to $R(T_i)$, (to be identified to $L^2(0,\tau;R(T_i))$, $i = 0,1$, so that $\tilde{T}_i$ has a bounded inverse $\tilde{T}_i^{-1} \in \mathcal{L}(L^2(0,\tau;R(T_i))))$, then the new change of variable $\tilde{T}_i(1-P_i)\eta_i = \sigma_i$ allows to deduce that (2.15) is equivalent to the system

$$P_i\eta_i = P_iG_i, \quad i = 0, 1, \tag{2.16}$$

$$(\epsilon + \tilde{Q}_i)\tilde{T}_i(1-P_i)\eta_i + (1-P_i)\eta_i = (1-P_i)G_i, \quad i = 0, 1. \tag{2.17}$$

Now (2.17) admits, for $j = 0, 1$, the unique solution

$$(1-P_j)\eta_j = \tilde{T}_j^{-1}\sigma_j = \frac{1}{2\pi i}\int_\gamma \tilde{T}_j^{-1}(z + \tilde{T}_j^{-1})^{-1}(\lambda - \epsilon - \tilde{Q}_j)^{-1}(1-P_j)G_j\, dz,$$

where γ is a suitable closed contour in the complex plane.

Furthermore, the following estimate holds

$$\|(1-P_i)\eta_i\|_{L^2(0,\tau;R(T_i))} \le C(1+\epsilon)^{-1}\|(1-P_i)G_i\|_{L^2(0,\tau;R(T_i))}, \tag{2.18}$$

for $i = 0, 1$. Therefore, taking into account (2.16) too, we have shown that (2.15) has the unique solution

$$\eta = (\eta_1, \eta_2) = (P_0G_0 + \tilde{T}_0^{-1}\sigma_0, P_1G_1 + \tilde{T}_1^{-1}\sigma_1) = PG + \tilde{T}^{-1}S_\epsilon(1-P)G, \tag{2.19}$$

where S_ϵ is a bounded linear operator from $L^2(0,\tau;R(T))$ into itself such that

$$\|S_\epsilon\|_{\mathcal{L}(L^2(0,\tau;R(T)))} \le C'(1+\epsilon)^{-1}. \tag{2.20}$$

This concludes the proof of the lemma. □

We are now in the position to prove the main result.

Theorem 2.1. *Let L_i, M_i, B_i, $i = 0, 1$, be closed linear operators in H, $B_i \in \mathcal{L}(H)$, $0 \in \rho(L_i)$, $D(L_i) \subseteq D(M_i)$, and let 0 be a simple pole of $(z + T_i)^{-1}$, where $T_i = M_i L_i^{-1}$. Moreover, let $h_0, h_1 > 0$ such that*

$$|L_0 u| \geq h_0 |u|, \quad |L_i v| \geq h_1 |v|, \quad (u, v) \in D(L_0) \times D(L_1), \tag{2.21}$$

where

$$\|P_0\|_{\mathcal{L}(H)} \|B_0\|_{\mathcal{L}(H)} h_1^{-1} + \|P_1\|_{\mathcal{L}(H)} \|B_1\|_{\mathcal{L}(H)} h_0^{-1} < 1. \tag{2.22}$$

Then there is $\epsilon_0 \geq 0$ such that for all $\epsilon \geq \epsilon_0$ problem (1.1)–(1.3) has a (unique) solution for all $f, g \in L^2(0, \tau; H)$ and any $y_0 \in D(L_0)$, $u_\tau \in D(L_1)$.

Remark 2.1. Assumption (2.22) is to be compared with assumption (3.2) in the paper [5] by J.M. Cooper, where the case $M_0 = M_1 = B_0 = B_1 = I$ is examined by the help of a method quite different from ours.

Remark 2.2. A condition like (2.22) seems unavoidable if we do not make further assumptions on the operators B_0, B_1 (to this purpose, see Da Prato [1], Theorem 4.8, p. 229, too), as the following trivial counterexample shows.

Consider the system

$$\begin{aligned} x' &= -x + u + 3v + f(t), \\ 0 &= -y + u + v + g(t), \\ -u' &= -u + 3y + h(t), \\ 0 &= -v + x + y - g(t), \end{aligned}$$

where $f, g, h \in L^2(0, \tau)$. Obviously,

$$M_0 = \begin{bmatrix} 1 & 0 \\ 0 & 0 \end{bmatrix} = M_1, \quad L_0 = I = L_1, \quad B_0 = \begin{bmatrix} -1 & -3 \\ -1 & -1 \end{bmatrix}, \quad B_1 = \begin{bmatrix} 0 & 3 \\ 1 & 1 \end{bmatrix},$$

and $z = 0$ is a simple pole for $(z + M_i)^{-1}$, $i = 0, 1$, (2.21) holds ($h_0 = h_1 = 1$), but (2.22) is not verified. Indeed, we see that necessarily

$$y = u + v + g(t) = u + x + y,$$

and thus $u + x \equiv 0$. Hence

$$\begin{aligned} x' &= -2x + 3v + f(t) = x + 3y + h(t) = x + 3(v - x + g(t)) + h(t) \\ &= -2x + 3v + h(t) + 3g(t). \end{aligned}$$

It follows that if $f \neq h + 3g$, no solution to the problem exists. This negative result is of course related to the non-positivity of the operators B_0 and B_1.

Remark 2.3. The method of proof of the Theorem is inspired by the paper [8] by Favini and Venni, in relation to the regular case. It must be carefully modified because the possible degeneration of the operator coefficients does not allow the needed decreasing of the resolvents as $\epsilon \longrightarrow +\infty$.

Proof of Theorem 2.1. After the change of variable $Rw = \eta$, equation (2.12) reads

$$(\epsilon + Q)T\eta + \eta + SR^{-1}\eta = F. \tag{2.23}$$

Taking into account Lemma 2.1, we look for a solution η to (2.23) under the form $\eta = PG + \tilde{T}^{-1}S_\epsilon(1-P)G$. Then (2.23) reduces to

$$G + SR^{-1}(PG + \tilde{T}^{-1}S_\epsilon(1-P)G) = F. \tag{2.24}$$

SR^{-1} is the operator in $L^2(0,\tau;X)$ induced by

$$(B_0L_1^{-1}v, -B_1L_0^{-1}y) \quad (= SR^{-1}(y,v)).$$

If we project (2.24) onto $N(T)$ and $R(T)$, we decouple (2.24) into

$$PG + PSR^{-1}PG + PSR^{-1}\tilde{T}^{-1}S_\epsilon(1-P)G = PF, \tag{2.25}$$

$$(1-P)G + (1-P)SR^{-1}PG + (1-P)SR^{-1}\tilde{T}^{-1}S_\epsilon(1-P)G = (1-P)F. \tag{2.26}$$

Since the norm of S_ϵ in $\mathcal{L}(L^2(0,\tau;R(T)))$ is arbitrarily small for ϵ suitably large, we can affirm that (2.26) has a unique solution $(1-P)G$ for given PG, i.e.,

$$(1-P)G = K_\epsilon[(1-P)F - (1-P)SR^{-1}PG],$$

where $K_\epsilon \in \mathcal{L}(L^2(0,\tau;R(T)))$, with the bound $\|K_\epsilon\|_{\mathcal{L}(L^2(0,\tau;R(T)))} \le h < \infty$ for all $\epsilon \ge \epsilon_0 \ge 0$. Then (2.25) becomes

$$PG + PSR^{-1}PG + PSR^{-1}\tilde{T}^{-1}S_\epsilon K_\epsilon(1-P)SR^{-1}PG = P\tilde{h}, \tag{2.27}$$

where $P\tilde{h} = PF - PSR^{-1}\tilde{T}^{-1}S_\epsilon K_\epsilon(1-P)F$.

Also observe that the norm of S_ϵ in $\mathcal{L}(L^2(0,\tau;R(T)))$ is small for large enough ϵ; hence we infer that (2.27) has a solution PG provided that

$$Ph + PSR^{-1}Ph = P\tilde{h} \tag{2.28}$$

is solvable for given $\tilde{h} \in L^2(0,\tau;X)$. On the other hand,

$$\begin{aligned}
&|P_0B_0L_1^{-1}P_1v|^2 + |P_1B_1L_0^{-1}P_0y|^2\\
&\le (\|P_0\|_{\mathcal{L}(H)}\|B_0\|^2_{\mathcal{L}(H)}\|L_1^{-1}\|^2_{\mathcal{L}(H)} + \|P_1\|_{\mathcal{L}(H)}\|B_1\|^2_{\mathcal{L}(H)}\|L_0^{-1}\|^2_{\mathcal{L}(H)})(|P_0y|^2 + |P_1v|^2)\\
&\le (\|P_0\|_{\mathcal{L}(H)}\|B_0\|_{\mathcal{L}(H)}h_1^{-1} + \|P_1\|_{\mathcal{L}(H)}\|B_1\|_{\mathcal{L}(H)}h_0^{-1})^2(|P_0y|^2 + |P_1v|^2)
\end{aligned}$$

and this finishes the proof in view of (2.22).

Alternatively, one could observe that under the assumptions above $PSR^{-1}P$ has a norm in $\mathcal{L}(L^2(0,\tau;X))$ less than 1, so that (2.28) has a unique solution. □

Let us remark that in view of (2.25), (2.26) the really basic assumption to solve the problem is $-1 \in \rho(PSR^{-1}P)$. If P_0 and P_1 are self-adjoint, Theorem 2.1 can be improved as follows.

Theorem 2.2. *Let L_i, M_i, B_i be operators as above, and suppose P_i to be self-adjoint, $i = 0,1$. If, in addition, for any $u \in D(L_0)$, $v \in D(L_1)$*

$$\Re e\{\langle B_0v, L_0u\rangle - \langle B_1u, L_1v\rangle\} \ge -c_0(|L_0u|^2 + |L_1v|^2) \tag{2.29}$$

with some constant $c_0 < 1$, then (1.1)–(1.3) has one solution (y,u) for every $f, g \in L^2(0,\tau;H)$ and any $y_0 \in D(L_0)$, $u_\tau \in D(L_1)$.

Proof. Since P_i is self-adjoint, so is P. Then for all $(x,y) \in N(T)$ one has

$$\begin{aligned}\Re e\langle PSR^{-1}P(x,y),(x,y)\rangle_X &= \Re e\langle PSR^{-1}P(x,y),P(x,y)\rangle_X \\ &= \Re e\langle SR^{-1}P(x,y),P(x,y)\rangle_X \geq \text{(in view of (2.29))} \\ &\geq -c_0(|x|^2+|y|^2).\end{aligned}$$

Remark 2.4. If $L_1 = L_0^*$ and $B_0 = B_1 = I$, then the left-hand side term in (2.29) equals

$$\Re e\{\langle v, L_0 u\rangle - \langle L_0 u, v\rangle\} = 0,$$

so that (2.29) holds with $c_0 = 0$.

Corollary 2.1. *Under the above assumptions on L_i, M_i, B_i, P_i as in Theorem 2.1, if $L_1 = L_0^*$, $B_1 = B_0^*$ and $D(L_0^*)$ is invariant under B_0 with $B_0L_0^* \subseteq L_0^*B_0$, then problem (1.1)–(1.3) has one solution (y,u) for every $f,g \in L^2(0,\tau;H)$ and any $y_0 \in D(L_0)$, $u_\tau \in D(L_1)$.*

Proof. For all $u \in D(L_0)$, $v \in D(L_0^*)$ we have

$$\begin{aligned}\langle B_0v, L_0u\rangle &= \langle L_0^*B_0v,u\rangle = \langle B_0L_0^*v,u\rangle = \langle L_0^*v, B_0^*u\rangle \\ &= \langle L_0^*v, B_1u\rangle = \overline{\langle B_1u, L_0^*v\rangle} = \overline{\langle B_1u, L_1v\rangle}.\end{aligned}$$

Hence (2.29) is verified with $c_0 = 0$ and the result follows from Theorem 2.2. □

If the projections P_i are self-adjoint, taking into account the proofs to Theorems 2.1 and 2.2, we can improve condition (2.22) a bit, too, as the following result shows.

Corollary 2.2. *Let L_i, M_i, B_i, $i = 0,1$, be closed linear operators in the Hilbert space H, $B_i \in \mathcal{L}(H)$, $0 \in \rho(L_i)$, $D(L_i) \subseteq D(M_i)$, $z = 0$ being a simple pole for $(z+T_i)^{-1}$, $T_i = M_iL_i^{-1}$. If, in addition, P_i (the projection onto $N(T_i)$) is self-adjoint,* (2.21) *holds and*

$$\|B_0\|_{\mathcal{L}(H)}h_1^{-1} + \|B_1\|_{\mathcal{L}(H)}h_0^{-1} < 2, \tag{2.30}$$

then (1.1)–(1.3) has one solution for any $f,g \in L^2(0,\tau;H)$ and all $y_0 \in D(L_0)$, $u_\tau \in D(L_1)$.

Proof. Under the present assumptions, for all $(u,v) \in D(L_0)\times D(L_1)$

$$\begin{aligned}&\Re e\{\langle B_0v, L_0u\rangle - \langle B_1u, L_1v\rangle\} \geq -|B_0v||L_0u| - |B_1u||L_1v| \\ \geq\ & -\{\|B_0L_1^{-1}\|_{\mathcal{L}(H)} + \|B_1L_0^{-1}\|_{\mathcal{L}(H)}\}|L_0u||L_1v| \\ \geq\ & -\frac{1}{2}(\|B_0\|_{\mathcal{L}(H)}h_1^{-1} + \|B_1\|_{\mathcal{L}(H)}h_0^{-1})(|L_0u|^2+|L_1v|^2).\end{aligned}$$

The bound (2.30) allows us to obtain an estimate (2.29) with

$$c_0 = 2^{-1}(\|B_0\|_{\mathcal{L}(H)}h_1^{-1} + \|B_1\|_{\mathcal{L}(H)}h_0^{-1}).$$

□

3. Examples

We shall indicate some possible applications of the abstract results to concrete differential equations. All examples could be generalized to more complicated situations, but we only have in mind to clarify the range of our theory, so we will confine to simple cases.

Example 3.1. Here we will consider optimal control of a trivial algebraic-differential equation, that is nevertheless enlighting the general treatment. Let us consider the system (1.4), (1.5), with cost functional (1.6), where

$$M = \begin{bmatrix} 0 & 0 \\ 0 & 1 \end{bmatrix}, \quad L = \begin{bmatrix} 1 & 0 \\ 0 & 1 \end{bmatrix}, \quad B = \begin{bmatrix} 1 & -1 \\ 1 & 0 \end{bmatrix},$$

$$C = \begin{bmatrix} 1 & 1 \\ 0 & 0 \end{bmatrix}, \quad N = \begin{bmatrix} 2 & 0 \\ 0 & 1 \end{bmatrix}.$$

The state variables are denoted (x, y), the controls are (u, v), while the initial condition reads $y(0) = y_0 \in R$. If (p, q) are the co-state variables, system (1.8)–(1.10) has the form

$$\frac{d}{dt}\begin{bmatrix} 0 & 0 \\ 0 & 1 \end{bmatrix}\begin{bmatrix} x \\ y \end{bmatrix} + \begin{bmatrix} x \\ y \end{bmatrix} + \frac{1}{2}\begin{bmatrix} 3 & 1 \\ 1 & 1 \end{bmatrix}\begin{bmatrix} p \\ q \end{bmatrix} = \begin{bmatrix} 0 \\ 0 \end{bmatrix}, \quad 0 < t < \tau, \tag{3.1}$$

$$-\frac{d}{dt}\begin{bmatrix} 0 & 0 \\ 0 & 1 \end{bmatrix}\begin{bmatrix} p \\ q \end{bmatrix} + \begin{bmatrix} p \\ yq \end{bmatrix} = \begin{bmatrix} 1 & 1 \\ 1 & 1 \end{bmatrix}\begin{bmatrix} x \\ y \end{bmatrix}, \quad 0 < t < \tau, \tag{3.2}$$

$$y(0) = y_0, \quad q(\tau) = 0. \tag{3.3}$$

Then $I + RS$, according to the notation in Section 2, is the operator associated to the matrix

$$\begin{bmatrix} 1 & 0 & 1/4 & -1/4 \\ 0 & 1 & -1/4 & -1/4 \\ 1/4 & -1/4 & 1 & 0 \\ -1/4 & -1/4 & 0 & 1 \end{bmatrix},$$

that is definite positive. Therefore Theorem 2.2 applies immediately. Of course, if we attempt to solve (3.1)–(3.3) directly, it is very easy to verify that $x = -\frac{3}{5}y - \frac{1}{5}q$, $p = \frac{2}{5}y - \frac{1}{5}q$ and the pair (y, q) satisfies the non-degenerate two-point problem (see Lions [10])

$$\frac{dy}{dt} = -\frac{6}{5}y - \frac{2}{5}q, \tag{3.4}$$

$$-\frac{dq}{dt} = -\frac{6}{5}q + \frac{2}{5}y, \tag{3.5}$$

together with the boundary conditions (3.3). We also point out that another possible direct treatment of problem (1.4)–(1.6) in this case would yield to minimize the integral

$$\int_0^\tau \{(u-v+y)^2+2u^2+v^2\}\, dt$$

subject to the differential constraint

$$\frac{dy}{dt}=-y+u, \quad y(0)=y_0, \quad u,v\in R.$$

Then the corresponding Euler equations are (see e.g. Burghes and Graham [2])

$$\frac{dr}{dt}=r-2u+2v-2y, \qquad 6u-2v+2y+r=0, \qquad -2u+4v-2y=0,$$

with the final condition $r(\tau)=0$. One then recognizes that all is reduced to the system

$$\frac{dy}{dt}=-\frac{6}{5}y-\frac{r}{5}, \qquad -\frac{dr}{dt}=-\frac{6}{5}r+\frac{4}{5}y.$$

But if $p=2r$, we immediately see that system (3.4), (3.5) is in fact obtained.

Example 3.2. Let Ω be a bounded domain in R^n, $n\geq 2$, with a smooth boundary $\partial\Omega$. In the cylinder $\Omega\times(0,\tau)$ we consider the initial-boundary value problem

$$\frac{\partial}{\partial t}(\lambda_0-\Delta)y=\alpha\Delta y-\beta\Delta^2 y+u, \quad (x,t)\in\Omega\times(0,\tau),$$

$$(\lambda_0-\Delta)y(x,0)=(\lambda_0-\Delta)y_0(x), \quad x\in\Omega,$$

$$y(x,t)=\Delta y(x,t)=0, \quad (x,t)\in\partial\Omega\times(0,\tau),$$

where λ_0 is the first negative eigenvalue of the Laplacian Δ with Dirichlet boundary conditions, $\alpha,\beta>0$, $y_0\in H_0^1(\Omega)\cap H^2(\Omega)$ is a given initial condition and $u\in L^2(\Omega\times(0,\tau))$ is the control. Similar equations model the evolution of a free surface of the filtered fluid. Sviridyuk and Efremov [12] investigated the spectral properties of the involved operators in $H=L^2(\Omega)$. If L, M are the operators in H associated to $-\alpha\Delta+\beta\Delta^2$ and $\lambda_0-\Delta$, respectively, so that

$$D(L)=\{u\in H_0^1(\Omega)\cap H^2(\Omega);\ \Delta u\in H_0^1(\Omega)\cap H^2(\Omega)\},\ D(M)=H_0^1(\Omega)\cap H^2(\Omega),$$

then, a reasoning as in Sviridyuk and Efremov [12], p. 1888 (see also Favini and Yagi [9], pp. 161–162) proves that $z=0$ is a simple pole for $L(zL+M)^{-1}$. Consider the cost functional

$$J(u)=\int_0^\tau \|y(t)\|^2_{L^2(\Omega)}\, dt+\int_0^\tau \|u(t)\|^2_{L^2(\Omega)}\, dt.$$

According to the notation in (1.8), (1.9), $B=B^*=N=C=I$. Moreover, since the operators M and L are self-adjoint and commute, we have $P=P^*$. Hence (2.29) is verified with $c_0=0$. Theorem 2.2 then affirms that the unique optimal control u^* for the minimum problem above is given by $u^* = -p$, where (y, p) is the (unique) solution to

$$\begin{aligned}
&\frac{\partial}{\partial t}(\lambda_0 - \Delta)y = \alpha\Delta y - \beta\Delta^2 y - p, \quad 0 < t < \tau,\ x \in \Omega,\\
&-\frac{\partial}{\partial t}(\lambda_0 - \Delta)p = \alpha\Delta p - \beta\Delta^2 p + y, \quad 0 < t < \tau,\ x \in \Omega,\\
&(\lambda_0 - \Delta)y(\cdot, 0) = (\lambda_0 - \Delta)y_0,\\
&(\lambda_0 - \Delta)p(\cdot, \tau) = 0,\\
&y(x,t) = \Delta y(x,t) = p(x,t) = \Delta p(x,t) = 0, \quad 0 < t < \tau,\ x \in \partial\Omega,
\end{aligned}$$

with $y_0 \in D(L)$.

Example 3.3. Let us consider the system

$$\begin{aligned}
&\frac{\partial}{\partial t}(1 + \frac{\partial^2}{\partial x^2})y = \frac{\partial^2 y}{\partial x^2} - q(x)y + m(x)u, \quad 0 < t < \tau,\ 0 < x < \pi,\\
&y(0,t) = y(\pi,t) = 0, \quad 0 < t < \tau,\\
&(1 + \frac{\partial^2}{\partial x^2})y(x,0) = (1 + \frac{\partial^2}{\partial x^2})y_0(x), \quad 0 < x < \pi.
\end{aligned}$$

Here we take $H = L^2(0,\pi)$, $D(L) = H_0^1(0,\pi) \cap H^2(0,\pi)$, $Ly = -\frac{d^2y}{dx^2} + qy$, $D(M) = D(L)$, $My = y + \frac{d^2y}{dx^2}$. It is assumed that $q, m \in C([0,\pi])$ and $q(x)$ has a non negative minimum over $[0,\pi]$. We have

$$\langle Ly, y\rangle = \int_0^\pi |y'(x)|^2\, dx + \int_0^\pi q(x)|y(x)|^2\, dx, \quad y \in D(L).$$

Moreover, it is readily seen that for all $y \in D(L)$

$$\int_0^\pi |y(x)|^2\, dx \le \frac{\pi^2}{2}\|y'\|_{L^2}^2.$$

Hence $\langle Ly, y\rangle \ge \frac{2}{\pi^2}\|y\|_{L^2}^2 + (\min_{[0,\pi]} q)\|y\|_{L^2}^2$, so that

$$\|Ly\|_{L^2} \ge (\frac{2}{\pi^2} + \min_{[0,\pi]} q)\|y\|_{L^2} \quad \text{and hence} \quad h_0 = h_1 = \frac{2}{\pi^2} + \min_{[0,\pi]} q.$$

It is well known that $z = 0$ is a simple pole for $(z + M)^{-1}$. Since L and M are self-adjoint and commute, the projection P_0 relative to $T_0 = ML^{-1}$ is self-adjoint, too. Let us associate to the problem above the cost functional

$$J(u) = \int_0^\tau \int_0^\pi c(x)y(t,x)^2\, dx\, dt + \int_0^\tau \int_0^\pi u(t,x)^2\, dx\, dt,$$

where c is a non negative continuous function on $[0,\pi]$. Since, see (2.1), (2.2), B_0 is multiplication by $-m(x)^2$ and $B_1 = C^*C$ is multiplication by $c(x)$, condition (2.30) reads

$$\max_{[0,1]} m(x)^2 + \max_{[0,1]} c(x) < 2\left(\frac{2}{\pi^2} + \min_{[0,1]} q(x)\right) = \frac{4}{\pi^2} + 2\min_{[0,1]} q(x). \tag{3.6}$$

Under these conditions the optimal control u^* minimizing $J(u)$ on $L^2((0,\pi)\times(0,\tau))$ exists and it is given by $u^* = -m^2 p$, where the pair (y,p) satisfies (1.8)–(1.10), with $N = I$. Clearly, the same argument extends to the n-dimensional case, where, for example, $L = -\Delta + q$ with Dirichlet boundary conditions, $M = \Delta + \lambda_0$, λ_0 an eigenvalue of the operator $-\Delta$.

References

[1] V. Barbu, A. Favini, *Control of degenerate differential systems,* Control and Cybernetics 28 (1999), 397–420.

[2] D. Burghes, A. Graham, "Introduction to Control Theory including Optimal Control", J. Wiley & Sons, New York-Chichester-Brisbane-Toronto, 1980.

[3] S. L. Campbell, "Singular Systems of Differential Equations II", Research Notes Math. 61, Pitman, San Francisco-London-Melbourne, 1982.

[4] J. D. Cobb, *Descriptor variable systems and optimal state regulation,* IEEE Trans. Aut. Control AC-28 (1983), 601–611.

[5] J. M. Cooper, *Two-point problems for abstract evolution equations,* J. Diff. Eqns. 9 (1971), 453–495.

[6] L. Dai, "Singular Control Systems", Lecture Notes in Control and Information Sciences 118, Springer-Verlag, Berlin-Heidelberg-New York-London-Paris-Tokyo, 1989.

[7] G. Da Prato, *Weak solutions for linear abstract differential equations in Banach spaces,* Advances in Math. 5 (1970), 181–245.

[8] A. Favini, A. Venni, *On a two-point problem for a system of abstract differential equations,* Numer. Funct. Appl. Optimiz. 2(4) (1980) 301–322.

[9] A. Favini, A. Yagi, "Degenerate Differential Equations in Banach Spaces", Monographs & Textbooks in Pure and Applied Math. 215, M. Dekker, New York-Basel-Hong Kong, 1999.

[10] J. L. Lions, "Optimal Control of Systems Governed by Partial Differential Equations", Die Grundlehren math. Wissenschaften in Einzeldarstellungen 170, Springer-Verlag, Berlin-Heidelberg-New York, 1971.

[11] L. Pandolfi, *On the regulator problem for linear degenerate control systems,* J. Opt. Theory Appl. 33 (1981), 241–254.

[12] G. A. Sviridyuk, A. A. Efremov, *Optimal control of Sobolev-type linear equations with relatively p-sectorial operators,* Diff. Uravn. 31 (1995), 1912–1916 ; English translation: Diff. Eqns. 31 (1995), 1882–1890.

Viorel Barbu
University of Iaşi
6600 Iaşi, Romania
e-mail: barbu@uaic.ro

Angelo Favini
Dipartimento di Matematica
Università degli Studi di Bologna
Piazza di Porta S. Donato 5
I-40126 Bologna, Italy
e-mail: favini@dm.unibo.it

Progress in Nonlinear Differential Equations
and Their Applications, Vol. 50, 39–48

Bounded Solutions for a Class of Quasi-linear Parabolic Problems with a Quadratic Gradient Term

Lucio Boccardo and Maria Michaela Porzio

1. Introduction and main results

In this paper we prove an existence result for a class of quasi-linear parabolic problems whose prototype is the following

$$\begin{cases} u_t - \operatorname{div}[\alpha(u)\nabla u] = \beta(u)|\nabla u|^2 + f(x,t) & \text{in} \quad \Omega_T, \\ u(x,t) = 0 & \text{on} \quad \partial\Omega \times (0,T), \\ u(x,0) = u_0 & \text{on} \quad \Omega. \end{cases} \tag{1}$$

Here $\Omega_T = \Omega \times (0,T)$, $T \in (0,+\infty)$, and Ω is a bounded open set of $\Re^N$ $(N \geq 3)$. We assume that α and β are positive continuous functions satisfying

$$\alpha \notin \mathrm{L}^1([0,+\infty)), \tag{2}$$

$$\frac{\beta}{\alpha} \in \mathrm{L}^1(\Re), \tag{3}$$

and

$$\sup_{\Re} \alpha \in \Re. \tag{4}$$

Notice that (3) and (4) imply that also β is in $\mathrm{L}^1(\Re)$. On the data u_0 and f we assume that there are non-negative functions belonging, respectively, to $\mathrm{L}^\infty(\Omega)$ and $\mathrm{L}^m(\Omega_T)$ with $m > 1 + N/2$.
An example of α and β is

$$\alpha(s) = \frac{1}{\sqrt{1+s^2}} \quad \text{and} \quad \beta(s) = \frac{1}{\sqrt{(1+s^2)^3}}. \tag{5}$$

Notice that the assumption done on β does not guarantee its boundedness. For example in (5) it is also possible to choose

$$\beta(s) = \frac{1}{\sqrt{(1+s^2)^3}} + \gamma(s), \tag{6}$$

with

$$\gamma(s) = \sum_{n=1}^{+\infty} \frac{n}{a_n} \left\{ \chi_{[n,n+\frac{a_n}{2}]}(s)(s-n) - 2\chi_{[n+\frac{a_n}{2},n+a_n]}(s)[s-(n+a_n)] \right\},$$

where $a_n = \frac{1}{2^{n+1}(n+1)^2}$ and $\chi_A(s)$ is the characteristic function of the set A.

There is an extensive literature concerning parabolic problems with a nonlinear term having a natural growth with respect to $|\nabla u|$ (i.e., of order $|\nabla u|^2$). Several papers impose further assumptions on it like a "controlled" growth or a sign condition. The existence of a bounded weak solution under rather general assumptions can be found in [1], [4], and [5]. Substantially the hypotheses done in these papers carried to the example (1) are the following

$$0 < \alpha_0 \leq \alpha(s) \leq \sup_{\Re} \alpha \in \Re, \tag{7}$$

$$\sup_{\Re} \beta \in \Re. \tag{8}$$

Here we have removed both these conditions. The main difficulties that arise in the proofs of the existence results are two.
The first one is that if we do not impose (7) we can have a lack of coerciveness (i.e $\inf_{\Re} \alpha$ can be zero).
Moreover, as just noticed in the example (6), we can have $\sup_{\Re} \beta = +\infty$ and this implies that the term

$$\beta(u)|\nabla u|^2 \leq B_u |\nabla u|^2,$$

is only a summable term when $u \in \mathrm{L}^\infty(\Omega_T) \cap \mathrm{L}^2(0,T;\mathrm{H}_0^1(\Omega))$ where a further difficulty is given by the fact that the constant B_u changes with u (and cannot be the same constant for every u in $\mathrm{L}^\infty(\Omega_T)$).
An open problem is to understand if it is possible to remove also the condition $\sup_{\Re} \alpha \in \Re$. We recall that such hypotheses arise to handle the part in time "u_t" and is not necessary in the stationary case (see [2]).
We state now our results more in details. Let us consider the following problem

$$\begin{cases} u_t - \operatorname{div}[a(x,t,u) \cdot \nabla u] + b(x,t,u,\nabla u) = f(x,t) & \text{in} \quad \Omega_T, \\ u(x,t) = 0 & \text{on} \quad \partial\Omega \times (0,T), \\ u(x,0) = u_0 & \text{on} \quad \Omega, \end{cases} \tag{9}$$

where $\Omega_T = \Omega \times (0,T)$, $T \in (0,+\infty)$, and Ω is a bounded open set of $\Re^N$ ($N \geq 3$). We assume that $a : \Omega_T \times \Re \to \Re^{N^2}$ and $b : \Omega_T \times \Re \times \Re^N \to \Re$ are Caratheodory functions satisfying the following structure hypotheses

$$\sup_{|s|\leq k} |a(x,t,s)| \in L^\infty(\Omega_T), \quad \forall k > 0, \tag{10}$$

$$[a(x,t,s) \cdot \xi] \cdot \xi \geq \alpha(s)|\xi|^2, \tag{11}$$

$$|b(x,t,s,\xi)| \leq \beta(s)|\xi|^2, \tag{12}$$

where, as before, α and β are positive continuous functions verifying

$$\alpha \notin \mathrm{L}^1([0,+\infty)), \tag{13}$$

$$\frac{\beta}{\alpha} \in \mathrm{L}^1(\Re), \tag{14}$$

and

$$\sup_{\Re} \alpha \in \Re. \tag{15}$$

Hence if we define

$$\gamma(s) = \int_0^s \frac{\beta(\tau)}{\alpha(\tau)} d\tau, \tag{16}$$

and

$$A(s) = \int_0^s \alpha(\tau) d\tau, \tag{17}$$

it follows that the function γ is bounded, while the function A satisfies

$$\lim_{s \to +\infty} A(s) = +\infty.$$

Definition 1.1. *We will say that a function* $u \in \mathrm{C}(0,T;\mathrm{L}^2(\Omega)) \cap \mathrm{L}^2(0,T;\mathrm{W}_0^{1,2}(\Omega))$ *is a weak solution of* (9) *if* $a(x,t,u)\cdot\nabla u \in (\mathrm{L}^2(\Omega_T))^N$, $b(x,t,u,\nabla u) \in \mathrm{L}^1(\Omega_T)$ *and*

$$\int_\Omega u\varphi dx|_{t_1}^{t_2} + \int_{t_1}^{t_2}\int_\Omega \{-u\varphi_t + [a(x,t,u)\cdot\nabla u]\cdot\nabla\varphi\}$$
$$+ \int_{t_1}^{t_2}\int_\Omega b(x,t,u,\nabla u)\varphi = \int_{t_1}^{t_2}\int_\Omega f\varphi, \tag{18}$$

for every $\varphi \in \mathrm{W}^{1,2}(0,T;\mathrm{L}^2(\Omega)) \cap \mathrm{L}^\infty(\Omega_T) \cap \mathrm{L}^2(0,T;\mathrm{W}_0^{1,2}(\Omega))$, *and for every* $[t_1,t_2] \subseteq [0,T]$.

We will also say that φ *may be taken as a test function in* (18) *(or* φ *is an admissible test function) if the functions* $[a(x,t,u)\cdot\nabla u]\cdot\nabla\varphi$, $b(x,t,u,\nabla u)\varphi$ *and* $f\varphi$ *belong to* $\mathrm{L}^1(\Omega_T)$ *and* (18) *holds true.*

We have the following result

Theorem 1.1. *Let* (10)–(15) *hold true. Assume that*

$$u_0 \in \mathrm{L}^\infty(\Omega), \quad u_0 \geq 0, \tag{19}$$

and

$$f \in \mathrm{L}^m(\Omega_T), \quad m > 1 + \frac{N}{2}, \quad f \geq 0. \tag{20}$$

Then there exists a nonnegative bounded weak solution of (9).

Remark 1.1. *We notice that in* (20) *we have done the same summability assumption that is done in the classical case (i.e., when the function* $a(x,t,s)\cdot\xi$ *satisfies the Leray-Lions conditions) to guarantee bounded solutions.*

Remark 1.2. *The sign assumptions on* f *and* u_0 *are done to have nonnegative solutions.*

The proof of Theorem 1.1 is in Section 4 and the main step is an a priori $\mathrm{L}^\infty(\Omega_T)$ estimate (see Section 3). To derive such an estimate we need some a priori technical inequalities proved in Section 2.

2. Technical inequalities

As said above, we prove here some integral inequalities that will be essential to prove an a priori $L^\infty(\Omega_T)$ estimate on the solutions of (9).

Lemma 2.1. *Let u be a weak nonnegative solution of (9) and let (10)–(15), (19) and (20) be satisfied. Assume that Ψ is a locally Lipschitz continuous and increasing real function such that $\Psi(0) = 0$. If $\Psi(u) \in L^\infty(\Omega_T) \cap L^2(0,T;W_0^{1,2}(\Omega))$; then it results*

$$\sup_{\tau\in[0,T]} \int_\Omega \psi(u(\tau))dx + \iint_{\Omega_T} e^{\gamma(u)}\Psi'(u)\alpha(u)|\nabla u|^2$$
$$\leq \int_\Omega \psi(u_0)dx + \iint_{\Omega_T} f e^{\gamma(u)}\Psi(u), \tag{1}$$

where $\psi(s) = \int_0^s e^{\gamma(z)}\Psi(z)dz$, γ is as in (16) and all the integrals that appear in (1) are finite.

Proof. Let us define

$$\varphi = e^{\gamma(T_k(u))}\Psi(T_k(u))$$

where $T_k(u)$ is the usual truncation function at levels $\pm k$, that is

$$T_k(s) = \max\{-k, \min\{k,s\}\}. \tag{2}$$

Notice that $\varphi \in L^\infty(\Omega_T) \cap L^2(0,T;W_0^{1,2}(\Omega))$. Using φ as a test function in (18) (the use of φ as a test function can be made rigorous using the Steklov averaging process) with $[t_1,t_2] = [0,\tau]$, $\tau \leq T$, we obtain

$$\int_\Omega [\psi_k(u(\tau)) - \psi_k(u_0)]\,dx + \iint_{\Omega_\tau} [a(x,t,u)\cdot\nabla u]\cdot\nabla\varphi$$
$$+ \iint_{\Omega_\tau} b(x,t,u,\nabla u)\varphi = \iint_{\Omega_\tau} f\varphi, \tag{3}$$

where we have set $\psi_k(s) = \int_0^s e^{\gamma(T_k(z))}\Psi(T_k(z))dz$ and $\Omega_\tau = \Omega \times (0,\tau)$.
We estimate now the integrals in (3). We notice that as $e^{\gamma(T_k(z))} \leq e^{\|\gamma\|_\infty}$ and Ψ is an increasing function, we have

$$0 \leq \psi_k(u) \leq e^{\|\gamma\|_\infty}\Psi(u)u \in L^1(\Omega_T).$$

Hence we can apply the dominated convergence Theorem in the first integral in (3) to conclude that

$$\exists \lim_{k\to+\infty} \int_\Omega [\psi_k(u(\tau)) - \psi_k(u_0)]\,dx = \int_\Omega [\psi(u(\tau)) - \psi(u_0)]\,dx,$$

where $\psi(s) = \int_0^s e^{\gamma(z)}\Psi(z)dz$. For the second integral in (3) we have, recalling the definition of γ

$$\begin{aligned}
&\iint_{\Omega_T} [a(x,t,u)\cdot\nabla u]\cdot\nabla\varphi \\
&= \iint_{\Omega_T} e^{\gamma(T_k(u))}\frac{\beta(T_k(u))}{\alpha(T_k(u))}\Psi(T_k(u))[a(x,t,u)\cdot\nabla u]\cdot\nabla T_k(u) \\
&+ \iint_{\Omega_T} e^{\gamma(T_k(u))}\Psi'(T_k(u))[a(x,t,u)\cdot\nabla u]\cdot\nabla T_k(u). \qquad (4)
\end{aligned}$$

Notice that the first integral in the right-hand side can be written as follows

$$\iint_{\Omega_T\cap\{u<k\}} e^{\gamma(u)}\frac{\beta(u)}{\alpha(u)}\Psi(u)[a(x,t,u)\cdot\nabla u]\cdot\nabla u,$$

where the integrand function is nonnegative. Hence we can apply the monotone convergence Theorem and conclude that as $k\to+\infty$ it tends to

$$\iint_{\Omega_T} e^{\gamma(u)}\frac{\beta(u)}{\alpha(u)}\Psi(u)[a(x,t,u)\cdot\nabla u]\cdot\nabla u.$$

Notice that at the moment the previous integral can be a positive number or $+\infty$: we will prove that really it is a finite number. Observe that, as by assumption $\Psi(u)$ belongs to $\mathrm{L}^2(0,T;\mathrm{W}_0^{1,2}(\Omega))$ it follows that $\Psi'(u)\nabla u\in\left(\mathrm{L}^2(\Omega_T)\right)^N$. Moreover $e^{\gamma(u)}$ is bounded and $a(x,t,u)\cdot\nabla u\in\mathrm{L}^2(\Omega_T)$. Hence the following integral is finite

$$\iint_{\Omega_T} e^{\gamma(u)}\Psi'(u)[a(x,t,u)\cdot\nabla u]\cdot\nabla u.$$

Besides, being u nonnegative, it results

$$\begin{aligned}
&\exists\lim_{k\to+\infty}\iint_{\Omega_T} e^{\gamma(T_k(u))}\Psi'(T_k(u))[a(x,t,u)\cdot\nabla u]\cdot\nabla T_k(u) \\
&= \lim_{k\to+\infty}\iint_{\Omega_T\cap\{u<k\}} e^{\gamma(u)}\Psi'(u)[a(x,t,u)\cdot\nabla u]\cdot\nabla u \\
&= \iint_{\Omega_T} e^{\gamma(u)}\Psi'(u)[a(x,t,u)\cdot\nabla u]\cdot\nabla u.
\end{aligned}$$

Observe that as $\Psi(u)\in\mathrm{L}^\infty(\Omega_T)$, by assumption and using again the boundedness of the function $e^{\gamma(s)}$, we can apply the dominated convergence Theorem to deduce that the third term in (3) satisfies

$$\begin{aligned}
&\exists\lim_{k\to+\infty}\iint_{\Omega_T} b(x,t,u,\nabla u)e^{\gamma(T_k(u))}\Psi(T_k(u)) \\
&= \iint_{\Omega_T} b(x,t,u,\nabla u)e^{\gamma(u)}\Psi(u).
\end{aligned}$$

Analogously we have

$$\exists\lim_{k\to+\infty}\iint_{\Omega_T} fe^{\gamma(T_k(u))}\Psi(T_k(u)) = \iint_{\Omega_T} fe^{\gamma(u)}\Psi(u).$$

Putting together all the previous estimate and passing to the limit on k we can conclude that

$$\begin{aligned}&\int_\Omega \psi(u(\tau))dx + \iint_{\Omega_\tau} e^{\gamma(u)}\frac{\beta(u)}{\alpha(u)}\Psi(u)[a(x,t,u)\cdot\nabla u]\cdot\nabla u\\&+\iint_{\Omega_\tau} e^{\gamma(u)}\Psi'(u)[a(x,t,u)\cdot\nabla u]\cdot\nabla u + \iint_{\Omega_\tau} b(x,t,u,\nabla u)e^{\gamma(u)}\Psi(u)\\&=\int_\Omega \psi(u_0)dx + \iint_{\Omega_\tau} fe^{\gamma(u)}\Psi(u).\end{aligned}\tag{5}$$

Notice that now we can affirm that the second integral in (5) is finite as are finite all the remaining integrals in (5).
Finally using the assumptions (11) and (12) we have that

$$\begin{aligned}&\iint_{\Omega_\tau} e^{\gamma(u)}\Psi((u))\left[\frac{\beta(u)}{\alpha(u)}[a(x,t,u)\cdot\nabla u]\cdot\nabla u + b(x,t,u,\nabla u)\right]\\&\geq \iint_{\Omega_\tau} e^{\gamma(u)}\Psi((u))\left[\frac{\beta(u)}{\alpha(u)}\alpha(u)|\nabla u|^2 - \beta(u)|\nabla u|^2\right] = 0,\end{aligned}$$

which with (5) gives

$$\begin{aligned}&\int_\Omega \psi(u(\tau))dx + \iint_{\Omega_\tau} e^{\gamma(u)}\Psi'(u)[a(x,t,u)\cdot\nabla u]\cdot\nabla u\\&\leq \int_\Omega \psi(u_0)dx + \iint_{\Omega_\tau} fe^{\gamma(u)}\Psi(u).\end{aligned}$$

Now the result follows taking the supremum for $\tau \in (0,T]$ and using again the assumption (11) and the hypothesis of increasing function done on Ψ. □

Noticing that

$$0 < \inf_{\Re} e^{\gamma(s)} \leq e^{\gamma(s)} \leq \sup_{\Re} e^{\gamma(s)} \in \Re,$$

an immediate consequence of the previous Lemma is the following

Lemma 2.2. *Under the assumptions of Lemma* 2.1 *there exists a positive constant* c_0 *depending only on* γ *such that the following integral estimate holds*

$$\begin{aligned}&\sup_{\tau\in[0,T]}\int_\Omega\left[\int_0^{u(\tau)}\Psi(z)dz\right]dx + \iint_{\Omega_T}\Psi'(u)\alpha(u)|\nabla u|^2\\&\leq c_0\int_\Omega\left[\int_0^{u_0}\Psi(z)dz\right]dx + c_0\iint_{\Omega_T} f\Psi(u).\end{aligned}\tag{6}$$

3. $L^\infty(\Omega_T)$-estimates

Using the results of the previous section we prove here the following $L^\infty(\Omega_T)$ a priori estimate.

Theorem 3.1. *Assume that the hypotheses of Theorem* 1.1 *hold true. If u is a nonnegative weak solution of* (9) *such that $A(u) \in L^\infty(\Omega_T) \cap L^2(0,T;W_0^{1,2}(\Omega))$ (here $A(u)$ is as in* (17)*) then the following estimates hold true*

$$\|A(u)\|_{L^\infty(\Omega_T)} \le c_1, \tag{1}$$

where c_1 is a positive constant depending only on m, $\|f\|_{L^m(\Omega_T)}$, $|\Omega|$, γ, N, $\|u_0\|_{L^\infty(\Omega)}$ and $\sup_{\Re}\alpha$. Hence it results

$$\|u\|_{L^\infty(\Omega_T)} \le A^{-1}(c_1). \tag{2}$$

Proof. Let $G_k(z) = \operatorname{sgn}(s)(|s|-k)_+$, where $k > A(\|u_0\|_{L^\infty})$. Notice that $\Psi(s) = G_k(A(s))$ satisfies the assumptions of Lemma 2.2.

Moreover, being $A(u) \in L^\infty(\Omega_T) \cap L^2(0,T;W_0^{1,2}(\Omega))$, also $\Psi(u) = G_k(A(u))$ belongs to $L^\infty(\Omega_T) \cap L^2(0,T;W_0^{1,2}(\Omega))$. Hence we can apply Lemma 2.2 and deduce that

$$\sup_{\tau\in[0,T]} \int_\Omega \left[\int_0^{u(\tau)} G_k(A(z))dz\right] dx + \iint_{\Omega_T\cap\{A(u)>k\}} \alpha^2(u)|\nabla u|^2$$
$$\le c_0 \int_\Omega \left[\int_0^{u_0} G_k(A(z))dz\right] dx + c_0 \iint_{\Omega_T} fG_k(A(u)). \tag{3}$$

We estimate now the integrals in (3). We have

$$\int_0^{u(\tau)} G_k(A(z))dz = \int_0^{u(\tau)} (A(z)-k)_+dz = V(k)\int_{A^{-1}(k)}^{u(\tau)} (A(z)-k)dz$$
$$= V(k)\left\{\int_{A^{-1}(k)}^{u(\tau)} A(z)dz - k[u(\tau) - A^{-1}(k)]\right\},$$

where we have set

$$V(k) = \begin{cases} 1, & \text{if} \quad u(\tau) > A^{-1}(k), \\ 0, & \qquad \text{otherwise.}\end{cases}$$

Notice that it results

$$\int_{A^{-1}(k)}^{s} A(z)dz - k[s - A^{-1}(k)] \ge c_2[G_k(A(s))]^2,$$

for every $s \ge A^{-1}(k)$ if we choose $c_2 = (4\sup_{\Re}\alpha)^{-1}$ (really it is true for every $c_2 \le (2\sup_{\Re}\alpha)^{-1}$).

Hence we can conclude that

$$\sup_{\tau\in[0,T]} \int_\Omega \left[\int_0^{u(\tau)} G_k(A(z))dz\right] dx \ge c_2 \sup_{\tau\in[0,T]} \int_\Omega |G_k(A(u(\tau)))|^2 dx.$$

Moreover, since $k > A(\|u_0\|_{L^\infty})$, it results that

$$c_0 \int_\Omega \left[\int_0^{u_0} G_k(A(z))dz \right] dx = 0.$$

Putting together all the previous estimates we obtain

$$\begin{aligned} & c_2 \sup_{\tau \in [0,T]} \int_\Omega |G_k(A(u(\tau)))|^2 dx + \iint_{\Omega_T} |\nabla G_k(A(u))|^2 \\ & \leq c_0 \iint_{\Omega_T} |f| G_k(A(u)). \end{aligned} \tag{4}$$

From (4) it follows (see [3]) that $A(u)$ verifies the following estimate

$$\|A(u)\|_{L^\infty(\Omega_T)} \leq c_1,$$

where c_1 is a constant depending only on $\sup_{\Re} \alpha$, γ, $\|u_0\|_{L^\infty(\Omega)}$, N, $|\Omega|$, m and $\|f\|_{L^m(\Omega_T)}$. □

4. Proof of Theorem 1.1

Let us consider the following approximating problems

$$\begin{cases} (u_n)_t - \operatorname{div}[a_n(x,t,u_n) \cdot \nabla u_n] + b_n(x,t,u_n,\nabla u_n) = f, \text{ in } \Omega_T, \\ u_n(x,t) = 0, \quad \text{on} \quad \partial\Omega \times (0,T), \\ u_n(x,0) = u_0 \quad \text{in} \quad \Omega, \end{cases} \tag{1}$$

where

$$\begin{cases} a_n(x,t,s) \equiv a(x,t,T_n(s)) \\ \alpha_n(s) \equiv \alpha(T_n(s)) \\ \beta_n(s) \equiv \alpha_n(s) \frac{\beta(s)}{\alpha(s)} \end{cases} \tag{2}$$

and

$$b_n(x,t,s,\xi) \equiv \min\left\{T_n(\beta_n(s)|\xi|^2), \max[-T_n(\beta_n(s)|\xi|^2), b(x,t,s,\xi)]\right\}. \tag{3}$$

Notice that from assumption (10) it follows that $|a_n|$ is bounded from above, i.e., there exists a positive constant Λ_n such that

$$|a_n(x,t,s)| \leq \Lambda_n.$$

Moreover, being α a continuous function, there is a positive constant λ_n such that

$$\alpha_n(s) \geq \lambda_n.$$

Hence it results

$$\lambda_n|\xi|^2 \leq \alpha_n(s)|\xi|^2 \leq [a_n(x,t,s) \cdot \xi] \cdot \xi \leq \Lambda_n|\xi|^2. \tag{4}$$

On the other hand, the function b_n is bounded and verifies

$$|b_n(x,t,s,\xi)| \leq T_n(\beta_n(s)|\xi|^2) \leq \beta_n(s)|\xi|^2.$$

Thus by the classical theory for every $n \in N$ there exists $u_n \in C(0,T;L^2(\Omega)) \cap L^2(0,T;W_0^{1,2}(\Omega))$ nonnegative weak solution of (1). Besides every solution of (1) is bounded in Ω_T (see Theorem 2.1 in [3]). Notice that $\frac{\beta_n}{\alpha_n}$ and β_n are $L^1(\Re)$ functions. As a matter of fact we have

$$\frac{\beta_n}{\alpha_n} = \frac{\beta}{\alpha}, \quad \text{and} \quad \beta_n \leq [\max_{|s|\leq n} \alpha(s)] \frac{\beta}{\alpha}.$$

Moreover the functions $A_n(s) = \int_0^s \alpha_n(\tau)d\tau$ are continuous and strictly increasing. Notice that

$$\alpha_n(s) = \alpha(T_n s) \leq \sup_{\Re} \alpha. \tag{5}$$

From the boundedness of u_n and (5) it follows that $A_n(u_n)$ belongs to $L^\infty(\Omega_T) \cap L^2(0,T; W_0^{1,2}(\Omega))$. Hence we can apply Theorem 3.1 and conclude that

$$\|u_n\|_{L^\infty(\Omega_T)} \leq A_n^{-1}(c_1),$$

where c_1 is as in Theorem 3.1, that is a constant independent on n. Notice that the sequence $A_n^{-1}(c_1)$ converges to $A^{-1}(c_1)$ as n goes to infinity. Thus there exists a constant c_3, independent on n, such that

$$\|u_n\|_{L^\infty(\Omega_T)} \leq c_3. \tag{6}$$

Notice that when $|s| \leq n$ it results

$$\begin{cases} a_n(x,t,s) = a(x,t,s) \\ \alpha_n(s) = \alpha(s) \\ \beta_n(s) = \beta(s) \\ b_n(x,t,s,\xi) = b(x,t,s,\xi). \end{cases} \tag{7}$$

Hence by (6) it follows that for every $n \geq c_3$ our approximating problem (1) becomes

$$\begin{cases} (u_n)_t - \operatorname{div}[a(x,t,u_n) \cdot \nabla u_n] + b(x,t,u_n,\nabla u_n) = f, \text{ in } \Omega_T, \\ u_n(x,t) = 0, \quad \text{on} \quad \partial\Omega \times (0,T), \\ u_n(x,0) = u_0 \quad \text{in} \quad \Omega. \end{cases} \tag{8}$$

By (6) and the continuity of α it follows also that there exists a positive constant λ, independent on n, such that

$$\alpha(u_n) \geq \lambda, \quad \forall n \geq c_3,$$

and thus by (11)

$$[a(x,t,u_n) \cdot \nabla u_n] \cdot \nabla u_n \geq \lambda |\nabla u_n|^2. \tag{9}$$

Notice that by (10) and again (6) it follows that

$$|a(x,t,u_n)| \leq \Lambda, \tag{10}$$

where Λ is a positive constant independent on n. Obviously it results

$$[a(x,t,s) \cdot \xi - a(x,t,s) \cdot \xi'] \cdot (\xi - \xi') > 0, \text{ if } \xi \neq \xi'. \tag{11}$$

Finally, using (6), (12) and the continuity of the function β we deduce that there exists a positive constant B (independent on n) such that

$$|b(x,t,u_n,\nabla u_n)| \leq \beta(u_n)|\nabla u_n|^2 \leq B|\nabla u_n|^2. \tag{12}$$

Thanks to the properties (9)–(12) and the a priori $L^\infty(\Omega_T)$-estimate (6) it is possible to proceed exactly as in [5] and conclude that there exists a subsequence of u_n, that we denote again u_n, such that

$$\begin{cases} u_n \to u & \text{a.e.} \quad \text{in} \quad \Omega_T, \\ u_n \to u & \text{strong} \quad \text{in} \quad L^2(0,T;W_0^{1,2}(\Omega)), \\ \nabla u_n \to \nabla u & \text{a.e.} \quad \text{in} \quad \Omega_T. \end{cases}$$

Hence we can pass to the limit (as n tends to $+\infty$) in (8) obtaining that u is a nonnegative, bounded weak solution of (9). □

References

[1] L. Boccardo, F. Murat, J.P. Puel, *Existence results for some quasilinear parabolic equations*, Nonlinear Anal. T.M.A., **13**, (1989), 373–392.

[2] L. Boccardo, S. Segura de León, C. Trombetti, *Bounded and unbounded solutions for a class of quasi-linear elliptic problems with a quadratic gradient term*, preprint.

[3] O.A. Ladyzenskaja, V.A. Solonnikov, N.N. Ural'ceva, *Linear and quasi-linear equations of parabolic type*, Translations of Math. Monographs, Vol. 23, Providence 1968.

[4] A. Mokrane, *Existence of bounded solutions for some nonlinear parabolic equations*, Proc. Roy. Soc. Edinburgh, **107A**, (1987), 313–326.

[5] L. Orsina, M.M. Porzio, *$L^\infty(Q)$-estimates and existence of solutions for some nonlinear parabolic equations*, Boll. U.M.I. (7), **6-B**, (1992), 631–647.

Lucio Boccardo
Dipartimento di Matematica
Università di Roma 1
Piazza A. Moro 2
I-00185 Roma
e-mail: boccardo@mat.uniroma1.it

Maria Michaela Porzio
Università del Sannio
Facoltà di Scienze MM. FF. NN.
via Port'Arsa 11
I-82100 Benevento
e-mail: porzio@unisannio.it.

Progress in Nonlinear Differential Equations
and Their Applications, Vol. 50, 49–78

Degenerate Evolution Systems Modeling the Cardiac Electric Field at Micro- and Macroscopic Level

Piero Colli Franzone and Giuseppe Savaré

Dedicated to the memory of Brunello Terreni.

1. Introduction

The aim of this paper is to study the reaction-diffusion systems arising from the mathematical models of the cardiac electric activity at the micro- and macroscopic level.

Recent theoretical and computational advanced studies in electrocardiology investigating the electrical behavior of the anisotropic cardiac tissue are based on the so-called "bidomain" model as a representation of the macroscopic tissue properties [41, 16, 23]. This approach conceives the cardiac muscle, despite its discrete structure, as the coupling of two anisotropic continuous superimposed domains. We refer to [24] for an extensive review on the use of the bidomain approach for simulating the electrical behavior of the cardiac tissue. Recently in [37, 30] it was presented a first formal derivation of the bidomain continuous representation of the cardiac tissue from a discrete model consisting of a periodic network of interconnected cells.

A distinctive feature of the mathematical description of both models, i.e. the macroscopic bidomain and the microscopic cellular models, lies in the structure of the coupling between the intra- and extra-cellular media. Both models are reaction-diffusion systems (see e.g. [50, 12]) but of degenerate parabolic type.

The microscopic model of the cardiac tissue

At a microscopic level the cardiac cellular structure of the tissue can be viewed as composed by two volumes: the intra-cellular space (inside the cells) and the extra-cellular space (outside) separated by the active membrane.

More specifically at the microscopic level the cardiac structure is composed of a collection of elongated cardiac cells, connected end-to-end and/or side-to-side by junctions, surrounded by the extra-cellular fluid. The end-to-end contacts

---

This work was partially supported by grants of M.U.R.S.T. (cofin1999–9901107579, cofin2000-MM01151559) and of the Institute of Numerical Analysis of the C.N.R., Pavia, Italy.

forms the long fibers structure of the cardiac muscle whereas the presence of lateral junctions establishes a connection between the elongated fibers. Since the interconnection between cells has junction resistance comparable to that of the intra-cellular volume, we can consider the cardiac tissue as a single isotropic intramural connected domain Ω_i separated from the extra-cellular fluid Ω_e by a membrane surface Γ_m.

THE GEOMETRY AND THE MAIN PHYSICAL QUANTITIES. Therefore

$$\begin{gathered}\Omega := \Omega_i \cup \Omega_e \cup \Gamma_m \subset \mathbb{R}^3 \text{ is the physical region occupied by the heart,}\\ \Omega_{i,e} \text{ are the } \textit{disjoint intra- and extra-cellular domains,}\\ \text{their interface } \overline{\Gamma}_m = \partial\Omega_i \cap \partial\Omega_e \text{ is the } \textit{active membrane.}\end{gathered} \tag{μ_1}$$

We denote by u_i, u_e

$$\begin{gathered}\text{the } \textit{intra- and extra-cellular electric potentials} \quad u_{i,e} : \overline{\Omega_{i,e}} \to \mathbb{R},\\ \text{whose difference is the } \textit{transmembrane potential} \quad v := u_i - u_e : \Gamma_m \to \mathbb{R},\end{gathered} \tag{μ_2}$$

and by σ_i, σ_e

$$\text{the } \textit{intra- and extra-cellular conductivities} \quad \sigma_{i,e} : \overline{\Omega_{i,e}} \to \mathbb{R}, \quad \inf_{\Omega_{i,e}} \sigma_{i,e} > 0. \tag{μ_3}$$

BASIC EQUATIONS. Due to the current conservation law, the normal current flux through the membrane is continuous: if ν_i, ν_e denote the unit exterior normals to the boundary of Ω_i and Ω_e respectively, satisfying $\nu_i = -\nu_e$ on Γ_m, we have

$$\sigma_i \nabla u_i \cdot \nu_i + \sigma_e \nabla u_e \cdot \nu_e = 0, \quad \text{on } \Gamma_m.$$

On the other hand, since the only active source elements lie on the membrane Γ_m, each flux equals the membrane current per unit area I_m, which consists of a capacitive and a ionic term (see [27]):

$$\sigma_i \nabla u_i \cdot \nu_i + I_m = -\sigma_e \nabla u_e \cdot \nu_e + I_m = 0, \quad I_m := \mathsf{C}_m \partial_t v + I_{ion}, \quad \text{on } \Gamma_m, \tag{μ_4}$$

where C_m is the surface capacitance of the membrane; in particular, Γ_m is a discontinuity surface for the potential.

Moreover, denoting by I_i^s, I_e^s the (given) stimulation currents applied to the intra- and extra-cellular space, we have

$$-\operatorname{div}(\sigma_i \nabla u_i) = I_i^s, \quad \text{in } \Omega_i, \qquad -\operatorname{div}(\sigma_e \nabla u_e) = I_e^s, \quad \text{in } \Omega_e. \tag{μ_5}$$

If suitable (lateral and initial) boundary conditions are provided, to complete the system (μ_4–μ_5) a further constitutive description of the ionic current I_{ion} is needed.

THE IONIC CURRENT: HODGKIN-HUXLEY MODEL. Dynamics structure of the ionic models have been developed extending the Hodgkin-Huxley formalism [25] to the ionic current source density of the cardiac membrane (see e.g. [20, 34, 18]) and are of the type

$$I_{ion} = I_{ion}(v, w_1, \dots, w_k) = \sum_n I_n(v, w_1, \dots, w_k), \tag{1.1}$$

where $w_1, \dots, w_k$ are additional *gating variables* taking their values between 0 and 1 and each I_n is a polynomial function of the form

$$I_n(v, w_1, \dots, w_k) := \mathsf{G}_n \left(\Pi_{j=1}^k w_j^{p_{j,n}}\right)(v - \mathsf{v}_n).$$

Here $p_{j,n}$ are integers (some of which may be zero), G_n is the conductance, and the equilibrium potential v_n depends on the intra- and extra-cellular concentration of the ionic species n.

Relation (1.1) should be coupled with the system of first order differential equations linking each gating variable to the transmembrane potential v: it has the form

$$\partial_t w_j = r_j(v, w_j) := \alpha_j(v)\,(1 - w_j) \; - \beta_j(v)\, w_j, \quad j = 1, \dots, k, \tag{1.2}$$

where α_j and β_j are in general *positive* rational functions of exponentials in v.

THE FITZHUGH-NAGUMO SIMPLIFICATION AND THE RECOVERY VARIABLE. Simplified models are often used for simulating the propagation of excitation wavefronts in large myocardial domains. In this work we first focus on a excitable model of the FitzHugh-Nagumo type, which was first introduced as a simplified membrane kinetic of the Hodgkin-Huxley equations in the description of the transmission of nervous electric impulses (see e.g. [14, 36]). It requires only one additional

$$\textit{recovery variable} \quad w : \Gamma_m \to \mathbb{R}, \tag{μ_6}$$

and (1.1), (1.2) reduce to

$$\begin{cases} I_{ion} = I_{ion}(v, w) := F(v) + \Theta w, \\ \partial_t w = r(v, w) := \eta v - \gamma w, \end{cases} \quad \text{on } \Gamma_m,$$
$$\text{where } \Theta, \eta, \gamma \geq 0 \text{ are given constants and} \tag{μ_7}$$
$$F \in C^1(\mathbb{R}) \text{ is a cubic-like function with } \inf_{\mathbb{R}} F' > -\infty.$$

We refer to (μ_1,. . . ,μ_7) as the *microscopic model*, together with Neumann boundary conditions imposed on u_i, u_e on the remaining part of the boundaries $\Gamma_{i,e} := \partial\Omega_{i,e} \setminus \Gamma_m$ respectively

$$\sigma_i \nabla u_i \cdot \nu_i = G_i \quad \text{on } \Gamma_i, \quad \sigma_e \nabla u_e \cdot \nu_e = G_e \quad \text{on } \Gamma_e, \tag{μ_8}$$

and with the (degenerate) initial Cauchy condition

$$v(x, 0) = v^0(x), \quad w(x, 0) = w^0(x), \quad \text{on } \Gamma_m. \tag{μ_9}$$

Thus our problem consists of two adjoining open domains with their boundaries partly intersecting, of a Poisson equation in each of them, and, on the common boundary, of two conditions connecting the fluxes and the potentials. In contrast to classical problems for the Poisson equation with a jump discontinuity across some surface, here the boundary conditions for the potentials is a dynamic boundary condition involving the assistant variable w.

The macroscopic "bidomain" model

At a macroscopic level, in spite of the discrete cellular structure, the cardiac tissue can be represented by a continuous model, called *bidomain model* (see e.g. [24, 17, 41] and also [30]), which attempts to describe the averaged electric potentials and current flows inside (intra-) and outside (extra-cellular space) the cardiac cells. The equations should result from an homogenization process as shown formally in [37] working on a scaled version of the cellular model on a periodic cardiac structure. For completeness we will present another formal derivation of the average model by using the standard two-scales method in the Appendix A and we defer to a forthcoming paper the rigorous mathematical justification of the model by the tools of Γ-convergence theory.

In this homogenized representation the heart domain coincides with the intra- (i) and extra-cellular (e) ones, which are two interpenetrating and superimposed continua connected at each point by the cardiac cellular membrane, i.e.

$$\Omega \equiv \Omega_i \equiv \Omega_e \subset \mathbb{R}^3 \text{ is the physical region occupied by the heart} \tag{M_1}$$

$$\begin{aligned} u_i, u_e : \Omega \to \mathbb{R} \quad &\text{are the intra- } (i) \text{ and extra-cellular } (e) \text{ electric potentials} \\ v := u_i - u_e : \Omega \to \mathbb{R} \quad &\text{is the transmembrane potential.} \end{aligned} \tag{M_2}$$

The anisotropy of the (i)–(e) media depends on the fiber structure of the myocardium. At the macroscopic level the fibers are regular curves, whose unit tangent vector at the point x is denoted by $\vec{a} = \vec{a}(x)$. Denoting by $\sigma^l_{i,e}(x)$, $\sigma^t_{i,e}(x)$ the conductivity coefficients along and across the fiber direction at a point x and always assuming axial symmetry for $\sigma^t_{i,e}$, the conductivity tensors $M_{i,e}$ in the media (i), (e), can be expressed by

$$M_{i,e}(x) = \sigma^t_{i,e} I + (\sigma^l_{i,e} - \sigma^t_{i,e})\vec{a} \otimes \vec{a},$$

and they are

$$\textit{symmetric, positive definite, continuous} \text{ tensors } M_{i,e} : \overline{\Omega} \to \mathbb{M}^{3\times 3}. \tag{M_3}$$

To the potentials u_i, u_e are associated the current densities $\vec{I}_{i,e} := -M_{i,e}\nabla u_{i,e}$; since induction effects are negligible, the current fields can be considered quasi-static and they are related to the membrane current per unit volume i_m and to the injected stimulating currents $i^s_{i,e}$ by the conservation laws

$$-\operatorname{div}(M_i \nabla u_i) = -i_m + i^s_i, \quad -\operatorname{div}(M_e \nabla u_e) = i_m + i^s_e, \quad \text{in } \Omega. \tag{M_4}$$

On the other hand the membrane current per unit volume i_m is the sum of a capacitance and a ionic term

$$i_m = \mathsf{c}_m \partial_t v + i_{ion} \quad \text{in } \Omega. \tag{M_5}$$

Again the FitzHugh-Nagumo model supply the form of the ionic current i_{ion} by means of a

$$\textit{recovery variable} \quad w : \Omega \to \mathbb{R}, \tag{M_6}$$

and of

$$\begin{cases} i_{ion} = i_{ion}(v, w) = f(v) + \theta w, \\ \partial_t w = r(v, w) = \eta v - \gamma w \end{cases} \quad \text{on } \Omega$$
$$\text{where } \theta, \eta, \gamma \geq 0 \text{ are given constants and} \tag{M_7}$$
$$f \in C^1(\mathbb{R}) \text{ is a cubic-like function with } \inf_{\mathbb{R}} f' > -\infty.$$

In fact, c_m and the forms of f, i_{ion} are related to the corresponding microscopic quantities by the relation

$$\mathsf{c}_m = \beta \mathsf{C}_m, \quad f(v) = \beta F(v), \quad \theta = \beta \Theta, \quad i_{ion}(v, w) = \beta I_{ion}(v, w),$$

where β is the ratio of membrane area per unit of tissue volume.

We complete the Reaction-Diffusion system (M_1,...,M_7) by imposing Neumann boundary conditions on u_i, u_e on $\partial\Omega$ respectively

$$M_i \nabla u_i \cdot \nu = g_i, \quad M_e \nabla u_e \cdot \nu = g_e, \tag{M_8}$$

and by assigning the (degenerate) initial Cauchy condition

$$v(x, 0) = v^0(x), \quad w(x, 0) = w^0(x), \quad \text{on } \Omega. \tag{M_9}$$

When $M_i = \lambda M_e$, with λ constant, the macroscopic system ($M_1 - M_9$) in the variables (u_i, u_e, w) is equivalent to a parabolic reaction-diffusion equation in $v = u_i - u_e$ coupled with the dynamics of the recovery variable w. This case is called in literature *equal anisotropic ratio* and this assumption is often used in modeling cardiac tissue. Experimental evidence indicates that this simplifying hypothesis is definitely not applicable to cardiac muscle. The degenerate structure of the mathematical model is primarily the result of differences in the intra- and extracellular anisotropy of the cardiac tissue. Moreover unequal anisotropic ratio makes possible more complex phenomena [52, 51] and can play an important role for the re-entrant excitation [53, 42].

Plan of the paper

In the two following sections we will propose a variational formulation of both the microscopic problem $(\mu_1, \ldots, \mu_9)$ and the macroscopic one $(M_1, \ldots, M_9)$. We will show that they have a common structure which fits into the abstract framework of (degenerate) evolution variational inequalities in Hilbert spaces; therefore, we will focus on the basic structural features of both the models, and we will state our main results about existence, uniqueness, and regularity of their solutions.

In § 4, we will present the abstract version of the evolution problems shared by the micro- and macroscopic models, we will discuss the links with the results available in the literature, and we will introduce an approach which allows us to eliminate the time degeneration; in this way it is possible to apply finer results which give a better insight into the properties of the solutions: in particular, general stability and approximation results could also be applied for the time discretization of both the systems.

The effort to enucleate a similar variational structure in the problems not only allows to reduce the length and the difficulty of the proofs, but also suggests a possible way to deduce the macroscopic setting from the microscopic one. However, the classical theory for convex evolution problems (see e.g. [1]) cannot be directly applied, due to the degeneracy and to the different kind of domains where the microscopic (Γ_m) and the macroscopic (Ω) quantities live. Nevertheless, we will show in a forthcoming work a possible way to overcome this difficulty; in the Appendix of the present paper, by developing formal asymptotic expansions using the two-scales method and neglecting the presence of stimulation currents, we will show how to convert the microscopic model of the cellular media into the averaged continuum "bidomain" representation of the cardiac tissue.

2. Variational formulation of the microscopic problem

In this section we will deal with the system $(\mu_1, \ldots, \mu_9)$ presented in the Introduction and we will rewrite it in a suitable variational form.

FIXING TECHNICAL ASSUMPTIONS AND NOTATION. Besides $(\mu_1, \ldots, \mu_9)$, we are also assuming that

$$\begin{aligned} \Omega_i, \Omega_e \quad & \text{are Lipschitz domains and at least one of them is connected,} \\ \Gamma_m \quad & \text{is a (not empty) Lipschitz hypersurface.} \end{aligned} \tag{μ_1'}$$

From the mathematical point of view, it is not always necessary to restrict the dimension of the ambient space, which we will denote by d. The regularity assumptions on F could be relaxed: so $F : \mathbb{R} \to \mathbb{R}$ could be a continuous function such that

$$F(0) = 0; \quad \exists\, \lambda_F \geq 0 : \quad \frac{F(x) - F(y)}{x - y} \geq -\lambda_F, \quad \forall\, x, y \in \mathbb{R}, \quad \text{with } x \neq y. \tag{μ_7'}$$

More general existence results in dimension $d = 3$ (see Proposition 2.4) will also require that F has a cubic growth at infinity, i.e.

$$0 < \liminf_{|s| \to +\infty} \frac{F(s)}{s^3} \leq \limsup_{|s| \to +\infty} \frac{F(s)}{s^3} < +\infty. \tag{μ_7''}$$

$]0, T[$ is the evolution time interval, and we define the associated space-time domains following the usual notation of [32]

$$Q_{i,e} := \Omega_{i,e} \times\,]0, T[, \qquad \Sigma_{i,e,m} := \Gamma_{i,e,m} \times\,]0, T[\,. \tag{2.1}$$

The formal statement of $(\mu_1 - \mu_9)$ is then:

Problem (μ). Given

$$I^s_{i,e} : Q_{i,e} \to \mathbb{R}, \quad G_{i,e} : \Sigma_{i,e} \to \mathbb{R}, \quad \text{and} \quad v^0, w^0 : \Gamma_m \to \mathbb{R}, \tag{2.2}$$

we seek

$$u_{i,e} : Q_{i,e} \to \mathbb{R}, \quad w : \Sigma_m \to \mathbb{R}, \quad v := u_i - u_e : \Sigma_m \to \mathbb{R},$$

satisfying the equations on $Q_{i,e}$ and $\Sigma_{i,e}$

$$\begin{cases} -\operatorname{div}(\sigma_{i,e}\nabla u_{i,e}) = I^s_{i,e}, & \text{in } Q_{i,e}, \\ \sigma_{i,e}\nabla u_{i,e}\cdot \nu_{i,e} = G_{i,e} & \text{on } \Sigma_{i,e}, \end{cases} \tag{2.3}$$

the evolution system on the surface Σ_m

$$\begin{cases} \mathsf{C}_m\partial_t v + F(v) + \Theta w = -\sigma_i \nabla u_i \cdot \nu_i = \sigma_e \nabla u_e \cdot \nu_e & \text{on } \Sigma_m, \\ \partial_t w + \gamma w - \eta v = 0 & \text{on } \Sigma_m, \\ v(x,0) = v^0(x), \quad w(x,0) = w^0(x) & \text{on } \Gamma_m. \end{cases} \tag{2.4}$$

Variational formulation

In order to write the variational formulation of Problem ($\boldsymbol{\mu}$), let us assume at first that for a.e. $t \in]0,T[$

$$I^s_{i,e}(\cdot,t) \in L^2(\Omega_{i,e}), \quad G_{i,e}(\cdot,t) \in H^{-1/2}(\Gamma_{i,e}), \tag{2.5}$$

$$u_{i,e}(\cdot,t),\ \partial_t u_{i,e}(\cdot,t) \in H^1(\Omega_{i,e}), \quad w(\cdot,t),\ \partial_t w(\cdot,t) \in L^2(\Gamma_m), \tag{2.6}$$

so that the trace operator $u_{i,e} \mapsto u_{i,e}|_{\Gamma_m}$ is well defined and maps continuously $H^1(\Omega_{i,e})$ onto $H^{1/2}(\Gamma_m)$ (cf. [32]). By using the simplified notation $v := u_i - u_e$ instead of $u_i|_{\Gamma_m} - u_e|_{\Gamma_m}$, we also suppose that

$$F(v(\cdot,t)) \in L^1(\Gamma_m) \cap H^{-1/2}(\Gamma_m). \tag{2.7}$$

We choose the test functions

$$\hat{u}_{i,e} \in H^1(\Omega_{i,e}), \quad \hat{w} \in L^2(\Gamma_m), \quad \text{with} \quad \hat{v} := \hat{u}_i - \hat{u}_e \in H^{1/2}(\Gamma_m), \tag{2.8}$$

and we multiply the two equations of the first row of (2.4) by the trace of $\hat{u}_i$ and $-\hat{u}_e$ respectively, and the next equation by $\rho\hat{w}$, ρ being the ratio Θ/η. Summing up them after an integration on Γ_m (we denote by $\mathcal{H}^{d-1}$ the usual $(d-1)$-dimensional Hausdorff measure) we get

$$\begin{aligned} &\int_{\Gamma_m} \big(\mathsf{C}_m\,\partial_t v\,\hat{v} + \rho\,\partial_t w\,\hat{w}\big)\,d\mathcal{H}^{d-1} \\ &\qquad + {}_{H^{-1/2}(\Gamma_m)}\langle F(v),\hat{v}\rangle_{H^{1/2}(\Gamma_m)} + \rho\gamma\int_{\Gamma_m} w\,\hat{w}\,d\mathcal{H}^{d-1} \\ &\qquad + \sum_{i,e}\int_{\Omega_{i,e}} \sigma_{i,e}\nabla u_{i,e}\cdot\nabla\hat{u}_{i,e}\,dx + \Theta\int_{\Gamma_m}(w\hat{v} - v\hat{w})\,d\mathcal{H}^{d-1} \\ &= \sum_{i,e}\int_{\Omega_{i,e}} I^s_{i,e}\,\hat{u}_{i,e}\,dx + \sum_{i,e} {}_{H^{-1/2}(\Gamma_{i,e})}\langle G_{i,e},\hat{u}_{i,e}\rangle_{H^{1/2}(\Gamma_{i,e})}, \end{aligned} \tag{2.9}$$

where we used (2.3) and the Green formulae

$$
\begin{aligned}
&{}_{H^{-1/2}(\Gamma_m)}\langle \sigma_{i,e}\nabla u_{i,e}\cdot \nu_{i,e}\, \hat u_{i,e}\rangle_{H^{1/2}(\Gamma_m)}\\
&\quad= \int_{\Omega_{i,e}} \Big(\sigma_{i,e}\nabla u_{i,e}\cdot\nabla \hat u_{i,e} + \operatorname{div}(\sigma_{i,e}\nabla u_{i,e})\,\hat u_{i,e}\Big)\,dx\\
&\qquad\qquad - {}_{H^{-1/2}(\Gamma_{i,e})}\langle \sigma_{i,e}\nabla u_{i,e}\cdot \nu_{i,e}, \hat u_{i,e}\rangle_{H^{1/2}(\Gamma_{i,e})}\\
&\quad= \int_{\Omega_{i,e}} \sigma_{i,e}\nabla u_{i,e}\cdot\nabla \hat u_{i,e}\,dx - \int_{\Omega_{i,e}} I^s_{i,e}\,\hat u_{i,e}\,dx - {}_{H^{-1/2}(\Gamma_{i,e})}\langle G_{i,e}, \hat u_{i,e}\rangle_{H^{1/2}(\Gamma_{i,e})}.
\end{aligned}
$$

which are justified by the usual arguments of [32]. Let us point out the particular structure of (2.9), which is common to more general situations.

"VECTOR" FUNCTIONAL SPACES. Denoting by boldface letters $\boldsymbol{u}, \hat{\boldsymbol{u}}, \dots$ the triple of functions (u_i, u_e, w), $(\hat u_i, \hat u_e, \hat w), \dots$, we introduce the product space

$$
\boldsymbol{X} := H^1(\Omega_i)\times H^1(\Omega_e)\times L^2(\Gamma_m), \tag{2.10}
$$

and the bilinear forms

$$
b(\boldsymbol{u},\hat{\boldsymbol{u}}) := \int_{\Gamma_m} \big[\mathsf{C}_m(u_i-u_e)(\hat u_i-\hat u_e) + \rho\, w\hat w\big]\,d\mathcal{H}^{d-1}, \tag{2.11}
$$

$$
\begin{aligned}
a(\boldsymbol{u},\hat{\boldsymbol{u}}) :=& \sum_{i,e}\int_{\Omega_{i,e}} \sigma_{i,e}\nabla u_{i,e}\cdot\nabla \hat u_{i,e}\,dx + \rho\gamma\int_{\Gamma_m} w\,\hat w\,d\mathcal{H}^{d-1}\\
&+\Theta\int_{\Gamma_m}\Big[w(\hat u_i-\hat u_e) - (u_i-u_e)\hat w\Big]\,d\mathcal{H}^{d-1},
\end{aligned} \tag{2.12}
$$

which are defined for every $\boldsymbol{u}, \hat{\boldsymbol{u}} \in \boldsymbol{X}$.

If (2.5) holds and v^0, w^0 belong to $L^2(\Gamma_m)$, then we can also associate to the right-hand member of (2.9) and to the initial data the time-depending family of linear functionals $\boldsymbol{L}(t)\in \boldsymbol{X}'$ and the linear functional $\boldsymbol{\ell}^0 \in \boldsymbol{X}'$ respectively, which act on a generic elements $\hat{\boldsymbol{u}} \in X$ in the following way

$$
\begin{cases}
\langle \boldsymbol{L}(t), \hat{\boldsymbol{u}}\rangle := \displaystyle\sum_{i,e}\int_{\Omega_{i,e}} I^s_{i,e}\,\hat u_{i,e}\,dx + \sum_{i,e} {}_{H^{-1/2}(\Gamma_{i,e})}\langle G_{i,e}, \hat u_{i,e}\rangle_{H^{1/2}(\Gamma_{i,e})},\\
\langle \boldsymbol{\ell}^0, \hat{\boldsymbol{u}}\rangle := \displaystyle\int_{\Gamma_m}\Big(\mathsf{C}_m\, v^0\,(\hat u_i-\hat u_e) + \rho\, w^0\,\hat w\Big)\,dx.
\end{cases} \tag{2.13}
$$

The remaining (non-linear term) $\mathfrak{F} : D(\mathfrak{F}) \subset \boldsymbol{X} \to \boldsymbol{X}'$ is defined by

$$
\langle \mathfrak{F}\boldsymbol{u}, \hat{\boldsymbol{u}}\rangle := {}_{H^{-1/2}(\Gamma_m)}\langle F(u_i-u_e), \hat u_i - \hat u_e\rangle_{H^{1/2}(\Gamma_m)}, \tag{2.14}
$$

for every $\boldsymbol{u}\in D(\mathfrak{F})$, $\hat{\boldsymbol{u}}\in \boldsymbol{X}$, with

$$
D(\mathfrak{F}) := \big\{\boldsymbol{u}\in \boldsymbol{X} : F(u_i-u_e)\in L^1(\Gamma_m)\cap H^{-1/2}(\Gamma_m)\big\}. \tag{2.15}
$$

We collect all these calculations in the following standard lemma.

Lemma 2.1. *Let $\boldsymbol{X}, b(\cdot,\cdot), a(\cdot,\cdot), \boldsymbol{L}, \mathfrak{F}$ be defined by* (2.10),...,(2.15) *respectively, and let us suppose that*

$$I^s_{i,e} \in L^2(Q) = L^2(0,T;L^2(\Omega_{i,e})), \quad G_{i,e} \in L^2(0,T;H^{-1/2}(\Gamma_{i,e})). \tag{2.16}$$

Then u_i, u_e, w satisfy (2.6), (2.7) *and solve Problem* ($\boldsymbol{\mu}$) *if and only if the function*

$$\boldsymbol{u}(t) := \big(u_i(\cdot,t), u_e(\cdot,t), w(\cdot,t)\big), \quad \textit{belongs to } W^{1,1}(0,T;\boldsymbol{X}), \tag{2.17}$$

and solves the abstract evolution equation

$$\begin{cases} b(\boldsymbol{u}',\hat{\boldsymbol{u}}) + a(\boldsymbol{u},\hat{\boldsymbol{u}}) + \langle \mathfrak{F}\boldsymbol{u},\hat{\boldsymbol{u}}\rangle = \langle \boldsymbol{L}(t),\hat{\boldsymbol{u}}\rangle, & \forall\,\hat{\boldsymbol{u}} \in \boldsymbol{X}, \quad \textit{a.e. in }]0,T[\,, \\ b(\boldsymbol{u}(0),\hat{\boldsymbol{u}}) = \langle \boldsymbol{\ell}^0,\hat{\boldsymbol{u}}\rangle, & \forall\,\hat{\boldsymbol{u}} \in X. \end{cases} \tag{2.18}$$

Remark 2.2. Different types of (variational) boundary conditions could also be considered in Problem ($\boldsymbol{\mu}$), by simply modifying $\boldsymbol{X}$ and $\boldsymbol{L}$. E.g., if we want to impose the homogeneous Dirichlet condition

$$u_e = 0, \quad \text{on } \Gamma_e, \tag{2.19}$$

instead of the corresponding Neumann one, we can replace $H^1(\Omega_e)$ by its closed subspace

$$H^1_{\Gamma_e}(\Omega_e) := \Big\{u \in H^1(\Omega_e) : u_{|\Gamma_e} = 0, \ \text{ in the sense of traces}\Big\}$$

in the definition (2.10) of $\boldsymbol{X}$.

Remark 2.3. We point out that (2.11), (2.12), and (2.14) are not modified by adding to $\boldsymbol{u}, \hat{\boldsymbol{u}}$ a constant vector $\mathbf{c}$ of the type

$$\mathbf{c} := (c,c,0), \quad \forall\, c \in \mathbb{R}. \tag{2.20}$$

Analogously, the solutions u_i, u_e, w of Problem ($\boldsymbol{\mu}$) can be determined at every time t up to this kind of additive constants. Therefore, it is natural to replace $\boldsymbol{X}$ by its quotient

$$\boldsymbol{V} := \Big(H^1(\Omega_i) \times H^1(\Omega_e) \times L^2(\Gamma_m)\Big)/\Big\{(c,c,0) : c \in \mathbb{R}\Big\} \tag{2.21}$$

and to look for $\boldsymbol{u}$ with values in $\boldsymbol{V}$; to simplify our notation, we still denote by (u_i, u_e, w) the corresponding equivalence class and take for granted to check the independence of the particular representative.
As usual, we can identify the dual $\boldsymbol{V}'$ of $\boldsymbol{V}$ with the closed subspace of $\boldsymbol{X}'$ whose elements $\boldsymbol{L}$ verify

$$\big\langle \boldsymbol{L},\mathbf{c}\big\rangle = 0, \quad \text{for every } \mathbf{c} \text{ given by (2.20)}. \tag{2.22}$$

It follows that we have to impose the compatibility condition

$$\sum_{i,e}\int_{\Omega_{i,e}} I^s_{i,e}\,dx + \sum_{i,e} {}_{H^{-1/2}(\Gamma_{i,e})}\langle G_{i,e},1\rangle_{H^{1/2}(\Gamma_{i,e})} = 0, \quad \text{a.e. in }]0,T[, \tag{2.23}$$

in order to have $\boldsymbol{L}(t) \in \boldsymbol{V}'$ in the case of (2.13). Finally, we note that $\mathfrak{F}$ maps $D(\mathfrak{F})$ into $\boldsymbol{V}'$ without any further restriction.

WHERE TO CHOOSE INITIAL DATA. In order to avoid extra regularity assumptions on the solution and to understand how to choose the correct functional framework for the initial data, we introduce the linear continuous operators $A, B : \boldsymbol{V} \to \boldsymbol{V}'$ associated to a, b respectively, i.e.

$$\big\langle A\boldsymbol{u}, \hat{\boldsymbol{u}}\big\rangle := a(\boldsymbol{u}, \hat{\boldsymbol{u}}), \quad \big\langle B\boldsymbol{u}, \hat{\boldsymbol{u}}\big\rangle := b(\boldsymbol{u}, \hat{\boldsymbol{u}}), \quad \forall\, \boldsymbol{u}, \hat{\boldsymbol{u}} \in \boldsymbol{V}. \tag{2.24}$$

Then we can rewrite (2.18) in the operator form

$$\big(B\boldsymbol{u}(t)\big)' + A\boldsymbol{u}(t) + \mathfrak{F}\boldsymbol{u}(t) = \boldsymbol{L}(t) \quad \text{in } \boldsymbol{V}', \quad \text{for a.e. } t \in]0, T[, \tag{2.25}$$

which suggests that (2.17) can be weakened by asking

$$\boldsymbol{u} \in L^2(0, T; \boldsymbol{V}), \quad B\boldsymbol{u} \in W^{1,1}_{loc}(0, T; \boldsymbol{V}') \cap C^0([0, T]; \boldsymbol{V}'), \tag{2.26}$$

so that we can give a meaning in $\boldsymbol{V}'$ to the initial value of $B\boldsymbol{u}$. Let us introduce the distribution spaces $\tilde{V}_b \subset \tilde{V}_b'$ on Γ_m

$$\tilde{V}_b := H^{1/2}(\Gamma_m) \times L^2(\Gamma_m), \quad \tilde{V}_b' := H^{-1/2}(\Gamma_m) \times L^2(\Gamma_m),$$

and let us observe that the operator $B : \boldsymbol{V} \to \boldsymbol{V}'$ admits the decomposition

$$B = \tilde{B}^* J_b \tilde{B}, \quad b(\boldsymbol{u}, \hat{\boldsymbol{u}}) = {}_{\boldsymbol{V}'}\langle B\boldsymbol{u}, \hat{\boldsymbol{u}}\rangle_{\boldsymbol{V}} = {}_{\tilde{V}_b'}\langle J_b \tilde{B}\boldsymbol{u}, \tilde{B}\hat{\boldsymbol{u}}\rangle_{\tilde{V}_b} \tag{2.27}$$

where $J_b : \tilde{V}_b \to \tilde{V}_b'$ is the inclusion map, $\tilde{B}$ is the linear surjection

$$\tilde{B} : \boldsymbol{V} \to \tilde{V}_b, \quad \tilde{B}\boldsymbol{u} := \Big(\sqrt{\mathsf{C}_m}\,(u_i - u_e), \sqrt{\rho}\, w\Big) \tag{2.28}$$

and $\tilde{B}^* : \tilde{V}_b' \to \boldsymbol{V}'$ is the transposed linear isomorphism

$$\begin{aligned}\big\langle \tilde{B}^*(v, w), \hat{\boldsymbol{u}}\big\rangle &:= {}_{H^{-1/2}(\Gamma_m)}\langle \sqrt{\mathsf{C}_m}\, v, \hat{v}\rangle_{H^{1/2}(\Gamma_m)} \\ &\quad + \int_{\Gamma_m} \sqrt{\rho}\, w\, \hat{w}\, d\mathcal{H}^{d-1}, \quad \forall\, \hat{\boldsymbol{u}} \in \boldsymbol{V}.\end{aligned} \tag{2.29}$$

We observe that (2.26) entails

$$(u_i - u_e, w) \in C^0([0, T]; \tilde{V}_b')$$

so that we have to require $(v^0, w^0) \in \tilde{V}_b'$ at least.

On the other hand, if $\mathfrak{F}$ would be linear, we could read from the equation that $\tilde{B}\boldsymbol{u}$ belongs also to $H^1(0, T; \tilde{V}_b') \cap L^2(0, T; \tilde{V}_b)$, and we could deduce by standard interpolation results (cf. [32]), the mapping

$$t \mapsto \big(v(t), w(t)\big) \quad \text{is uniformly continuous in} \quad \tilde{H}_b := L^2(\Gamma_m) \times L^2(\Gamma_m). \tag{2.30}$$

Even if $B\boldsymbol{u}$ does not belong to $H^1(0, T; \boldsymbol{V}')$ in the general nonlinear case, however (2.30) still holds, and we shall require

$$(v^0, w^0) \in L^2(\Gamma_m) \times L^2(\Gamma_m), \quad \boldsymbol{\ell}^0 \in H_b := \tilde{B}^*(\tilde{H}_b) \tag{2.31}$$

writing the initial condition as

$$\big(B\boldsymbol{u}\big)(0) = \boldsymbol{\ell}^0 \in H_b. \tag{2.32}$$

Main result

We have now all the elements to reformulate Problem ($\boldsymbol{\mu}$) in an abstract variational form and to state our main result about it.

Problem ($\boldsymbol{A\mu}$). Let us assume that (2.16), (2.23), and (2.31) hold. Then, if $\boldsymbol{V}$, b, a, L, $\mathfrak{F}$ are defined by (2.21), (2.11), ...,(2.14) respectively, we look for

$$\boldsymbol{u} \in L^2(0,T;\boldsymbol{V}), \quad \text{with} \quad B\boldsymbol{u} \in W^{1,1}_{loc}(0,T;\boldsymbol{V}') \cap C^0([0,T];\boldsymbol{V}') \tag{2.33}$$

satisfying the abstract Cauchy problem (2.25), (2.32).

Theorem 1. *In the framework of* $(\mu_1, \ldots, \mu_9)$, (μ_1', μ_7'), *let us assume that*

$$I^s_{i,e} \in W^{1,1}(0,T;L^2(\Omega_{i,e})), \quad G_{i,e} \in W^{1,1}(0,T;H^{-1/2}(\Gamma_{i,e})), \tag{2.34}$$

and (2.23), (2.31) *are satisfied. Then there exists a unique solution* $\boldsymbol{u}$ *of Problem* ($\boldsymbol{A\mu}$)*; in particular there exist a couple*

$$u_i, u_e \in L^2(0,T;H^1(\Omega_{i,e})) \cap C^0(]0,T];H^1(\Omega_{i,e}))$$

uniquely determined up to a family of additive constants $c(t)$ *and a unique couple* (v,w) *with*

$$v \in C^0([0,T];L^2(\Gamma_m)) \cap L^2(0,T;H^{1/2}(\Gamma_m)), \quad \partial_t v \in L^1_{loc}(0,T;H^{-1/2}(\Gamma_m)), \tag{2.35}$$

$$w, \partial_t w \in C^0([0,T];L^2(\Gamma_m)) \cap L^2(0,T;H^{1/2}(\Gamma_m)), \tag{2.36}$$

which solve the microscopic model $(\mu_1, \ldots, \mu_9)$, *as formulated by Problem* ($\boldsymbol{\mu}$), *in the standard distribution sense. Moreover, if*

$$v^0 \in H^{1/2}(\Gamma_m), \quad v^0\, F(v^0) \in L^1(\Gamma_m), \tag{2.37}$$

then

$$u_{i,e} \in C^0([0,T];H^1(\Omega_{i,e})), \quad \partial_t v \in L^2(\Sigma_m), \quad w \in C^0([0,T];H^{1/2}(\Gamma_m)). \tag{2.38}$$

Structural properties

The *proof* of this theorem follows from a direct application of Theorem 4 of § 4; now we limit ourselves to point out some distinctive properties of $a, b, \mathfrak{F}, \boldsymbol{L}$ which will guide us in the choice and the application of the abstract framework and which are common to different situations, as the macroscopic model. These properties will conclude the proof of Theorem 1.

P1. b is *symmetric* and the associated quadratic form (which we will denote by b again) is *nonnegative* but its kernel has infinite dimension, so that (2.18) is a *degenerate evolution equation.* However, by (μ_3) and Poincaré inequality, the sum of the quadratic forms associated to a and b is coercive on $\boldsymbol{V}$, i.e.

$$\exists\, \alpha > 0: \quad a(\boldsymbol{u},\boldsymbol{u}) + b(\boldsymbol{u},\boldsymbol{u}) \geq \alpha \|\boldsymbol{u}\|^2_{\boldsymbol{V}}, \quad \forall \boldsymbol{u} \in \boldsymbol{V}, \tag{2.39}$$

and a verifies

$$\begin{aligned}|a(\boldsymbol{u},\hat{\boldsymbol{u}})-a(\hat{\boldsymbol{u}},\boldsymbol{u})|^2 &= 4\Theta^2\left|\int_{\Gamma_m}(v\hat{w}-\hat{v}\,w)\,d\mathcal{H}^{d-1}\right|^2\\ &\le\frac{4\Theta^2}{\rho\mathsf{C}_m}\int_{\Gamma_m}(\mathsf{C}_m v^2+\rho w^2)\,d\mathcal{H}^{d-1}\int_{\Gamma_m}(\mathsf{C}_m\hat{v}^2+\rho\hat{w}^2)\,d\mathcal{H}^{d-1}\\ &\le\frac{4\Theta^2}{\rho\mathsf{C}_m}b(\boldsymbol{u},\boldsymbol{u})\;b(\hat{\boldsymbol{u}},\hat{\boldsymbol{u}}).\end{aligned}\tag{2.40}$$

P2. The nonlinear operator $\mathfrak{F}: D(\mathfrak{F})\subset\boldsymbol{V}\to\boldsymbol{V}'$ is a linear perturbation of a subdifferential one; in order to show this fact, let us set for $\lambda\in\mathbb{R}$

$$F_\lambda(v):=F(v)+\lambda v,\quad \phi_\lambda(v):=\int_0^v F_\lambda(s)\,ds=\frac{\lambda\mathsf{C}_m}{2}v^2+\int_0^v F(s)\,ds.\tag{2.41}$$

In this way, if $\lambda\ge\lambda_F/\mathsf{C}_m$, F_λ is an increasing function whose primitive ϕ_λ is convex. We introduce the convex l.s.c. functional on $\boldsymbol{V}$

$$\Phi(\boldsymbol{u}):=\begin{cases}\int_{\Gamma_m}\left(\phi_\lambda(u_i-u_e)+\frac{\lambda}{2}\rho w^2\right)d\mathcal{H}^{d-1}, & \text{if }\phi(u_i-u_e)\in L^1(\Gamma_m)\\ +\infty & \text{otherwise,}\end{cases}\tag{2.42}$$

and, adapting the results of [9], we know that a functional $\boldsymbol{\ell}\in\boldsymbol{V}'$ belongs to the subdifferential $\partial\Phi(\boldsymbol{u})$ if and only if

$$\begin{cases}F_\lambda(u_i-u_e)\in L^1(\Gamma_m)\cap H^{-1/2}(\Gamma_m), \quad \text{i.e. } \boldsymbol{u}\in D(\mathfrak{F}),\text{ and}\\ \langle\boldsymbol{\ell},\hat{\boldsymbol{u}}\rangle={}_{H^{-1/2}(\Gamma_m)}\langle F_\lambda(u_i-u_e),\hat{u}_i-\hat{u}_e\rangle_{H^{1/2}(\Gamma_m)}+\lambda\rho\int_{\Gamma_m}w\hat{w}\,d\mathcal{H}^{d-1}.\end{cases}\tag{2.43}$$

This formula shows that $\boldsymbol{\ell}\in\partial\Phi(\boldsymbol{u})$ is equivalent to

$$\begin{aligned}\langle\boldsymbol{\ell},\hat{\boldsymbol{u}}\rangle &=\langle\mathfrak{F}\boldsymbol{u},\hat{\boldsymbol{u}}\rangle+\lambda\int_{\Gamma_m}\Big(\mathsf{C}_m(u_i-u_e)\,(\hat{u}_i-\hat{u}_e)+\rho w\hat{w}\Big)\,d\mathcal{H}^{d-1}\\ &=\langle\mathfrak{F}\boldsymbol{u},\boldsymbol{u}\rangle+\lambda b(\boldsymbol{u},\hat{\boldsymbol{u}}),\end{aligned}$$

i.e.

$$\mathfrak{F}=\partial\Phi-\lambda B.\tag{2.44}$$

Let us remark that if $v\in L^2(\Gamma_m)$, the property $\phi(v)\in L^1(\Gamma_m)$ is equivalent to ask $vF(v)\in L^1(\Gamma_m)$, by the monotonicity of F_λ and (2.41).

P3. Let us denote by $\boldsymbol{K}_b\subset\boldsymbol{V}$ the kernel of B and $b(\cdot,\cdot)$, i.e.

$$\boldsymbol{K}_b:=\{\boldsymbol{u}\in\boldsymbol{V}: u_i|_{\Gamma_m}\equiv u_e|_{\Gamma_m},\ w\equiv 0\}=\{\boldsymbol{u}\in\boldsymbol{V}: b(\boldsymbol{u},\boldsymbol{u})=0\}.\tag{2.45}$$

The functional Φ is invariant with respect to the translation of $\boldsymbol{K}_b$, or equivalently, it admits the obvious decomposition

$$\Phi:=\tilde{\phi}_b\circ\tilde{B},\quad\text{where}\quad\tilde{\phi}_b:\tilde{V}_b\to[0,+\infty]\tag{2.46}$$

can be formally obtained from (2.42) by replacing u_i-u_e with v. Analogously, $\mathfrak{F}$ can be written as

$$\mathfrak{F}=\tilde{B}^*\circ\tilde{\mathfrak{F}}_b\circ\tilde{B},\quad\text{with}\quad\tilde{\mathfrak{F}}_b: D(\tilde{\mathfrak{F}}_b)\subset\tilde{V}_b\to\tilde{V}_b'.\tag{2.47}$$

We observe that, with respect to the duality between $\tilde{V}_b$ and $\tilde{V}_b'$, $\tilde{\Phi}_b$ and $\tilde{\mathfrak{F}}_b$ have the same properties of $\Phi, \mathfrak{F}$ with respect to $\boldsymbol{V}, \boldsymbol{V}'$.

P4. We can characterize the image $V_b' := \tilde{B}^*(\tilde{V}_b')$ as the closed subspace of $\boldsymbol{V}'$

$$V_b' := \big\{ \boldsymbol{\ell} \in \boldsymbol{V}' : \langle \boldsymbol{\ell}, \boldsymbol{u} \rangle = 0, \quad \forall \boldsymbol{u} \in \boldsymbol{K}_b \big\}. \tag{2.48}$$

Setting $V_b := \tilde{B}^* J_b(\tilde{V}_b) = B(\boldsymbol{V}) \subset V_b'$, endowed with the induced norm, $\tilde{B}^*$ establishes an isomorphism between the Hilbert triple $\tilde{V}_b, \tilde{H}_b, \tilde{V}_b'$ and V_b, H_b, V_b', where H_b is the intermediate interpolation space $H_b := (V_b, V_b')_{1/2,2}$. It is easy to translate all the properties holding in one of these triples (as in the next Theorem 4) into the corresponding ones in the other setting.

Extra regularity

In the case of a single parabolic equation of second order, further regularity properties can often be deduced by the usual maximum principle arguments (see [50]); since we are dealing with a parabolic system, this technique cannot be applied.

In any case, we are mostly interested to point out the abstract variational structure of the micro- and macroscopic problems, which is preserved by the asymptotic (formal) limit and which will play a crucial role in a rigorous asymptotic analysis. Postponing to a forthcoming paper this study, here we do not exploit further regularity properties, which are also influenced by the particular choice of F and by the singular character of the boundary conditions at the junction points of Γ_i, Γ_e, and Γ_m. Let us only mention another existence result under slightly weaker assumptions, which is important for the specific cardiac model we are referring to, and which is an immediate consequence of the abstract theory.

Proposition 2.4. *Suppose that (the space dimension) $d = 3$, Γ_m has a finite measure, and F has a cubic growth at infinity as in (μ_7''). Then* (2.35) *and* (2.36) *hold even if we assume* (2.16) *instead of the stronger* (2.34).

Proof. By the Sobolev Embedding Theorem, we know that $H^{1/2}(\Gamma_m)$ is continuously embedded into $L^4(\Gamma_m)$, and consequently

$$L^2(\Gamma_m) \subset L^{4/3}(\Gamma_m) \subset H^{-1/2}(\Gamma_m).$$

We deduce that, for $\lambda \geq \lambda_F$ and suitable constants C_i dependent only on Γ_m

$$\begin{aligned} \|F(v)\|_{H^{-1/2}(\Gamma_m)}^{4/3} &\leq C_1 \|F(v)\|_{L^{4/3}(\Gamma_m)}^{4/3} \leq C_2\Big(1 + \int_{\Gamma_m} v^4 \, d\mathcal{H}^2\Big) \\ &\leq C_3\Big(1 + \int_{\Gamma_m} \phi_\lambda(v) \, d\mathcal{H}^2\Big) \leq C_3\big(1 + \Phi(\boldsymbol{u})\big); \end{aligned} \tag{2.49}$$

thus we can apply the next Proposition 4.3. □

3. Variational formulation of the macroscopic problem

We now consider the variational formulation of the macroscopic bidomain model $(M_1, \dots, M_9)$. Because of the analogy with the procedure of the previous section, we will proceed on more quickly.

PRELIMINARIES. As we did before, we slightly reinforce (M_1, M_3, M_7) by requiring that

$$\Omega \text{ is a Lipschitz domain of } \mathbb{R}^d,\ \Gamma = \partial\Omega,\ \nu := \text{unitary exterior normal to } \Gamma, \quad (M_1')$$

As usual we set $Q := \Omega \times]0, T[$, $\Sigma := \Gamma \times]0, T[$.

We also suppose that $M_i(x), M_e(x)$, are measurable and satisfy the uniform ellipticity condition

$$\exists\, \alpha, m > 0: \quad \alpha|\xi|^2 \le M_{i,e}(x)\xi\cdot\xi \le m|\xi|^2, \quad \forall \xi \in \mathbb{R}^d,\ x \in \Omega. \qquad (M_3')$$

Finally, f is a continuous function with

$$f(0) = 0; \quad \exists\, \lambda_f \ge 0: \quad \frac{f(x) - f(y)}{x - y} \ge -\lambda_f, \quad \forall x, y \in \mathbb{R}, \quad \text{with } x \ne y; \quad (M_7')$$

sometimes we will also require that f has a cubic growth at infinity, i.e.

$$0 < \liminf_{|s|\to+\infty} \frac{f(s)}{s^3} \le \limsup_{|s|\to+\infty} \frac{f(s)}{s^3} < +\infty. \qquad (M_7'')$$

The following problem collects the discussion presented in the Introduction:

Problem (M). Given

$$i^s_{i,e} : Q \to \mathbb{R}, \quad g_{i,e} : \Sigma \to \mathbb{R}, \quad \text{and} \quad v^0, w^0 : \Omega \to \mathbb{R}, \qquad (3.1)$$

we seek

$$u_i, u_e, w : Q \to \mathbb{R}, \quad \text{with} \quad v := u_i - u_e,$$

which solve

$$\begin{cases} \mathrm{c}_m \partial_t v + f(v) + \theta w = \operatorname{div}\big(M_i \nabla u_i\big) + i^s_i & \text{in } Q, \\ \mathrm{c}_m \partial_t v + f(v) + \theta w = -\operatorname{div}\big(M_e \nabla u_e\big) - i^s_e & \text{in } Q, \\ \partial_t w + \gamma w - \eta v = 0 & \text{in } Q, \\ v(x,0) = v^0(x), \quad w(x,0) = w^0(x) & \text{in } \Omega, \\ M_{i,e}\nabla u_{i,e}\cdot\nu = g_{i,e} & \text{on } \Sigma. \end{cases} \qquad (3.2)$$

Arguing as in the previous section, we introduce the Hilbert spaces

$$\boldsymbol{X} := H^1(\Omega) \times H^1(\Omega) \times L^2(\Omega), \quad \boldsymbol{V} := X/\big\{(c, c, 0) : c \in \mathbb{R}\big\}, \qquad (3.3)$$

(whose generic elements (u_i, u_e, w) we denote by $\boldsymbol{u}$ again), the bilinear forms on $\boldsymbol{V} \times \boldsymbol{V}$ (here $\rho := \theta/\eta$)

$$b(\boldsymbol{u}, \hat{\boldsymbol{u}}) := \int_\Omega \left[\mathsf{c}_m(u_i - u_e)(\hat{u}_i - \hat{u}_e) + \rho\, w\hat{w}\right] dx, \tag{3.4}$$

$$a(\boldsymbol{u}, \hat{\boldsymbol{u}}) := \sum_{i,e} \int_\Omega M_{i,e}\nabla u_{i,e} \cdot \nabla \hat{u}_{i,e}\, dx + \theta \int_\Omega \Big[w(\hat{u}_i - \hat{u}_e) - (u_i - u_e)\hat{w}\Big]\, dx, \tag{3.5}$$

the functionals on X

$$\begin{cases} \langle \boldsymbol{L}(t), \hat{\boldsymbol{u}}\rangle := \displaystyle\sum_{i,e} \int_\Omega i^s_{i,e}\, \hat{u}_{i,e}\, dx + \sum_{i,e} {}_{H^{-1/2}(\Gamma)}\langle g_{i,e}, \hat{u}_{i,e}\rangle_{H^{1/2}(\Gamma)}, \\ \langle \boldsymbol{\ell}^0, \hat{\boldsymbol{u}}\rangle := \displaystyle\int_\Omega \Big(\mathsf{C}_m\, v^0\, \hat{v} + \rho\, w^0\, \hat{w}\Big)\, dx, \end{cases} \tag{3.6}$$

and the operator $\mathfrak{F} : D(\mathfrak{F}) \subset \boldsymbol{V} \to \boldsymbol{V}'$

$$\langle \mathfrak{F}\boldsymbol{u}, \hat{\boldsymbol{u}}\rangle := \int_\Omega \Big[f(u_i - u_e)(\hat{u}_i - \hat{u}_e)\Big]\, dx, \tag{3.7}$$

with domain

$$D(\mathfrak{F}) := \big\{\boldsymbol{u} \in \boldsymbol{V} : f(u_i - u_e) \in L^1(\Omega) \cap (H^1(\Omega))'\big\}; \tag{3.8}$$

this means that for every $\boldsymbol{u} \in D(\mathfrak{F})$ there exists a constant $C > 0$ such that

$$\int_\Omega f(u_i - u_e)\,\zeta\, dx \le C\|\zeta\|_{H^1(\Omega)}, \quad \forall \zeta \in H^1(\Omega) \cap L^\infty(\Omega). \tag{3.9}$$

Again we observe that all these definitions are compatible with the quotient space $\boldsymbol{V}$, provided that

$$\int_\Omega (i^s_i + i^s_e)\, dx + {}_{H^{-1/2}(\Gamma)}\langle g_i + g_e, 1\rangle_{H^{1/2}(\Gamma)} = 0, \qquad \text{for a.e. } t \in]0, T[. \tag{3.10}$$

Here is the precise statement of Problem $(\boldsymbol{M})$ in a variational abstract form.

Problem $(\boldsymbol{AM})$. Let us assume that

$$i^s_{i,e} \in L^2(Q), \quad g_{i,e} \in L^2(0, T; H^{-1/2}(\Gamma)), \quad v^0, w^0 \in L^2(\Omega), \tag{3.11}$$

and (3.10) is satisfied. Then, if $\boldsymbol{V}, b, a, L, \boldsymbol{\ell}^0, \mathfrak{F}$ are defined by (3.3),. . .,(3.7) respectively, and A, B are defined as in (2.24), we seek $\boldsymbol{u}(t) := \big(u_i(\cdot,t), u_e(\cdot,t), w(\cdot,t)\big)$, with

$$\boldsymbol{u} \in L^2(0, T; \boldsymbol{V}), \quad B\boldsymbol{u} \in W^{1,1}_{loc}(0, T; \boldsymbol{V}') \cap C^0([0, T]; \boldsymbol{V}'), \tag{3.12}$$

satisfying for

$$(B\boldsymbol{u})' + A\boldsymbol{u} + \mathfrak{F}\boldsymbol{u} = \boldsymbol{L}, \quad \text{in } \boldsymbol{V}', \quad \text{a.e. in }]0, T[, \tag{3.13}$$

together to the initial condition

$$(B\boldsymbol{u})(0) = \boldsymbol{\ell}^0. \tag{3.14}$$

Also for this problem we could repeat the same structural remarks **P1,...,4** of the previous section, the basic properties of $\boldsymbol{V}, b, a, \mathfrak{F}, \boldsymbol{L}$ being the same. We deduce the following statement.

Theorem 2. *In the framework of the macroscopic model* (M_1, ..., M_9), (M_1', M_3', M_7'), *let us assume that* (3.11) *and* (3.10) *are satisfied, together with*

$$i_i^s + i_e^s \in W^{1,1}(0,T;L^2(\Omega)), \quad g_{i,e} \in W^{1,1}(0,T;H^{-1/2}(\Gamma)). \tag{3.15}$$

Then Problem (***AM***) *admits a unique solution* $\boldsymbol{u}$. *In particular there exist a couple*

$$u_i, u_e \in L^2(0,T;H^1(\Omega)),$$

uniquely determined up to a family of additive constants $c(t)$, *and a unique couple* (v, w) *with*

$$v \in C^0([0,T];L^2(\Omega)) \cap L^2(0,T;H^1(\Omega)), \quad \partial_t v \in L^2_{loc}(0,T;L^2(\Omega)), \tag{3.16}$$

$$w, \partial_t w \in C^0([0,T];L^2(\Omega)),$$

which solve the macroscopic model $(M_1, \ldots, M_9)$ *in the sense of Problem* (***M***). *Moreover, if*

$$v^0 \in H^1(\Omega), \quad v^0 f(v^0) \in L^1(\Omega), \tag{3.17}$$

then

$$u_{i,e} \in C^0([0,T];H^1(\Omega)), \quad \partial_t v \in L^2(Q), \quad w \in C^0([0,T];H^1(\Omega)). \tag{3.18}$$

Proof. We argue as before by invoking Theorem 4 of the next session; in this case we can establish an isomorphism of the "abstract Hilbert triple" V_b, H_b, V_b' introduced there with the "concrete"

$$\tilde{V}_b := H^1(\Omega) \times L^2(\Omega), \quad \tilde{H}_b := L^2(\Omega) \times L^2(\Omega), \quad \tilde{V}_b' := \left(H^1(\Omega)\right)' \times L^2(\Omega)$$

via the transpose of the operator $\tilde{B}\boldsymbol{u} := (u_i - u_e, w)$. The first condition of (3.15) follows from the simple splitting

$$\int_\Omega (i_i^s\, u_i + i_e^s\, u_e)\, dx = \frac{1}{2}\int_\Omega (i_i^s - i_e^s)(u_i - u_e)\, dx + \frac{1}{2}\int_\Omega (i_i^s + i_e^s)(u_i + u_e)\, dx$$

□

Further results

We conclude this section by extending Theorem 2 in two different directions.

Proposition 3.1. *Suppose that* $d = 3$, *and* (M_7'') *holds; then, if* (3.11) *holds, Problem* (***M***) *admits a unique strong solution* $\boldsymbol{u}$. *Moreover, if* (3.15) *and* (3.17) *hold, then*

$$-\operatorname{div}\left(M_{i,e}\nabla u_{i,e}\right) \in L^2(Q). \tag{3.19}$$

Proof. It follows from Propositions 4.3 and 4.4 by the same arguments of Proposition 2.4. □

Remark 3.2. If

$$\Omega \text{ is of class } C^{1,1}, \quad M_{i,e} \text{ are Lipschitz in } \Omega, \tag{3.20}$$

and

$$g_{i,e} \in L^2(0,T;H^{1/2}(\Gamma)), \tag{3.21}$$

then (3.19) and standard regularity estimates for elliptic problems yield

$$u_{i,e} \in L^2(0,T;H^2(\Omega)). \tag{3.22}$$

Proposition 3.3. *Suppose that* (3.20) *holds and* $M_i\nu$ *and* $M_e\nu$ *have the same direction on* Γ. *If*

$$\partial_t i^s_{i,e} \in L^2(Q), \quad \partial_t g_{i,e} \in L^2(0,T;H^{-1/2}(\Gamma)) \tag{3.23}$$

and

$$v^0 \in H^2(\Omega), \quad f(v^0) \in L^2(\Omega), \quad g_{i,e}(0) \in H^{1/2}(\Gamma) \tag{3.24}$$

then, for every $\varepsilon > 0$,

$$v \in W^{1,\infty}(0,T;L^2(\Omega)) \cap H^{3/2-\varepsilon}(0,T;L^2(\Omega)), \quad u_{i,e} \in H^1(0,T;H^1(\Omega)). \tag{3.25}$$

Proof. It follows from the third implication of Theorem 4.

4. The abstract theory

Let $\boldsymbol{V}$ be a (separable) Hilbert space with dual $\boldsymbol{V}'$, $\langle\cdot,\cdot\rangle$ denoting the duality pairing between them,

$$\text{let } a(\cdot,\cdot), \quad b(\cdot,\cdot) : \boldsymbol{V}\times\boldsymbol{V} \to \mathbb{R} \quad \text{be continuous bilinear forms,} \tag{A_1}$$

with the associated linear continuous operators

$$A, B : \boldsymbol{V} \to \boldsymbol{V}'; \quad \langle A\boldsymbol{u},\boldsymbol{w}\rangle = a(\boldsymbol{u},\boldsymbol{w}), \quad \langle B\boldsymbol{u},\boldsymbol{w}\rangle = b(\boldsymbol{u},\boldsymbol{w}), \quad \forall\,\boldsymbol{u},\boldsymbol{w}\in\boldsymbol{V}, \tag{A_2}$$

$$\text{let } \mathfrak{F} : D(\mathfrak{F}) \subset \boldsymbol{V} \to \boldsymbol{V}' \quad \text{be a (nonlinear) operator.} \tag{A_3}$$

We are interested to the following abstract Cauchy problem.

Problem $(\boldsymbol{A})$. Given $\boldsymbol{L} \in L^2(0,T;\boldsymbol{V}')$ and $\boldsymbol{\ell}^0 \in \boldsymbol{V}'$, find

$$\boldsymbol{u} \in L^2(0,T;\boldsymbol{V}), \quad \text{with} \quad B\boldsymbol{u} \in W^{1,1}_{loc}(0,T;\boldsymbol{V}') \cap C^0([0,T];\boldsymbol{V}')$$

such that

$$\begin{cases} (B\boldsymbol{u})' + A\boldsymbol{u} + \mathfrak{F}\boldsymbol{u} = \boldsymbol{L}, & \text{a.e. in }]0,T[, \\ (B\boldsymbol{u})(0) = \boldsymbol{\ell}^0. \end{cases} \tag{4.1}$$

Degenerate parabolic equations of this kind have been studied by many authors (see [11, 8, 19, 15], and [13, 49, 21] for an extensive bibliography) in a very general context: e.g. $\boldsymbol{V}$ could be a reflexive Banach space and A a (pseudo)monotone bounded operator; under suitable compatibility assumptions [19] also B could be nonlinear. In particular, adapting a result of [13] via the techniques developed in [7], an existence and uniqueness result could be given for a weak formulation of (4.1). Another possible way is to use a regularizing method as in [19], replacing the (possibly) degenerate bilinear form b with a coercive one b_ε and proving uniform estimates with respect to $\varepsilon \to 0^+$ in an appropriate functional setting. A natural choice in this approach is

$$b_\varepsilon(u,w) := b(u,w) + \varepsilon(u,w)_{\boldsymbol{V}}.$$

Main assumptions

In this paper, inspired by the structural properties **P1–4** we detailed in § 2, we follow a more direct way, exploiting the particular compatibility of the nonlinear operator $\mathfrak{F}$ with respect to the degeneration of b.

More precisely, we assume that b is symmetric and positive, but possibly degenerate on the closed subspace $\boldsymbol{K}_b$:

$$b(\boldsymbol{u},\boldsymbol{w}) = b(\boldsymbol{w},\boldsymbol{u}), \quad b(\boldsymbol{u},\boldsymbol{u}) \geq 0; \quad b(\boldsymbol{u},\boldsymbol{u}) = 0 \ \Leftrightarrow \ \boldsymbol{u} \in \boldsymbol{K}_b \quad \forall\, \boldsymbol{u},\boldsymbol{w} \in \boldsymbol{V}, \quad (A_4)$$

a is weakly coercive on $\boldsymbol{V}$ and "almost" symmetric:

$$\exists\, \alpha, c > 0: \quad \begin{cases} a(\boldsymbol{u},\boldsymbol{u}) + b(\boldsymbol{u},\boldsymbol{u}) \geq \alpha \|\boldsymbol{u}\|^2_{\boldsymbol{V}}, \\ |a(\boldsymbol{u},\boldsymbol{w}) - a(\boldsymbol{w},\boldsymbol{u})|^2 \leq c\, b(\boldsymbol{u},\boldsymbol{u})\, b(\boldsymbol{w},\boldsymbol{w}), \end{cases} \quad \forall\, \boldsymbol{u},\boldsymbol{w} \in \boldsymbol{V}. \quad (A_5)$$

We will suppose that $\mathfrak{F} = \partial\Phi - \lambda B$, $\lambda \geq 0$, is a linear perturbation of the subdifferential of

$$\begin{gathered} \Phi : \boldsymbol{V} \to [0,+\infty] \quad \text{proper, convex, l.s.c. function,} \\ \boldsymbol{\ell} = \mathfrak{F}\boldsymbol{u} \quad \Leftrightarrow \quad \langle \boldsymbol{\ell}, \boldsymbol{w}-\boldsymbol{u}\rangle + \lambda b(\boldsymbol{u},\boldsymbol{w}-\boldsymbol{u}) + \Phi(\boldsymbol{u}) \leq \Phi(\boldsymbol{w}), \quad \forall\, \boldsymbol{w} \in \boldsymbol{V}. \end{gathered} \quad (A_6)$$

Finally, we shall assume that Φ is invariant with respect to the translation of $\boldsymbol{K}_b$:

$$\begin{gathered} \Phi(\boldsymbol{u}+\boldsymbol{k}) = \Phi(\boldsymbol{u}) \quad \forall\, \boldsymbol{u} \in \boldsymbol{V},\ \boldsymbol{k} \in \boldsymbol{K}_b, \quad \text{or, equivalently,} \\ \langle \mathfrak{F}\boldsymbol{u}, \boldsymbol{k}\rangle = 0, \quad \mathfrak{F}(\boldsymbol{u}+\boldsymbol{k}) = \mathfrak{F}(\boldsymbol{u}), \quad \forall\, \boldsymbol{u} \in D(\mathfrak{F}),\ \forall\, \boldsymbol{k} \in \boldsymbol{K}_b. \end{gathered} \quad (A_7)$$

A reduction technique

We shall show that Problem $(\boldsymbol{A})$ can be reduced, by the "change of variable $v := B\boldsymbol{u}$", to a usual (non degenerate) evolution variational inequality of the type

$$v(0) = \boldsymbol{\ell}^0, \quad \big(v' + (A_b - \lambda)v - L_b, v - w\big)_b + \Phi_b(v) \leq \Phi_b(w), \qquad \forall\, w \in V_b \quad (4.2)$$

in a suitable Hilbert triple $V_b \subset H_b \subset V_b'$, with A_b, L_b, Φ_b related in a explicit way to A, L, Φ.

In this formulation, regularity and approximation results are easier and deeply studied (see [33, 10, 7, 31, 22, 2, 48]) and they give corresponding information about the original Problem $(\boldsymbol{A})$. In particular the study of the time discretization by the backward Euler method and of some further regularity properties of (4.1) are developed by [2, 43, 47, 48, 38]. In order to state our results, we introduce the Hilbert spaces

$$V_b = B(\boldsymbol{V}), \quad \text{with the norm} \quad \|v\|_b = \inf\big\{\|\boldsymbol{u}\|_{\boldsymbol{V}} :\ \boldsymbol{u} \in \boldsymbol{V},\, B\boldsymbol{u} = v\big\}, \quad (4.3)$$

and

$$V_b' = \Big\{L \in \boldsymbol{V}' :\ \langle L, \boldsymbol{k}\rangle = 0, \quad \forall\, \boldsymbol{k} \in \boldsymbol{K}_b\Big\}. \quad (4.4)$$

It is easy to see that V_b is included in V_b' and it is isomorphic to the quotient space $\boldsymbol{V}/\boldsymbol{K}_b$, whereas V_b' is isomorphic to its dual, so that our notation is correct. We denote by $\boldsymbol{J} : V_b \to \boldsymbol{V}$ a right inverse of B, defined by

$$\boldsymbol{J}v = \boldsymbol{u} \quad \Leftrightarrow \quad B\boldsymbol{u} = v, \quad \text{and} \quad a(\boldsymbol{u}, \boldsymbol{k}) = 0, \quad \forall\, \boldsymbol{k} \in \boldsymbol{K}_b. \tag{4.5}$$

By (A_5) $a(\cdot,\cdot)$ is coercive on $\boldsymbol{K}_b$; Lax-Milgram lemma ensures that $\boldsymbol{J}$ is a linear isomorphism. Observe that, when a is symmetric, (4.5) is equivalent to the minimization problem

$$B\boldsymbol{u} = v, \quad a(\boldsymbol{u}, \boldsymbol{u}) = \min\big\{a(\boldsymbol{w}, \boldsymbol{w}) : B\boldsymbol{w} = v\big\}. \tag{4.6}$$

Each element $\boldsymbol{u} \in \boldsymbol{V}$ admits the linear decomposition

$$\boldsymbol{u} = \boldsymbol{J}v + \boldsymbol{k} : \quad v = B\boldsymbol{u}, \quad \boldsymbol{k} = \boldsymbol{u} - \boldsymbol{J}B\boldsymbol{u} \in \boldsymbol{K}_b. \tag{4.7}$$

We define the duality pairing between V_b' and V_b as

$$\big(\ell, v\big)_b := {}_{\boldsymbol{V}'}\langle \ell, \boldsymbol{J}v\rangle_{\boldsymbol{V}} = {}_{\boldsymbol{V}'}\langle \ell, \boldsymbol{u}\rangle_{\boldsymbol{V}}, \quad \forall\, \ell \in V_b',\ v \in V_b,\ \boldsymbol{u} \in B^{-1}v. \tag{4.8}$$

It is easy to see that $(\cdot,\cdot)_b$ restricted to $V_b \times V_b$ is a scalar product, associated to the intermediate norm

$$|v|_b^2 := (v, v)_b = b(\boldsymbol{J}v, \boldsymbol{J}v). \tag{4.9}$$

By the standard duality theory, we can identify the completion H_b of V_b with respect to this norm with the space

$$H_b' = \Big\{\ell \in V_b' : \sup_{w \in \boldsymbol{V}\setminus\boldsymbol{K}_b} \frac{\langle \ell, w\rangle}{\sqrt{b(w,w)}} = \sup_{v \in V_b\setminus\{0\}} \frac{(\ell, v)_b}{|v|_b} < +\infty\Big\}. \tag{4.10}$$

In this way V_b, $H_b \equiv H_b'$, V_b' becomes a standard Hilbert triple and defining

$$\Phi_b(v) := \Phi(\boldsymbol{J}v), \qquad a_b(v, w) := a(\boldsymbol{J}v, \boldsymbol{J}w), \tag{4.11}$$

we can consider the following evolution variational inequality:

Problem $(\boldsymbol{A}_b)$. Given $L_b \in L^2(0, T; V_b')$ and $\boldsymbol{\ell}^0 \in V_b'$, find

$$v \in L^2(0, T; V_b) \cap W^{1,1}_{loc}(0, T; V_b') \cap C^0([0, T]; V_b')$$

such that:

$$\begin{aligned} &\big(v' + (A_b - \lambda)v - L_b, v - w\big)_b + \Phi_b(v) \le \Phi_b(w), \quad \forall w \in V_b, \text{ a.e. in }]0, T[, \\ &v(0) = \boldsymbol{\ell}^0. \end{aligned} \tag{4.12}$$

The link with Problem $(\boldsymbol{A})$ is given by the following result.

Theorem 3. *The function u is a strong solution of Problem $(\boldsymbol{A})$ if and only if it admits the decomposition*

$$\boldsymbol{u} = u_{\boldsymbol{L}} + \boldsymbol{J}v, \tag{4.13}$$

where $u_{\boldsymbol{L}}$ solves

$$u_{\boldsymbol{L}}(t) \in \boldsymbol{K}_b, \quad a(u_{\boldsymbol{L}}(t), \boldsymbol{k}) = \langle \boldsymbol{L}(t), \boldsymbol{k}\rangle, \quad \forall\, \boldsymbol{k} \in \boldsymbol{K}_b, \text{ a.e. in }]0, T[, \tag{4.14}$$

and v is a solution of Problem $(\boldsymbol{A}_b)$ with respect to

$$L_b(t) := \boldsymbol{L}(t) - Au_{\boldsymbol{L}}(t). \tag{4.15}$$

Proof. Observe that $L_b(t) \in V_b'$ for a.e. $t \in]0,T[$ so that (4.12) makes sense. Let $\boldsymbol{u}$ be a strong solution of Problem $(\boldsymbol{A})$ and $v := B\boldsymbol{u}$; by (A_7) we have $\mathfrak{F}\boldsymbol{u} \in V_b'$ so that, taking the duality of each member of equation (4.1) by a generic element $\boldsymbol{k} \in \boldsymbol{K}_b$ we get

$$a(\boldsymbol{u}(t), \boldsymbol{k}) = \langle \boldsymbol{L}(t), \boldsymbol{k}\rangle, \quad \forall \boldsymbol{k} \in \boldsymbol{K}_b, \quad \text{for a.e. } t \in]0,T[. \tag{4.16}$$

This implies that

$$a(\boldsymbol{u}(t) - u_{\boldsymbol{L}}(t), \boldsymbol{k}) = 0, \quad B(\boldsymbol{u}(t) - u_{\boldsymbol{L}}(t)) = v(t), \quad \forall \boldsymbol{k} \in \boldsymbol{K}_b, \text{ for a.e. } t \in]0,T[,$$

i.e. $\boldsymbol{u} - u_{\boldsymbol{L}} = \boldsymbol{J}v$ by definition (4.5).

Coming back to equation (4.1), since $\mathfrak{F}$ is invariant by $\boldsymbol{K}_b$-translations, we read that

$$v' + A(\boldsymbol{J}v) + \mathfrak{F}(\boldsymbol{J}v) = (B\boldsymbol{u})' + A(\boldsymbol{u} - u_{\boldsymbol{L}}) + \mathfrak{F}(\boldsymbol{u} - u_{\boldsymbol{L}}) = \boldsymbol{L} - Au_{\boldsymbol{L}} = L_b. \tag{4.17}$$

Taking the duality of (4.17) with $\boldsymbol{J}(v - w)$, $w \in V_b$, and recalling (4.11), we get

$$\begin{aligned}(v', v - w)_b + a_b(v, v - w) - (L_b, v - w)_b = \langle \mathcal{F}(\boldsymbol{J}v), w - v\rangle \le \\ \le \Phi(\boldsymbol{J}w) - \Phi(\boldsymbol{J}v) - \lambda b(\boldsymbol{J}v, \boldsymbol{J}(w - v)) = \Phi_b(w) - \Phi_b(v) - \lambda(v, w - v)_b\end{aligned}$$

by (A_6), (4.5), (4.8), and (4.11); therefore, v solves (4.12).

Let now v be a solution of (4.12), $u_{\boldsymbol{L}}$ be given by (4.14), and $\boldsymbol{u} := u_{\boldsymbol{L}} + \boldsymbol{J}v$. By (4.11) and (4.15) we deduce

$$\langle B(\boldsymbol{J}v)' + A(\boldsymbol{J}v) - L_b, \boldsymbol{J}(v - w)\rangle \le \Phi(\boldsymbol{J}w) - \Phi(\boldsymbol{J}v) - \lambda\langle B\boldsymbol{J}v, \boldsymbol{J}(w - v)\rangle, \quad \forall w \in V_b.$$

By (4.14), (4.5), and (A_7) we have

$$\begin{aligned}&\langle B\boldsymbol{J}v' + A(\boldsymbol{J}v) - (\boldsymbol{L} - Au_{\boldsymbol{L}}), \boldsymbol{J}v + u_{\boldsymbol{L}} - (\boldsymbol{J}w + \boldsymbol{k})\rangle \le \\ &\Phi(\boldsymbol{J}w + \boldsymbol{k}) - \Phi(\boldsymbol{J}v + u_{\boldsymbol{L}}) - \lambda\langle B\boldsymbol{J}v, \boldsymbol{J}w + \boldsymbol{k} - (\boldsymbol{J}v + u_{\boldsymbol{L}})\rangle, \ \forall w \in V_b, \ \forall \boldsymbol{k} \in \boldsymbol{K}_b.\end{aligned}$$

Since $Bu_{\boldsymbol{L}} \equiv 0$, (4.15) yields

$$\langle (B\boldsymbol{u})' + A\boldsymbol{u} - \boldsymbol{L}, \boldsymbol{u} - (\boldsymbol{J}w + \boldsymbol{k})\rangle \le \Phi(\boldsymbol{J}w + \boldsymbol{k}) - \Phi(\boldsymbol{u}) - \lambda\langle B\boldsymbol{u}, \boldsymbol{J}w + \boldsymbol{k} - \boldsymbol{u}\rangle,$$

for every choice of $w \in V_b$, $\boldsymbol{k} \in \boldsymbol{K}_b$. Since $\boldsymbol{V} = \boldsymbol{J}(V_b) + \boldsymbol{K}_b$ thanks to (4.7), by (A_6) we deduce that

$$(B\boldsymbol{u})' + A\boldsymbol{u} - \boldsymbol{L} = -\mathfrak{F}\boldsymbol{u}$$

i.e. (4.1). □

Main result

We have now all the elements to state our main result.

Theorem 4. *Let us assume that*

$$\boldsymbol{\ell}^0 \in \overline{D(\Phi_b)}^{H_b}, \quad \boldsymbol{L} \in W^{1,1}(0,T;\boldsymbol{V}') + L^2(0,T;H_b'). \tag{4.18}$$

Then there exists a unique strong solution $\boldsymbol{u}$ *of Problem* $(\boldsymbol{A})$ *with*

$$v := B\boldsymbol{u} \in H^1_{loc}(0,T;H_b) \cap C^0([0,T];H_b), \quad \boldsymbol{u} \in C^0(]0,T];V). \tag{4.19}$$

If

$$\boldsymbol{\ell}^0 \in V_b, \quad \text{and} \quad \exists\, \boldsymbol{u}^0 \in D(\Phi) : \; B\boldsymbol{u}^0 = \boldsymbol{\ell}^0, \tag{4.20}$$

then

$$v := B\boldsymbol{u} \in H^1(0,T;H_b), \quad \boldsymbol{u} \in C^0([0,T];V). \tag{4.21}$$

Finally, if

$$\boldsymbol{L} \in H^1(0,T;\boldsymbol{V}'), \quad \boldsymbol{\ell}^0 \in V_b, \; \boldsymbol{J}\boldsymbol{\ell}^0 \in D(\mathfrak{F}), \quad A(\boldsymbol{J}\boldsymbol{\ell}^0) + \mathfrak{F}(\boldsymbol{J}\boldsymbol{\ell}^0) \in H_b' \tag{4.22}$$

then

$$\boldsymbol{u} \in H^1(0,T;\boldsymbol{V}) \quad \text{and} \quad v = B\boldsymbol{u} \in W^{1,\infty}(0,T;H_b) \cap H^{3/2-\varepsilon}(0,T;H_b), \tag{4.23}$$

for every $\varepsilon > 0$.

Proof. Let us observe that $(A_4,\ldots,A_7)$ imply that

$$a_b(\cdot,\cdot) - \lambda(\cdot,\cdot)_b \text{ is weakly coercive on } V_b, \quad \Phi_b \text{ is l.s.c. and convex on } V_b. \tag{4.24}$$

By the general theory of evolution variational inequalities (see e.g. [6], Proposition II.2 and [48], Theorem 3), if (4.22) holds then Problem $(\boldsymbol{A}_b)$ admits a unique strong solution

$$v \in H^1(0,T;V_b) \cap W^{1,\infty}(0,T;H_b) \cap H^{3/2-\varepsilon}(0,T;H_b), \tag{4.25}$$

and Theorem 3 gives (4.23). Moreover, the mapping $(L_b, \boldsymbol{\ell}^0) \mapsto v$ is Lipschitz with respect to the norm (cf. [48, 4.12]) of

$$\Big(L^2(0,T;V_b') + L^1(0,T;H_b')\Big) \times H_b$$

with values in $L^\infty(0,T;H_b) \cap L^2(0,T;V_b)$. This regular dependence of the solution from the data and a standard density argument allow us to prove the other two results (which are already known, in a slightly different form, when $a(\cdot,\cdot)$ is symmetric; cf. [6, cor. II.2] by simply showing the corresponding a priori estimates. To get (4.21), we write (4.12) as a differential equation governed by the subdifferential of Φ_b in H_b and multiply it by v', obtaining

$$\begin{aligned} |v'|_b^2 + \frac{d}{dt}\Big[\frac{1}{2}a_b(v,v) + \Phi_b(v)\Big] &= (L_b, v')_b - \frac{1}{2}\big[a_b(v,v') - a_b(v',v)\big] + \lambda(v,v')_b \\ &\le (L_b, v')_b + (c+\lambda)|v|_b\,|v'|_b, \end{aligned}$$

thanks to (A_5). By integrating in time and applying Gronwall lemma, we get the estimate

$$\int_0^T |v'|_b^2\,dt + \sup_{t\in[0,T]}\Big[a_b(v,v)+\Phi_b(v)\Big] \\ \le C\Big[\|\boldsymbol{\ell}^0\|_{V_b}^2 + \Phi_b(\boldsymbol{\ell}^0) + \|L_b\|_{W^{1,1}(0,T;V_b')+L^2(0,T;H_b)}\Big]$$

with C independent of the data. Since $\Phi_b(\boldsymbol{\ell}^0) = \Phi(u^0)$, (4.21) follows by (4.20).

The first implication follows by the same technique, multiplying the equation by tv' (cf. [6]). □

Remark 4.1. Analogous results hold if $\mathfrak{F}$ is a multivalued (subdifferential) operator; we considered the single-valued case, in order to simplify our exposition.

Remark 4.2. We could give an existence and uniqueness result for a suitable weak formulation of Problems $(\boldsymbol{A})$, $(\boldsymbol{A}_b)$ (see [6, Remark II.5], and [48, Remark 1.1]), which requires

$$\boldsymbol{L} \in L^2(0,T;\boldsymbol{V}') + L^1(0,T;H_b), \quad \boldsymbol{\ell}^0 \in \overline{D(\Phi_b)}^{H_b} \tag{4.26}$$

and gives $\boldsymbol{u} \in L^2(0,T;\boldsymbol{V})$ with $B\boldsymbol{u} \in C^0([0,T];H_b)\cap H^{1/2-\varepsilon}(0,T;H_b)$, for every $\varepsilon > 0$, together to the a priori estimate (cf. [48, 3.1 and Theorem 3])

$$\sup_{[0,T]} b(\boldsymbol{u},\boldsymbol{u}) + \int_0^T \Big[\|\boldsymbol{u}\|_{\boldsymbol{V}}^2 + \Phi(\boldsymbol{u})\Big]\,dt \le C\Big[\|L_b\|^2_{L^2(0,T;\boldsymbol{V}')+L^1(0,T;H_b)} + |\boldsymbol{\ell}^0|_b^2\Big]. \tag{4.27}$$

We give two other simple regularity results, under some additional assumptions on $\mathfrak{F}$.

Proposition 4.3. *Assume that* $D(\mathfrak{F}) \equiv D(\Phi)$ *and that there exists* $C > 0$, $p \in]1,2]$ *such that*

$$\|\mathfrak{F}\boldsymbol{w}\|_{\boldsymbol{V}'}^p \le C(1+\|\boldsymbol{w}\|_{\boldsymbol{V}}^2 + \Phi(\boldsymbol{w})), \quad \forall \boldsymbol{w} \in D(\Phi). \tag{4.28}$$

If (4.26) *holds (instead of the stronger* (4.18)*), then Problem* $(\boldsymbol{A})$ *admits a unique strong solution.*

Proof. Starting from the strongest result of Theorem 4 and the a priori estimate (4.27), we deduce

$$\|v'\|^p_{L^p(0,T;\boldsymbol{V}')} + \|\mathfrak{F}\boldsymbol{u}\|^p_{L^p(0,T;\boldsymbol{V}')} \le C\Big[T + \int_0^T \Big(\|\boldsymbol{u}\|_{\boldsymbol{V}}^2 + \Phi(\boldsymbol{u})\Big)\,dt\Big] \\ \le C\Big[1 + \|\boldsymbol{L}\|^2_{L^2(0,T;\boldsymbol{V}')+L^1(0,T;H_b)} + |\boldsymbol{\ell}^0|_b^2\Big] \tag{4.29}$$

with C independent of the data. By the usual density arguments, this is sufficient to obtain a strong solution of the Problems $(\boldsymbol{A})$,$(\boldsymbol{A}_b)$ under (4.26). □

Proposition 4.4. *Assume that* $D(\mathfrak{F}) \equiv \boldsymbol{V}$, $\mathfrak{F}(\boldsymbol{V}) \subset H_b$ *and there exists* $C > 0$, $p \geq 1$ *such that*

$$\|\mathfrak{F}\boldsymbol{w}\|_{H_b} \leq C(1 + \|\boldsymbol{w}\|_{\boldsymbol{V}}^p) \quad \forall \boldsymbol{w} \in \boldsymbol{V}. \tag{4.30}$$

If (4.20) *holds, we get*

$$A\boldsymbol{u} - \boldsymbol{L} \in L^2(0, T; H_b) \tag{4.31}$$

Proof. It is an immediate consequence of (4.21) and the equation (4.1). □

Approximation results

We conclude this section with an *example* of the possible approximation results, which follow by applying the theory developed in [48, 38] to the abstract formulation $(\boldsymbol{A}_b)$: other *a priori* and *a posteriori* estimates are available and it should not be difficult to translate them in the framework of the micro- and macroscopic Problems $(\boldsymbol{\mu},\boldsymbol{M})$; this is also the starting point for a complete space-time discretization, as studied in [46] (for other approximation results in the context of reaction-diffusion problems, see e.g. [26, 28, 35]; a different abstract approach has been developed in [4].)

We choose a partition $\mathcal{P}$ of the time interval $[0, T]$ into N subintervals

$$\mathcal{P} := \{0 = t_0 < t_1 < \ldots < t_{N-1} < t_N = T\}, \tag{4.32}$$

with variable step $\tau_n := t_n - t_{n-1}$, and we consider the sequence $\{U_n\}_{n=0}^N$ whose first term is given and satisfies $BU^0 = \boldsymbol{\ell}^0$ and the other ones are recursively defined for $1 \leq n \leq N$ by

$$\frac{1}{\tau_n} B(U^n - U^{n-1}) + A(U^n) + \mathfrak{F}U^n = L^n, \quad \text{for } n = 1, \ldots, N,$$

where L^n is a suitable approximation of $\boldsymbol{L}$ in the time interval $(t_{n-1}, t_n]$, e.g. $L^n := \boldsymbol{L}(t_n)$.

Hereafter, $\boldsymbol{U}$ is the piecewise linear interpolant of the values $\{U_n\}_{n=0}^N$ on the grid $\mathcal{P}$ and $|\tau| := \max_{1 \leq n \leq N} \tau_n$ denotes the maximum of the time-step sizes. Combining Theorems 8 and 9 of [38] (see also [48, Thm. 3] in the case of a uniform mesh) we get

Theorem 5. *Assume that* (4.22) *holds. Then we have the estimate*

$$\|\boldsymbol{u} - \boldsymbol{U}\|_{L^2(0,T;\boldsymbol{V})} + \max_{t \in [0,T]} b(\boldsymbol{u} - \boldsymbol{U}, \boldsymbol{u} - \boldsymbol{U}) \leq C|\tau|, \tag{4.33}$$

with C *independent of the partition* $\mathcal{P}$.

Appendix A. The derivation of the "bidomain" model

For completeness in this appendix, using the two-scales method, we develop formal asymptotic expansions and we convert a microscopic model of the cellular media into an averaged continuum representation of the cardiac tissue, when the presence of stimulation currents are neglected and $\sigma_{i,e}$ do not depend on the position x. With respect to the procedure followed in [37], based on current balances expressed by means of integral identities, the formal derivation here presented is shorter and more standard. For the electric potentials u_i, u_e we can consider two characteristic length scales, the microscopic one related to a typical dimension d_c of the cells (e.g. the cell diameter is of order $10^{-3}cm$) and the other macroscopic one determined by a suitable length constant of the tissue. Following [37] we consider the dimensionless parameter

$$\varepsilon^2 = \mathsf{d}_c/(\mathsf{R}_m\sigma_i) \quad \text{with} \quad \mathsf{R}_m^{-1} = \partial_v I_{ion}(0,0)$$

assuming that $v = 0$, $w = 0$ is the equilibrium point for the Problem $(\boldsymbol{\mu})$.

We convert the cellular problem into a non-dimensional form, by scaling space and time with the macroscopic units of length $\mathsf{L} = \mathsf{d}_c/\varepsilon$ and with respect to the membrane constant $\mathsf{T} = \mathsf{R}_m\ \mathsf{C}_m$, i.e.

$$\widehat{x} = x/\mathsf{L}, \quad \widehat{t} = t/\mathsf{T}$$

the microscopic space variable measured in unit cell is defined by

$$\xi := \widehat{x}/\varepsilon = x/\mathsf{d}_c.$$

Disregarding the presence of applied current terms, rescaling the equations (μ_4), (μ_5) in the intra- and extra-cellular potentials we obtain:

$$\begin{cases} -\Delta u_{i,e} = 0 & \text{in } \Omega^\varepsilon_{i,e}, \\ \varepsilon\ (\partial_t v + I(v,w)) = -\nabla u_i \cdot \nu_i = \alpha_e\ \nabla u_e \cdot \nu_e & \text{on } \Gamma^\varepsilon_m, \\ \partial_t w = r(v,w) & \text{on } \Gamma^\varepsilon_m, \end{cases} \tag{A.1}$$

where $\alpha_e := \sigma_e/\sigma_i$, $I = \mathsf{R}_m\, I_{ion}$, and for convenience, the superscripts $\widehat{\ }$ of the dimensionless variables are omitted.

In order to establish a formal relationship between a cellular model and the macroscopic bidomain structure we consider an ideal geometry and interconnection model of the cardiac cells assuming a periodic organization of the cellular tissue i.e. a periodic network of interconnected cells similar to a regular lattice of interconnected cylinders.

More precisely Ω^ε_i and Ω^ε_e are two open, disjoint, connected, periodic domains covering the entire space i.e. $\overline{\Omega^\varepsilon_i} \cup \overline{\Omega^\varepsilon_i} = \mathbb{R}^3$ with their common boundary $\partial\Omega^\varepsilon_i \cap \partial\Omega^\varepsilon_e = \Gamma^\varepsilon_m$ locally Lipschitz.

For $\alpha > 0$, we set

$$Y^\varepsilon = [0,\varepsilon]^2 \times [0,\alpha\,\varepsilon], \quad Y^\varepsilon_{i,e} = \Omega^\varepsilon_{i,e} \cap Y^\varepsilon, \quad \partial Y^\varepsilon_i \cap \partial Y^\varepsilon_e = S^\varepsilon$$

which defines the elementary unit ε-sized box with its intra- and extra-cellular parts, whose periodic repetition cover the whole space $\mathbb{R}^3$, while

$$Y, \quad Y_i, \quad Y_e, \quad S$$

denote the corresponding transformed sets expressed in the microscopic variable $\xi = x/\varepsilon$. When the reference parallelepiped Y^ε represents a volume box including a cardiac cell unit, then α can be interpreted as the ratio between the length and the diameter of the elongated cardiac cells.

To investigate solutions of (A.1) we use the two-scales method (see [3, 5, 45, 44, 29, 40, 29]) and we seek a solution u_i and u_e having the following asymptotic form in powers of ε of the type (the indexes i and e are omitted):

$$u = u_0(x,\xi,t) + \varepsilon u_1(x,\xi,t) + \varepsilon^2 u_2(x,\xi,t) + \cdot \tag{A.2}$$

where x denotes the slow macroscopic variable and $\xi = x/\varepsilon$ the fast microscopic one. The slow and fast variables correspond respectively to the global and local structure of the field and the coefficients u_k are 1-periodic function of ξ.

Considering the full derivative operators,

$$\nabla u = \varepsilon^{-1}\nabla_\xi u + \nabla_x u,$$

$$\Delta u = \varepsilon^{-2}\Delta_{\xi\xi}u + \varepsilon^{-1}\operatorname{div}_\xi \nabla_x u + \varepsilon^{-1}\operatorname{div}_x \nabla_\xi u + \Delta_{xx}u,$$

substituting the asymptotic form for $u = u_i$ into the original equation (A.1) and equating the coefficients of the powers $-1, 0, 1,$ of ε to zero, we obtain the following equations for the functions $u_k(x,\xi,t)$, $k = 0,1,2$:

$$\begin{cases} \Delta_{\xi\xi}u_0 = 0 & \text{in } Y_i, \\ \nabla_\xi \cdot \nu_\xi u_0 = 0 & \text{on } S, \end{cases} \tag{A.3}$$

$$\begin{cases} \Delta_{\xi\xi}u_1 = -2\operatorname{div}_x\nabla_\xi u_0 & \text{in } Y_i, \\ \nabla_\xi u_1 \cdot \nu_\xi + \nabla_x u_0 \cdot \nu_\xi = 0 & \text{on } S, \end{cases} \tag{A.4}$$

$$\begin{cases} \Delta_{\xi\xi}u_2 = -\operatorname{div}_x \nabla_\xi u_1 - \operatorname{div}_\xi \nabla_x u_1 - \Delta_{xx}u_0 & \text{in } Y_i, \\ \nabla_\xi u_2 \cdot \nu_\xi + \nabla_x u_1 \cdot \nu_\xi = -(\partial_t v_0 + I(v_0,w_0)) & \text{on } S, \\ v_0 = u_{i,0} - u_{e,0}, \qquad \partial_t w_0 = r(v_0,w_0). \end{cases} \tag{A.5}$$

In the previous problems x appears as a parameter and we seek 1-periodic solutions in ξ on the reference cell Y_i. We recall the following well-known result [39, 3].

Lemma A.1. *Let Ω be a bounded domain with Lipschitz boundary and S be a subset of $\partial\Omega$. If $f_k(\xi)$, $k = 0,1,2,3$ e $g(\xi)$ are 1-periodic bounded, measurable functions in ξ then the problem*

$$\begin{cases} \text{Find } u \in H^1(\Omega) \text{ periodic such that:} \\ -\Delta u = f_0 - \operatorname{div} \boldsymbol{f} & \text{in } \Omega, \\ \nabla u \cdot \nu = \nu_\xi \cdot \boldsymbol{f} + g & \text{on } S, \end{cases}$$

admits a unique solution apart from an additive constant if and only if

$$\int_\Omega f_0 - \int_S g = 0.$$

Applying this lemma to Problem (A.3) we obtain that the 1-periodic solution u_0 is independent of ξ; since u_0 depends only on the macroscopic variable x then it represents a potential average over Y_i.

Problem (A.4) becomes:

$$\Delta_{\xi\xi}u_1 = 0 \quad \text{in } Y_i, \qquad \nabla_\xi u_1 \cdot \nu_\xi = -\nabla_x u_0 \cdot \nu_\xi \quad \text{on } S, \tag{A.6}$$

and Lemma A.1 guarantees its solvability. An easy check shows that the solution of (A.4) can be represented as:

$$u_1(x,\xi,t) = -\boldsymbol{w}(\xi)\cdot\nabla_x u_0 + \tilde{u}_1(x,t), \tag{A.7}$$

where $\boldsymbol{w} = (w^1(\xi), w^2(\xi), w^3(\xi))^T$ satisfies:

$$\Delta_{\xi\xi} w^k = 0, \quad \text{in } Y_i, \qquad \nabla_\xi w^k \cdot \nu_\xi = n_{\xi_k} \quad \text{on } S, \quad k = 1,2,3. \tag{A.8}$$

These problems are solvable by Lemma A.1 apart an additive constant, which can be fixed e.g. by the condition $\int_S w^k = 0$.

Problem (A.5) becomes:

$$\begin{cases} \Delta_{\xi\xi} u_2 = \operatorname{div}_\xi \nabla_x (\nabla_x u_0 \cdot \boldsymbol{w}) + \operatorname{div}_x \nabla_\xi (\nabla_x u_0 \cdot \boldsymbol{w}) - \Delta_{xx} u_0 & \text{in } Y_i, \\ \nabla_\xi u_2 \cdot \nu = \nabla_\xi (\nabla_x u_0 \cdot \boldsymbol{w}) \cdot \nu - (\partial_t v_0 + I(v_0, w_0)) & \text{on } S. \end{cases} \tag{A.9}$$

Applying always Lemma A.1 for the solvability of Problem (A.9) for the class of 1-periodic functions we obtain:

$$-\int_{Y_i} \operatorname{div}_x \nabla_\xi (\nabla_x u_0 \cdot \boldsymbol{w})\, d\xi - \int_{Y_i} \Delta_{xx} u_0 \, d\xi + \int_S (v_{0,t} + I(v_0, w_0))\, d\sigma_\xi = 0.$$

Considering that u_0 is independent of ξ and substituting in (A.7) it follows:

$$\operatorname{div}_x \left[\int_{Y_i} \nabla_\xi \boldsymbol{w}\, d\xi - |Y_i| I \right] \cdot \nabla_x u_0 + |S| (\partial_t v_0 + I(v_0, w_0)) = 0, \tag{A.10}$$

where $\nabla_\xi \boldsymbol{w} = [\nabla_\xi w^1, \nabla_\xi w^2, \nabla_\xi w^3]$ and $|Y_i|$, $|S|$ denote the volume and the area of Y_i and S respectively.

Let $\widehat{\beta} = |S|/|Y|$ be the ratio between the surface membrane and the volume of the unit cell; then we have:

$$\operatorname{div}_x \left[\frac{1}{|Y|} \left(\int_{Y_i} \nabla_\xi \boldsymbol{w}\, d\xi - |Y_i| I \right) \nabla_x u_0 \right] + \widehat{\beta} (\partial_t v_0 + I(v_0, w_0)) = 0,$$

with reference to medium (i) $\boldsymbol{w}_i = \boldsymbol{w}$ and we set

$$D_i = \frac{1}{|Y|} \left\{ |Y_i| I - \int_{Y_i} \nabla_\xi \boldsymbol{w}_i \, d\xi \right\}. \tag{A.11}$$

Applying the Green formula and taking into account the periodicity of $\boldsymbol{w}_i$ we have the following expression for the macroscopic tensor of intra-cellular conductivity:

$$D_i = \frac{1}{|Y|} \left\{ |Y_i| I - \int_S \nu_i \otimes \boldsymbol{w}_i \, d\sigma_\xi \right\}.$$

Hence we obtain the following *"averaged equation"* for the intra-cellular potential:

$$\operatorname{div} D_i \nabla_x u_{i,0} = \widehat{\beta} \left(\partial_t v_0 + I(v_0, w_0) \right)).$$

Proceeding similarly for $u = u_e$ we obtain the following averaged equation for the extra-cellular potential:

$$\operatorname{div} D_e \nabla_x u_{e,0} = -\widehat{\beta} \left(\partial_t v_0 + I(v_0, w_0) \right)),$$

where

$$D_e = \frac{\alpha_e}{|Y|}\left\{|Y_e|I - \int_{Y_e} \nabla_\xi \boldsymbol{w}_e d\xi\right\}$$
$$= \frac{\alpha_e}{|Y|}\left\{|Y_e|I - \int_S \nu_e \otimes \boldsymbol{w}_e \, d\sigma_\xi, \right\}, \qquad \alpha_e = \sigma_e/\sigma_i, \tag{A.12}$$

and w_e^k, $k = 1, 2, 3$, are solutions of:

$$\Delta_{\xi\xi} w_e^k = 0, \quad \text{in } Y_e, \qquad \nabla_\xi w_e^k \cdot \nu_e = \nu_e^k \quad \text{on } S, \quad k = 1, 2, 3.$$

Finally we rescale the previous dimensionless equations using $x = \widehat{x}\cdot\mathsf{L}$ and $t = \widehat{t}\cdot\mathsf{T}$, and we denote by Y^d, Y_i^d, Y_e^d, S^d the dimensional sets in the variable $x = \xi \cdot \mathsf{d}_c$ corresponding to Y, Y_i, Y_e, S, respectively; moreover we set $\beta = |S^d|/|Y^d| = \widehat{\beta}/\mathsf{d}_c$.

Summarizing, for a periodic network of interconnected cells the governing dimensional equations, at the zero order in ε, of the macroscopic intra- and extra-cellular potentials are given by:

$$\begin{cases} \operatorname{div} M_i \nabla_x u_i = \beta\left(\partial_t v + I_{ion}(v, w)\right), \\ \operatorname{div} M_e \nabla_x u_e = -\beta\left(\partial_t v + I_{ion}(v, w)\right), \\ v = u_i - u_e, \quad \partial_t w = r(v, w), \end{cases} \tag{A.13}$$

where the effective conductivity tensors are given by:

$$M_{i,e} = \frac{\sigma_{i,e}}{|Y^d|}\left\{|Y_{i,e}^d|I - \int_{S^d} \nu_{i,e} \otimes \boldsymbol{w}_{i,e} \, d\sigma_\xi\right\},$$

with $\nu_{i,e}$ unit normal to S^d pointing outside $Y_{i,e}^d$ and w_i^k and w_e^k solutions of the cellular problems:

$$\Delta \boldsymbol{w}_i = 0 \quad \text{in } Y_i^d, \qquad \nabla_\xi \boldsymbol{w}_i \cdot \nu_i = \nu_i \quad \text{on } S^d,$$
$$\Delta \boldsymbol{w}_e = 0 \quad \text{in } Y_e^d, \qquad \nabla_\xi \boldsymbol{w}_e \cdot \nu_e = \nu_e \quad \text{on } S^d.$$

Remark A.2. Following [5], it is easy to verify that the macroscopic conductivity tensors of the intra-cellular M_i and extra-cellular M_e spaces are symmetric and positive definite.

References

[1] H. Attouch, *Variational convergence for functions and operators*, Pitman (Advanced Publishing Program), Boston, MA, 1984.

[2] C. Baiocchi, *Discretization of evolution variational inequalities*, Partial differential equations and the calculus of variations, Vol. I (F. Colombini, A. Marino, L. Modica, and S. Spagnolo, eds.), Birkhäuser Boston, Boston, MA, 1989, pp. 59–92.

[3] N. Bakhvalov and G. Panasenko, *Homogenisation: averaging processes in periodic media*, Kluwer Academic Publishers Group, Dordrecht, 1989, Mathematical problems in the mechanics of composite materials, Translated from the Russian by D. Leĭtes.

[4] F. Bassetti, *Variable time-step discretization of degenerate evolution equations in Banach spaces*, Tech. report, IAN-CNR, Pavia, 2002.

[5] A. Bensoussan, J.-L. Lions, and G. Papanicolaou, *Asymptotic analysis for periodic structures*, North-Holland Publishing Co., Amsterdam, 1978.

[6] H. Brézis, *Monotonicity methods in Hilbert spaces and some applications to nonlinear partial differential equations*, Contribution to Nonlinear Functional Analysis, Proc. Sympos. Math. Res. Center, Univ. Wisconsin, Madison, 1971, Academic Press, New York, 1971, pp. 101–156.

[7] H. Brézis, *Opérateurs maximaux monotones et semi-groupes de contractions dans les espaces de Hilbert*, North-Holland Publishing Co., Amsterdam, 1973, North-Holland Mathematics Studies, No. 5. Notas de Matemática (50).

[8] H. Brezis, *On some degenerate nonlinear parabolic equations*, Nonlinear Functional Analysis (Proc. Sympos. Pure Math., Vol. XVIII, Part 1, Chicago, Ill., 1968), Amer. Math. Soc., Providence, R.I., 1970, pp. 28–38.

[9] H. Brézis, *Intégrales convexes dans les espaces de Sobolev*, Proceedings of the International Symposium on Partial Differential Equations and the Geometry of Normed Linear Spaces (Jerusalem, 1972), vol. 13, 1972, pp. 9–23 (1973).

[10] H. Brézis, *Problèmes unilatéraux*, J. Math. Pures Appl. (9) **51** (1972), 1–168.

[11] H. Brézis C. Bardos, *Sur une classe de problèmes d'evolution non linéaires*, J. Differential Equations **6** (1969), 345–394.

[12] N. F. Britton, *Reaction-diffusion equations and their applications to biology*, Academic Press Inc. [Harcourt Brace Jovanovich Publishers], London, 1986.

[13] R. W. Carroll and R. E. Showalter, *Singular and degenerate Cauchy problems*, Academic Press [Harcourt Brace Jovanovich Publishers], New York, 1976, Mathematics in Science and Engineering, Vol. 127.

[14] R. G. Casten, H. Cohen, and P. A. Lagerstrom, *Perturbation analysis of an approximation to the Hodgkin-Huxley theory*, Quart. Appl. Math. **32** (1974/75), 365–402.

[15] P. Colli and A. Visintin, *On a class of doubly nonlinear evolution equations*, Comm. Partial Differential Equations **15** (1990), no. 5, 737–756.

[16] P. Colli Franzone and L. Guerri, *Spreading of excitation in 3-d models of the anisotropic cardiac tissue*, Math. Biosc. **113** (1993), 145–209.

[17] P. Colli Franzone, L. Guerri, and S. Rovida, *Wavefront propagation in an activation model of the anisotropic cardiac tissue: asymptotic analysis and numerical simulations*, J. Math. Biol. **28** (1990), no. 2, 121–176.

[18] J. Cronin, *Mathematics of cell electrophysiology*, Marcel Dekker Inc., New York, 1981.

[19] E. DiBenedetto and R. E. Showalter, *Implicit degenerate evolution equations and applications*, SIAM J. Math. Anal. **12** (1981), no. 5, 731–751.

[20] L. Ebihara and E. A. Johnson, *Fast sodium current in cardia muscle*, Biophys. J. **32** (1980), 779–790.

[21] A. Favini and A. Yagi, *Degenerate differential equations in Banach spaces*, Marcel Dekker Inc., New York, 1999.

[22] R. Glowinski, J.-L. Lions, and R. Trémolières, *Numerical analysis of variational inequalities*, North-Holland Publishing Co., Amsterdam, 1981, Translated from the French.

[23] C. S. Henriquez, A. L. Muzikant, and C. K. Smoak, *Anisotropy, fiber curvature, and bath loading effects on activation in thin and thick cardiac tissue preparations:*

simulations in a three dimensional bidomain model, J. Cardiovasc. Electrophysiol. **7** (1996), 424–444.

[24] C. S. Henriquez, *Simulating the electrical behavior of cardiac tissue using the bidomain model*, Crit. Rev. Biomed. Engr. **21** (1993), 1–77.

[25] A. L. Hodkin and A. F. Huxley, *A quantitative description of membrane current and its application to conduction and excitation in nerve*, J. Physiol. **117** (1952), 500–544.

[26] D. Hoff, *Stability and convergence of finite difference methods for systems of nonlinear reaction-diffusion equations*, SIAM J. Numer. Anal. **15** (1978), no. 6, 1161–1177.

[27] J. J. B. Jack, D. Noble, and R. W. Tsien, *Electric current flow in excitable cells*, Clarendon Press, Oxford, 1983.

[28] J. W. Jerome, *Convergence of successive iterative semidiscretizations for FitzHugh-Nagumo reaction diffusion systems*, SIAM J. Numer. Anal. **17** (1980), no. 2, 192–206.

[29] V. V. Jikov, S. M. Kozlov, and O. A. Oleĭnik, *Homogenization of differential operators and integral functionals*, Springer-Verlag, Berlin, 1994, Translated from the Russian by G. A. Yosifian [G. A. Iosif′yan].

[30] J. Keener and J. Sneyd, *Mathematical physiology*, Springer-Verlag, New York, 1998.

[31] J.-L. Lions, *Quelques méthodes de résolution des problèmes aux limites non linéaires*, Dunod, Gauthier-Villars, Paris, 1969.

[32] J.-L. Lions and E. Magenes, *Non-homogeneous boundary value problems and applications. Vol. I-II*, Springer-Verlag, New York, 1972, Translated from the French by P. Kenneth, Die Grundlehren der mathematischen Wissenschaften, Band 182.

[33] J.-L. Lions and G. Stampacchia, *Variational inequalities*, Comm. Pure Appl. Math. **20** (1967), 493–519.

[34] C. H. Luo and Y. Rudy, *A dynamic model of the cardiac ventricular action potential. i. simulations of ionic currents and concentration changes*, Circ. Res. **74** (1994), 1071–1096.

[35] M. Mascagni, *The backward Euler method for numerical solution of the Hodgkin-Huxley equations of nerve conduction*, SIAM J. Numer. Anal. **27** (1990), no. 4, 941–962.

[36] R. M. Miura, *Accurate computation of the stable solitary wave for the FitzHugh-Nagumo equations*, J. Math. Biol. **13** (1981/82), no. 3, 247–269.

[37] J. S. Neu and W. Krassowska, *Homogenization of syncitial tissues*, Crit. Rev. Biom. Engr. **21** (1993), 137–199.

[38] R. H. Nochetto, G. Savaré, and C. Verdi, *A posteriori error estimates for variable time-step discretizations of nonlinear evolution equations*, Comm. Pure Appl. Math. **53** (2000), no. 5, 525–589.

[39] O. A. Oleĭnik, A. S. Shamaev, and G. A. Yosifian, *Mathematical problems in elasticity and homogenization*, North-Holland Publishing Co., Amsterdam, 1992.

[40] O. A. Oleĭnik and T. Shaposhnikova, *On homogenization problems for the Laplace operator in partially perforated domains with Neumann's condition on the boundary of cavities*, Atti Accad. Naz. Lincei Cl. Sci. Fis. Mat. Natur. Rend. Lincei (9) Mat. Appl. **6** (1995), no. 3, 133–142.

[41] B. J. Roth, *How the anisotropy of the intracellular and extracellular conductivities influence stimulation of cardiac muscle*, J. Math. Biol. **30** (1992), 633–646.

[42] B. J. Roth and W. Krassowska, *The induction of reentry in cardiac tissue. The missing link: how electric fields alter transmembrane potential*, Chaos **8** (1998), 204–219.

[43] J. Rulla, *Error analysis for implicit approximations to solutions to Cauchy problems*, SIAM J. Numer. Anal. **33** (1996), 68–87.

[44] E. Sánchez-Palencia and A. Zaoui (eds.), *Homogenization techniques for composite media*, Springer-Verlag, Berlin, 1987, Papers from the course held in Udine, July 1–5, 1985.

[45] E. Sánchez-Palencia, *Nonhomogeneous media and vibration theory*, Springer-Verlag, Berlin, 1980.

[46] S. Sanfelici, *Convergence of the galerkin approximation of a degenerate evolution problem in electrocardiology*, Numer. Methods for Partial Differential Equations **18** (2002), 218–240.

[47] G. Savaré, *Approximation and regularity of evolution variational inequalities*, Rend. Acc. Naz. Sci. XL Mem. Mat. **XVII** (1993), 83–111.

[48] G. Savaré, *Weak solutions and maximal regularity for abstract evolution inequalities*, Adv. Math. Sci. Appl. **6** (1996), 377–418.

[49] R. E. Showalter, *Monotone operators in Banach space and nonlinear partial differential equations*, American Mathematical Society, Providence, RI, 1997.

[50] J. Smoller, *Shock waves and reaction-diffusion equations*, second ed., Springer-Verlag, New York, 1994.

[51] N. Trayanova, K. Skouibine, F. Aguel, *The role of cardiac tissue structure in defibrillation*, Chaos **8** (1998), 221–253.

[52] J. P. Wikswo, *Tissue anisotropy, the cardiac bidomain, and the virtual cathod effect*, Cardiac Electrophysiology: From Cell to Beside (D. P. Zipes and J. Jalife, eds.), W. B. Saunders Co., Philadelphia, 1994, pp. 348–361.

[53] A. L. Wit, S. M. Dillon, and J. Coromilas, *Anisotropy reentry as a cause of ventricular tachyarhythmias*, Cardiac Electrophysiology: From Cell to Beside (D. P. Zipes and J. Jalife, eds.), W. B. Saunders Co., Philadelphia, 1994, pp. 511–526.

Piero Colli Franzone
Dipartimento di Matematica "F. Casorati"
Università di Pavia
Via Ferrata, 1
I-27100 Pavia, Italy
e-mail: `colli@ian.pv.cnr.it`

Giuseppe Savaré
Dipartimento di Matematica "F. Casorati"
Università di Pavia
Via Ferrata, 1
I-27100 Pavia, Italy
e-mail: `savare@ian.pv.cnr.it`
`http://www.ian.pv.cnr.it/~savare`

Progress in Nonlinear Differential Equations
and Their Applications, Vol. 50, 79–96

Singular Limits for Nonlinear Hyperbolic Systems

Donatella Donatelli and Pierangelo Marcati

In ricordo di Brunello Terreni.

1. Introduction

In this paper we are interested in the study of the singular limits for the following hyperbolic nonlinear system of partial differential equation of the form

$$W_t + A(x, W, D)W = F(W), \tag{1.1}$$

where $W = W(x,t)$ takes values in $\mathbb{R}^N$ and denotes the density vector of some physical quantities over the space variable, $x \in \mathbb{R}^d$ and $A(x, W, D)$ is a first order differential or a pseudodifferential operator which acts on W, $F(W)$ denotes the nonhomogeneous term which in physical systems usually represents the external forces. Some physical problems often have inner relaxation parameters (internal scales) and a typical form for these systems is given by

$$W_t + A(x, W, D)W = \frac{1}{\varepsilon}F(W), \tag{1.2}$$

where $\varepsilon > 0$ represents possibly a relaxation parameter (e.g. relaxation time) depending on the involved physical quantities. We wish to investigate the relaxation phenomena leading to relaxed equilibrium states described by means of parabolic type systems. An elementary example for this kind of problems is given by the following damped wave equation

$$\varepsilon w_{tt} - w_{xx} + w_t = 0.$$

This equation can be converted to the form of (1.2) in the usual way, by putting

$$u = w_x \qquad v = w_t$$

to get

$$\begin{cases} u_t - v_x = 0 \\ v_t - \varepsilon^{-1} u_x = -\varepsilon^{-1} v \end{cases}$$

and as $\varepsilon \downarrow 0$, u converge in suitable norms to the solution of the heat equation

$$u_t = u_{xx}.$$

An important motivation for studying these kind of problems lies in the strong similarity with the limiting analysis of the Boltzmann equation, the so-called hydrodynamical limit. Let us denote by $f(x,\xi,t)$ the density of gas particles as a function of the time t, the position x and the velocity ξ. The Boltzmann equation has the following form

$$\nu f_t + \xi\nabla\cdot f = \frac{1}{\varepsilon}Q(f,f),$$

ν and ε are called respectively the Mach number and the Knudsen number. By averaging $f, \xi f, \xi^2 f$ we get macroscopic quantities. The hydrodynamical limits regards the behaviour of those quantities as ε goes to zero. The convergence is different if the Mach number is of the same order of the Knudsen number or if it is different. In the former case the limit can be described by the Navier Stokes equation, in the latter case by the Euler equation. A very simply model of this kind of situation is given by the two velocity model, namely the so-called Carlemann's equation,

$$\begin{cases} f_{1t} + \dfrac{1}{\varepsilon}f_{1x} = \dfrac{1}{\varepsilon^2}(f_2^2 - f_1^2) \\ f_{2t} - \dfrac{1}{\varepsilon}f_{2x} = \dfrac{1}{\varepsilon^2}(f_1^2 - f_2^2), \end{cases} \tag{1.3}$$

where we take $\xi \in \{-1,1\}$ and $f_1 = f(x,1,t)$, $f_2 = f(x,-1,t)$. Setting $\rho = f_1 + f_2$ and $m = \dfrac{f_1 - f_2}{\varepsilon}$ we get ρ, m satisfy

$$\begin{cases} \rho_t + m_x = 0 \\ \varepsilon^2 m_t + \rho_x = -2\rho m. \end{cases} \tag{1.4}$$

As $\varepsilon \downarrow 0$ ρ satisfies the nonlinear diffusion equation

$$\rho_t = \frac{1}{2}(\log\rho)_{xx}.$$

This model with a more complicated right-hand side was further investigated by P.L. Lions and Toscani [11] by using the methods proposed in the paper of Marcati and Milani [14]. Relaxation phenomena appear also in many physical situations, for example, in the equation describing the porous media flow [14], in the nonlinear heat conduction [15], in the semiconductor physics.

Here we will focus our attention on the convergence problem. To obtain rigorous results of convergence it is needed to deal with tools involving weak convergence methods. In particular some models were studied by using the methods of compensated compactness [20], [22], [23], while for others it was necessary the use of the more sophisticated theory of Microlocal Defect Measures. In this article we outline some advances on these topics and we explain with more details the results on multi-D semilinear hyperbolic systems. The plan of this paper is as follows. Section 2 begins with the fast-slow scales asymptotic expansions by following the approach of Lattanzio and Yong [10] in the quasilinear case of (1.1), then we show examples of the kind of limits arising in some physics problems. In Section 3 we follow the general framework proposed by Marcati and Rubino [18], to study the 2×2 nonlinear hyperbolic systems moreover some results on 1-D semilinear systems are provided [18], [5]. Section 4 is enterely devoted to the mathematical theory of semilinear hyperbolic systems, with nonconstant coefficients, in several space-D.

2. Asymptotic analysis and examples

Now we will give various methods and examples to investigate these type of relaxation phenomena all of these methods regard singular limits with multiscale expansions with respect to a small parameter. Another way to understand these phenomena is to consider the large time behavior of the system (1.1) and to look at the asymptotic profile as the relaxed equilibria, but we will not discuss this point of view here.

2.1. Asymptotic expansions

Since, as we said before, the limit involves parabolic equations, we can expect it is sufficiently smooth. Hence it is reasonable to construct the solution by formal asymptotic approximation, in particular close to equilibrium. Following [10], it is possible to consider

$$W_t + \sum_{j=1}^{d} \overline{A}_j(W)W_{x_j} + \frac{1}{\varepsilon}\sum_{j=1}^{d} A_j(\varepsilon W)W_{x_j} = \frac{Q(W)}{\varepsilon^2}, \tag{2.1}$$

$$W(x,0) = W_0(x;\varepsilon),$$

where $Q(W) = \begin{pmatrix} 0 \\ q(W) \end{pmatrix}$.

Corresponding to the partition of Q it is possible to write

$$Q(W) = \begin{pmatrix} u \\ v \end{pmatrix} \qquad \overline{A}_j = \begin{pmatrix} \overline{A}_j^{11} & \overline{A}_j^{12} \\ \overline{A}_j^{21} & \overline{A}_j^{22} \end{pmatrix} \qquad A_j = \begin{pmatrix} A_j^{11} & A_j^{12} \\ A_j^{21} & A_j^{22} \end{pmatrix}.$$

One has to look for a solution of the form

$$W = \sum_{k=0}^{\infty} \varepsilon^k O_k(x,t) + \sum_{k=0}^{\infty} \varepsilon^k I_k(x, t/\varepsilon^2).$$

Therefore it is possible to prove the following theorem in the H^s framework.

Theorem 2.1. [10] *Under the parabolic structure condition*

PS *$q(u,v) = 0$ if and only if $v = 0$, $q_v(u,0)$ is invertible for any u under consideration and $A_j^{11}(0) = 0$ for all j,*

and the following stability conditions

(i) *system (2.1) is symmetrizable,*

(ii) *there exists a positive definite $r \times r$ matrix $A_0^\varepsilon \equiv A_0(\varepsilon W)$ such that*

$$A_0^\varepsilon(W)(\overline{A}_j(W), A_j(\varepsilon W)) = (\overline{A^*}_j(W), A^*{}_j(\varepsilon W))(\overline{A}_j(W), A_j(\varepsilon W)).$$

Moreover assume suitable regularity, then there exists a constant $K > 0$, independent if ε such that

$$\|W_\varepsilon^m(t) - W_\varepsilon\|_{H^s} \le K\varepsilon^m$$

for ε sufficiently small, $m > 2$ and t in a "uniform" time interval.

2.2. Models from mathematical physics and engineering

We investigate here some known models from physics and engineering that are known to have nonsmooth weak solutions. Moreover we will not restrict to small perturbations therefore for those reasons we need to use different mathematical frameworks such as Compensated Compactness [20],[22], [23] and Microlocal Defect Measures [7]. Some of the models investigated in the sequel, because of their physical meaning, have internal relaxation parameters.

Example 2.2. (Nonlinear heat conduction) *We recall the nonlinear wave equation proposed by Cattaneo [1] to describe nonlinear heat conduction*

$$\phi_{\tau\tau} + \phi_\tau - \sigma(\phi_y)_y = 0 \tag{2.2}$$

where $\sigma \in C^2(\mathbb{R})$ verifies the conditions

$$\begin{aligned} \sigma'(\lambda) &> 0 && \text{for all } \lambda > 0 \\ \sigma''(\lambda)\lambda &> 0 && \text{for all } \lambda \neq 0. \end{aligned}$$

Now setting $z = s_\tau$, $w = s_y$ and rescaling as

$$\begin{cases} w(y,t) = u^\varepsilon(\varepsilon y, \varepsilon^2\tau) \\ z(y,t) = \varepsilon v^\varepsilon(\varepsilon y, \varepsilon^2\tau) \end{cases} \tag{2.3}$$

the equation (2.2) becomes

$$\begin{cases} u_t^\varepsilon - v_x^\varepsilon = 0 \\ \varepsilon^2 v_t^\varepsilon - \sigma(u^\varepsilon)_x = -v^\varepsilon. \end{cases} \tag{2.4}$$

In [15] was proved that as $\varepsilon \downarrow 0$:

$$\begin{aligned} v^\varepsilon &\rightharpoonup v && \text{weakly in } L^2 \\ v^\varepsilon &\to v && \text{strongly in } L^p_{loc} \text{ for all } p \ge 1 \end{aligned}$$

where (u, v) verifies the system

$$\begin{cases} u_t - v_x = 0 \\ v - \sigma(u)_x = 0. \end{cases} \tag{2.5}$$

In particular u satisfies in the H^{-1}_{loc} sense the nonlinear heat equation

$$u_t - \sigma(u)_{xx}.$$

Example 2.3. (Porous media equation) *Let us consider a dumped compressible Euler flow in one space dimension*

$$\begin{cases} \rho_\tau + (\rho n)_y = 0 \\ (\rho n)_\tau + (\rho n^2 + p(\rho))_y = -kn\,, \end{cases} \tag{2.6}$$

where $k > 0$ *and* $p(\rho) = \frac{1}{\gamma}\rho^\gamma$, $\gamma > 1$. *If we denote by* $\rho = w$ *and* $\rho = z$, *system* (2.6) *can be written in the following form*

$$\begin{cases} w_\tau + z_y = 0 \\ z_\tau + (\dfrac{z^2}{w} + p(w))_y = -k\dfrac{z}{w} . \end{cases} \tag{2.7}$$

By applying the scaling (2.3) *the system* (2.7) *becomes*

$$\begin{cases} u_t^\varepsilon + v_x^\varepsilon = 0 \\ \varepsilon^2 v_t^\varepsilon + (\varepsilon^2 \dfrac{(v^\varepsilon)^2}{u} + p(u^\varepsilon))_x = -k\dfrac{v^\varepsilon}{u^\varepsilon} . \end{cases} \tag{2.8}$$

In [14] *was proved that there exists* (u, v) *such that as* $\varepsilon \downarrow 0$:

$$\begin{array}{ll} v^\varepsilon \rightharpoonup v & \textit{weakly in } L^2_{loc} \\ \varepsilon v^\varepsilon \to 0 & \textit{strongly in } L^p_{loc} \textit{ for all } p \geq 1 \\ u^\varepsilon \to u & \textit{strongly in } L^p_{loc} \textit{ for all } p \geq 1. \end{array}$$

The limit function u *is a weak solution of the porous media equation*

$$u_t - \frac{1}{k}(up(u)_x)_x = 0$$

and (u, v) *satisfy the Darcy's law* $v = -\frac{1}{k}up(u)_x$.

These two examples have been generalized to the $2 - D$ case by Marcati and Rubino in [18]. Finally we conclude this section with a model arising in the physics of semiconductor devices. In this case we have to point out that there exists a natural relaxation time $\tau > 0$.

Example 2.4. (Drift diffusion) *We consider the Euler-Poisson equation for semiconductors*

$$\begin{cases} n_s + j_y = 0 \\ j_s + (j^2/n + p(n))_y = nE - j/\tau \\ E_y = n - c(y). \end{cases} \tag{2.9}$$

Let us rescale the variables by putting $N(x,t) = n(x, t/\tau)$ *and* $\tau J(x,t) = j(x, t/\tau)$, *then it follows*

$$\begin{cases} N_t^\tau + J_x^\tau = 0 \\ \tau^2 J_t^\tau + (\tau^2 (J^\tau)^2/N^\tau + p(N^\tau))_x = N^\tau \overline{E} - J^\tau \\ \overline{E}_x = N^\tau - c(x). \end{cases} \tag{2.10}$$

As investigated in [16], [17], *the limit profile* $(N, \overline{E})$ *as* τ *goes to zero satisfy the drift diffusion equation*

$$\begin{cases} N_t + (N\overline{E} - p(N)_x)_x = 0 \\ \overline{E}_x = n - c(x). \end{cases}$$

3. 1-D Mathematical Theory

In this section we will report the main ideas for a general mathematical theory in the 1-D case.

3.1. Nonlinear 2×2 systems

We follow here the approach for 2×2 nonlinear systems due to Marcati and Rubino in [18]. Consider the following quasilinear nonhomogeneous 2×2 hyperbolic system

$$\begin{cases} w_s + f(w,z)_y = 0 \\ z_s + g(w,z)_y = h(w,z), \end{cases} \tag{3.1}$$

where $y \in \mathbb{R}$, $s \geq 0$ and assume the following hypotheses holds

(B.1) $f, g, h:\ \mathbb{R}^2 \longrightarrow \mathbb{R}$ are continuously differentiable functions such that

$$f(w,0) = 0, \qquad\qquad h(w,0) = 0,$$

(B.2) if we denote by $\lambda_{\pm}(w,z)$ the characteristic speeds of (3.1) (namely the eigenvalues of $\frac{\partial(f,g)}{\partial(w,z)}$), the system (3.1) is strictly hyperbolic, namely $\lambda_- < \lambda_+$, that is equivalent to assume

$$\Delta(w,z) := (f_w + g_z)^2 + 4g_w f_z > 0 \qquad\qquad \text{for all } (w,z),$$

(B.3) denote by

$$f^*(w) := \frac{\partial f}{\partial z}(w,0), \quad h^*(w) := \frac{\partial h}{\partial z}(w,0), \quad g^0(w) := g(w,0).$$

then, for all $w \in \mathbb{R}$,

$$h^*(w) < 0, \quad (g^0(w))' f^*(w) > 0.$$

We can implement two different type of scaling depending whether or not there exists some internal relaxation time. For system (3.1), it is possible to rescale both space and time by setting

$$\begin{cases} u^\varepsilon(x,t) = w\left(\frac{x}{\sqrt{\varepsilon}}, \frac{t}{\varepsilon}\right) \\ v^\varepsilon(x,t) = \dfrac{1}{\sqrt{\varepsilon}} z\left(\frac{x}{\sqrt{\varepsilon}}, \frac{t}{\varepsilon}\right) \end{cases} \tag{3.2}$$

for all $\varepsilon > 0$, $x \in \mathbb{R}$, $t \geq 0$. The rescaled system is given by

$$\begin{cases} u^\varepsilon_t + \dfrac{1}{\sqrt{\varepsilon}} f\left(u^\varepsilon, \sqrt{\varepsilon} v^\varepsilon\right)_x = 0 \\ \varepsilon v^\varepsilon_t + g\left(u^\varepsilon, \sqrt{\varepsilon} v^\varepsilon\right)_x = \dfrac{1}{\sqrt{\varepsilon}} h\left(u^\varepsilon, \sqrt{\varepsilon} v^\varepsilon\right) \end{cases} \tag{3.3}$$

supplemented by the initial conditions

$$\begin{cases} u^\varepsilon(x,0) = u^\varepsilon_0(x) \\ v^\varepsilon(x,0) = v^\varepsilon_0(x). \end{cases} \tag{3.4}$$

The second type of scaling which can be considered is related with systems which possess an internal relaxation time. In this case, the limiting behavior as $\tau \downarrow 0$, of the system

$$\begin{cases} w_s + f(w,z)_x = 0 \\ z_s + g(w,z)_x = h\left(w, \frac{1}{\tau}z\right) \end{cases} \tag{3.5}$$

under the initial conditions

$$\begin{cases} w(x,0) = w_0(x) \\ z(x,0) = z_0(x) \end{cases} \tag{3.6}$$

can be investigated by using the following scaling

$$\begin{cases} u^\tau(x,t) = w\left(x, \dfrac{t}{\tau}\right) \\ v^\tau(x,t) = \dfrac{1}{\tau} z\left(x, \dfrac{t}{\tau}\right), \end{cases} \tag{3.7}$$

where $x \in \mathbb{R}$, $t \geq 0$, $\tau > 0$. Then we end up to study the following system

$$\begin{cases} u_t^\tau + \dfrac{1}{\tau} f(u^\tau, \tau v^\tau)_x = 0 \\ \tau^2 v_t^\tau + g(u^\tau, \tau v^\tau)_x = h(u^\tau, v^\tau) \end{cases} \tag{3.8}$$

under the initial conditions

$$\begin{cases} u^\tau(x,0) = w(x,0) = w_0(x) \\ v^\tau(x,0) = \dfrac{1}{\tau} z(x,0) = \dfrac{1}{\tau} z_0(x). \end{cases}$$

In order to deal with this case it is necessary to modify (B.3) in the following more restrictive condition:

(B.3′) assume that f^*, g^0 are defined as in the condition (B.3) and for all $w, z \in \mathbb{R}$,

$$\frac{\partial h}{\partial z}(w,z) < 0, \qquad\qquad (g^0(w))' f^*(w) > 0.$$

On the other hand, condition (B.1) can be modified in the less restrictive:

(B.1′) $f, g, h : \mathbb{R}^2 \longrightarrow \mathbb{R}$ are continuously differentiable functions such that $f(w,0) = 0$.

In [18], finally it is possible to prove the following theorems

Theorem 3.1. *Assume that the hypotheses (B.1), (B.2), (B.3) hold and the initial datum u_0^ε satisfies the condition*

$$u_0^\varepsilon \overset{*}{\rightharpoonup} u \qquad \text{in } L^\infty \text{ as } \varepsilon \downarrow 0;$$

moreover assume that

(B.4) *there exists a convex entropy, entropy-flux, pair* (η^*, q^*), *for the system* (3.1), *and a continuous function* $\psi \in C^2$, $\psi^{-1}(\{0\}) = \{0\}$ *and a constant* $\gamma > 0$ *such that* $\psi'' > \gamma$ *and we have*

$$\eta_z^*(w,z)h(w,z) \leq -\psi(z) \qquad \text{for all } (w,z);$$

(B.5) $(u^\varepsilon, v^\varepsilon)$ *is a solution to* (3.3), *limit of the vanishing viscosity method, such that* $(u^\varepsilon, \sqrt{\varepsilon}v^\varepsilon)$ *is uniformly bounded in* L^∞.

Then one has, as $\varepsilon \downarrow 0$

$$u^\varepsilon \longrightarrow u \qquad \text{strongly in } L^p_{loc},\ p < +\infty,$$
$$v^\varepsilon \rightharpoonup v \qquad \text{weakly in } L^2,$$

and it follows that (u,v) *verifies the relations*

$$u_t + (f^*(u)\,v)_x = 0 \tag{3.9}$$
$$g^0(u)_x = h^*(u)\,v \tag{3.10}$$

where (3.9) *is verified in the sense of distributions and* (3.10) *in* L^2.

In a similar way it can be investigated the relaxation problem for the system (3.5).

Theorem 3.2. *Assume that the hypotheses* (B.1′), (B.2), (B.3′), (B.4) *hold and*

(B.5′) (u^τ, v^τ) *is a solution to* (3.8), *limit of the vanishing viscosity method, such that* $(u^\tau, \tau v^\tau)$ *is uniformly bounded in* L^∞.

Then one has, as $\tau \downarrow 0$,

$$u^\tau \longrightarrow u \qquad \text{strongly in } L^p_{loc},\ p < +\infty,$$
$$v^\tau \rightharpoonup v \qquad \text{weakly in } L^2,$$

and it follows that (u,v) *verifies the relations*

$$u_t + (f^*(u)\,v)_x = 0 \tag{3.11}$$
$$g^0(u)_x = h(u,v) \tag{3.12}$$

where (3.11) *is verified in the sense of distributions and* (3.12) *in* L^2.

Remark 3.3. *The two cases are actually very similar from a mathematical point of view. The basic ideas used in the proof of the theorems involves the methods of compensated compactness as they have been created by Tartar* [22], [23] *and the exponential entropies of Di Perna* [2], [3].

We want also point out that although the results in Theorem 3.1 *and Theorem* 3.2 *are apparently similar, there is a big difference on the hypotheses for the initial data in the Cauchy problem. The rescaling* (3.7) *is only in time while the initial data remain unchanged under the rescaling. The situation is very different for Theorem* 3.1. *Here the system* (3.3) *is obtained from the initial system* (3.1) *via the rescaling* (3.2), *which acts also on the space scale. In this situation we fix the initial data after the change of scale.*

3.2. Semilinear systems

In this section we show some basic results on 1-D semilinear hyperbolic systems. Let us consider the system

$$\begin{cases} \partial_s \overline{U} + \sum\limits_{j=1}^{d} K_j^1 \partial_j \overline{V} = 0 \\ \partial_s \overline{V} + \sum\limits_{j=1}^{d} H_j^2 \partial_j \overline{U} + \sum\limits_{j=1}^{d} K_j^2 \partial_j \overline{V} = R\left(\overline{U}, \overline{V}\right). \end{cases} \tag{3.13}$$

With a scaling of the first type (3.2) we get

$$\begin{cases} \partial_t U^\varepsilon + K^1 \partial_x V^\varepsilon = 0 \\ \varepsilon \partial_t V^\varepsilon + H^2 \partial_x U^\varepsilon + \varepsilon^{1/2} K^2 \partial_x V^\varepsilon = \varepsilon^{-1/2} R\left(U^\varepsilon, \varepsilon^{1/2} V^\varepsilon\right). \end{cases} \tag{3.14}$$

In order to obtain a convergence result for the solutions of the system (3.14), we make on (3.14) the following hypotheses:

- **(C.1)** $K^2 \in \mathcal{M}_{(N-k)\times(N-k)}$ is symmetric,
- **(C.2)** there exists a symmetric positive definite matrix $B_0 \in \mathcal{M}_{k\times k}$ such that $\left(K^1\right)^T B_0 = H^2$,
- **(C.3)** there exists $\lambda_1 > 0$ such that for any $U \in \mathbb{R}^k$, $R(U,0) \equiv 0$ and $\left\|R_V(U,0)\right\|_\infty \leq \lambda_1$,
- **(C.4)** there exist $\lambda_2, \lambda_3 > 0$ such that for any $(U,Z) \in \mathbb{R}^k \times \mathbb{R}^{N-k}$, $R_V(U,Z) \leq -\lambda_2 Z$ and $\left\|R_{VV}(U,Z)\right\|_\infty \leq \lambda_3$
- **(C.5)** $\det\left(\left(H^2\right)^T H^2\right) \neq 0$.

Since $H^2 \in \mathcal{M}_{(N-k)\times k}$, from elementary linear algebra we deduce condition (C.5) is violated whenever $k > N/2$. The following convergence result will be shown by using only the information provided by the energy estimates.

Theorem 3.4. *Let us consider the Cauchy problem for the system* (3.14). *Assume that the hypotheses (C.1), (C.2), (C.3), (C.4) and (C.5) hold. Then there exists* $\left(U^0, V^0\right) \in L^2$ *such that, as* $\varepsilon \downarrow 0$, *one has*

$$U^\varepsilon \longrightarrow U^0 \qquad \text{strongly in } L^2_{loc}, \tag{3.15}$$

$$V^\varepsilon \rightharpoonup V^0 \qquad \text{weakly in } L^2, \tag{3.16}$$

$$\sqrt{\varepsilon} V^\varepsilon \longrightarrow 0 \qquad \text{strongly in } L^2_{loc} \tag{3.17}$$

and the limit profile $\left(U^0, V^0\right)$ *verifies in* $\mathcal{D}'$ *the system*

$$\begin{cases} \partial_t U^0 + K^1 \partial_x V^0 = 0 \\ H^2 \partial_x U^0 = R_V^0\left(U^0\right) V^0. \end{cases}$$

This result was further generalized in [5]. We consider there the following semilinear system with variable coefficients.

$$\begin{cases} \partial_s \overline{U} + K^{(1)}(y)\partial_y \overline{V} = F^{(1)}(x,\overline{U},\overline{V}) \\ \partial_s \overline{V} + H^{(2)}(y)\partial_y \overline{U} + K^{(2)}(y)\partial_y \overline{V} = \varepsilon^{-1}R(y,\overline{U})\overline{V} + F^{(2)}(x,\overline{U},\overline{V}), \end{cases} \tag{3.18}$$

where $(y,s) \in \mathbb{R}\times\mathbb{R}_+$, $\overline{U} = \overline{U}(y,s) \in \mathbb{R}^k, \overline{V} = \overline{V}(y,s) \in \mathbb{R}^{N-k}$, $K^{(1)}(y)$, $H^{(2)}(y)$, $K^{(2)}(y)$, $R(y,\overline{U})$ are matrices such that $K^{(1)}(y) \in \mathcal{M}_{k\times(N-k)}$, $H^{(2)}(y) \in \mathcal{M}_{(N-k)\times k}$, $K^{(2)}(y) \in \mathcal{M}_{(N-k)\times(N-k)}$, $R(y,\overline{U}) \in \mathcal{M}_{(N-k)\times(N-k)}$, $F^{(1)}(y,\overline{U},\overline{V}) \in \mathbb{R}^k$, $F^{(2)}(y,\overline{U},\overline{V}) \in \mathbb{R}^{N-k}$. The scaling that is needed is the following: for any $\varepsilon > 0$ we set

$$\begin{aligned} y &= x, & s &= \frac{t}{\varepsilon}, \\ U^\varepsilon(x,t) &= \overline{U}\left(x,\frac{t}{\varepsilon}\right), & V^\varepsilon(x,t) &= \frac{1}{\varepsilon}\overline{V}\left(x,\frac{t}{\varepsilon}\right). \end{aligned} \tag{3.19}$$

With previous position the system (3.18) transforms into

$$\begin{cases} \partial_t U^\varepsilon + K^{(1)}(x)\partial_x V^\varepsilon = F^{(1)}(x,U^\varepsilon,\varepsilon V^\varepsilon) \\ \varepsilon^2\partial_t V^\varepsilon + H^{(2)}(x)\partial_x U^\varepsilon + \varepsilon K^{(2)}(x)\partial_x V^\varepsilon = R(x,U^\varepsilon)V^\varepsilon(x,t) \\ \qquad\qquad\qquad\qquad\qquad\qquad\qquad\qquad +F^{(2)}(x,U^\varepsilon,\varepsilon V^\varepsilon). \end{cases} \tag{3.20}$$

We make on (3.20) the following hypotheses

- **(E.1)** $\overline{U}(x,0) = \overline{U}_0(x) \in \left[L^2(\mathbb{R})\right]^k$, $\overline{V}(x,0) = \overline{V}_0(x) \in \left[L^2(\mathbb{R})\right]^{N-k}$,
- **(E.2)** there exist symmetric positive definite matrices $B_0(x) \in \mathcal{M}_{k\times k}$, $D_0(x) \in \mathcal{M}_{(N-k)\times(N-k)}$ such that $(K^{(1)}(x))^T B_0(x) = D_0(x)H^{(2)}(x)$, $\forall x \in \mathbb{R}$, $|B_0(x)| \le \gamma$, $|D_0(x)| \le \gamma$, $\operatorname{meas}\{x \mid \det B_0(x) = 0\} = 0$,
- **(E.3)** $R(x,U) \in C(\mathbb{R}\times\mathbb{R}^k, \mathcal{M}_{(N-k)\times(N-k)})$, $D_0(x)R(x,U) + R(x,U)^T D_0(x)$ is negative definite for all $x \in \mathbb{R}$, namely there exists $\lambda \in \mathbb{R}$, $\lambda > 0$ such that $D_0(x)R(x,U) + R(x,U)^T D_0(x) \le -\lambda I \quad$ for all $x \in \mathbb{R}$,
- **(E.4)** $K^{(1)} \in C^1(\mathbb{R}, \mathcal{M}_{k\times(N-k)})$, $K^{(1)}(x)$ is bounded for all $x \in \mathbb{R}$ and $\det\left[K^{(1)}(x)(K^{(1)}(x))^T\right] \neq 0$ $\forall x \in \mathbb{R}$,
- **(E.5)** $H^{(2)} \in C^1(\mathbb{R}, \mathcal{M}_{(N-k)\times k})$, $H^{(2)}(x)$ is bounded for all $x \in \mathbb{R}$ and we set $M = \sup_{x\in\mathbb{R}}\left(\left(D_0(x)H^{(2)}(x)\right)_x\right)$,
- **(E.6)** $K^{(2)} \in C^1(\mathbb{R}, \mathcal{M}_{(N-k)\times(N-k)})$, for all $x \in \mathbb{R}$, $D_0(x)K^{(2)}(x) = (K^{(2)}(x))^T D_0(x)$ and there exists $N \in \mathbb{R}$, such that $|\mu_j(x)| \le N$, $\forall j = 1,\dots,m$, where $\mu_j(x)$ are the eigenvalues of $\left(D_0(x)K^{(2)}(x)\right)_x + \left(\left(D_0(x)K^{(2)}(x)\right)_x\right)^T$,
- **(E.7)** $F(x,U,V) = \left(F^{(1)}(x,U,V), F^{(2)}(x,U,V)\right) \in \mathbb{R}^N$, is α-lipschitz function of (U,V), $\alpha \in \mathbb{R}$ moreover $F(x,0,0) = 0$ for all $x \in \mathbb{R}$, $F^{(1)}(x,U,0) = 0$ for all $U \in \mathbb{R}^k$.

Most of the previous hypotheses can be obtained if we suppose the system strictly hyperbolic. The main result in this case is

Theorem 3.5. *Let us consider the solution* $\{U^\varepsilon\}$, $\{V^\varepsilon\}$ *of the Cauchy problem for system* (3.20). *Assume the hypotheses (E.1), (E.2), (E.3), (E.4), (E.5), (E.6), (E.7) and moreover*

(E.8) $R(x,U)$ *is bounded on* U.

Then for any $T > 0$, *independent from* ε, $U^0 \in [L^2(\mathbb{R}\times\mathbb{R}_+)]^k$, $V^0 \in [L^2(\mathbb{R}\times\mathbb{R}_+)]^{N-k}$ *such that, as* $\varepsilon \downarrow 0$, *one has (extracting eventually subsequences),*

$$V^\varepsilon \rightharpoonup V^0 \qquad \text{weakly in } L^2(\mathbb{R}\times[0,T]) \tag{3.21}$$

$$\varepsilon V^\varepsilon \longrightarrow 0 \qquad \text{strongly in } L^2_{loc}(\mathbb{R}\times\mathbb{R}_+) \tag{3.22}$$

$$\{\varepsilon^2 V^\varepsilon_t\} \longrightarrow 0 \qquad \text{in } H^{-1}_{loc}(\mathbb{R}\times\mathbb{R}_+). \tag{3.23}$$

$$U^\varepsilon \longrightarrow U^0 \qquad \text{a.e in } \mathbb{R}\times\mathbb{R}_+ \tag{3.24}$$

$$R(x,U^\varepsilon) \longrightarrow R(x,U^0) \qquad \text{strongly in } L^p_{loc}(\mathbb{R}\times\mathbb{R}_+) \tag{3.25}$$

$$F^{(1)}(x,U^\varepsilon,\varepsilon V^\varepsilon) \longrightarrow F^{(1)}(x,U^0,0) \qquad \text{strongly in } L^2_{loc}(\mathbb{R}\times\mathbb{R}_+) \tag{3.26}$$

$$F^{(2)}(x,U^\varepsilon,\varepsilon V^\varepsilon) \longrightarrow F^{(2)}(x,U^0,0) \qquad \text{strongly in } L^2_{loc}(\mathbb{R}\times\mathbb{R}_+)\ . \tag{3.27}$$

moreover (U^0, V^0) *verifies, in the sense of distributions, the following system*

$$\begin{cases} \partial_t U^0(x,t) + K^{(1)}(x)\partial_x V^0(x,t) = F^{(1)}(x,U^0(x,t),0) \\ H^{(2)}(x)\partial_x U^0(x,t) = R(x,U^0(x,t))V^0(x,t) + F^{(2)}(x,U^0(x,t),0) \end{cases} \tag{3.28}$$

that in the sense of distribution has the equivalent formulation

$$U^0_t + K^{(1)}(x)\left(R(x,U^0)^{-1}H^{(2)}(x)U^0_x - R(x,U^0)^{-1}F^{(2)}(x,U^0,0)\right)_x = F^{(1)}(x,U^0,0). \tag{3.29}$$

The proof of the Theorem 3.5 follows from the arguments in [5]. The difference in the case of variable coefficients lies in the technique used to get strong convergence of the sequence $\{U^\varepsilon\}$. While in the constant coefficient case we can use the classical Compensated Compactness ([22], [23]), in the variable coefficient case it is not sufficient. It is then necessary the use of the generalized compensated compactness due to Tartar [24] and Gérard [7]. How to use this tool will be better explained in the next section.

4. Multi-D mathematical theory

Here we consider the following semilinear multidimensional hyperbolic system with a small parameter $\varepsilon > 0$

$$W_t(x,t) + \frac{1}{\varepsilon}A(x,D)W(x,t) = \frac{1}{\varepsilon^2}B(x,W(x,t)) + \frac{1}{\varepsilon}D(W(x,t)), \tag{4.1}$$

where $W = W(x,t)$ takes values in $\mathbb{R}^N$, $x \in \mathbb{R}^d$, $t \geq 0$, $A(x,D)$ is a pseudodifferential operator. System (4.1) includes also the case of hyperbolic differential operators of the form

$$W_t(x,t) + \frac{1}{\varepsilon}\sum_{j=1}^{d} A_j(x)\partial_j W(x,t) = \frac{1}{\varepsilon^2}B(x, W(x,t)) + \frac{1}{\varepsilon}D(W(x,t)), \qquad (4.2)$$

where $A_j(x)$, $j = 1,\dots,d$ are $N \times N$ matrices for any $x \in \mathbb{R}^d$. Our aim is to understand the limiting behaviour of the system (4.1) as ε goes to zero. We will look for structure condition in order that (4.1) will approximate a second order parabolic system. With respect to the previous paragraph the problem here is more complicated since the symmetrizers are pseudodifferential operators which make more difficult the proof of the a priori estimates. We will give here only the principal steps of the path to follow in this situation. We consider now

$$W_t(x,t) + A(x,D)W(x,t) = B(x, W(x,t)) + D(W(x,t)), \qquad (4.3)$$

and we assume the following hypotheses hold.

- **(A.1)** $B(x,W), D(W) \in \mathbb{R}^N$,
- **(A.2)** $A(x,D) \in OPS^1$, the system (4.3) is hyperbolic, namely the principal symbol of $A(x,D)$ is the matrix $A(x,\xi) \in \mathcal{M}_{N\times N}$ whose eigenvalues for $(x,\xi) \in \mathbb{R}^d \times \mathbb{R}^d$, $\xi \neq 0$ are pure imaginary,
- **(A.3)** denote by $S = span\,\{B(x,W) \mid W \in \mathbb{R}^N\}$ then dim $S = N - k$, $0 < k < N$.

Considering the hypothesis (A.3) there exists a matrix $P^I \in \mathcal{M}_{k\times N}$ such that $P^I B(W) = 0$ and $Z^I = P^I W$ is the conserved vector. Therefore, we can construct an invertible matrix

$$P = \begin{bmatrix} P^I \\ P^{II} \end{bmatrix}, \qquad P^{II} \in \mathcal{M}_{(N-k)\times N}.$$

Now we set, for any $W \in \mathbb{R}^N$

$$Z^I = P^I W, \qquad Z^{II} = P^{II} W,$$

$$Z = \begin{bmatrix} Z^I \\ Z^{II} \end{bmatrix}, \qquad Q(Z^I, Z^{II}) = P^{II} B(P^{-1}Z)$$

$$D^I(Z^I, Z^{II}) = P^I D(P^{-1}Z) \qquad D^{II}(Z^I, Z^{II}) = P^{II} D(P^{-1}Z)$$

$$PA(x,\xi)P^{-1} = \begin{bmatrix} M^{11}(x,\xi) & M^{12}(x,\xi) \\ M^{21}(x,\xi) & M^{22}(x,\xi) \end{bmatrix} = \overline{M}(x,\xi) \qquad M^{ij}(x,D) = OPM^{ij}(x,\xi),$$

hence we can rewrite the system (4.3) in the following way

$$\begin{cases} Z^I_t + M^{11}(x,D)Z^I + M^{12}(x,D)Z^{II} = D^I(Z^I, Z^{II}) \\ Z^{II}_t + M^{21}(x,D)Z^I + M^{22}(x,D)Z^{II} = Q(x, Z^I, Z^{II}) + D^{II}(Z^I, Z^{II}), \end{cases} \qquad (4.4)$$

by construction $Z^I = Z^I(x,t) \in \mathbb{R}^k$, $Z^{II} = Z^{II}(x,t) \in \mathbb{R}^{N-k}$ and $M^{11}(x,\xi) \in \mathcal{M}_{k\times k}$, $M^{12}(x,\xi) \in \mathcal{M}_{k\times(N-k)}$, $M^{21}(x,\xi) \in \mathcal{M}_{(N-k)\times k}$, $M^{22}(x,\xi) \in \mathcal{M}_{(N-k)\times(N-k)}$, $D^I(Z^I, Z^{II}) \in \mathbb{R}^k$, $D^{II}(Z^I, Z^{II}) \in \mathbb{R}^{N-k}$, $Q(x, Z^I, Z^{II}) \in \mathbb{R}^{N-k}$. We remark that system (4.4) is still hyperbolic. In this way we have decoupled (4.1) in order to single out the conserved quantities from the others. In order to perform our analysis on system (4.4) we need the following structural assumption:

(S.1) $M^{11}(x,\xi) = 0$, for any $(x,\xi) \in \mathbb{R}^d \times \mathbb{R}^d$.

The condition (S.1) is essential otherwise the only relaxation process would be the trivial one which relaxes on the null solution. Now, considering (4.4), by using the previous notations and by denoting

$$M(x,D) = OP\left\{-\begin{pmatrix} 0 & M^{12}(x,\xi) \\ M^{21}(x,\xi) & M^{22}(x,\xi) \end{pmatrix}\right\} \tag{4.5}$$

we can rewrite the system (4.4) in the following form

$$Z_t - M(x,D)Z = Q(x, Z^I, Z^{II}) + D(Z^I, Z^{II}), \tag{4.6}$$

where $D(Z^I, Z^{II}) = (D^I(Z^I, Z^{II}), D^{II}(Z^I, Z^{II}))$.
We define the scaling needed in this situation

$$Z^I(x,t) = U^I(x,t), \qquad\qquad Z^{II}(x,t) = \varepsilon U^{II}(x,t). \tag{4.7}$$

The previous scaling, as seen in Section 2.1 has an equivalent interpretation as a scaling of the time variable setting $\partial_\tau = \varepsilon\partial_t$. With previous position and the decoupling (4.4) the system (4.1) transforms into

$$\begin{cases} U_t^I + M^{12}(x,D)U^{II} = \dfrac{1}{\varepsilon}D^I(U^I, \varepsilon U^{II}) \\ \varepsilon^2 U_t^{II} + M^{21}(x,D)U^I + \varepsilon M^{22}(x,D)U^{II} = \dfrac{1}{\varepsilon}Q(x, U^I, \varepsilon U^{II}) + D^{II}(U^I, \varepsilon U^{II}). \end{cases} \tag{4.8}$$

In order to establish a priori estimates we assume here the existence of a symmetrizer for the system (4.6) (Taylor [25]), namely

(A.4) there exists $R(x,D) \in OPS^0_{1,0}$ such that
$R(x,D)M(x,D) + (R(x,D)M(x,D))^* \in OPS^0_{1,0}$ and its symbol $R(x,\xi)$ is a positive definite matrix for $|\xi| > 1$.

The next structure condition regards the structure of a symmetrizer for the system (4.6). We assume here a special "block structure" which is natural for strictly hyperbolic systems (see for instance the seminal paper of Kreiss [8] or of Majda and Osher [13] and Ralston [21]). The block structure follows also for non-strictly hyperbolic systems having constant multiplicity, by a deep result due to Métivier [19]. Here we are not assuming anyone of the previous conditions but only the "block structure" of the symmetrizer $R(x,D)$.

(S.2) the symbol of $R(x,D)$ has the following form

$$R(x,\xi) = \begin{pmatrix} R_{11}(x,\xi) & 0 \\ 0 & R_{22}(x,\xi) \end{pmatrix},$$

where $R_{11}(x,\xi) \in \mathcal{M}_{k\times k}$, $R_{22}(x,\xi) \in \mathcal{M}_{(N-k)\times(N-k)}$ are symmetric positive definite matrices and $R(x,D)M(x,D) + (R(x,D)M(x,D))^* \in OPS^0_{1,0}$.

Moreover we assume the following hypotheses

(G.1) $M^{12}(x,\xi), M^{21}(x,\xi), M^{22}(x,\xi) \in S^1$,

(G.2) $\det\left[\left(M^{21}(x,\xi)\right)^T M^{21}(x,\xi)\right] \neq 0$,

(G.3) $D{=}(D^I(Z^I,Z^{II}), D^{II}(Z^I,Z^{II})){\in}\, C^1(\mathbb{R}^N;\mathbb{R}^k\times\mathbb{R}^{N-k})$, $D^I(Z^I,0)=0$, $D^{II}(0,0)=0$, and D^I_ν is bounded in (Z^I,Z^{II}), D^{II} is a α-lipschitz function in (Z^I,Z^{II}),

(D) $Q(x,Z^I,Z^{II})$ has the following form

$$Q(x,Z^I,Z^{II}) = Q_0(x,Z^I,Z^{II}) + Q_1(x,Z^I,Z^{II})$$

and $Q(x,Z^I,0)=0$ for any $(x,Z^I)\in\mathbb{R}^d\times\mathbb{R}^k$ moreover

(d1) $Q_0(x,Z^I,Z^{II}) \in C^1(\mathbb{R}^{N+d};\mathbb{R}^{N-k})$, $Q_{0\nu}(x,Z^I,Z^{II})$ is bounded in (x,Z^I,Z^{II}) and $[Q_{0\nu}, R_{22}^{1/2}(x,D)] = 0$. There exists $\lambda_0>0$ such that for any $x\in\mathbb{R}^d$, $(Z^I,Z^{II})\in\mathbb{R}^k\times\mathbb{R}^{N-k}$, $Q_{0\nu}(x,Z^I,Z^{II}) \leq -\lambda_0 I$,

(d2) $Q_1(x,Z^I,Z^{II}) \in C^1(\mathbb{R}^{N+1};\mathbb{R}^{N-k})$ $Q_{1\nu}(x,Z^I,Z^{II})$ is bounded in (x,Z^I,Z^{II}) and the operator $R_{22}(x,D)Q_{1\nu}$ has norm $\|R_{22}(x,D)Q_{1\nu}\|_{\mathcal{L}(L^2)} \leq \lambda_1$, $\lambda_1>0$, $\lambda_1\leq\lambda_0/2$.

We remark that since $M^{21}(x,\xi)\in\mathcal{M}_{(N-k)\times k}$ from elementary linear algebra we deduce condition (B.3) is violated whenever $k>\frac{N}{2}$. Now we are going to establish a priori estimates. In this step the block structure of the symmetrizer plays an important rule. It allows us, in fact to single out the estimates for U^I and U^{II}. So we are able to prove the following result

Theorem 4.1. *Let us consider the solution $\{U^I\}$, $\{U^{II}\}$ of the Cauchy problem for system (4.8). Assume that the hypotheses (A.4), (S.2), (G.1), (G.2), (G.3), (D) hold. Then for ε small enough, one has*

(i) *for any $T>0$ there exists $M(T)>0$, independent of ε, such that $\|U^{II}\|_{L^2(\mathbb{R}^d\times[0,T])}\leq M(T)$ and $\sup_{[0,T]}\|\varepsilon U^{II}(\cdot,t)\|_{L^2(\mathbb{R}^d)}\leq M(T)$,*

(ii) $\{\varepsilon^2 U^{II}_t\}$ *is relatively compact in $H^{-1}_{loc}(\mathbb{R}^d\times\mathbb{R}_+)$,*

(iii) *$\{U^I\}$ is uniformly bounded, with respect to ε, in $L^\infty(\mathbb{R}_+, L^2(\mathbb{R}))$, namely for any $T>0$ there exists $M(T)>0$, independent of ε, such that $\sup_{[0,T]}\|U^I(\cdot,t)\|_{L^2(\mathbb{R}^d)}\leq M(T)$.*

The proof is done by using the previous hypotheses and the block structure of the symmetrizer in order to obtain energy estimates which leads to (i), (ii), (iii). Details are contained in the theory developed in [6]. In order to get the relaxation result we need the convergence of our sequence of solutions. A simple consequence of (i) and (ii) of Theorem (4.1) is the following theorem.

Theorem 4.2. *Let us consider the solution $\{U^{II}\}$ of the Cauchy problem for system (4.8). Assume the hypotheses (A.4), (S.2), (G.1), (G.2), (G.3), (D) hold. Then there exists $U^{II0} \in [L^2(\mathbf{R}^d \times [0,T])]^{N-k}$, such that, as $\varepsilon \downarrow 0$ one has (extracting eventually subsequences)*

$$U^{II} \rightharpoonup U^{II0} \qquad \text{weakly in } L^2(\mathbb{R}^d \times [0,T]) \tag{4.9}$$

$$\varepsilon U^{II} \longrightarrow 0 \qquad \text{strongly in } L^2_{loc}(\mathbb{R}^d \times \mathbb{R}_+) \tag{4.10}$$

$$\{\varepsilon^2 U_t^{II}\} \longrightarrow 0 \qquad \text{in } H^{-1}_{loc}(\mathbb{R}^d \times \mathbb{R}_+). \tag{4.11}$$

Our next step is to prove strong convergence for the sequence $\{U^I\}$ in $L^2_{loc}(\mathbb{R}^d \times \mathbb{R}_+)$. To this end we will use only the estimate obtained in the previous theorems and the following theorem taken from Gérard [7].

Theorem 4.3.
(Generalized Compensated Compactness via Microlocal Defect Measures)
Let be Ω an open subset of $\mathbb{R}^d$, H and $H^\sharp$ Hilbert spaces. Take $P \in OPS^m$ with principal symbol $p(x,\xi)$ and $\{u_k\}$ be a bounded sequence of $L^2_{loc}(\Omega, H)$, such that $u_k \rightharpoonup u$. Assume that there exists a dense subset $D \in H^\sharp$ such that, for any $h \in D$, the sequence $(\langle Pu_k, h\rangle)$ is relatively compact in $H^{-m}_{loc}(\Omega)$. Moreover, let $q \in C(\Omega, \mathcal{K}(H))$.
If for all $(x,\xi,h) \in \Omega \times S^{n-1} \times H$, one has

$$(p(x,\xi)h = 0) \Rightarrow (\langle q(x)h, h\rangle = 0)$$

then

$$\langle q(x)u_k, u_k\rangle \quad \text{converges to} \quad \langle q(x)u, u\rangle \qquad \text{in } \mathcal{D}'(\Omega).$$

We have now

Theorem 4.4. *Let us consider the solution $\{U^I\}$, of the Cauchy problem for system (4.8). Assume the hypotheses (A.4), (S.2), (G.1), (G.2), (G.3), (D) hold. Then there exists $U^{I0} \in [L^2(\mathbb{R}^d \times \mathbb{R}_+)]^k$, such that, as $\varepsilon \downarrow 0$, one has (extracting eventually subsequences)*

$$U^I \longrightarrow U^{I0} \qquad \text{strongly in } L^2_{loc}(\mathbb{R}^d \times \mathbb{R}_+). \tag{4.12}$$

Proof. By using the hypothesis (D) and the estimate of Theorem 4.1 we conclude

$$\begin{pmatrix} U_t^I + M^{12}(x,D)U^{II} \\ M^{21}(x,D)U^I \end{pmatrix} \qquad \text{is relatively compact in } (H^{-1}_{loc})^2.$$

In order to fit into the framework of the Theorem 4.3 we set

$$P\begin{bmatrix} U^I \\ U^{II} \end{bmatrix} = \begin{bmatrix} I_{k\times k} & 0 \\ 0 & 0 \end{bmatrix} \partial_t \begin{bmatrix} U^I \\ U^{II} \end{bmatrix} + \begin{bmatrix} 0 & M^{12}(x,D) \\ M^{21}(x,D) & 0 \end{bmatrix} \begin{bmatrix} U^I \\ U^{II} \end{bmatrix}$$

Let us denote by $p(x,\xi)$, for $\xi = (\xi_0, \xi') \in \mathbb{R}^{d+1}$, $|\xi| = 1$, the principal symbol of P, then we have

$$\left\{ (x,\xi,\lambda,\mu) \quad \text{such that} \quad p(x,\xi)\begin{bmatrix}\lambda\\ \mu\end{bmatrix} = 0 \right\} \subset \{\lambda \mid \lambda = 0\}.$$

If we define $q(x) = \begin{bmatrix} I_{k\times k} & 0 \\ 0 & 0 \end{bmatrix}$ for all $\xi \neq 0$, $\xi = (\xi_0, \xi')$ we have that

$$p(x,\xi)\begin{bmatrix}\lambda\\ \mu\end{bmatrix} = 0 \qquad \text{implies} \qquad \langle q(x)\begin{bmatrix}\lambda\\ \mu\end{bmatrix}, \begin{bmatrix}\lambda\\ \mu\end{bmatrix}\rangle = 0$$

for all $\lambda \in \mathbf{R}^k$, $\mu \in \mathbb{R}^{N-k}$. Now we can apply the Theorem (4.3) and we conclude

$$U^I \longrightarrow U^{I0} \qquad \text{strongly in } L^2_{loc}(\mathbb{R}^d \times \mathbb{R}_+),$$

where U^{I0} denotes, in view of the Theorem (4.1) the weak limit of U^I in $L^2(\mathbb{R}^d \times \mathbb{R}_+)$. □

In this way it follows our main theorem.

Theorem 4.5. *Assume that the hypotheses of Theorems* (4.2), (4.4) *hold, then* (U^{I0}, U^{II0}) *verifies, in the sense of distributions, the following system*

$$\begin{cases} U_t^{I0} + M^{12}(x,D)U^{II0} = D^I_\nu(U^{I0},0)U^{II0} \\ M^{21}(x,D)U^{I0} = Q_\nu(x,U^{I0},0)U^{II0} + D^{II}(U^{I0},0). \end{cases} \tag{4.13}$$

In the case we consider a differential operator of the form (4.2) an equivalent formulation of (4.13) is given by (where we set $U = U^{I0}$)

$$\begin{aligned} U_t + \sum_{j=1}^d M_j^{12}(x)\partial_j \left(Q_\nu^{-1}(x,U,0) \sum_{k=1}^d M_k^{21}(x)\partial_k U \right) \\ = \sum_{j=1}^d M_j^{12}(x)\partial_j (Q_\nu^{-1}(x,U,0)D^{II}(U,0)) \\ + D^I_\nu(U,0)Q_\nu^{-1}(x,U,0)\left[\sum_{k=1}^d M_k^{21}(x)\partial_k U - D^{II}(U,0)\right]. \end{aligned} \tag{4.14}$$

The parabolicity of (4.14) follows from the relations between the blocks of the symmetrizer and the coefficients of the system (4.8).

References

[1] C. Cattaneo, *Sulla Conduzione del calore*, Atti Sem. Mat. Fisico Univ. Modena, **3**, 1948.

[2] R.J. DiPerna, *Convergence of approximate solutions to conservation laws*, Arch. Rational Mech. Anal., **82** (1983), no. 1, 27–70.

[3] R.J. DiPerna. *Compensated compactness and general systems of conservation laws*, Trans. Amer. Math. Soc., **292** (1985), no.2, 383–420. 83–101.

[4] D. Donatelli, P. Marcati, *Relaxation of semilinear hyperbolic systems with variable coefficients*, Ricerche di Matematica, **48** (1999), suppl., 295–310.

[5] D. Donatelli, P. Marcati, $1-\mathcal{D}$ *Relaxation from hyperbolic to parabolic systems with variable coefficients*, Rendiconti dell'Istituto di Matematica dell'Università di Trieste, **31** (2000), suppl., 63–85.

[6] D. Donatelli, P. Marcati, *Convergence of singular limits for multi-D semilinear hyperbolic systems to parabolic systems*, Preprint di Matematica, n.12, della Scuola Normale Superiore di Pisa (Aprile 2000).

[7] P. Gérard, *Microlocal defect measures*, Comm. Partial Differential Equations, **16** (1991), no. 11, 1761–1794.

[8] H.-0. Kreiss, *Initial-boundary Value Problems for Hyperbolic systems*, Comm. on Pure and Applied Math., **23** (1970), 277–298.

[9] H.-0. Kreiss, J. Lorenz, *Initial-boundary Value Problems and the Navier-Stokes Equations*, Academic Press (1989).

[10] C. Lattanzio, W.-A. Yong, *Hyperbolic-Parabolic singular limits for first order nonlinear systems*, Comm. Partial Differential Equations **26** (2001), no. 5–6, 939–964.

[11] P.L. Lions, G. Toscani, *Diffusive limit for finite velocity Boltzmann kinetic models*, Rev. Mat. Iberoamericana, **13** (1997), no. 3, 473–513.

[12] A. Majda, *Compressible Fluid Flow and Systems of Conservation Laws in Several Space Dimensions*, Appl. Math. Sci., **53** (1984), Springer-Verlag.

[13] A. Majda and S. Osher, *Initial-boundary Value Problems for hyperbolic Equations with Uniformly Characteristics Boundary*, Comm. on Pure and Applied Math., **28** (1975), no. 5, 607–675.

[14] P. Marcati, A. Milani, *The one-dimensional Darcy's law as the limit of a compressible Euler flow*, J. Differential Equations, **84** (1990), no. 1, 129–147.

[15] P. Marcati, A. Milani, P. Secchi, *Singular convergence of weak solutions for a quasilinear nonhomogeneous hyperbolic system*, Manuscripta Math.,**60** (1988), no. 1, 49–69.

[16] P. Marcati, R. Natalini, *Weak solutions to a hydrodynamic model for semiconductors and relaxation to the drift-diffusion equation*, Arch. Rational Mech. Anal., **129** (1995), no. 2, 129–145.

[17] P. Marcati, R. Natalini, *Weak solutions to a hydrodynamic model for semiconductors: the Cauchy problem*, Proc. Roy. Soc. Edinburgh Sect. A, **125** (1995), no. 1, 115–131.

[18] P. Marcati, B. Rubino, *Hyperbolic to parabolic relaxation theory for quasilinear first order systems*, J. Differential Equations, **162** (2000), no. 2, 359–399.

[19] G. Métivier, *The block structure condition for symmetric hyperbolic systems*, Bull. London Math. Soc., **32** (2000), no. 6, 689–702.

[20] F. Murat, *Compacité par compensation*, Ann. Scuola Norm. Sup. Pisa Cl. Sci.(4), **5** (1978), no. 3, 489–507.

[21] J. V. Ralston, *Note on a paper of Kreiss*, Comm. on Pure and Applied Math., **24** (1971), no. 6, 759–762.

[22] L. Tartar, *Compensated compactness and applications to partial differential equations*, Research Notes in Math., **39** (1979), 136–210.

[23] L. Tartar, *The compensated compactness method applied to partial differential equations*, Systems of Nonlinear Partial Differential Equations, Reidel, Dordrecht, 1983. NATO ASI.

[24] L. Tartar, *H-measures, a new approach for studying homogenization and concentration effects in partial differential equations*, Proc. Roy. Soc. Edinburg Sect. A, **115** (1990), no. 3–4, 193–230.

[25] M.E. Taylor, *Pseudodifferential Operators*, Princeton mathematical series, **34** (1981), Princeton University Press, Princeton New Jersey.

[26] M.E. Taylor, *Pseudodifferential Operators and Nonlinear PDE*, Progress in Mathematics, **100** (1991), Birkhäuser.

Donatella Donatelli
Dip. di Matematica Pura ed Applicata
Università degli Studi dell'Aquila
I-67100 L'Aquila, Italy
e-mail: `donatell@univaq.it`

Pierangelo Marcati
Dip. di Matematica Pura ed Applicata
Università degli Studi dell'Aquila
I-67100 L'Aquila, Italy
e-mail: `marcati@univaq.it`

Progress in Nonlinear Differential Equations
and Their Applications, Vol. 50, 97–114

Bounded Perturbations of Ornstein-Uhlenbeck Semigroups

Giuseppe Da Prato

Dedicated to Brunello Terreni

1. Introduction

Let H be a separable Hilbert space (norm $|\cdot|$, inner product $\langle\cdot,\cdot\rangle$) and $L(H)$ the Banach algebra of all linear bounded operators from H into H endowed with the norm

$$\|T\| = \sup\{|Tx|,\ x\in H,\ |x|\le 1\}.$$

We shall denote by $L_1^+(H)$ the subset of $L(H)$ of all symmetric nonnegative operators of trace class. If $T\in L_1^+(H)$ then the trace of T is defined by

$$\operatorname{Tr} T = \sum_{k=1}^{\infty}\langle Te_k, e_k\rangle,$$

where (e_k) is any complete orthonormal basis on H.

For any $x\in H$ and any $Q\in L_1^+(H)$ we shall denote by $N_{x,Q}$ the Borel Gaussian measure in H with mean x and covariance operator Q. If $x=0$ we shall write $N_{x,Q}=N_Q$ for short.

We are given a linear closed operator $A: D(A)\subset H\to H$, such that

Hypothesis 1.1.

(i) *A is the infinitesimal generator of a strongly continuous semigroup e^{tA} in H of negative type. That is, there are $M>0$ and $\omega>0$ such that*

$$\|e^{tA}\|\le Me^{-\omega t},\ t\ge 0. \tag{1.1}$$

(ii) *We have*

$$\int_0^{+\infty}\operatorname{Tr}\,[e^{tA}e^{tA^*}]dt<+\infty, \tag{1.2}$$

where A^* is the adjoint of A.

If Hypothesis 1.1 holds then the linear operator Q:

$$Qx=\int_0^{+\infty} e^{sA}e^{sA^*}xds,\ t\ge 0,\ x\in H,$$

is of trace class. An important rôle will be played in the following by the Gaussian measure $\mu = N_Q$.

Let us now introduce the *Ornstein-Uhlenbeck* generator, that is the differential operator

$$L_0\varphi = \frac{1}{2}\,\mathrm{Tr}\,[D^2\varphi] + \langle x, A^*D\varphi\rangle,\ \varphi \in \mathcal{E}_A(H).$$

Here $D\varphi$ and $D^2\varphi$ represent respectively the first and second Fréchet derivatives of φ and $\mathcal{E}_A(H)$ the linear span of all the real parts of functions φ_h, $x \in D(A^*)$ of the form

$$\varphi_h(x) := e^{i\langle h,x\rangle},\ x \in H.$$

Clearly L_0 is well defined on $\mathcal{E}_A(H)$. For instance if $h \in D(A^*)$ and

$$\varphi(x) = a\sin\langle h, x\rangle + b\cos\langle h, x\rangle,$$

we have

$$L_0\varphi(x) = -\frac{1}{2}\,a|h|^2\varphi(x) + (a\cos\langle h, x\rangle - b\sin\langle h, x\rangle)\,\langle x, A^*h\rangle, x \in H.$$

L_0 is related (see §2 below) to the following semigroup of linear operators on $C_b(H)$ [1]:

$$R_t\varphi(x) = \int_H \varphi(e^{tA}x + y)N_{Q_t}(dy),\ \varphi \in C_b(H),\ t \geq 0, \tag{1.3}$$

where N_{Q_t} is the Gaussian measure on H with mean 0 and covariance operator

$$Q_tx = \int_0^t e^{tA}e^{tA^*}x dt,\ x \in H.$$

Notice that, in view of Hypothesis 1.1, Q_t is a trace class operator, so that the measure N_{Q_t} is well defined.

R_t possesses a unique invariant measure $\mu = N_Q$, see §2. This means that

$$\int_H R_t\varphi d\mu = \int_H \varphi d\mu,\ \forall\,\varphi \in C_b(H). \tag{1.4}$$

Consequently R_t can be uniquely extended to a strongly continuous semigroup of contractions on $L^p(H,\mu)$ which we denote by R_t^p. The infinitesimal generator of R_t^p will be denoted by L_p.

The goal of this paper is to describe perturbations of L_0 of the form

$$N_0\varphi = L_0\varphi + \langle F(x), D\varphi\rangle,\ \varphi \in \mathcal{E}_A(H),$$

where $F : H \to H$ is a Borel bounded mapping. This perturbation is very special, however it is useful as a first step in order to approximate more general perturbations.

Let us describe the content of the paper.

---

[1] $C_b(H)$ is the Banach space of all uniformly continuous and bounded mappings $\varphi : H \to \mathbb{R}$ endowed with the norm $\|\varphi\|_0 = \sup_{x\in H} |\varphi(x)|$

Section §2 is devoted to recall several properties of the Ornstein-Uhlenbeck semigroup both in the space $C_b(H)$ where we denote the generator of R_t by L and in the spaces $L^p(H,\mu)$ where we denote the generator of R_t by L_p.

In §3 we prove that when F is continuous and bounded the linear operator

$$N\varphi = L\varphi + \langle F(x), D\varphi\rangle,\ \varphi \in D(L),$$

is m-dissipative on $C_b(H)$.

In §4 we consider the operator

$$N_2\varphi = L_2\varphi + \langle F(x), D\varphi\rangle,\ \varphi \in D(L_2).$$

We prove that N_2 generates a strongly continuous semigroup on $L^2(H,\mu)$, denoted by P_t^2. Moreover we show that P_t^2 has an invariant measure ν absolutely continuous with respect to μ and that the density $\rho = \frac{d\nu}{d\mu}$ belongs to $W^{1,2}(H,\mu)$ together with $\sqrt{\rho}$.

Finally, under additional assumptions, we give an explicit characterization of the adjoint N_2^* of N_2. This characterization provides additional regularity properties of ρ.

Finally in §5 we show that the operator N is dissipative on $L^2(H,\nu)$ and that its closure is m-dissipative on $L^2(H,\nu)$.

Several results presented here are known but scattered in different papers. Some proofs and some results are new.

2. Ornstein-Uhlenbeck semigroups

2.1. Space of continuous fuctions

We are here concerned with the family of linear operators

$$R_t\varphi(x) = \int_H \varphi(e^{tA}x + y)N_{Q_t}(dy) = \int_H \varphi(y)N_{x,Q_t}(dy),\ t \geq 0, \tag{2.1}$$

defined for all $\varphi : H \to \mathbb{R}$ having polynomial growth. In this subsection we shall be essentially interested to consider R_t acting on $C_b(H)$. However it is also convenient to consider R_t acting on the space $C_{b,1}(H)$ of all mappings $\varphi : H \to \mathbb{R}$ such that the function $H \to \mathbb{R} :\ \ x \to \frac{\varphi(x)}{1+|x|}$, belongs to $C_b(H)$. $C_{b,1}(H)$, endowed with the norm,

$$\|\varphi\|_{b,1} := \sup_{x\in H} \frac{|\varphi(x)|}{1+|x|}.$$

is a Banach space.

It is clear that R_t maps $C_b(H)$ into $C_b(H)$ and $C_{b,1}(H)$ into $C_{b,1}(H)$. We shall see in a moment that R_t is a semigroup of linear operators both in $C_b(H)$ and in $C_{b,1}(H)$.

For this the following elementary result about approximation by exponential functions is useful, see [13].

Lemma 2.1. *For all* $\varphi \in C_b(H)$ *(resp.* $C_{b,1}(H)$*) there exists a sequence* $(\zeta_n) = (\zeta_{n_1,n_2,n_3}) \subset \mathcal{E}_A(H)$ *such that*

$$\lim_{n\to\infty} \zeta_n = \lim_{n_1\to\infty} \lim_{n_2\to\infty} \lim_{n_3\to\infty} \zeta_{n_1,n_2,n_3}(x) = \varphi(x), \ x \in H. \tag{2.2}$$

$$\|\zeta_n\|_0 \le \|\varphi\|_0, \ (\text{resp.} \|\zeta_n\|_{b,1} \le \|\varphi\|_{b,1}), \ \forall\, n_1, n_2, n_3 \in \mathbb{N}. \tag{2.3}$$

We show now that the semigroup law for R_t holds. We have in fact for any $h \in D(A^*)$ and any $t, s \ge 0$,

$$R_s R_t \varphi_h(x) = \exp\{i\langle e^{(t+s)A}x, h\rangle - \frac{1}{2}\langle (Q_t + e^{tA}Q_s e^{tA^*})h, h\rangle\}.$$

Since

$$Q_t + e^{tA}Q_s e^{tA^*} = Q_{t+s},$$

we have

$$R_s R_t \varphi_h = R_{t+s}\varphi_h.$$

In view of Lemma 2.1 the semigroup law $R_{t+s}\varphi = R_t R_s \varphi$, follows for any $\varphi \in C_{b,1}(H)$.

It is well known that R_t is not strongly continuous neither in $C_b(H)$ nor in $C_{b,1}(H)$. However we can define, following [4], the infinitesimal generator of R_t through its resolvent

$$R(\lambda, L)f(x) = \int_0^{+\infty} e^{-\lambda t} R_t f(x)dt, \ x \in H, \ \lambda > 0, \ f \in C_b(H). \tag{2.4}$$

$$R(\lambda, L_{b,1})f(x) = \int_0^{+\infty} e^{-\lambda t} R_t f(x)dt, \ x \in H, \ \lambda > 0, \ f \in C_{b,1}(H). \tag{2.5}$$

Notice that

$$\mathcal{E}_A(H) \subset D(L_{b,1}).$$

Now we want to show that any function $\varphi \in D(L_{b,1})$ can be approximated pointwise in the graph norm by functions in $\mathcal{E}_A(H)$ with uniformly bounded norm on $C_{b,1}(H)$. The following result is proved in [13].

Proposition 2.2. *For any* $\varphi \in D(L_{b,1})$ *there exists a multi-sequence* $(\varphi_n) \subset \mathcal{E}_A(H)$ *and* $C_\varphi > 0$ *such that for all* $x \in H$ *we have*

$$\lim_{n\to\infty} \varphi_n(x) = \varphi(x), \ \lim_{n\to\infty} L_{b,1}\varphi_n(x) = L_{b,1}\varphi(x), \ \lim_{n\to\infty} D\varphi_n(x) = D\varphi(x) \tag{2.6}$$

and

$$\|\varphi_n\|_{b,1} + \|L_{b,1}\varphi_n\|_{b,1} + \|D\varphi_n\|_{b,1} \le C_\varphi. \tag{2.7}$$

Proposition 2.3. *For all* $\varphi \in C_b(H)$ *and for all* $t > 0$ *we have* $R_t\varphi \in C_b^1(H)$ (2) *and*

$$|DR_t\varphi(x)| \le t^{-1/2}\, \|\varphi\|_0, \ x \in H. \tag{2.8}$$

---

(2) $C_b^1(H)$ is the set of all $\varphi \in C_b(H)$ which are continuously differentiable with bounded derivatives.

Proof. Let $\varphi \in C_b(H)$ and $t > 0$. Then, by using the Cameron-Martin formula, we see that $R_t\varphi \in C_b^1(H)$ and we have, see [14],

$$\langle DR_t\varphi(x), h\rangle = \int_H \langle \Gamma(t)h, Q_t^{-1/2}y\rangle \varphi(e^{tA}x + y)N_{Q_t}(dy), \tag{2.9}$$

where $\Gamma(t) = Q_t^{-1/2}e^{tA}$. One can show that $\|\Gamma(t)\| \le t^{-1/2}$, see [14]. Now by the Hölder inequality it follows that

$$\begin{aligned}|\langle DR_t\varphi(x), h\rangle|^2 &\le \int_H |\langle \Gamma(t)h, Q_t^{-1/2}y\rangle|^2 \int_H \varphi^2(e^{tA}x + y)N_{Q_t}(dy)\\ &= |\Gamma(t)h|^2 R_t(\varphi^2) \le \frac{1}{t}\, |h|^2 R_t(\varphi^2).\end{aligned}$$

Due to the arbitrariness of h it follows

$$|DR_t\varphi(x)|^2 \le \frac{1}{t}\, R_t(\varphi^2),\ x \in H,\ t > 0. \tag{2.10}$$

Thus (2.8) is proved. □

Proposition 2.4. *For all $f \in C_b(H)$ and all $\lambda > 0$ we have $R(\lambda, L)f \in C_b^1(H)$ and*

$$|DR(\lambda, L)f(x)| \le \sqrt{\frac{\pi}{\lambda}}\, \|f\|_0. \tag{2.11}$$

Consequently $D(L) \subset C_b^1(H)$ with continuous embedding.

Proof. The conclusion follows from (2.8) by taking Laplace transform. □

The following result is proved in [11].

Proposition 2.5. *If $\varphi \in D(L)$ we have $\varphi^2 \in D(L)$ and*

$$L(\varphi^2) = 2\varphi L\varphi + |D\varphi|^2. \tag{2.12}$$

Proof. **Step 1**. For any $\varphi \in \mathcal{E}_A(H)$ we have

$$L_{b,1}(\varphi^2) = 2\varphi L_{b,1}\varphi + |D\varphi|^2. \tag{2.13}$$

We have in fact

$$D(\varphi^2) = 2\varphi D\varphi,\quad D^2(\varphi^2) = 2D\varphi \otimes D\varphi + 2\varphi D^2\varphi.$$

It follows that

$$\mathrm{Tr}\,[D^2(\varphi^2)] = 2|D\varphi|^2 + 2\varphi\,\mathrm{Tr}\,[D^2\varphi],$$

and (2.13) follows.

Step 2. Let $\varphi \in \mathcal{E}_A(H)$ and $f = \varphi - L_{b,1}\varphi$. Then we have

$$\varphi^2 = R(2, L_{b,1})(2\varphi f + |D\varphi|^2). \tag{2.14}$$

We have in fact

$$\varphi^2 - \varphi L_{b,1}\varphi = \varphi f,$$

from which, taking into account (2.13),

$$2\varphi^2 - L_{b,1}(\varphi^2) = 2\varphi f + |D\varphi|^2,$$

that yields (2.14).

Step 3. Conclusion.

Let $\varphi \in D(L)$. By Proposition 2.2 there exists $(\varphi_n) \subset \mathcal{E}_A(H)$ such that (2.6) and (2.7) hold. Now, setting $f_n = \varphi_n - L_{b,1}\varphi_n$, we have by (2.14),

$$\varphi_n^2 = R(2, L_{b,1})(2\varphi_n f_n + |D\varphi_n|^2). \tag{2.15}$$

Letting n tend to ∞ we find

$$\varphi^2 = R(2, L_{b,1})(2\varphi f + |D\varphi|^2).$$

Since $2\varphi f + |D\varphi|^2 \in C_b(H)$ it follows

$$\varphi^2 = R(2, L)(2\varphi f + |D\varphi|^2).$$

□

2.2. Spaces $L^p(H, \mu)$

Let us prove that $\mu = N_Q$ where

$$Qx = \int_0^{+\infty} e^{sA}e^{sA^*}x ds, \ t \geq 0, \ x \in H, \tag{2.16}$$

is an invariant measure for R_t. For this it is enough to check, in view of Lemma 2.1, that for any $\varphi = \varphi_h$ we have

$$\int_H R_t\varphi_h(x)\mu(dx) = \int_H \varphi_h(x)\mu(dx). \tag{2.17}$$

In fact (2.17) is equivalent to

$$\langle Qe^{tA^*}h, e^{tA^*}h\rangle + \langle Q_t h, h\rangle = \langle Qh, h\rangle, \ h \in H,$$

which is also equivalent to

$$e^{tA}Qe^{tA^*} + Q_t = Q,$$

that it is obviously fulfilled.

Consequently R_t can be uniquely extended to a strongly continuous semigroup of contractions R_t^p on $L^p(H, \mu)$. We shall denote by L_p its infinitesimal generator. The following result holds.

Proposition 2.6. *For all $p \geq 1$, L_p is the closure of L_0 on $L^p(H, \mu)$.*

Proof. It is enough to show that $\mathcal{E}_A(H)$ is a core for L_p. In fact $\mathcal{E}_A(H)$ is dense in $L^p(H, \mu)$, and it is invariant for R_t^p. Now the conclusion follows from a classical result, see e.g. [15]. □

2.3. Sobolev spaces

Let Q be the operator defined by (2.16). Then Q is of trace class by Hypothesis 1.1–(ii). Moreover Ker $Q = \{0\}$; in fact if $x_0 \in H$ is such that $Qx_0 = 0$ we have

$$0 = \langle Qx_0, x_0\rangle = \int_0^{+\infty} |e^{sA^*}x_0|^2 ds, \ t \geq 0,$$

that yields $x_0 = 0$. Therefore there exists an orthonormal basis (e_k) in H and a sequence of positive numbers (λ_k) such that

$$Qe_k = \lambda_k e_k, \ \ k \in \mathbb{N}.$$

We set $x_k = \langle x, e_k \rangle, \ k \in \mathbb{N}$ and we denote by D_k the partial derivative with respect to $x_k, \ k \in \mathbb{N}$.

It is important to notice that Q fulfills, as easily checked, the following *Lyapunov* equation:

$$2 < Qx, A^*x >= -|x|^2, \quad x \in D(A^*). \tag{2.18}$$

Moreover the following integration by parts formula holds.

Proposition 2.7. *Assume that $\varphi, \psi \in \mathcal{E}_A(H)$ and that $k \in \mathbb{N}$. Then we have*

$$\int_H (D_k\varphi)\psi d\mu + \int_H \varphi(D_k\psi)d\mu = \frac{1}{\lambda_k}\int_H x_k\varphi\psi d\mu. \tag{2.19}$$

Proof. It is enough to check (2.19) for $\varphi(x) = e^{i\langle h,x\rangle}$ and $\psi(x) = e^{i\langle k,x\rangle}, \ h,k \in D(A^{*)}$. But this follows by a straightforward computation. □

Proposition 2.8. *For all $k \in \mathbb{N}$, D_k is closable in $L^2(H,\mu)$.*

Proof. Let $k \in \mathbb{N}$, $(\varphi_n) \subset \mathcal{E}_A(H)$ and $\psi \in L^2(H,\mu)$ be such that

$$\varphi_n \to 0, \quad D_k\varphi_n \to \psi \text{ in } L^2(H,\mu).$$

We have to show that $\psi = 0$. In fact if $\zeta \in \mathcal{E}_A(H)$, then by (2.19) we have

$$\int_H (D_k\varphi_n)\zeta d\mu = -\int_H \varphi_n(D_k\zeta)d\mu + \frac{1}{\lambda_k}\int_H x_k\varphi_n\zeta d\mu.$$

As $n \to \infty$ the first integral tends to $\int_H \psi\zeta d\mu$, the second tends to 0, and the third tends also to 0, since $x_k\zeta$ belongs to $L^2(H,\mu)$ as easily checked. Therefore $\int_H \psi\zeta d\mu = 0$ for all $\zeta \in \mathcal{E}_A(H)$, so that $\psi = 0$ as required. □

We shall still denote by D_k the closure of D_k on $L^2(H,\mu)$.

Corollary 2.9. *The linear operator*

$$D: \ \mathcal{E}_A(H) \subset L^p(H,\mu) \to L^p(H,\mu;H),$$

is closable on $L^p(H,\mu)$; we still denote by D its closure.

Proof. Let $(\varphi_n) \subset \mathcal{E}_A(H)$ be such that

$$\lim_{n\to\infty} \varphi_n = 0 \text{ in } L^2(H,\mu)$$

$$\lim_{n\to\infty} D\varphi_n = G \text{ in } L^2(H,\mu;H).$$

Then, for any $k \in \mathbb{N}$, we have clearly

$$\lim_{n\to\infty} D_k\varphi_n = \lim_{n\to\infty} \langle D\varphi_n, e_k\rangle = \langle G, e_k\rangle \text{ in } L^2(H,\nu).$$

Since D_k is closable this implies $\langle G, e_k\rangle = 0$ for any $k \in \mathbb{N}$, and so $G = 0$ as required. □

If φ belongs to the domain of D on $L^p(H,\mu)$ we shall say that $D\varphi$ belongs to $L^p(H,\mu;H)$.

We define $W^{1,p}(H,\mu)$ as the linear space of all functions $\varphi \in L^p(H,\mu)$ such that $D\varphi \in L^p(H,\mu;H)$ and

$$\int_H |D\varphi(x)|^p \mu(dx) < +\infty.$$

$W^{1,p}(H,\mu)$, endowed with the norm

$$\|\varphi\|_{W^{1,p}(H,\mu)} = \left[\|\varphi\|_{W^{1,p}(H,\mu)} + \int_H |D\varphi(x)|^p \mu(dx)\right]^{1/p},$$

is a Banach space.

Moreover $W^{1,2}(H,\mu)$, endowed with the inner product

$$\langle \varphi, \psi \rangle_{W^{1,2}(H,\mu)} = \langle \varphi, \psi \rangle_{L^2(H,\mu)} + \int_H \langle D\varphi, D\psi \rangle d\mu,$$

is a Hilbert space.

If $\varphi \in W^{1,2}(H,\mu)$ we set

$$D\varphi(x) = \sum_{k=1}^{\infty} D_k\varphi(x) e_k, \ \ x \in H, \mu - \text{a.s.}$$

Since

$$|D\varphi(x)|^2 = \sum_{k=1}^{\infty} |D_k\varphi(x)|^2, \ \ x \in H, \mu - \text{a.s.},$$

this series above is convergent for almost all $x \in H$.

The following result is proved in [12] and [20].

Proposition 2.10. *The embedding $W^{1,2}(H,\mu) \subset L^2(H,\mu)$ is compact.*

Let us define $W^{2,2}(H,\mu)$. First we notice that for any $h,k \in \mathbb{N}$ the linear operator D_hD_k, defined in $\mathcal{E}_A(H)$, is closable; we shall still denote by D_hD_k the closure of D_hD_k.

Now we define $W^{2,2}(H,\mu)$ as the linear space of all functions $\varphi \in L^2(H,\mu)$ such that $D_hD_k\varphi \in L^2(H,\mu)$ for all $h,k \in \mathbb{N}$ and

$$\sum_{h,k=1}^{\infty} \int_H |D_hD_k\varphi(x)|^2 \mu(dx) < +\infty.$$

Endowed with the inner product

$$\begin{aligned}\langle \varphi, \psi \rangle_{W^{2,2}(H,\mu)} &= \langle \varphi, \psi \rangle_{L^2(H,\mu)} + \sum_{k=1}^{\infty} \int_H (D_k\varphi)(D_k\psi) d\mu \\ &+ \sum_{h,k=1}^{\infty} \int_H (D_hD_k\varphi(x))(D_hD_k\psi(x))\mu(dx),\end{aligned}$$

$W^{2,2}(H,\mu)$ is a Hilbert space.

If $\varphi \in W^{2,2}(H,\mu)$ we define $D^2\varphi$ as follows

$$\langle D^2\varphi(x)z,z\rangle = \sum_{h,k=1}^{\infty} D_h D_k\varphi(x)\langle z,e_h\rangle\langle z,e_k\rangle,\ x,z\in H, \mu-\text{a.s.}$$

It is easy to see that $D^2\varphi(x)$ is a Hilbert-Schmidt operator for almost all $x \in H$.

It is not difficult to check that $\mathcal{E}_A(H)$ is dense in $L^2(H,\mu)$ and in $W^{1,2}(H,\mu)$.

2.4. Some properties of R_t^p

The following propositions are known, see [14], [9],[17],[3]. However we give some sketches of proofs for the reader's convenience.

Proposition 2.11. *For any $\varphi \in C_b(H)$ we have*

$$\lim_{t\to+\infty} R_t\varphi(x) = \int_H \varphi(y)\mu(dy), \tag{2.20}$$

so that μ is ergodic.

Proof. In view of Lemma 2.1 it is enough to check (2.20) for $\varphi = \varphi_h,\ h \in D(A^*)$. In this case we have

$$R_t\varphi_h(x) = e^{i\langle e^{tA}x,h\rangle} \to 1, \text{ as } t \to +\infty,\ x \in H,$$

and the conclusion follows. □

Proposition 2.12. *For all $\varphi \in L^p(H,\mu)$ and for all $t > 0$ we have $R_t^p\varphi \in W^{1,p}(H,\mu)$ and*

$$\int_H |DR_t\varphi|^p d\mu \le \frac{1}{t^{p/2}} \int_H |\varphi|^p d\mu. \tag{2.21}$$

Proof. Since $C_b(H)$ is dense in $L^p(H,\mu)$ for all $p \ge 1$, it is enough to show (2.21) for $\varphi \in C_b(H)$. Let first $p = 2$. Then, integrating (2.10) with respect to μ, and taking into account the invariance of μ, (2.21) follows. Finally the case with a general p follows by interpolation. □

Proposition 2.13. *For all $f \in L^p(H,\mu)$ and all $\lambda > 0$ we have $R(\lambda, L_p)f \in W^{1,p}(H,\mu)$ and*

$$\int_H |DR(\lambda,L_p)f|^p d\mu \le \left(\frac{\pi}{\lambda}\right)^p \int_H |f|^p d\mu. \tag{2.22}$$

Moreover $D(L_p) \subset W^{1,p}(H,\mu)$ with continuous embedding.

Proof. The first statement follows from (2.21) taking the Laplace transform. The second one follows easily. □

The following result is proved in [3], [17].

Proposition 2.14. *For any $\varphi \in D(L_2)$, $\psi \in W^{1,2}(H,\mu)$ we have*

$$\int_H (L\varphi)\psi d\mu = \int_H \langle AQD\psi, D\varphi\rangle d\mu, \tag{2.23}$$

and

$$\int_H (L\varphi)\varphi d\mu = -\frac{1}{2}\int_H |D\varphi|^2 d\mu. \tag{2.24}$$

Note that (2.15) is meaningful since $AQ \in L(H)$ by [8].

Proof. It is enough to check (2.23) for $\varphi(x) = e^{i\langle h,x\rangle}$ and $\psi(x) = e^{i\langle k,x\rangle}$, $h,k \in D(A^*)$. This follows by a straightforward computation.

Now (2.24) follows setting $\varphi = \psi$ in (2.23) and taking into account the Lyapunov equation (2.18). □

When $A = A^*$ one can characterize the domain of L_2. In fact the following result holds, [9].

Proposition 2.15. *Assume that $A = A^*$. Then*

$$D(L_2) = \left\{\varphi \in W^{2,2}(H,\mu) : \int_H |(-A)^{1/2}|D\varphi|^2 d\mu < +\infty\right\}. \tag{2.25}$$

3. Perturbations on $C_b(H)$

In this section we want to show that the linear operator

$$N\varphi = L\varphi + \langle F, D\varphi\rangle,\ \varphi \in D(L),$$

is m-dissipative on $C_b(H)$.

Here $F : H \to H$ is continuous and bounded. We set

$$\|F\|_0 = \sup_{x\in H} |F(x)|.$$

Lemma 3.1. *Let $F : H \to H$ be continuous and bounded. Then for any $\lambda > \lambda_0 := \pi\|F\|_0^2$ and any $f \in C_b(H)$, there is a unique solution $\varphi \in D(L) \cap C_b^1(H)$ of the equation*

$$\lambda\varphi - L\varphi - \langle F(x), D\varphi\rangle = f. \tag{3.1}$$

Therefore N is m-dissipative $C_b(H)$.

Proof. Let $\lambda > \lambda_0$. Then, setting $\psi = \lambda\varphi - L\varphi$, equation (3.1) becomes

$$\psi - T_\lambda\psi = f, \tag{3.2}$$

where T_λ is defined by

$$T_\lambda\psi(x) = \langle F(x), DR(\lambda, L)\psi(x)\rangle,\ \psi \in C_b(H),\ x \in H. \tag{3.3}$$

Recalling the estimate (2.11), we see that

$$\|T_\lambda\psi\|_0 \le \sqrt{\frac{\pi}{\lambda}}\ \|F\|_0\ \|\psi\|_0.$$

Therefore if $\lambda > \lambda_0$ (3.1) has a unique solution φ. Moreover the resolvent set of the linear operator N contains $(\lambda_0, +\infty)$ and we have

$$\|R(\lambda N)f\|_0 \le \frac{1}{\lambda - \lambda_0}\ \|f\|_0.$$

The conclusion follows from the Hille-Yosida theorem. □

Now we are going to show that N is m-dissipative on $C_b(H)$. For this we first consider the case when F is in addition of class C^1.

Lemma 3.2. *Let $F : H \to H$ be bounded and of class C^1. Then N is m-dissipative.*

Proof. It is convenient to introduce for any $h > 0$ an approximating operator

$$N_h\varphi = L\varphi + \Delta_h\varphi, \quad \varphi \in D(L),$$

where

$$\Delta_h\varphi(x) = \frac{1}{h}\left(\varphi(\eta(h,x)) - \varphi(x)\right), \tag{3.4}$$

and η is the solution to

$$\eta_t(t,x) = F(\eta(t,x)), \quad \eta(0,x) = x \in H. \tag{3.5}$$

Clearly for any $\varphi \in C_b^1(H)$ we have

$$\lim_{h\to 0} \Delta_h\varphi = \langle F, D\varphi\rangle, \text{ in } C_b(H). \tag{3.6}$$

Now, given $\lambda > 0$ and $f \in C_b(H)$, we consider the equation

$$\lambda\varphi_h - L\varphi_h - \Delta_h\varphi_h = f. \tag{3.7}$$

Clearly equation (3.7) can be solved as before and, by a standard fixed point argument depending on the parameter h, we see that

$$\lim_{h\to 0} \varphi_h = \varphi \text{ in } C_b(H). \tag{3.8}$$

Now by (3.7) we find

$$\left(\lambda + \frac{1}{h}\right)\varphi_h - L\varphi_h = f + \frac{1}{h}\,\varphi(\eta(h,x)). \tag{3.9}$$

It follows

$$\|\varphi_h\|_0 \le \frac{1}{\lambda + \frac{1}{h}}\left(\|f\|_0 + \|\varphi_h\|_0\right),$$

that yields

$$\|\varphi_h\|_0 \le \frac{1}{\lambda}\,\|f\|_0.$$

Consequently, letting h tend to 0 gives

$$\|\varphi\|_0 \le \frac{1}{\lambda}\,\|f\|_0.$$

Therefore N is m-dissipative as required. □

Finally we consider the general case.

Proposition 3.3. *Let $F : H \to H$ be continuous and bounded. Then N is m-dissipative.*

Proof. Let (F_n) be a sequence of C^1 bounded functions such that

(i) $\lim_{n\to\infty} F_n(x) = F(x)$, for all $x \in H$.

(ii) $\|F_n\|_0 \le \|F\|_0$ for all $n \in \mathbb{N}$.

Given $\lambda \ge \lambda_0 = \pi\|F\|_0^2$ and $f \in C_b(H)$, let consider the equation

$$\lambda\varphi_n - L\varphi_n - \langle F_n(x), D\varphi_n\rangle = f, \tag{3.10}$$

that can be solved as in Lemma 3.1 by successive approximations. Moreover it is not difficult to see that these approximations converge pointwise as $n \to \infty$ and consequently that

$$\lim_{n\to\infty} \varphi_n(x) = \varphi(x),\ x \in H,$$

where $\varphi = R(\lambda, N)f$. By Lemma 3.2 it follows that

$$|\varphi_n(x)| \le \frac{1}{\lambda}\,\|f\|_0.$$

Therefore

$$|\varphi(x)| \le \frac{1}{\lambda}\,\|f\|_0,$$

and consequently N is m-dissipative. □

4. Perturbations on $L^2(H,\mu)$

Let N_2 be the linear operator

$$N_2\varphi = L_2\varphi + \langle F(x), D\varphi\rangle,\ \varphi \in D(L_2). \tag{4.1}$$

Proposition 4.1. *N_2 is the infinitesimal generator of a strongly continuous semigroup P_t^2 on $L^2(H,\mu)$. Moreover its resolvent $R(\lambda, N_2)$ is given by*

$$R(\lambda, N_2) = R(\lambda, L_2)(1 - T_\lambda)^{-1},\ \lambda > \lambda_0, \tag{4.2}$$

where

$$\lambda_0 = \pi\|F\|_0^2 = \pi \sup_{x\in H} |F(x)|, \tag{4.3}$$

and

$$T_\lambda\psi(x) = \langle F(x), DR(\lambda, L_2)\psi(x)\rangle,\ \psi \in L^2(H,\mu),\ x \in H. \tag{4.4}$$

Finally $\mathcal{E}_A(H)$ is a core for N_2.

Proof. Let $\lambda > 0$, $f \in L^2(H,\mu)$. Consider the equation

$$\lambda\varphi - L_2\varphi - \langle F(x), D\varphi\rangle = f. \tag{4.5}$$

Setting $\psi = \lambda\varphi - L_2\varphi$ equation (4.5) becomes

$$\psi - T_\lambda\psi = f, \tag{4.6}$$

where T_λ is defined by (4.4). Taking into account (2.22) we find

$$\|T_\lambda\psi\|_{L^2(H,\mu)} \le \sqrt{\frac{\pi}{\lambda}}\,\|F\|_0\,\|\psi\|_{L^2(H,\mu)}.$$

Therefore if $\lambda > \lambda_0$ (4.6) has a unique solution and (4.2) follows.

It remains to prove that $\mathcal{E}_A(H)$ is a core for N_2. Let $\varphi \in D(N_2) = D(L_2)$. Since $\mathcal{E}_A(H)$ is a core for L_2 there exists a sequence $(\varphi_n) \subset \mathcal{E}_A(H)$ such that

$$\varphi_n \to \varphi, \quad L_2\varphi_n \to L_2\varphi, \text{ in } L^2(H,\mu).$$

By Proposition 2.13 it follows that $\varphi \in W^{1,2}(H,\mu)$ and

$$\lim_{n\to\infty} \int_H |D\varphi - D\varphi_n|^2 d\mu = 0.$$

Consequently

$$\lim_{n\to\infty} N_2\varphi_n = L_2\varphi + \langle F, D\varphi\rangle = N_2\varphi,$$

and the conclusion follows. □

We now consider the adjoint semigroup $(P_t^2)^*$; we denote by N_2^* its infinitesimal generator, and by Σ^* the set of all its stationary points:

$$\Sigma^* = \left\{\varphi \in L^2(H,\mu) : \ (P_t^2)^*\varphi = \varphi, \ t \geq 0\right\}.$$

Lemma 4.2. $(P_t^2)^*$ *has the following properties:*

(i) *For all $\varphi \geq 0$ μ–a.e, one has $(P_t^2)^*\varphi \geq 0$ μ–a.e.*
(ii) *Σ^* is a lattice, that is if $\varphi \in \Sigma^*$ then $|\varphi| \in \Sigma^*$.*

Proof. Let $\psi_0 \geq 0$ μ–a.e. Then for all $\varphi \geq 0$ μ–a.e and all $t > 0$ we have

$$\int_H P_t\varphi\psi_0 \, d\mu = \int_H \varphi(P_t^2)^*\psi_0 \, d\mu \geq 0.$$

This implies that $\psi_0 \geq 0$ μ–a.e, and (i) is proved.

Let us prove (ii). Assume that $\varphi \in \Sigma^*$, so that $\varphi(x) = (P_t^2)^*\varphi(x)$. Then we have

$$|\varphi(x)| = |(P_t^2)^*\varphi(x)| \leq (P_t^2)^*(|\varphi|)(x). \tag{4.7}$$

We claim that

$$|\varphi(x)| = (P_t^2)^*(|\varphi|)(x), \ x - \mu \text{ a.s.}$$

Assume in contradiction that there is a Borel subset $I \subset H$ such that $\mu(I) > 0$ and

$$|\varphi(x)| < (P_t^2)^*(|\varphi|)(x), \ x \in I.$$

Then we have

$$\int_H |\varphi(x)|\mu(dx) < \int_H (P_t^2)^*(|\varphi|)(x)\mu(dx). \tag{4.8}$$

On the other hand

$$\int_H (P_t^2)^*(|\varphi|)d\mu = \langle (P_t^2)^*(|\varphi|), 1\rangle_{L^2(H,\mu)} = \langle |\varphi|, 1\rangle_{L^2(H,\mu)} = \int_H |\varphi| d\mu,$$

which in contradiction with (4.8). □

We prove now a regularity result for the domain $D(N_2^*)$ of the adjoint semigroup $(P_t^2)^*$.

Proposition 4.3. *We have $D(N_2^*) \subset W^{1,2}(H,\mu)$.*

Proof. Let $\lambda > \pi\|F\|_0$, and let $f \in L^2(H,\mu)$. Let us consider the bilinear form $b : W^{1,2}(H,\mu) \times W^{1,2}(H,\mu) \to \mathbb{R}$ defined as

$$b(\psi, v) = \lambda \int_H \psi v d\mu - \int_H \langle AQD\psi, Dv\rangle d\mu - \int_H \langle F, Dv\rangle \psi d\mu.$$

Clearly b is continuous since $AQ \in L(H)$ and

$$|b(\psi, v)| \leq \lambda\|\psi\|_{L^2(H,\mu)}\|v\|_{L^2(H,\mu)} + \|AQ\|\|D\psi\|_{L^2(H,\mu)}\|Dv\|_{L^2(H,\mu)}$$

$$+ \|F\|_0\|Dv\|_{L^2(H,\mu)}\|\psi\|_{L^2(H,\mu)}.$$

Moreover b is coercive since, recalling the Lyapunov equation, we have

$$b(\psi, \psi) = \lambda \int_H \psi^2 d\mu + \frac{1}{2} \int_H |D\psi|^2 d\mu - \int_H \langle F, Dv\rangle \psi d\mu.$$

By the Lax-Milgram theorem there exists $\psi \in W^{1,2}(H,\mu)$ such that

$$b(\psi, v) = \int_H f v d\mu, \quad \forall\, v \in W^{1,2}(H,\mu).$$

Choosing $v \in D(L_2)$ we have

$$\lambda \int_H \psi v d\mu - \int_H (N_2 v)\psi d\mu = \int_H f v d\mu.$$

Consequently $\psi \in D(N_2^*)$ and $\lambda\psi - N_2^*\psi = f$. The conclusion follows from the arbitrariness of f. □

The following result is proved in [14]. We present here a different simpler proof.

Proposition 4.4. *There exists an invariant measure ν of P_t^2 absolutely continuous with respect to μ. If ν_1 is another invariant measure of P_t^2 absolutely continuous with respect to μ, we have $\nu_1 = \nu$.*

Proof. Let $\lambda > 0$ be fixed and let φ_0 be the functions identically equal to 1. Clearly $\varphi_0 \in D(N_2)$ and we have $N_2\varphi_0 = 0$. Consequently $1/\lambda$ is an eigenvalue of $R(\lambda, N_2)$ since

$$R(\lambda, N_2)\varphi_0 = \frac{1}{\lambda}\, \varphi_0.$$

Moreover $1/\lambda$ is a simple eigenvalue because μ is ergodic. Since the embedding $W^{1,2}(H,\mu) \subset L^2(H,\mu)$ is compact and $D(L_2) \subset W^{1,2}(H,\mu)$ by Proposition 2.13, it follows that $R(\lambda, N_2)$ is compact as well for any $\lambda > 0$. Therefore $R(\lambda, N_2^*)$ is compact and $1/\lambda$ is a simple eigenvalue for $R(\lambda, N_2^*)$. Consequently there exists $\rho \in L^2(H,\mu)$ such that

$$R(\lambda, N_2^*)\rho = \frac{1}{\lambda}\, \rho. \tag{4.9}$$

It follows that $\rho \in D(N_2^*)$ and $N_2^*\rho = 0$. Since Σ^* is a lattice, ρ can be chosen to be nonnegative and such that $\int_H \rho d\mu = 1$.

Now set

$$\nu(dx) = \rho(x)\mu(dx), \; x \in H.$$

We claim that ν is an invariant measure for P_t^2. In fact taking the inverse Laplace transform in (4.9) we find

$$(P_t^2)^*\rho = \rho$$

that implies for any $\varphi \in L^2(H,\mu)$

$$\int_H P_t^2\varphi d\nu = \int_H P_t^2\varphi\ \rho d\mu = \int_H \varphi(P_t^2)^*\rho d\mu = \int_H \varphi d\nu.$$

It remains to show uniqueness. Let ν_1 be another invariant measure of P_t^2, and assume that $\nu_1 << \mu$ and $\rho_1 = \frac{d\nu_1}{d\mu}$. Then we have $P_t^2\rho_1 = \rho_1,\ t \geq 0$, and consequently $R(\lambda, N_2)\rho_1 = \frac{1}{\lambda}\,\rho_1$. Therefore $\rho = \rho_1$ since $1/\lambda$ is a simple eigenvalue of $R(\lambda, N_2^*)$. □

Remark 4.5. If $(P_t^2)^*1 = 1$ then $\nu = \mu$.

Remark 4.6. One can show that P_t is irreducible and strong Feller, so that the invariant measure ν is unique, see [14].

We now study the regularity of the density ρ. First notice that, since $\rho \in D(N_2^*)$, then by Proposition 4.3 we have that $\rho \in W^{1,2}(H,\mu)$.

The following result was proved in [2] when A is self-adjoint.

Proposition 4.7. *For all $\varphi \in W^{1,2}(H,\mu)$ we have*

$$\frac{1}{2}\int_H |D\rho|^2 d\mu = \int_H \langle F, D\rho\rangle \rho\, d\mu. \tag{4.10}$$

Moreover $\sqrt{\rho} \in W^{1,2}(H,\mu)$ and we have

$$\frac{1}{2}\int_H |D\sqrt{\rho}|^2\, d\mu \leq \int_H |\langle F, D\rho\rangle| d\mu. \tag{4.11}$$

Proof. Since ν is an invariant measure for P_t^2 we have

$$\int_H (N_2\varphi)\rho\, d\mu = \int_H (L_2\varphi)\rho\, d\mu + \int_H \langle F, D\varphi\rangle \rho\, d\mu = 0,\ \forall\, \varphi \in D(L).$$

Since $\rho \in W^{1,2}(H,\mu)$ we have by Proposition 2.14

$$\int_H \langle AQD\rho, D\varphi\rangle d\mu + \int_H \langle F, D\varphi\rangle \rho d\mu = 0, \tag{4.12}$$

for all $\varphi \in D(L_2)$. Since $D(L_2)$ is dense in $W^{1,2}(H,\mu)$ we can conclude that (4.12) holds for all $\varphi \in W^{1,2}(H,\mu)$. Finally, $\varphi = \rho$ and recalling the Lyapunov equation (2.18) we obtain (4.10).

Moreover, setting in (4.12) $\varphi = \log(\rho + \varepsilon)$, with $\varepsilon > 0$, and using again the Lyapunov equation, we find

$$2\int_H |D\sqrt{\rho+\varepsilon}|^2 d\mu = \int_H \langle F, D\rho\rangle \frac{\rho}{\rho+\varepsilon} d\mu \leq \int_H |\langle F, D\rho\rangle| d\mu.$$

Now (4.12) follows letting ε tend to zero. □

When F is sufficiently regular we can give an explicit expression for N_2^*. Let $F \in C_b(H;H)$. We say that F has a finite *divergence* if for any $x \in H$ the series

$$\text{div } F(x) := \sum_{k=1}^{\infty} D_k F_k(x), \; x \in H,$$

where $F_k(x) = \langle F(x), e_k \rangle$, is convergent and moreover div $F \in C_b(H)$. If in addition the function

$$Q(X) \to H, \;\; x \to \langle Q^{-1}x, F(x) \rangle,$$

is extendible to an uniformly continuous and bounded function, we says that F has a finite divergence with respect to μ, and we set

$$\text{div}_\mu \, F(x) = \text{div } F(x) - \langle Q^{-1}x, F(x) \rangle, \; x \in H.$$

The following result is an easy consequence of the integration by parts formula (2.19).

Lemma 4.8. *Assume that $F \in C_b(H;H)$ has a finite divergence with respect to μ. Then for any $\varphi, \psi \in W^{1,2}(H,\mu)$ we have*

$$\int_H \langle F, D\varphi \rangle \psi d\mu = -\int_H \varphi \langle F, D\psi \rangle d\mu - \int_H \text{div}_\mu \, F(x) \varphi \psi d\mu. \tag{4.13}$$

Proof. It is enough to consider $\varphi, \psi \in \mathcal{E}_A(H)$. In this case taking into account (2.19) we find

$$\begin{aligned}
\int_H \langle F, D\varphi \rangle \psi d\mu &= \sum_{k=1}^{\infty} \int_H F_k D_k \varphi \, \psi d\mu \\
&= -\sum_{k=1}^{\infty} \int_H \varphi \, (D_k F_k \psi + F_k D_k \psi) \, d\mu - \sum_{k=1}^{\infty} \frac{1}{\lambda_k} \int_H x_k F_k \varphi \psi d\mu \\
&= -\int_H \varphi \psi \text{ div } F \, d\mu - \int_H \langle F, D\psi \rangle \varphi d\mu - \int_H \langle Q^{-1}x, F \rangle \varphi \psi d\mu,
\end{aligned}$$

and the conclusion follows. □

Proposition 4.9. *Assume that $A = A^*$ and that $F \in C_b(H;H)$ has a finite divergence with respect to μ. Then $D(N_2^*) = D(L_2)$ and*

$$N_2^* \psi = L_2 \varphi - \langle F(x), D\psi \rangle - \text{ div}_\mu \, F(x) \psi, \; \psi \in D(N_2^*) = D(L_2). \tag{4.14}$$

Proof. Let $\varphi, \psi \in D(L_2)$. Then, taking into account (2.23) and (4.13), we have

$$\begin{aligned}
\int_H N_2 \varphi \, \psi d\mu &= \int_H L_2 \varphi \, \psi d\mu + \int_H \langle F, D\varphi \rangle \psi d\mu \\
&= \int_H \varphi \, L_2 \psi d\mu - \int_H \varphi \langle F, D\psi \rangle d\mu - \int_H \text{div}_\mu F \varphi \psi d\mu.
\end{aligned}$$

□

Corollary 4.10. *Under the assumptions of Proposition 4.9 we have $\rho \in D(L_2)$. If in particular $A = A^*$ then $\rho \in W^{2,2}(H,\mu)$ and $|(-A)^{1/2} D\rho| \in L^2(H,\mu)$.*

5. The semigroup P_t on $L^2(H,\nu)$

Let ν be the invariant measure of P_t. Let us start with an integration by parts formula.

Proposition 5.1. *For any $\varphi \in D(L)$ we have*

$$\int_H (N\varphi)\varphi d\nu = -\frac{1}{2}\int_H |D\varphi|^2 d\nu. \tag{5.1}$$

Consequently N is dissipative on $L^2(H,\nu)$.

Proof. Let $\varphi \in D(L)$. Then by Proposition 2.5 it follows that $\varphi^2 \in D(L)$ and

$$N(\varphi^2) = 2\varphi N\varphi + |D\varphi|^2.$$

Integrating with respect to ν and taking into account the invariance of ν, yields the conclusion. □

Since N is dissipative on $L^2(H,\nu)$, it is closable. We shall denote by N_2^ν its closure.

Theorem 5.2. *N_2^ν is m-dissipative on $L^2(H,\nu)$.*

Proof. Let $\lambda > \lambda_0$. Since the range of $\lambda - N_2$ contains $C_b(H)$, which is dense in $L^2(H,\nu)$, we have that the range of $\lambda - N_2$ is dense on $L^2(H,\nu)$. Now the conclusion follows from a classical result due to Lumer and Phillips. □

References

[1] D. Bakry, *L'hypercontractivité et son utilisation en théorie des semi-groupes,* Lectures on Probability Theory, LNM 1581, Springer, (1994).

[2] V. Bogachev, G. Da Prato and M. Röckner, *Regularity of invariant measures for a class of perturbed Ornstein-Uhlenbeck,* Nonlinear Differential Equations and Applications, Vol 3, No. 2, 261–268, (1996).

[3] V. Bogachev, M. Röckner and B. Schmuland, *Generalized Mehler semigroups and applications*, Probability and Related Fields, **114**, 193–225, (1996).

[4] S. Cerrai, *A Hille-Yosida theorem for weakly continuous semigroups,* Semigroup Forum, **49**, 349–367, (1994).

[5] S. Cerrai, *Weakly continuous semigroups in the space of functions with polynomial growth,* Dynamic Systems and Applications, 4, 351–372, (1995).

[6] S. Cerrai and F. Gozzi, *Strong solutions of Cauchy problems associated to weakly continuous semigroups,* Differential Integral Equations, 465–486, (1995).

[7] A.Chojnowska-Michalik and B.Goldys, *Existence, uniqueness and invariant measures for stochastic semilinear equations on Hilbert spaces,* Probab. Theory Rel. Fields 102 (1995), 331–356

[8] G. Da Prato, *Null controllability and strong Feller property of Markov transition semigroups,* Nonlinear Analysis TMA, **25**, 9–10, 941–949, (1995).

[9] G. Da Prato, *The Ornstein-Uhlenbeck generator perturbed by the gradient of a potential,* Bollettino UMI, (8) I-B, 501–519, (1998).

[10] G. Da Prato, *Lipschitz perturbations of Ornstein-Uhlenbeck semigroups,* submitted.

[11] G. Da Prato, A. Debussche and B. Goldys, *Invariant measures of non symmetric dissipative stochastic systems*, to appear in Probab. Related Fields.

[12] G. Da Prato, P. Malliavin and D. Nualart, *Compact families of Wiener functionals,* C. R. Acad. Sci. Paris, 315, 1287–1291, (1992).

[13] G. Da Prato and L.Tubaro, *Some results about dissipativity of Kolmogorov operators,* Czechoslovak Mathematical Journal, **51** 126, 685–699, 2001.

[14] G. Da Prato and J. Zabczyk, *Ergodicity for Infinite Dimensional Systems,* London Mathematical Society Lecture Notes, n. 229, Cambridge University Press, (1996).

[15] E. B. Davies,(1980) ONE PARAMETER SEMIGROUPS, Academic Press.

[16] J.D. Deuschel and D. Stroock, *Large Deviations,* Academic Press, (1984).

[17] M. Fuhrman, *Analyticity of transition semigroups and closability of bilinear forms in Hilbert spaces,* Studia Mathematica, **115**, 53–71, (1995).

[18] B. Goldys and M. Kocan, *Diffusion semigroups in spaces of continuous functions with mixed topology,* Journal of Differential Equations, **173**, no 1, 17–39, 2001.

[19] Z. M. Ma and M. Röckner, *Introduction to the Theory of (Non Symmetric) Dirichlet Forms,* Springer-Verlag, (1992).

[20] S. Peszat, *On a Sobolev space of function of infinite numbers of variables,* Bull. Polish. Acad. Sci, **40**, 55–60, (1993).

[21] E. Priola, *π-Semigroups and applications,* Preprint no. 9 of the Scuola Normale Superiore di Pisa, (1998).

[22] E. Priola, *On a class of Markov type semigroups in spaces of uniformly continuous and bounded functions, Studia Mathematica,* 136 (3), 271–295.

Giuseppe Da Prato
Scuola Normale Superiore di Pisa
Piazza dei Cavalieri 7
I-56126, Pisa, Italy
e-mail: `daprato@sns.it`

Progress in Nonlinear Differential Equations
and Their Applications, Vol. 50, 115–135

Degenerate Integrodifferential Equations of Volterra Type in Banach Space

Angelo Favini, Alfredo Lorenzi and Hiroki Tanabe

This paper is concerned with the following degenerate integrodifferential equations of parabolic type

$$\begin{cases} \dfrac{d}{dt}(M(t)u(t)) + L(t)u(t) + \displaystyle\int_0^t K(t,s)u(s)ds = f(t), \quad 0 < t \le T, \\ M(t)u(t)|_{t=0} = M(0)u_0. \end{cases} \tag{1}$$

This type of equations without the integral terms are discussed in great detail in Chapters III and IV of the book A. Favini and A. Yagi [2] based on the theory of analytic semigroups generated by multivalued linear operators. In Section 1 making extensive use of the results of [2; Chapter IV] we show the existence and uniqueness of a solution to the nonautonomous equation (1). We use the idea of M. G. Crandall and J. A. Nohel [1; Proposition 1] to deal with the integral term. Section 2 is devoted to the autonomous case based on the results of Chapter III of [2]. In the nonautonomous case rather restrictive assumptions are required for the constants α and β which appear in the hypotheses for the operators $M(t)$ and $L(t)$. In the autonomous case this restriction is considerably relaxed. Finally in Section 3 we consider the case in which the assumption (P) of [2; p. 92] is satisfied with $\alpha = \beta = 1$. In this case using the method of J. Prüss [3] we show the existence and uniqueness of a function satisfying (1) with the integral term understood in the improper sense under a weaker assumption of the initial data.

1. General Case

We state the hypotheses.

(I) $M(t)$, $L(t)$ are closed linear operators in X such that $D(L(t)) \subset D(M(t))$ for each $t \in [0,T]$. There exist positive numbers α and β satisfying

$$0 < \beta \le \alpha \le 1, \quad 2\alpha + \beta > 2$$

and positive constants c_0 and C_0 such that for any $t \in [0,T]$ and complex number λ in the region

$$\Sigma \equiv \{\lambda \in \mathbf{C};\ \mathrm{Re}\lambda \ge -c_0(|\mathrm{Im}\lambda| + 1)^\alpha\}$$

$\lambda M(t) + L(t)$ has an everywhere defined bounded inverse and the inequality

$$\|M(t)(\lambda M(t) + L(t))^{-1}\| \leq \frac{C_0}{(|\lambda| + 1)^\beta}$$

holds.

Since $0 \in \Sigma$, $L(t)$ has an everywhere defined bounded inverse. However, this restriction is not essential.

Let $A(t)$ be the possibly multivalued linear operator defined by $A(t) = L(t)M(t)^{-1}$. It is shown in Chapter III of Favini and Yagi [2] that $A(t)$ generates an analytic semigroup $e^{-\tau A(t)}$, $\tau > 0$, by the formula

$$e^{-\tau A(t)} = \frac{1}{2\pi i}\int_\Gamma e^{\lambda\tau}(\lambda + A(t))^{-1} d\lambda = \frac{1}{2\pi i}\int_\Gamma e^{\lambda\tau} M(t)(\lambda M(t) + L(t))^{-1} d\lambda,$$

where Γ is the contour

$$\Gamma = \{\lambda = -c_0(|\eta| + 1)^\alpha + i\eta; -\infty < \eta < \infty\}.$$

It is shown in [2] that the following inequalities hold:

$$\left\|e^{-\tau A(t)}\right\| \leq C\tau^{(\beta-1)/\alpha}, \quad \left\|\frac{\partial}{\partial\tau} e^{-\tau A(t)}\right\| \leq C\tau^{(\beta-2)/\alpha}. \tag{2}$$

(II) For $\lambda \in \Sigma$, $0 \leq s < t \leq T$,

$$\|A(t)^\circ(\lambda - A(t))^{-1}\{A(t)^{-1} - A(s)^{-1}\}\| \leq \frac{C_1|t-s|^\mu}{(|\lambda| + 1)^\nu}$$

with $0 < \mu \leq 1$ and $0 < \nu \leq 1$, where $A(t)^\circ$ is a linear section of $A(t)$;

(III) $2(\alpha + \beta) + \alpha\mu + \nu > 5$.

Several sufficient conditions in order that (II) and (III) be satisfied are found in terms of $M(t)$ and $L(t)$ in [2; Chapter IV]. Also a number of examples which satisfy these conditions are given in [2].

For the integral kernel K we assume that

(IV) $D(K(t,s)) \supset D(L(s))$ for $0 \leq s \leq t \leq T$;

(V) $K(t,s)L(s)^{-1}$ is continuous in $0 \leq s \leq t \leq T$ in the uniform operator topology, and Hölder continuous in t of order $\rho > (2 - \alpha - \beta)/\alpha$:

$$\|K(t,s)L(s)^{-1} - K(t',s)L(s)^{-1}\| \leq C_2|t - t'|^\rho.$$

By a solution of (1) we mean a function u satisfying

$$\begin{aligned}
&u \in C(]0,T]; X),\ u(t) \in D(M(t))\ \forall t \in [0,T],\\
&Mu \in C([0,T]; X) \cap C^1(]0,T]; X),\\
&t^\gamma d(Mu)(t)/dt \text{ is bounded in }]0,T[\text{ for some } \gamma < (\alpha + \nu - 1)/\alpha + \mu,\\
&u(t) \in D(L(t))\ \forall t \in]0,T],\ Lu \in C(]0,T]; X) \cap L^1(0,T; X),\\
&\text{and (1) holds.}
\end{aligned} \tag{3}$$

Theorem 1 *Suppose that the assumptions* (I)–(V) *are satisfied. Then, for any* $f \in C^{\rho}([0,T];X)$, $(2-\alpha-\beta)/\alpha < \rho \le 1$, *and* $u_0 \in D(M(0))$ *satisfying one of the following two conditions:*

$$M(0)u_0 \in X^{\theta}_{A(0)} = \{u \in X; \sup_{\xi>0} \xi^{\theta} \|L(0)(\xi M(0) + L(0))^{-1}u\| < \infty\}, \tag{4}$$

$$M(0)u_0 \in (X, D(A(0)))_{\theta,\infty} \tag{5}$$

with $2-\alpha-\beta < \theta < 1$, *a solution of problem* (1) *exists and is unique.*

It is shown in Favini and Yagi [2; Chapter IV] that under the assumptions (I), (II), (III) the fundamental solution $U(t,s)$ to the problem

$$\frac{d}{dt}v(t) + A(t)v(t) \ni f(t), \quad v(0) = v_0,$$

exists, and satisfies

$$\|U(t,s)\| \le C(t-s)^{(\beta-1)/\alpha}, \quad \left\|\frac{\partial}{\partial t}U(t,s)\right\| \le C(t-s)^{(\beta-2)/\alpha}. \tag{6}$$

The convolution of an operator valued function $A(t,s)$ and a function $f(t)$ is defined by

$$(A * f)(t) = \int_0^t A(t,s)f(s)ds,$$

and that of two operator valued functions $A(t,s)$ and $B(t,s)$ is by

$$(A * B)(t,s) = \int_s^t A(t,r)B(r,s)dr.$$

Set $H(t,s) = K(t,s)L(s)^{-1}$. Then the equation (1) is rewritten as

$$\frac{d}{dt}Mu + Lu + H * Lu = f. \tag{7}$$

Let $R(t,s)$ be the solution to the integral equation

$$R + H + H * R = 0. \tag{8}$$

This equation can be solved by successive approximations:

$$R = -H + H * H - H * H * H + \cdots,$$

and the solution R also satisfies

$$R + H + R * H = 0. \tag{9}$$

It follows from (8) and (V) that $R(t,s)$ is Hölder continuous in t of order ρ:

$$\|R(t,s) - R(t',s)\| \le C|t-t'|^{\rho}. \tag{10}$$

Suggested by M. G. Crandall and J. A. Nohel [1; Proposition 1] we derive the integral equation to be satisfied by $(Mu)'$. Starting from the solution to this integral equation we will construct the desired solution to (1).

Let u be a solution of (1). Addition of (7) and

$$R * (Mu)' + R * Lu + R * H * Lu = R * f$$

which is obtained by convoluting R and (7) yields

$$(Mu)' + Lu = f + R * f - R * (Mu)'. \tag{11}$$

Therefore, the new unknown function $v(t) = M(t)u(t)$ should be the solution to the initial value problem

$$\begin{aligned} &\frac{d}{dt}v(t) + A(t)v(t) \ni f(t) + (R * f)(t) - (R * v')(t), \\ &v(0) = v_0, \end{aligned} \tag{12}$$

where $v_0 = M(0)u_0$. Since $t^\gamma v'(t) = t^\gamma (Mu)'(t)$ is bounded in $]0,T[$ for some $\gamma < (\alpha+\nu-1)/\alpha+\mu$, one can apply Theorem 4.3 of [2] to transform the problem (12) into

$$v(t) = U(t,0)v_0 + \int_0^t U(t,s)\{f(s) + (R * f)(s) - (R * v')(s)\}ds. \tag{13}$$

Define the function g by

$$g(t) = U(t,0)v_0 + \int_0^t U(t,s)\left\{f(s) + (R * f)(s)\right\} ds. \tag{14}$$

By a change of the order of integration

$$\begin{aligned} \int_0^t U(t,s)(R * v')(s)ds &= \int_0^t U(t,s) \int_0^s R(s,\tau)v'(\tau)d\tau ds \\ &= \int_0^t \int_\tau^t U(t,s)R(s,\tau)ds\, v'(\tau)d\tau. \end{aligned} \tag{15}$$

Then (13) is rewritten as

$$v(t) = g(t) - \int_0^t \int_s^t U(t,\tau)R(\tau,s)d\tau\, v'(s)ds. \tag{16}$$

Since $2-\alpha-\beta < \theta < 1$ implies $1-\beta < \theta < 1$, one has by Theorem 3.3 of [2] that

$$\lim_{t\to 0} e^{-tA(0)}v_0 = v_0 \tag{17}$$

if $v_0 \in X^\theta_{A(0)}$. In case $v_0 \in (X, D(A(0)))_{\theta,\infty}$, one can show that (17) is also true by using

$$\|A(0)^\circ(\lambda - A(0))^{-1}u\|_X \le C|\lambda|^{1-\beta-\theta}\|u\|_{(X,D(A(0)))_{\theta,\infty}}$$

in place of (3.4) in the proof of Theorem 3.3 of [2]. It follows from (4.8) of Favini and Yagi [2] and (17) that

$$U(t,0)v_0 - v_0 = \left(U(t,0)v_0 - e^{-tA(0)}v_0\right) + \left(e^{-tA(0)}v_0 - v_0\right) \to 0 \tag{18}$$

as $t \to 0$. This shows that

$$\lim_{t\to 0} g(t) = v_0 \text{ and } g \in C([0,T];X). \tag{19}$$

Next we show that g is differentiable and obtain the estimate of the derivative g'. For that purpose we make some preparations. Let $A_n(t)$ be the Yosida approximation of $A(t)$, and $U_n(t,s)$ the associated fundamental solution:

$$\frac{\partial}{\partial t}U_n(t,s) + A_n(t)U_n(t,s) = 0, \qquad U_n(s,s) = I.$$

It is shown in [2] that $U_n(t,s) \to U(t,s)$ strongly as $n \to \infty$, and that the inequalities

$$\|U_n(t,s)\| \le C(t-s)^{(\beta-1)/\alpha}, \quad \|A_n(t)U_n(t,s)\| \le C(t-s)^{(\beta-2)/\alpha}, \tag{20}$$

hold with some constant C independent of n. In view of Lemma 4.2 of Favini and Yagi [2]

$$\|A_n(t)e^{-(t-s)A_n(t)} - A_n(s)e^{-(t-s)A_n(s)}\| \le C(t-s)^{(\beta+\nu-3)/\alpha+\mu}. \tag{21}$$

Let $W(t,s)$ and $W_n(t,s)$ be the operator valued functions defined by

$$W(t,s) = \frac{\partial}{\partial t}U(t,s) + \frac{\partial}{\partial s}e^{-(t-s)A(t)}, \tag{22}$$

$$\begin{aligned} W_n(t,s) &= \frac{\partial}{\partial t}U_n(t,s) + \frac{\partial}{\partial s}e^{-(t-s)A_n(t)} \\ &= -A_n(t)U_n(t,s) + A_n(t)e^{-(t-s)A_n(t)}. \end{aligned} \tag{23}$$

Then the argument of pp. 94–95 of Favini and Yagi [2] shows that

$$\|W_n(t,s)\| \le C(t-s)^{\{2(\alpha+\beta)+\alpha\mu+\nu-5\}/\alpha-1} \tag{24}$$

with some constant C independent of n, and $W_n(t,s) \to W(t,s)$ strongly as $n \to \infty$. This and (24) imply that

$$\|W(t,s)\| \le C(t-s)^{\{2(\alpha+\beta)+\alpha\mu+\nu-5\}/\alpha-1}. \tag{25}$$

In view of Proposition 3.4 of [2] and

$$\left\|\frac{d}{dt}e^{-tA(0)}u\right\| \le \begin{cases} Ct^{(\beta-2)/\alpha}\|u\| & \text{for } u \in X, \\ Ct^{(\beta-1)/\alpha}\|u\|_{D(A(0))} & \text{for } u \in D(A(0)), \end{cases}$$

one observes that

$$\left\|\frac{d}{dt}e^{-tA(0)}v_0\right\| \le \begin{cases} C_\theta t^{(\beta+\theta-2)/\alpha}\|v_0\|_{X^\theta_{A(0)}}, \\ C_\theta t^{(\beta+\theta-2)/\alpha}\|v_0\|_{(X,D(A(0)))_{\theta,\infty}} \end{cases} \tag{26}$$

according as (4) or (5) is satisfied. By virtue of (21) and (24) and noting the inequality $\{2(\alpha+\beta)+\alpha\mu+\nu-5\}/\alpha - 1 \le (\beta+\nu-3)/\alpha+\mu$ we obtain

$$\begin{aligned} &\left\|\frac{d}{dt}U_n(t,0)v_0\right\| \\ &= \left\|W_n(t,0)v_0 - \{A_n(t)e^{-tA_n(t)} - A_n(0)e^{-tA_n(0)}\}v_0 - A_n(0)e^{-tA_n(0)}v_0\right\| \\ &\le Ct^{\{2(\alpha+\beta)+\alpha\mu+\nu-5\}/\alpha-1}\|v_0\| + \left\|\frac{d}{dt}e^{-tA_n(0)}v_0\right\|. \end{aligned} \tag{27}$$

Letting $n \to \infty$ in (27) and using (26) one gets

$$\left\| \frac{d}{dt} U(t,0)v_0 \right\| \le C\left\{ t^{\{2(\alpha+\beta)+\alpha\mu+\nu-5\}/\alpha-1} \|v_0\| + t^{(\beta+\theta-2)/\alpha} \|v_0\|_\theta \right\}, \tag{28}$$

where $\|v_0\|_\theta = \|v_0\|_{X^\theta_{A(0)}}$ or $\|v_0\|_{(X,D(A(0)))_{\theta,\infty}}$. From the assumption on f and (10) it follows that $f + R * f \in C^\rho([0,T];X)$, $\rho > (2-\alpha-\beta)/\alpha$. Hence by virtue of Theorem 4.3 of [2] and its Remark g is differentiable and

$$\begin{aligned} g'(t) &= \frac{d}{dt} U(t,0)v_0 + e^{-tA(t)}\{f(t) + (R*f)(t)\} \\ &+ \int_0^t \frac{\partial}{\partial t} U(t,s)\{f(s) + (R*f)(s) - f(t) - (R*f)(t)\}ds \\ &+ \int_0^t W(t,s)\{f(s) + (R*f)(s)\}ds. \end{aligned} \tag{29}$$

With the aid of (2), (6), (25), (28) and (29) and noting $(\beta-1)/\alpha > (\beta+\theta-2)/\alpha$ one obtains

$$\|g'(t)\| \le C\{t^{\{2(\alpha+\beta)+\alpha\mu+\nu-5\}/\alpha-1} + t^{(\beta+\theta-2)/\alpha}\}. \tag{30}$$

The right-hand side of (30) is integrable in $(0,T)$ in view of (III) and $(\beta+\theta-2)/\alpha > -1$ by assumption. Thus we have proved that

$$g \in C([0,T];X) \cap C^1(]0,T];X), \quad g' \in L^1(0,T;X), \quad g(0) = v_0. \tag{31}$$

In view of the Remark to Theorem 4.3 of Favini and Yagi [2] one has

$$\begin{aligned} \frac{\partial}{\partial t} \int_s^t U(t,\tau)R(\tau,s)d\tau &= e^{-(t-s)A(t)}R(t,s) \\ &+ \int_s^t \frac{\partial}{\partial t} U(t,\tau)\cdot(R(\tau,s) - R(t,s))d\tau + \int_s^t W(t,\tau)R(t,s)d\tau. \end{aligned} \tag{32}$$

From (2), (6), (10), (25) and (32) it follows that

$$\left\| \frac{\partial}{\partial t} \int_s^t U(t,\tau)R(\tau,s)d\tau \right\| \le C(t-s)^{(\beta-1)/\alpha}. \tag{33}$$

Hence differentiation of both sides of (16) yields

$$v'(t) = g'(t) - \int_0^t \frac{\partial}{\partial t} \int_s^t U(t,\tau)R(\tau,s)d\tau\, v'(s)ds. \tag{34}$$

This is the integral equation to be satisfied by $(Mu)'$. In view of (30) and (33) the integral equation

$$w(t) = g'(t) - \int_0^t \frac{\partial}{\partial t} \int_s^t U(t,\tau)R(\tau,s)d\tau\, w(s)ds \tag{35}$$

has a unique solution w satisfying

$$\|w(t)\| \le C\{t^{\{2(\alpha+\beta)+\alpha\mu+\nu-5\}/\alpha-1} + t^{(\beta+\theta-2)/\alpha}\}. \tag{36}$$

First we define the function w as the solution of the integral equation (35), and then noting that $w \in L^1(0,T;X)$ define the function v by

$$v(t) = v_0 + \int_0^t w(s)ds.$$

Then $v \in C([0,T];X) \cap C^1(]0,T])$ and $v' = w$, and hence

$$\|v'(t)\| \le C\{t^{\{2(\alpha+\beta)+\alpha\mu+\nu-5\}/\alpha-1} + t^{(\beta+\theta-2)/\alpha}\} \tag{37}$$

and (34) holds. If we set

$$\gamma = \max\left\{1 - \frac{2(\alpha+\beta)+\alpha\mu+\nu-5}{\alpha}, \frac{2-\beta-\theta}{\alpha}\right\}, \tag{38}$$

then

$$0 < \gamma < \min\left\{\frac{\alpha+\nu-1}{\alpha} + \mu, 1\right\}, \tag{39}$$

and $t^\gamma v'(t)$ is bounded. Integrating both sides of (34) from 0 to t we obtain (16) or (13). The inequality (10) implies that for $0 \le t' < t \le T$

$$\begin{aligned}
&\|(R*v')(t) - (R*v')(t')\| \\
&= \left\| \int_{t'}^t R(t,s)v'(s)ds + \int_0^{t'} (R(t,s) - R(t',s))v'(s)ds \right\| \\
&\le C(t-t')t'^{-\gamma} + C(t-t')^\rho \int_0^t \|v'(s)\|ds. \qquad (40)
\end{aligned}$$

Since $0 < \gamma < 1$ and $f + R*f \in C^\rho([0,T];X)$ the proof of Theorem 4.3 of [2] shows that v satisfies (12).

Define the function u by

$$u(t) = L(t)^{-1}\{f(t) + (R*f)(t) - (R*v')(t) - v'(t)\}. \tag{41}$$

Then in view of (12)

$$\begin{aligned}
L(t)u(t) &= f(t) + (R*f)(t) - (R*v')(t) - v'(t) \\
&\in A(t)v(t) = L(t)M(t)^{-1}v(t). \qquad (42)
\end{aligned}$$

Since $L(t)$ is an injection, (42) implies that $u(t) \in M(t)^{-1}v(t)$ or $M(t)u(t) = v(t)$. Substituting this in the first half of (42) we get

$$(Mu)' + Lu + R*(Mu)' = f + R*f. \tag{43}$$

Adding (43) and

$$H*(Mu)' + H*Lu + H*R*(Mu)' = H*f + H*R*f$$

which is obtained by convoluting H and (43), and using (8) one concludes

$$(Mu)' + Lu + H*Lu = f,$$

which is the first part of (1). Since $\lim_{t\to0} M(t)u(t) = \lim_{t\to0} v(t) = v_0 = M(0)u_0$, we have proved that u is the desired solution of (1) satisfying (3) with γ defined by (38).

Finally we prove the uniqueness of the solution to (1). Let u be a solution to (1) with $f \equiv 0$ and $u_0 = 0$. Using the argument by which we derived (34) from (7) we reach the relation (34) with $g'(t) \equiv 0$. Hence $v'(t) \equiv 0$. Combination of this and $v(0) = 0$ implies $M(t)u(t) = v(t) \equiv 0$. Substituting this in (7) one gets $Lu + H * Lu = 0$. Hence $Lu \equiv 0$, which implies $u(t) \equiv 0$, since $L(t)$ is invertible.

Example 1 Let $X = L^p(\Omega), 1 < p < \infty$, where Ω is a bounded domain in R^n with smooth boundary, and M be the multiplication operator by a function $m(x)$ such that $0 \le m \in L^\infty(\Omega)$. Let

$$\mathcal{L}(t) = -\sum_{i,j=1}^{n} \frac{\partial}{\partial x_j}\left(a_{ij}(x,t)\frac{\partial}{\partial x_i}\right) + a_0(x,t) \tag{44}$$

be a linear second order differential operator for each $t \in [0,T]$ whose coefficients a_{ij} and a_0 are real valued functions satisfying

$$a_{ij},\ \frac{\partial}{\partial x_j}a_{ij},\ a_0 \in C(\overline{\Omega} \times [0,T]),$$

$$\{a_{ij}(x,t)\} \text{ is a positive definite symmetric matrix for each } (x,t) \in \overline{\Omega} \times [0,T],$$

$$a_{ij},\ \frac{\partial}{\partial x_j}a_{ij},\ a_0 \text{ are uniformly Hölder continuous in } t:$$

$$\left.\begin{array}{l} |a_{ij}(x,t) - a_{ij}(x,s)|,\ \left|\dfrac{\partial}{\partial x_j}a_{ij}(x,t) - \dfrac{\partial}{\partial x_j}a_{ij}(x,s)\right| \\ |a_0(x,t) - a_0(x,s)| \end{array}\right\} \le C|t-s|^\mu,\ 0 < \mu \le 1;$$

$$a_0(x,t) > 0 \text{ for each } (x,t) \in \overline{\Omega} \times [0,T].$$

Let $L(t)$ be the realization of $\mathcal{L}(t)$ in $L^p(\Omega)$ under the Dirichlet boundary condition. The argument of Favini and Yagi [2; p. 80 and p. 106] shows that the assumptions (I) and (II) are satisfied with $\alpha = 1$ and $\beta = \nu = 1/p$. Hence, if $3/p + \mu > 3$, (III) is also satisfied. Let $K(t,s)$, $(t,s) \in \Delta = \{(t,s); 0 \le s \le t \le T\}$, be the differential operator

$$K(t,s) = \sum_{i,i=1}^{n} k_{ij}(x,t,s)\frac{\partial^2}{\partial x_i \partial x_j} + \sum_{i=1}^{n} k_i(x,t,s)\frac{\partial}{\partial x_i} + k_0(x,t,s), \tag{45}$$

with coefficients k_{ij}, k_i, k_0 belonging to $C(\overline{\Omega} \times \Delta)$ and which are uniformly Hölder continuous in t of order $\rho > 1 - 1/p$. Then (IV) and (V) are also satisfied.

Example 2 Let $X = H^{-1}(\Omega)$, where Ω is a bounded domain in R^n with smooth boundary, and M be the multiplication operator by a function $m(x)$ such that $0 \le m \in L^\infty(\Omega)$. Let $\mathcal{L}(t)$ be the differential operator (44) such that a_{ij} and a_0

are real valued functions satisfying

$a_{ij},\ a_0 \in C(\overline{\Omega} \times [0,T])$, and $a_{ij},\ a_0$ are uniformly Hölder continuous in t:

$|a_{ij}(x,t) - a_{ij}(x,s)|,\ |a_0(x,t) - a_0(x,s)| \le C|t-s|^{\mu},\ 0 < \mu \le 1;$

$\{a_{ij}(x,t)\}$ is a positive definite symmetric matrix for each $(x,t) \in \overline{\Omega} \times [0,T]$,

$a_0(x,t) \ge 0$ for each $(x,t) \in \overline{\Omega} \times [0,T]$.

Let $L(t) : H_0^1(\Omega) \to H^{-1}(\Omega)$ be the mapping defined by $L(t)u = \mathcal{L}(t)u$ for $u \in H_0^1(\Omega)$. The argument of Favini and Yagi [2; Example 3.3 and p. 106] shows that the assumptions (I) and (II) are satisfied with $\alpha = \beta = \nu = 1$. Hence, (III) is also satisfied. Let $K(t,s)$ be the differential operator (45) such that $k_{ij}, \partial k_{ij}/\partial x_j, k_i, k_0$ belong to $C(\overline{\Omega} \times \Delta)$ and are uniformly Hölder continuous in t. Then (IV) and (V) are also satisfied.

Remark The above method can be applied to the equation treated in Section 4 of J. Prüss [3], and it can be shown that the solution of the equation exists and is unique if the initial value u_0 belongs to the subspace $(X, D(A(0)))_{\theta,p}, 0 < \theta < 1$.

2. Autonomous equations

In Example 1 assumption (III) is not satisfied if $p = 2$, since $\alpha = 1$ and $\beta = 1/2$ in this case. However, the autonomous equation

$$\begin{cases} \dfrac{d}{dt}Mu(t) + Lu(t) + \displaystyle\int_0^t K(t-s)u(s)ds = f(t), \quad 0 < t \le T, \\ Mu(0) = Mu_0. \end{cases} \tag{46}$$

can be solved under weaker assumptions for α and β. Since the details will be published in [4], only an outline is given here.

We state the assumptions.

(I′) M, L are closed linear operators in X such that $D(L) \subset D(M)$. There exist positive numbers α and β satisfying

$$0 < \beta \le \alpha \le 1 \quad 2\alpha + \beta > 2 \tag{47}$$

and positive constants c_0 and C_0 such that for any complex number

$$\lambda \in \Sigma \equiv \{\lambda \in \mathbf{C};\ \mathrm{Re}\lambda \ge -c_0(|\mathrm{Im}\lambda| + 1)^{\alpha}\}$$

$\lambda M + L$ has an everywhere defined bounded inverse and the inequality

$$\|M(\lambda M + L)^{-1}\| \le \frac{C_0}{(|\lambda| + 1)^{\beta}}$$

holds.

(IV′) K is a linear operator valued function such that $D(K(t)) \supset D(L)$.

(V′)

$$K \in C^{\rho}([0,T]; \mathcal{L}(D(L), X)), \quad \rho > \frac{2-\alpha-\beta}{\alpha}.$$

Theorem 2 *Suppose that the assumptions* (I′), (IV′) *and* (V′) *are satisfied. Then, for any* $f \in C^{\rho}([0,T]; X)$, $0 < (2-\alpha-\beta)/\alpha < \rho \le 1$, *and* $u_0 \in D(M)$ *satisfying one of the following two conditions:*

$$Mu_0 \in X_A^{\theta} = \{u \in X; \sup_{\xi > 0} \xi^{\theta} \|L(\xi M + L)^{-1} u\| < \infty\},$$
$$Mu_0 \in (X, D(A))_{\theta,\infty}$$

with $2 - \alpha - \beta < \theta < 1$, *a solution to problem* (46) *exists and is unique.*

In this section the convolution $a * b$ of two functions a and b is defined by

$$(a * b)(t) = \int_0^t a(t-s) b(s) ds.$$

Let $H(t) = K(t)L^{-1}$ and R be the solution to the integral equation

$$R(t) + H(t) + (H * R)(t) = R(t) + H(t) + \int_0^t H(t-s) R(s) ds = 0.$$

By assumptions (IV′) and (V′) H and R belong to $C^{\rho}([0,T]; \mathcal{L}(X))$. The integral equations (13) and (35) to be satisfied by $v = Mu$ and v' are expressed respectively as

$$v(t) = e^{-tA} v_0 + \int_0^t e^{-(t-s)A} \{f(s) + (R * f)(s) - (R * v')(s)\} ds,$$

and

$$w(t) = g'(t) - \int_0^t \frac{\partial}{\partial t} \int_s^t e^{-(t-\tau)A} R(\tau - s) d\tau w(s) ds, \tag{48}$$

where $A = LM^{-1}$ and

$$g(t) = e^{-tA} v_0 + \int_0^t e^{-(t-s)A} \{f(s) + (R * f)(s)\} ds.$$

If we define the operator valued function $Q(t)$ by

$$Q(t) = -\frac{d}{dt} \int_0^t e^{-(t-s)A} R(s) ds,$$

then

$$\|Q(t)\| = \left\| -e^{-tA} R(t) + \int_0^t \frac{\partial}{\partial t} e^{-(t-s)A} \cdot (R(s) - R(t)) ds \right\| \le C t^{(\beta-1)/\alpha}, \tag{49}$$

and

$$\frac{\partial}{\partial t} \int_s^t e^{-(t-\tau)A} R(\tau - s) d\tau = \frac{\partial}{\partial t} \int_0^{t-s} e^{-(t-s-\tau)A} R(\tau) d\tau = -Q(t-s).$$

Hence (48) is rewritten as

$$w(t) = g'(t) + \int_0^t Q(t-s)w(s)ds. \tag{50}$$

In view of (49) we can proceed as in the previous section starting with (50).

3. Case $\alpha = \beta = 1$

Suppose that the assumptions of Section 1 are satisfied with $\alpha = \beta = 1$ and with (V) replaced by the following stronger condition:

(VI) $\|K(t,s)L(s)^{-1} - K(t',s')L(s')^{-1}\| \le C\left(|t-t'|^\rho + |s-s'|^\rho\right), \ 0 < \rho \le 1.$

We are going to show that problem (1) has a unique solution for any initial value such that $u(0) \in D(M(0))$ and $M(0)u(0) \in \overline{D(A(0))}$ if we understand the integral $\int_0^t K(t,s)u(s)ds$ in the improper sense $\lim_{\epsilon\to 0}\int_\epsilon^t K(t,s)u(s)ds$. We make extensive use of the method of J. Prüss [3].

In this section we mean by a solution to (1) a function u satisfying the following conditions:

$u \in C(]0,T];X), u(t) \in D(M(t))\ \forall t \in [0,T], Mu \in C([0,T];X) \cap C^1(]0,T];X)$, $\|td(Mu)(t)/dt\|$ is bounded in $]0,T]$, $u(t) \in D(L(t))\ \forall t \in]0,T], Lu \in C(]0,T];X)$, $\int_\epsilon^t K(t,s)u(s)ds$ is uniformly bounded for $0 < \epsilon < t \le T$, and the limit

$$\int_{+0}^t K(t,s)u(s)ds = \lim_{\epsilon\to 0}\int_\epsilon^t K(t,s)u(s)ds \tag{51}$$

exists, and (1) holds with the integral $\int_0^t K(t,s)u(s)ds$ understood in the improper sense (51).

Note that since we are assuming $\alpha = \beta = 1$, the conditions (III) and $(2-\alpha-\beta)/\alpha < \rho \le 1$ in the previous section become

$$\mu + \nu > 1 \ \text{ and } \ 0 < \rho \le 1 \tag{52}$$

respectively.

Theorem 3 *Suppose that the assumptions* (I)–(IV) *and* (VI) *are satisfied with* $\alpha = \beta = 1$. *Then, for any* $f \in C^\rho([0,T];X)$, $0 < \rho \le 1$, *and* $u_0 \in D(M(0))$ *such that* $M(0)u_0 \in \overline{D(A(0))}$ *a solution to problem* (1) *exists and is unique.*

Since we are assuming $\alpha = \beta = 1$, we have

$$\left\|e^{-\tau A_n(t)}\right\| \le C, \quad \left\|A_n(t)e^{-\tau A_n(t)}\right\| \le C/\tau, \tag{53}$$

and (20) and (24) become

$$\|U_n(t,s)\| \le C, \ \|A_n(t)U_n(t,s)\| \le C(t-s)^{-1}, \ \|W_n(t,s)\| \le C(t-s)^{\mu+\nu-2}. \tag{54}$$

Following [3; Section 4] the fundamental solution $S_n(t,s)$ to the initial value problem for the equation

$$\frac{d}{dt}v_n(t) + A_n(t)v_n(t) + \int_0^t H(t,s)A_n(s)v_n(s)ds = f(t) \tag{55}$$

can be constructed as follows:

$$S_n = U_n + A_n^{-1}V_n, \tag{56}$$

$$V_n = Z_n + K_{1,n} * V_n, \quad K_{1,n} = -A_nU_n * H, \quad Z_n = -A_nU_n * H * A_nU_n. \tag{57}$$

The uniform estimates for these operators are established following the proof of Lemma 1 of p. 226 of [3]. $K_{1,n}(t,s)$ is expressed as

$$\begin{aligned} K_{1,n}(t,s) &= -\int_s^t A_n(t)U_n(t,\tau)(H(\tau,s) - H(t,s))d\tau \\ &\quad + \int_s^t W_n(t,\tau)H(t,s)d\tau - \left(I - e^{-(t-s)A_n(t)}\right)H(t,s). \end{aligned} \tag{58}$$

Hence

$$\|K_{1,n}(t,s)\| \le C\int_s^t (t-\tau)^{\rho-1}d\tau + C\int_s^t (t-\tau)^{\mu+\nu-2}d\tau + C \le C. \tag{59}$$

For $0 \le s' < s < t \le T$ we write, noting (58),

$$K_{1,n}(t,s) - K_{1,n}(t,s') = \mathrm{I} + \mathrm{II} + \mathrm{III}, \tag{60}$$

where

$$\begin{aligned} \mathrm{I} &= -\int_s^t A_n(t)U_n(t,\tau)(H(\tau,s) - H(t,s))d\tau \\ &\qquad + \int_{s'}^t A_n(t)U_n(t,\tau)(H(\tau,s') - H(t,s'))d\tau, \\ \mathrm{II} &= \int_s^t W_n(t,\tau)H(t,s)d\tau - \int_{s'}^t W_n(t,\tau)H(t,s')d\tau \\ \mathrm{III} &= -\left(I - e^{-(t-s)A_n(t)}\right)H(t,s) + \left(I - e^{-(t-s')A_n(t)}\right)H(t,s'). \end{aligned}$$

Using the inequality

$$\|H(\tau,s) - H(t,s) - H(\tau,s') + H(t,s')\| \le C(t-\tau)^{\rho/2}(s-s')^{\rho/2} \tag{61}$$

we easily obtain

$$\begin{aligned}
\|\mathrm{I}\| &= \Big\| -\int_s^t A_n(t)U_n(t,\tau)(H(\tau,s)-H(t,s)-H(\tau,s')+H(t,s'))d\tau \\
&+ \int_{s'}^s A_n(t)U_n(t,\tau)(H(\tau,s')-H(t,s'))d\tau \Big\| \\
&\le C\int_s^t (t-\tau)^{\rho/2-1}(s-s')^{\rho/2}d\tau + C\int_{s'}^s (t-\tau)^{\rho-1}d\tau \\
&\le C\left\{(t-s)^{\rho/2}(s-s')^{\rho/2}+(s-s')^{\rho}\right\}, \qquad (62)\\
\|\mathrm{II}\| &= \left\| \int_s^t W_n(t,\tau)d\tau(H(t,s)-H(t,s')) - \int_{s'}^s W_n(t,\tau)d\tau H(t,s')\right\| \\
&\le C\int_s^t (t-\tau)^{\mu+\nu-2}d\tau(s-s')^{\rho} + C\int_{s'}^s (t-\tau)^{\mu+\nu-2}d\tau \\
&\le C\left\{(t-s)^{\mu+\nu-1}(s-s')^{\rho}+(s-s')^{\mu+\nu-1}\right\}, \qquad (63)\\
\|\mathrm{III}\| &= \Big\| -\left(I-e^{-(t-s)A_n(t)}\right)(H(t,s)-H(t,s')) \\
&+ \left(e^{-(t-s)A_n(t)}-e^{-(t-s')A_n(t)}\right)H(t,s')\Big\| \\
&\le C\left\{(s-s')^{\rho}+\log\frac{t-s'}{t-s}\right\}. \qquad (64)
\end{aligned}$$

It follows from (60), (62), (63) and (64) that

$$\|K_{1,n}(t,s)-K_{1,n}(t,s')\| \le C\left\{(s-s')^{\rho/2}+(s-s')^{\mu+\nu-1}+\log\frac{t-s'}{t-s}\right\} \qquad (65)$$

for $0 \le s' < s < t \le T$. Therefore, by expressing $Z_n(t,s)$ as

$$\begin{aligned}
Z_n(t,s) &= (K_{1,n} * A_nU_n)(t,s) \\
&= \int_s^t (K_{1,n}(t,\tau)-K_{1,n}(t,s))A_n(\tau)U_n(\tau,s)d\tau \\
&-K_{1,n}(t,s)(U_n(t,s)-I), \qquad (66)
\end{aligned}$$

one can show

$$\begin{aligned}
\|Z_n(t,s)\| &\le C\int_s^t \left\{(\tau-s)^{\rho/2}+(\tau-s)^{\mu+\nu-1}+\log\frac{t-s}{t-\tau}\right\}\frac{d\tau}{\tau-s}+C \\
&\le C\left\{(t-s)^{\rho/2}+(t-s)^{\mu+\nu-1}+\int_0^1 \log\frac{1}{1-r}\frac{dr}{r}\right\}+C \le C. \qquad (67)
\end{aligned}$$

From (57), (59) and (67) it follows that

$$\|V_n(t,s)\| \le C. \qquad (68)$$

A familiar argument shows

$$(H * A_n U_n)(t,s) = \int_s^t H(t,\tau) A_n(\tau) U_n(\tau,s) d\tau$$
$$= \int_s^t (H(t,\tau) - H(t,s)) A_n(\tau) U_n(\tau,s) d\tau - H(t,s)(U_n(t,s) - I). \quad (69)$$

With the aid of (61) and (69) it can be shown without difficulty that the following inequalities hold:

$$\|(H * A_n U_n)(t,s)\| \le C, \quad (70)$$
$$\|(H * A_n U_n)(t,s) - (H * A_n U_n)(t',s)\| \le C \left\{ (t-t')^{\rho/2} + \log \frac{t-s}{t'-s} \right\} \quad (71)$$

for $s < t' < t$. Using (70), (71) and

$$\frac{\partial}{\partial t} \int_s^t U_n(t,\tau)(H * A_n U_n)(\tau,s) d\tau$$
$$= \int_s^t \frac{\partial}{\partial t} U_n(t,\tau) \left\{ (H * A_n U_n)(\tau,s) - (H * A_n U_n)(t,s) \right\} d\tau$$
$$+ \int_s^t W_n(t,\tau)(H * A_n U_n)(t,s) d\tau + e^{-(t-s)A_n(t)} (H * A_n U_n)(t,s)$$

one easily shows that

$$\left\| \frac{\partial}{\partial t} (U_n * H * A_n U_n)(t,s) \right\| = \left\| \frac{\partial}{\partial t} \int_s^t U_n(t,\tau)(H * A_n U_n)(\tau,s) d\tau \right\| \le C. \quad (72)$$

It is easier to show that

$$\left\| \frac{\partial}{\partial t} (U_n * H * V_n)(t,s) \right\| = \left\| \frac{\partial}{\partial t} \int_s^t U_n(t,\tau)(H * V_n)(\tau,s) d\tau \right\| \le C. \quad (73)$$

From (72) and (73) it follows that

$$\left\| \frac{\partial}{\partial t} \left(A_n(t)^{-1} V_n(t,s) \right) \right\|$$
$$= \left\| -\frac{\partial}{\partial t} (U_n * H * A_n U_n)(t,s) - \frac{\partial}{\partial t} (U_n * H * V_n)(t,s) \right\| \le C, \quad (74)$$

and hence that

$$\|(\partial/\partial t) S_n(t,s)\| \le C/(t-s). \quad (75)$$

Letting $n \to \infty$ in (58), (65) and (66) one obtains

$$K_{1,n}(t,s) \to K_1(t,s) = \int_s^t \frac{\partial}{\partial t} U(t,\tau)(H(\tau,s) - H(t,s))d\tau$$
$$+ \int_s^t W(t,\tau)H(t,s)d\tau - \left(I - e^{-(t-s)A(t)}\right) H(t,s), \tag{76}$$
$$\|K_1(t,s) - K_1(t,s')\|$$
$$\leq C\left\{(s-s')^{\rho/2} + (s-s')^{\mu+\nu-1} + \log\frac{t-s'}{t-s}\right\}, \ 0 \leq s' < s < t \leq T, \tag{77}$$
$$Z_n(t,s) \to Z(t,s) = -\int_s^t (K_1(t,\tau) - K_1(t,s))\frac{\partial}{\partial \tau} U(\tau,s)d\tau \tag{78}$$
$$-K_1(t,s)(U(t,s) - I).$$

Hence $V_n(t,s) \to V(t,s)$, where V is the solution to the integral equation

$$V = Z + K_1 * V, \tag{79}$$

and

$$S_n(t,s) \to S(t,s) = U(t,s) + A(t)^{-1}V(t,s), \tag{80}$$
$$A_n(t)S_n(t,s) = A_n(t)U_n(t,s) + V_n(t,s)$$
$$\to -\frac{\partial}{\partial t}U(t,s) + V(t,s) = -W(t,s) + \frac{\partial}{\partial s}e^{-(t-s)A(t)} + V(t,s). \tag{81}$$

As is easily seen $K_1(t,s)$ and $Z(t,s)$ are bounded and continuous in $0 \leq s < t \leq T$, and hence so are $V(t,s)$ and $S(t,s)$. We show that

$$\lim_{t \to s} S(t,s)w = w \ \text{ for } \ w \in \overline{D(A(s))}. \tag{82}$$

In view of (79), (80) and the uniform boundedness of $Z(t,s)$ it suffices to show that

$$\lim_{t \to s} Z(t,s)w = 0 \ \text{ for } \ w \in D(A(s)). \tag{83}$$

One can show analogously to (28) that

$$\left\|\frac{\partial}{\partial \tau}U(\tau,s)w\right\| \leq C(\tau-s)^{\mu+\nu-2}\|w\| + C\|w\|_{D(A(s))} \leq C(\tau-s)^{\mu+\nu-2}\|w\|_{D(A(s))}.$$

This and (77) imply

$$\left\|\int_s^t (K_1(t,\tau) - K_1(t,s))\frac{\partial}{\partial \tau}U(\tau,s)wd\tau\right\|$$
$$\leq C\int_s^t \left\{(\tau-s)^{\rho/2} + (\tau-s)^{\mu+\nu-1} + \log\frac{t-s}{t-\tau}\right\}(\tau-s)^{\mu+\nu-2}\|w\|_{D(A(s))}d\tau$$
$$\leq C(t-s)^{\mu+\nu-1}\|w\|_{D(A(s))} \to 0 \tag{84}$$

as $t \to s$. It follows from (78), (84), the uniform boundedness of $K_1(t,s)$ and $\lim_{t\to s} U(t,s)w = w$ that (83) holds.

Let v_n be the solution to (55) satisfying the initial condition $v_n(0) = v_0$, where $v_0 = M(0)u_0$. Then

$$\begin{aligned} v_n(t) &= S_n(t,0)v_0 + \int_0^t S_n(t,s)f(s)ds \\ &= U_n(t,0)v_0 + A_n(t)^{-1}V_n(t,0)v_0 + \int_0^t \{U_n(t,s) + A_n(t)^{-1}V_n(t,s)\}f(s)ds, \end{aligned}$$

$$v_n(t) \to v(t) = S(t,0)v_0 + \int_0^t S(t,s)f(s)ds. \tag{85}$$

Since $v_0 \in \overline{D(A(0))}$ by assumption, (82) and (85) imply that $\lim_{t\to 0} v(t) = v_0$. Therefore

$$v \in C([0,T];X) \text{ and } v(0) = v_0. \tag{86}$$

With the aid of a familiar argument we have

$$\begin{aligned} A_n(t)v_n(t) &= A_n(t)U_n(t,0)v_0 + V_n(t,0)v_0 \\ &\quad + A_n(t)\int_0^t U_n(t,s)f(s)ds + \int_0^t V_n(t,s)f(s)ds \\ &= A_n(t)U_n(t,0)v_0 + Y_n(t), \end{aligned} \tag{87}$$

where

$$\begin{aligned} Y_n(t) &= V_n(t,0)v_0 + \int_0^t A_n(t)U_n(t,s)(f(s) - f(t))ds \\ &\quad - \int_0^t W_n(t,s)f(t)ds + (I - e^{-tA_n(t)})f(t) + \int_0^t V_n(t,s)f(s)ds. \end{aligned}$$

Since $Y_n(t)$ is uniformly bounded and

$$\begin{aligned} Y_n(t) \to Y(t) &= V(t,0)v_0 - \int_0^t \frac{\partial}{\partial t}U(t,s)\cdot(f(s) - f(t))ds \\ &\quad - \int_0^t W(t,s)f(t)ds + (I - e^{-tA(t)})f(t) + \int_0^t V(t,s)f(s)ds \end{aligned}$$

as $n \to \infty$, one has

$$A_n(t)v_n(t) \to -\frac{\partial}{\partial t}U(t,0)v_0 + Y(t) \equiv \psi(t). \tag{88}$$

It follows from (85) and

$$v_n(t) = A_n(t)^{-1}A_n(t)v_n(t) \to A(t)^{-1}\psi(t), \tag{89}$$

that

$$\psi(t) \in A(t)v(t). \tag{90}$$

It follows from (54), (87), (88) and the uniform boundedness of $Y_n(t)$ that

$$\|A_n(t)v_n(t)\| \le C/t, \qquad \|\psi(t)\| \le C/t. \tag{91}$$

It is easily seen that

$$(H * A_n v_n)(t) = \int_0^t H(t,\tau)A_n(\tau)U_n(\tau,0)v_0 d\tau + \int_0^t H(t,\tau)Y_n(\tau)d\tau$$
$$= \int_0^t (H(t,\tau) - H(t,0))A_n(\tau)U_n(\tau,0)v_0 d\tau$$
$$-H(t,0)(U_n(t,0)v_0 - v_0) + \int_0^t H(t,\tau)Y_n(\tau)d\tau$$
$$\to -\int_0^t (H(t,\tau) - H(t,0))\frac{\partial}{\partial\tau}U(\tau,0)v_0 d\tau - H(t,0)(U(t,0)v_0 - v_0)$$
$$+\int_0^t H(t,\tau)Y(\tau)d\tau. \tag{92}$$

Using (88) and noting $\lim_{t\to 0} U(t,0)v_0 = v_0$ which follows from $v_0 \in \overline{D(A(0))}$ the last side of (92) is equal to

$$(H * \psi)(t) = \int_{+0}^t H(t,\tau)\psi(\tau)d\tau.$$

Hence

$$(H * A_n v_n)(t) \to (H * \psi)(t). \tag{93}$$

It is easily seen that $(H * A_n v_n)(t)$ is uniformly bounded. Hence it follows from (55), (88), (91) and (93) that

$$\|v_n'(t)\| \le C/t, \quad v_n'(t) \to v'(t), \quad \|v'(t)\| \le C/t, \tag{94}$$

and

$$v' + \psi + H * \psi = f. \tag{95}$$

Let R be the solution to the integral equation

$$R + H + H * R = 0.$$

Then $R(t,s)$ is Hölder continuous in (t,s) of order ρ:

$$\|R(t,s) - R(t',s')\| \le C\left(|t-t'|^\rho + |s-s'|^\rho\right). \tag{96}$$

It follows from (55) and

$$R * v_n' + R * A_n v_n + R * H * A_n v_n = R * f$$

that

$$v_n' + A_n v_n = f + R * f - R * v_n'. \tag{97}$$

In view of (VI), (85), (95) and (86)

$$\begin{aligned}(H * v_n')(t) &= \int_0^t H(t,\tau)v_n'(\tau)d\tau \\ &= \int_0^t (H(t,\tau) - H(t,0))v_n'(\tau)d\tau + H(t,0)(v_n(t) - v_0) \\ &\to \int_0^t (H(t,\tau) - H(t,0))v'(\tau)d\tau + H(t,0)(v(t) - v_0) \\ &= \int_{+0}^t H(t,\tau)v'(\tau)d\tau = (H * v')(t). \end{aligned} \tag{98}$$

Analogously using (96) instead of (VI) one shows

$$(R * v_n')(t) \to \int_{+0}^t R(t,\tau)v'(\tau)d\tau = (R * v')(t). \tag{99}$$

Noting that

$$\int_\epsilon^t R(t,s)v'(s)ds = \int_\epsilon^t (R(t,s) - R(t,0))v'(s)ds + R(t,0)(v(t) - v(\epsilon))$$

is uniformly bounded and tends to $(R * v')(t)$ as $\epsilon \to 0$, we have

$$\begin{aligned}\lim_{\epsilon\to 0}\int_\epsilon^t (R(t,s) + H(t,s))v'(s)ds &= -\lim_{\epsilon\to 0}\int_\epsilon^t (H * R)(t,s)v'(s)ds \\ &= -\lim_{\epsilon\to 0}\int_\epsilon^t \int_s^t H(t,\tau)R(\tau,s)d\tau v'(s)ds \\ &= -\lim_{\epsilon\to 0}\int_\epsilon^t H(t,\tau)\int_\epsilon^\tau R(\tau,s)v'(s)dsd\tau = -\int_0^t H(t,\tau)(R * v')(\tau)d\tau \\ &= -(H * (R * v'))(t). \end{aligned} \tag{100}$$

This means that

$$R * v' + H * v' = -H * (R * v'). \tag{101}$$

Letting $n \to \infty$ in (97) and using (88), (94), (99) one gets

$$v' + \psi = f + R * f - R * v'. \tag{102}$$

Define the function u by

$$u = L^{-1}(f + R * f - R * v' - v'). \tag{103}$$

Then by (102) and (90)

$$Lu = f + R * f - R * v' - v' = \psi \in Av = LM^{-1}v. \tag{104}$$

This implies $Mu = v$. Since $\int_\epsilon^t H(t,s)v'(s)ds$ is uniformly bounded and converges as $\epsilon \to 0$, one shows with the aid of (104) that $\int_\epsilon^t H(t,s)L(s)u(s)ds$ is also uniformly bounded, and tends to

$$H * Lu = H * f + H * R * f - H * (R * v') - H * v'. \tag{105}$$

Adding (104) and (105), and using (101) we conclude

$$Lu + H * Lu = f - v' = f - (Mu)',$$

and the existence part of the proof is complete.

Next, we show the uniqueness. Let u be the solution to (1) with $f = 0$ and $u_0 = 0$. By virtue of

$$\begin{aligned}
&\int_\epsilon^t H(t,s)L(s)u(s)ds + \int_\epsilon^t R(t,s)L(s)u(s)ds \\
&\quad + \int_\epsilon^t R(t,\tau)\int_\epsilon^\tau H(\tau,s)L(s)u(s)dsd\tau \\
&\quad = \int_\epsilon^t H(t,s)L(s)u(s)ds + \int_\epsilon^t R(t,s)L(s)u(s)ds \\
&\quad + \int_\epsilon^t \int_s^t R(t,\tau)H(\tau,s)d\tau L(s)u(s)ds \\
&\quad = \int_\epsilon^t \{H(t,s) + R(t,s) + (R * H)(t,s)\}L(s)u(s)ds = 0,
\end{aligned}$$

$\int_\epsilon^t R(t,s)L(s)u(s)ds$ is uniformly bounded and converges as $\epsilon \to 0$, and

$$H * Lu + R * Lu + R * (H * Lu) = 0. \tag{106}$$

Since

$$(R * (Mu)')(t) = \int_0^t (R(t-s) - R(t))(Mu)'(s)ds + R(t)M(t)u(t) \tag{107}$$

exists, we obtain from (1) with $f = 0$

$$R * (Mu)' + R * Lu + R * (H * Lu) = 0. \tag{108}$$

Adding (1) with $f = 0$ and (108), and using (106) we get

$$(Mu)' + Lu + R * (Mu)' = 0. \tag{109}$$

Hence, $v = Mu$ satisfies

$$v' + Av \ni -R * v', \quad v(0) = 0.$$

By the second half of Theorem 4.3 of [2] this implies

$$v(t) = -\int_0^t U(t,\tau)(R * v')(\tau)d\tau.$$

Let for $0 < \epsilon < t \leq T$

$$v_\epsilon(t) = -\int_\epsilon^t U(t,\tau)\int_\epsilon^\tau R(\tau,s)v'(s)dsd\tau.$$

Then $v_\epsilon(t) \to v(t)$ as $\epsilon \to 0$, and by the change of the order of integration

$$v_\epsilon(t) = -\int_\epsilon^t \int_s^t U(t,\tau)R(\tau,s)d\tau v'(s)ds.$$

Differentiation yields

$$v'_\epsilon(t) = -\int_\epsilon^t P(t,s)v'(s)ds, \tag{110}$$

where

$$P(t,s) = \frac{\partial}{\partial t}\int_s^t U(t,\tau)R(\tau,s)d\tau.$$

We write

$$v'_\epsilon(t) = -\int_\epsilon^t (P(t,s)-P(t,0))v'(s)ds - P(t,0)(v(t)-v(\epsilon)), \tag{111}$$

and will estimate $\|P(t,s)-P(t,0)\|$. Remark to Theorem 4.3 in [2] shows

$$\begin{aligned} P(t,s) &= e^{-(t-s)A(t)}R(t,s) \\ &\quad + \int_s^t \frac{\partial}{\partial t}U(t,\tau)\cdot(R(\tau,s)-R(t,s))d\tau + \int_s^t W(t,\tau)R(t,s)d\tau. \end{aligned} \tag{112}$$

In view of

$$\left\|e^{-(t-s)A(t)} - e^{-tA(t)}\right\| = \left\|\int_{t-s}^t \frac{\partial}{\partial r}e^{-rA(t)}dr\right\| \le C\int_{t-s}^t \frac{dr}{r} = C\log\frac{t}{t-s},$$

and (96) one has

$$\left\|e^{-(t-s)A(t)}R(t,s) - e^{-tA(t)}R(t,0)\right\| \le C\left(\log\frac{t}{t-s} + s^\rho\right). \tag{113}$$

One also gets without difficulty

$$\begin{aligned} &\left\|\int_s^t \frac{\partial}{\partial t}U(t,\tau)\cdot(R(\tau,s)-R(t,s))d\tau - \int_0^t \frac{\partial}{\partial t}U(t,\tau)\cdot(R(\tau,0)-R(t,0))d\tau\right\| \\ &\quad = \left\|\int_s^t \frac{\partial}{\partial t}U(t,\tau)\cdot(R(\tau,s)-R(t,s)-R(\tau,0)+R(t,0))d\tau \right.\\ &\quad\quad \left. - \int_0^s \frac{\partial}{\partial t}U(t,\tau)\cdot(R(\tau,0)-R(t,0))d\tau\right\| \\ &\quad \le C\int_s^t (t-\tau)^{\rho/2-1}d\tau\cdot s^{\rho/2} + C\int_0^s (t-\tau)^{\rho-1}d\tau \\ &\quad \le C(t-s)^{\rho/2}s^{\rho/2} + Cs^\rho \le Cs^{\rho/2}, \end{aligned} \tag{114}$$

$$\begin{aligned} &\left\|\int_s^t W(t,\tau)R(t,s)d\tau - \int_0^t W(t,\tau)R(t,0)d\tau\right\| \\ &\quad = \left\|\int_s^t W(t,\tau)d\tau(R(t,s)-R(t,0)) - \int_0^s W(t,\tau)d\tau R(t,0)\right\| \\ &\quad \le C\int_0^s (t-\tau)^{\mu+\nu-2}d\tau + Ct^{\mu+\nu-1}s^\rho \le Cs^{\mu+\nu-1} + Ct^{\mu+\nu-1}s^\rho. \end{aligned} \tag{115}$$

It follows from (112), (113), (114) and (115) that

$$\|P(t,s)-P(t,0)\| \le C\left(\log\frac{t}{t-s}+s^{\rho/2}+s^{\mu+\nu-1}\right). \tag{116}$$

Using (111) and (116) one obtains that

$$\begin{aligned}\|v_\epsilon'(t)\| &\le C\int_0^t \left(\log\frac{t}{t-s}+s^{\rho/2}+s^{\mu+\nu-1}\right)\frac{ds}{s}+C\\ &\le C\left(\int_0^1 \log\frac{1}{1-s}\cdot\frac{ds}{s}+t^{\rho/2}+t^{\mu+\nu-1}\right)+C\le C,\end{aligned}$$

and

$$v_\epsilon'(t)\to -\int_0^t (P(t,s)-P(t,0))v'(s)ds-P(t,0)v(t)\equiv w(t)$$

as $\epsilon\to 0$. As is easily seen $v'(t)=w(t)$ is uniformly bounded. Letting $\epsilon\to 0$ in (110) yields

$$v'(t)=-\int_0^t P(t,s)v'(s)ds.$$

This implies $v'(t)\equiv 0$, and hence $Lu+H*Lu=0$. Therefore $Lu(t)$ is bounded, and the uniqueness of this integral equation implies $Lu=0$, and hence $u=0$.

References

[1] M. G. Crandall and J. A. Nohel: *An abstract functional differential equation and a related nonlinear Volterra equation*, Israel J. Math. **29** (1978), 313–328.

[2] A. Favini and A. Yagi: Degenerate Differential Equations in Banach Spaces, Marcel Dekker, New York · Basel · Hong Kong, 1998.

[3] J. Prüss: *On resolvent operators for linear integrodifferential equations of Volterra type*, J. Integral Equations **5** (1983), 211–236.

[4] H. Tanabe: *On degenerate integrodifferential equations of parabolic type*, The 6th International Conference on Nonlinear Functional Analysis and Applications, Masan and Chinju, Korea, September 1–5, 2000.

Angelo Favini
Dipartimento di Matematica
Università degli Studi di Bologna
Piazza di Porta S. Donato 5
I-40126 Bologna, Italy
e-mail: `favini@dm.unibo.it`

Alfredo Lorenzi
Departimento di Matematica
Università di Milano
Milano, Italia
e-mail: `Alfredo.Lorenzi@mat.unimi.it`

Hiroki Tanabe
Department of Economics
Otemon Gakuin University
Osaka, Japan
e-mail: `h7tanabe@jttk.zaq.ne.jp`

Progress in Nonlinear Differential Equations
and Their Applications, Vol. 50, 137–154

On a Class of Quasi Linear Equations in Infinite-dimensional Spaces

Marco Fuhrman

1. Introduction

1.1. Statement of the problem and aims of this note

In this paper we consider the following differential problem:

$$\begin{cases} \dfrac{\partial}{\partial t} v(t,x) &= \mathcal{L}v(t,x) + f(x, v(t,x), \nabla v(t,x)), \\ v(0,x) &= \phi(x) \qquad t \geq 0,\ x \in H. \end{cases} \tag{1.1}$$

Here H is a real separable Hilbert space, $v : [0,\infty) \times H \to \mathbb{R}$ is the unknown function, $\mathcal{L}$ is a linear second order differential operator, satisfying suitable ellipticity conditions, $f : H \times \mathbb{R} \times H \to \mathbb{R}$ and $\phi : H \to \mathbb{R}$ are given functions, ∇ denotes derivative with respect to x.

When H is the space $\mathbb{R}^n$, this is a classical parabolic problem. Recently, the general, infinite-dimensional case has deserved the attention of many researchers: it is the purpose of this paper to give an account of the developments of this area. Although still in expansion, the research activity in this field has already produced a large number of results; moreover, a variety of techniques has been used, ranging from methods of partial differential equations to semigroup theory, stochastic analysis etc. At the same time, applications to different areas have been given, including quantum physics, stochastic control, particle systems, fluid dynamics and many others.

We will focus attention on solutions of equation (1.1), and some of its variants, in spaces of continuous functions. The equation (1.1) can also be studied in L^p-spaces with respect to a suitably chosen measure on H (as well as in other function spaces), and the approach via Dirichlet forms has also turned out to be very effective: we refer the reader to [35] and [26].

Even in the framework of continuous solutions the existing literature is very large and for reasons of brevity we will be able to report only on a part of it. This note has to be understood as a reader's guide to the main results. We will try to warn the reader on important subjects and theories that are not given adequate treatment here. Proofs are often not given; sometimes they are sketched, in order to indicate to the reader some of the techniques used. When a theorem admits several variants, we will select only one of them and state it in a precise way, referring the reader to the original papers. Occasionally this prevents us to state a result in its full generality, but helps to avoid a very lenghty and technical exposition, due to the complexity of the infinite-dimensional framework.

Although this is mainly an expository article, we will also include some new results contained in a forthcoming joint paper [19].

1.2. Motivations

As mentioned earlier, results on problem (1.1) can be applied to a variety of different subjects. Among them we mention the connections with stochastic evolution equations. Indeed, in many cases, the solution v can be represented in the form $v(t,x) = P_t\phi(x)$, where (P_t) is the transition semigroup of a Markov stochastic process in H, constructed as the solution of a differential stochastic equation of Ito type. Knowledge about the solution of (1.1) can give information about the semigroup (P_t); conversely, properties of (P_t) can be used to prove existence and regularity of the solution v.

Another motivation for the study of (1.1) comes from the optimal control theory for stochastic evolution equations; in many cases the so-called *value function* of the optimal control problem is a solution of the *Hamilton-Jacobi-Bellman equation*, which turns out to be of the form (1.1) in many situations.

1.3. Standing assumptions on the operator $\mathcal{L}$. Plan of the paper

Let us fix some notations. H stands for a real separable Hilbert space, with scalar product $\langle \cdot, \cdot \rangle$. Let K be a Banach space; $C^k(H;K)$ denotes the space of functions from H to K that are continuous together with their (Fréchet) derivatives of order $1, \ldots, k$. $C_b^k(H;K)$ (respectively, $UC_b^k(H;K)$) is the subspace of the elements of $C^k(H;K)$ that are bounded with bounded derivatives of order $1, \ldots, k$ (respectively, uniformly continuous and bounded, with uniformly continuous and bounded derivatives). $\mathcal{B}_b(H)$ is the space of bounded measurable functions $\phi : H \to \mathbb{R}$. For functions $v : [0,T] \times H \to \mathbb{R}$, the notation $v \in UC_b^{1,2}([0,T] \times H; \mathbb{R})$ means that v is differentiable with respect to t, twice differentiable with respect to x, and it is bounded and uniformly continuous, with respect to all arguments, together with its derivatives. Spaces like $C_b^{0,1}([0,T] \times H; \mathbb{R})$ etc. are defined in a similar way.

$L(H,K)$ is the space of linear bounded operators from H to K. We set $L(H) = L(H,H)$ for brevity. If K is a Hilbert space, $L_{HS}(H,K)$ is the subspace of $L(H,K)$ consisting of Hilbert-Schmidt operators, with the Hilbert-Schmidt norm. Norms are denoted $|\cdot|$, with a subscript to denote the space if necessary.

We specify the form of the operator $\mathcal{L}$ appearing in (1.1). For $\phi \in C^2(H;\mathbb{R})$ we set

$$\mathcal{L}\phi(x) = \frac{1}{2}\mathrm{Trace}\left(G(x)G(x)^*\nabla^2\phi(x)\right) + \langle Ax + F(x), \nabla\phi(x)\rangle,$$

where $\nabla\phi(x)$ and $\nabla^2\phi(x)$ are the Fréchet derivatives of ϕ at $x \in H$; they are identified with elements of H and $L(H)$ respectively.

The following are standing assumptions in the rest of the paper.

Hypothesis 1.

- *A is the generator of a strongly continuous semigroup of linear operators e^{tA}, $t \geq 0$, in H;*
- *$F : H \to H$ belongs to $C_b^1(H;H)$;*
- *G is a mapping from H to $L(H)$ such that $e^{tA}G(x) \in L_{HS}(H,H)$ for $x \in H$ and $t > 0$, the mapping $e^{tA}G : H \to L_{HS}(H,H)$ belongs to the space $C_b^1(H; L_{HS}(H,H))$ and there exist $C > 0$ and $\gamma \in [0,1/2)$ such that*

$$\begin{aligned} &|G(x)|_{L(H)} \leq C\,(1+|x|),\\ &|e^{tA}G(x)|_{L_{HS}(H,H)} \leq C\,t^{-\gamma}(1+|x|),\\ &|\nabla e^{tA}G(x)|_{L(H,L_{HS}(H,H))} \leq C\,t^{-\gamma}, \qquad t \in (0,1],\ x \in H. \end{aligned}$$

If $G(x) = G$ is constant then the last assumption reduces to

$$\text{Trace}\left(e^{tA}GG^*e^{tA^*}\right) \leq Ct^{-2\gamma}.$$

This implies in particular that the nonnegative operators Q_t defined by

$$Q_t x = \int_0^t e^{sA}GG^*e^{sA^*}x\,ds, \qquad x \in H, \tag{1.2}$$

have finite trace.

The special form of the operator $\mathcal{L}$ is motivated by applications. The assumptions turn out to be natural, as explained in the following sections.

We note that $\mathcal{L}$ is elliptic in the sense that $G(x)G(x)^* \geq 0$. The condition of uniform ellipticity:

$$G(x)G(x)^* \geq \epsilon I, \quad \text{for some } \epsilon > 0,$$

is not assumed in general.

The plan of the paper is as follows. In Section 2 we study the linear problem

$$\begin{cases} \dfrac{\partial}{\partial t}v(t,x) &= \mathcal{L}v(t,x),\\ v(0,x) &= \phi(x) \qquad t \geq 0,\ x \in H. \end{cases} \tag{1.3}$$

In Section 3 we consider equation (1.1) whereas Section 4 is devoted to applications to stochastic control.

2. The linear equation $\frac{\partial}{\partial t}v = \mathcal{L}v$

2.1. A special case: the operator $\mathcal{L}^0$

We first consider the following special case:

$$\frac{\partial}{\partial t}v(t,x) = \mathcal{L}^0 v(t,x), \quad v(0,x) = \phi(x), \tag{2.1}$$

where

$$\mathcal{L}^0\phi(x) = \frac{1}{2}\text{Trace}\left(GG^*\nabla^2\phi(x)\right) + \langle x, A^*\nabla\phi(x)\rangle.$$

This is obtained by setting $F = 0$ and $G(x) = G$, a constant operator. A candidate solution is given by the so-called *Mehler formula*, which is used as the definition of a semigroup P_t^0, $t \geq 0$:

$$v(t,x) = P_t^0\phi(x) = \int_H \phi(e^{tA}x + y)\,\mathcal{N}_{Q_t}(dy),$$

where $\mathcal{N}_{Q_t}$ denotes the centered gaussian measure in H with covariance operator Q_t: compare (1.2); $\mathcal{N}_{Q_t}$ is well defined, since Q_t is nonnegative and our assumptions imply that it has finite trace. In the Mehler formula we assume that ϕ belongs to the space $UC_b(H,\mathbb{R})$. It can be proved that (P_t^0) acts in $UC_b(H,\mathbb{R})$ as a contraction semigroup, called *Ornstein-Uhlenbeck semigroup*; if $\dim H = \infty$, it can be proved that it is not strongly continuous, in general.

The following result, taken from [37], Theorem 5.1, shows that v is indeed the required solution under some additional conditions.

Theorem 1. *If* Trace $GG^* < \infty$, $\phi \in UC_b^2(H;\mathbb{R})$, $A^*\nabla\phi$ *is well defined and belongs to* $C_b(H;H)$, *then for all* $x \in H$, $t \geq 0$, *we have* $v(t,\cdot) \in UC_b^2(H;\mathbb{R})$, $A^*\nabla v(t,\cdot) \in C_b(H;H)$, $v(\cdot,x)$ *is continuously differentiable and*

$$\begin{cases} \dfrac{\partial}{\partial t}v(t,x) &=& \dfrac{1}{2}\mathrm{Trace}\left(GG^*\nabla^2 v(t,x)\right) + \langle x, A^*\nabla v(t,x)\rangle, \\ v(0,x) &=& \phi(x). \end{cases}$$

If ϕ only belongs to $UC_b(H;\mathbb{R})$ then $v(t,x) = P_t^0\phi(x)$ is called a *generalized* solution. It can be proved that it is the pointwise limit of a sequence (v_n) of solutions corresponding to properly chosen initial conditions (ϕ_n) that converge pointwise to ϕ: see [37], Theorem 5.4.

Note that the Mehler formula makes sense for arbitrary $\phi \in \mathcal{B}_b(H)$. In this case regularity properties of the corresponding generalized solution (equivalently, of the Ornstein-Uhlenbeck semigroup) reflect the elliptic character of the operator $\mathcal{L}^0$. One may ask which nondegeneracy conditions are needed in order to have $v(t,\cdot) = P_t^0\phi$ smooth. The following result, taken from [15], Theorem 7.2.1, gives a very precise answer.

Theorem 2. $P_t^0\phi = v(t,\cdot)$ *belongs to* $C^0(H;\mathbb{R})$ *for every* $\phi \in \mathcal{B}_b(H)$ *and every* $t > 0$ *if and only if*

$$\text{Image } e^{tA} \subset \text{Image } Q_t^{1/2}, \qquad t > 0.$$

If this happens then $v(t,\cdot) \in C_b^\infty(H;\mathbb{R})$. *If there exist* $C > 0$ *and* $\alpha < 1$ *such that*

$$|Q_t^{-1/2}e^{tA}| \leq C\,t^{-\alpha}, \qquad t \in (0,1],$$

where $Q_t^{-1/2}$ *denotes the pseudoinverse of the operator* $Q_t^{1/2}$, *then, for some* $c > 0$,

$$|\nabla v(t,x)| \leq c\,t^{-\alpha}. \qquad t \in (0,1].$$

The Ornstein-Uhlenbeck semigroup (P_t^0) can be considered in different function spaces than $UC_b(H;\mathbb{R})$. Properties of (P_t^0) have been deeply investigated in spaces of continuous functions (possibly unbounded). A detailed study of its continuity properties and a characterization of the generator can be found in the papers [5], [6], [7], [33], [34]. There the reader will also find results on the inhomogeneous Cauchy problem associated to (2.1) and on elliptic (stationary) equations for the operator $\mathcal{L}^0$, both in H and in subsets of H.

Under suitable conditions, (P_t^0) admits an extension to a space $L^p(H,\mu)$, where μ a properly chosen, gaussian measure on H. Properties of (P_t^0) in the space $L^p(H,\mu)$ are studied in [8], [9], [10], [11], [12], [20], [2] [16], [17].

Finally, we mention that the Ornstein-Uhlenbeck semigroup has a central position in the analysis of gaussian spaces of random variables, and is connected with the so-called Malliavin calculus: we refer the reader to [29].

For generalizations of the Mehler formula see also [2], [18].

2.2. The general linear case

Next we consider the general linear equation

$$\begin{cases} \dfrac{\partial}{\partial t} v(t,x) & = & \mathcal{L}v(t,x) \\ & = & \dfrac{1}{2}\operatorname{Trace}\left(G(x)G(x)^*\nabla^2 v(t,x)\right) + \langle Ax + F(x), \nabla v(t,x)\rangle, \\ v(0,x) & = & \phi(x), \qquad t \in [0,T],\ x \in H. \end{cases} \tag{2.2}$$

We would like to proceed as in the previous section and look for a semigroup (P_t) acting on functions ϕ belonging to $UC_b(H;\mathbb{R})$ (or to some other function space) such that $v(t,x) = P_t\phi(x)$ gives a candidate solution, and P_t is given by a more or less explicit formula. It turns out that this is possible, by means of a probabilistic representation formula.

We consider the stochastic differential equation

$$\begin{aligned} dX_\tau &= AX_\tau\, d\tau + F(X_\tau)\, d\tau + G(X_\tau)\, dW_\tau, \\ X_0 &= x \in H, \end{aligned}$$

where W_τ, $\tau \geq 0$, is a standard (cylindrical) Wiener process in H defined in a probability space $(\Omega, \mathcal{F}, \mathbb{P})$. The unknown function $X : [0,T] \times \Omega \to H$ is a stochastic process, also denoted X_τ, $\tau \in [0,T]$. The precise meaning of the stochastic equation is as follows: X is called *mild solution* if

(i) X_τ is $\mathcal{F}_\tau$-measurable, where

$$\mathcal{F}_\tau = \sigma\{W_r \ : \ r \leq \tau\} \vee \mathcal{N}, \tag{2.3}$$

is the σ-algebra generated by the random variables W_r, $r \leq \tau$, and by the family $\mathcal{N}$ of $\mathbb{P}$-null sets of $\mathcal{F}$;

(ii) $X(\cdot,\omega) \in C([0,T];H)$ for $\mathbb{P}$-a.a. $\omega \in \Omega$ and $\mathbb{E}\sup_{\tau\in[0,T]} |X_\tau|^p < \infty$ for every $p \in [1,\infty)$;

(iii) $\mathbb{P}$-a.s. we have for every $\tau \in [0,T]$,

$$X_\tau = e^{\tau A}x + \int_0^\tau e^{(\tau-\sigma)A}F(X_\sigma)\, d\sigma + \int_0^\tau e^{(\tau-\sigma)A}G(X_\sigma)\, dW_\sigma.$$

The last integral is an Ito stochastic integral; we refer to [14] for a detailed account of infinite-dimensional stochastic integration. The assumptions made in Hypothesis 1 guarantee that there exists a unique mild solution. Indeed, these assumptions are standard in the literature, and this is the reason why we assume them to hold throughout the paper.

We write $X(\tau, x)$ instead of X_τ when we need to stress dependence on the initial condition. We define

$$P_t\phi(x) = \mathbb{E}\,\phi(X(t,x)). \tag{2.4}$$

(P_t) is called *transition semigroup* of the process X. It is indeed a semigroup on $C_b(H;\mathbb{R})$. Somewhat imprecisely, we call (P_t) the semigroup "generated" by $\mathcal{L}$, without specifying the exact meaning of this terminology. We set $v(t,x) = P_t\phi(x)$. The following theorem combines an existence result in [14] Theorem 9.17 with a regularity statement in [37] Theorem 6.9.

Theorem 3. *If* $G \in C_b^2(H; L_{HS}(H,H))$, $\phi \in C_b^2(H;\mathbb{R})$, $F \in C_b^2(H;H)$, *then* $v \in C^{0,2}([0,T] \times H;\mathbb{R})$, *and for* $x \in \mathrm{dom}(A)$ *and* $t \in [0,T]$ *we have* $v(\cdot, x) \in C^1([0,T];\mathbb{R})$ *and*

$$\left\{ \begin{array}{lcl} \dfrac{\partial}{\partial t} v(t,x) & = & \dfrac{1}{2}\mathrm{Trace}\left(G(x)G(x)^*\nabla^2 v(t,x)\right) + \langle Ax + F(x), \nabla v(t,x)\rangle, \\ v(0,x) & = & \phi(x). \end{array} \right. \tag{2.5}$$

The fact that the stochastic representation formula (2.4) gives the desired solution is natural: equation (2.5) is the usual backward Kolmogorov equation for the Markov process defined by the stochastic equation. However, since this may look strange for the reader unfamiliar with stochastic differential equations, we would like to sketch below an argument which shows that, under additional conditions, the solution of (2.5) is necessarily given by the formula (2.4).

The argument is based on the *Ito formula*: for $u \in UC_b^{1,2}([0,T] \times H;\mathbb{R})$,

$$d\,u(\tau, X_\tau) = \frac{\partial}{\partial t}u(\tau, X_\tau)d\tau + \mathcal{L}u(\tau,X_\tau)d\tau + \langle \nabla u(\tau, X_\tau), G(X_\tau)dW_\tau\rangle,$$

provided e.g. $A \in L(H)$ (see for example [14], Theorem 4.17). The differential notation used in the last formula has the following precise meaning:

$$\begin{aligned} u(\tau, X_\tau) - u(0, X_0) &= \int_0^\tau [\frac{\partial}{\partial t}u(\sigma, X_\sigma) + \mathcal{L}u(\sigma, X_\sigma)]\, d\sigma \\ &\quad + \int_0^\tau \langle \nabla u(\sigma, X_\sigma), G(X_\sigma)dW_\sigma\rangle, \quad \tau \in [0,T]. \end{aligned}$$

Now assume that v is a strict solution in $UC_b^{1,2}([0,T] \times H;\mathbb{R})$ and that $A \in L(H)$. We fix $t \in [0,T]$ and we apply the Ito formula to the process $v(t-\tau, X_\tau)$, $\tau \in [0,t]$:

$$\begin{aligned} d\,v(t-\tau, X_\tau) &= -\frac{\partial}{\partial t}v(t-\tau, X_\tau)d\tau + \mathcal{L}v(t-\tau, X_\tau)d\tau \\ &\quad + \langle \nabla v(t-\tau, X_\tau), G(X_\tau)dW_\tau\rangle \\ &= \langle \nabla v(t-\tau, X_\tau), G(X_\tau)dW_\tau\rangle, \end{aligned}$$

since $\frac{\partial}{\partial t}u = \mathcal{L}u$. This means

$$v(0, X_t) - v(t, X_0) = \int_0^t \langle \nabla v(t-\tau, X_\tau), G(X_\tau)dW_\tau \rangle.$$

If we take expectation the stochastic integral vanishes and we obtain

$$\mathbb{E}\, v(0, X_t) = \mathbb{E}\, v(t, X_0),$$

and since $v(0,x) = \phi(x)$ and $X_0 = x$ we conclude that $v(t,x) = \mathbb{E}\,\phi(X_t)$, i.e. v is given by the formula (2.4).

We mention that this argument can be refined to remove the additional restriction on v and A, and leads to uniqueness statements for the solution of (2.5): see [14] Theorem 9.17 and [37] Theorem 7.2.

Finally, we do not address the problem of smoothness of $v(t,x) = P_t\phi(x)$ when we only assume $\phi \in \mathcal{B}_b(H;\mathbb{R})$. Generalizations of Theorem 2 can be proved: see [22] and [32].

3. The nonlinear equation

We look for $v : [0,T] \times H \to \mathbb{R}$ solution of

$$\begin{cases} \frac{\partial}{\partial t} v(t,x) = \mathcal{L}v(t,x) + f(x, v(t,x), \nabla v(t,x)), \\ v(0,x) = \phi(x), \qquad t \in [0,T],\ x \in H, \end{cases} \tag{3.1}$$

where $\phi : H \to \mathbb{R}$ and $f : H \times \mathbb{R} \times H \to \mathbb{R}$ are given functions. We may define a solution of this equation in a classical sense and prove generalizations of the statements of the previous section. However, very stringent assumptions were required for the existence of a solution, even in the simpler linear case. In order to relax these assumptions, we introduce a different, weaker concept of solution, already used in the literature.

If (P_t) is a semigroup "generated" (in some sense) by $\mathcal{L}$ on $UC_b(H;\mathbb{R})$ (or another space of real functions on H) then we call v a *mild solution* if for $x \in H$ and $t \in [0,T]$,

$$v(t,x) = P_t\phi(x) + \int_0^t P_{t-s}[f(\cdot, v(s,\cdot), \nabla v(s,\cdot))](x)\, ds. \tag{3.2}$$

Of course, this is motivated by the variation of constants formula.

3.1. Nonlinear perturbations of the operator $\mathcal{L}^0$

Let us recall the operator $\mathcal{L}^0$ introduced in the previous section:

$$\mathcal{L}^0\phi(x) = \frac{1}{2}\mathrm{Trace}\left(GG^*\nabla^2\phi(x)\right) + \langle Ax, \nabla\phi(x)\rangle,$$

and consider the following special instance of equation (3.1):

$$\begin{cases} \dfrac{\partial}{\partial t}v(t,x) &= \mathcal{L}^0 v(t,x) + g(x) + \psi_0(\nabla v(t,x)) + \langle F(x), \nabla v(t,x)\rangle, \\ v(0,x) &= \phi(x), \qquad t \in [0,T],\ x \in H, \end{cases}$$

where $g, \psi_0 : H \to \mathbb{R}$ and $F : H \to H$ are given functions. This equation arises in connection with optimal control of stochastic evolution equations driven by additive noise, and has been deeply studied. We discuss application to control theory in the next section, therefore we postpone references to the literature. Here we just report the following theorem due to Gozzi, [21] Theorem 4.9, as an example of the kind of results that can be proved.

Theorem 4. *Assume $\psi_0 : H \to \mathbb{R}$ is Lipschitz, $g \in UC_b(H;\mathbb{R})$, $\phi \in UC_b(H;\mathbb{R})$, $F \in UC_b(H;H)$ and there exist $C > 0$ and $\alpha < 1$ such that*

$$|Q_t^{-1/2} e^{tA}| \leq C\, t^{-\alpha}, \qquad t \in (0,1]. \tag{3.3}$$

Then there exists a unique mild solution v with following properties:

- *for every $\tau \in (0,T)$, and every compact set $K \subset H$, the maps $v(\cdot,x) : [0,T] \to \mathbb{R}$, $\nabla v(\cdot,x) : [\tau,T] \to H$ are uniformly continuous, uniformly with respect to $x \in K$;*
- *the map $v(t,\cdot) : H \to \mathbb{R}$ is uniformly continuous, uniformly with respect to $t \in [0,T]$;*
- *for every $\tau \in (0,T)$, the map $\nabla v(t,\cdot) : H \to H$ is uniformly continuous, uniformly with respect to $t \in [\tau,T]$;*
- *there exists $c > 0$ such that*

$$|\nabla v(t,x)| \leq c\, t^{-\alpha}, \qquad t \in (0,T],\ x \in H. \tag{3.4}$$

As in the discussion preceeding Theorem 2, the hypothesis (3.3) is a kind of nondegeneracy assumption for the elliptic operator $\mathcal{L}^0$.

3.2. Nonlinear perturbations of the operator $\mathcal{L}$

Now we consider the case of the operator $\mathcal{L}$. We look for $v : [0,T] \times H \to \mathbb{R}$ solution of

$$\begin{cases} \dfrac{\partial}{\partial t}v(t,x) &= \mathcal{L}v(t,x) + \psi(x, v(t,x), G(x)^* \nabla v(t,x)) \\ v(0,x) &= \phi(x), \qquad t \in [0,T],\ x \in H, \end{cases} \tag{3.5}$$

where

$$\mathcal{L}\phi(x) = \frac{1}{2}\mathrm{Trace}\left(G(x)G(x)^* \nabla^2 \phi(x)\right) + \langle Ax + F(x), \nabla \phi(x)\rangle,$$

and $\psi : H \times \mathbb{R} \times H \to \mathbb{R}$, $\phi : H \to \mathbb{R}$. We still assume that Hypothesis 1 holds.

Note the appearance of G in the nonlinear term: the method used below to solve the equation imposes this restriction on the class of equation under consideration.

Theorem 5. *If $\psi \in C^1(H \times \mathbb{R} \times H; \mathbb{R})$ and $\phi \in C^1(H;\mathbb{R})$ have bounded derivatives, then there exists a unique $v \in C^{0,1}([0,T] \times H; \mathbb{R})$ with bounded derivative such that for $x \in H$ and $t \in [0,T]$,*

$$v(t,x) = P_t\phi(x) + \int_0^t P_{t-s}[\psi(\cdot, v(s,\cdot), G(\cdot)^* \nabla v(s,\cdot)](x)\, ds.$$

Remark 6. In this theorem all the derivatives are understood in the sense of Gâteaux, and continuity of the derivatives is understood in the strong sense: for instance in the definition of the space $C_b^1(H;H)$ the continuity requirement on the Gâteaux derivative, which is a function from H to $L(H)$, is to be understood by endowing $L(H)$ of the strong operator topology (it is well known that continuity of the Gâteaux derivative in the sense of the norm operator topology implies Fréchet differentiability). This remark is important in view of applications to stochastic partial differential equation: indeed, typical nonlinear operators in L^2 spaces on a euclidean domain are the so-called Nemytskii operators (also known as evaluation or superposition operators), that are not Fréchet differentiable except in trivial cases.

Theorem 5 is proved in [19]. We refer the reader to this paper for a detailed proof and for various extensions. Here we will just outline the method used for the proof. In analogy with the previous section, it turns out that there exists a formula for the solution v, obtained by means of a stochastic differential system. Let us consider again the process X_τ solution of

$$\begin{aligned} dX_\tau &= AX_\tau\, d\tau + F(X_\tau)\, d\tau + G(X_\tau)\, dW_\tau, \\ X_0 &= x \in H, \quad \tau \geq 0. \end{aligned} \tag{3.6}$$

For fixed $t \in [0,T]$, let us consider the stochastic equation

$$\begin{aligned} dY_\tau &= Z_\tau\, dW_\tau - \psi(X_\tau, Y_\tau, Z_\tau)\, d\tau, \\ Y_t &= \phi(X_t), \quad \tau \in [0,t]. \end{aligned} \tag{3.7}$$

This is called a *backward* equation, since the value of the process (Y_τ) is specified at the final time $\tau = t$. The precise meaning is as follows: a pair of $(\mathcal{F}_\tau)$-adapted processes Y_τ, Z_τ, $\tau \in [0,t]$, is called solution of (3.7) if

$$Y_\tau + \int_\tau^t Z_\sigma\, dW_\sigma = \int_\tau^t \psi(X_\sigma, Y_\sigma, Z_\sigma)\, d\sigma + \phi(X_t), \quad \tau \in [0,t].$$

The equations (3.6)–(3.7) are called a *forward-backward system.* One can prove that there exists a unique pair of adapted, square integrable processes Y_τ, Z_τ, $\tau \in [0,t]$, solving the backward equation. These processes depend on the values of x and t, fixed in advance:

$$Y_\tau = Y_\tau(t,x), \quad Z_\tau = Z_\tau(t,x).$$

We set

$$v(t,x) = Y_0(t,x), \tag{3.8}$$

and we note that $v(t,x)$ is deterministic, since it is $\mathcal{F}_0$-measurable. One can show that v is the required solution.

Backward stochastic equations have been intensively studied in recent years, starting from the paper by E. Pardoux and S. Peng [30], and the connection of forward-backward systems with quasi-linear partial differential equations is now fairly well understood in the finite-dimensional case dim $H < \infty$. We refer the reader to the book by J. Ma and J. Yong [27].

In the infinite-dimensional case (dim $H = \infty$) there are not many papers on stochastic backward equations: we are only aware of [25], [36], [28], [38]. The connection between forward-backward systems and partial differential equations in infinitely many variables was not investigated before. Theorem 5 is close to the results obtained in [31] for dim $H < \infty$, but the extension to the Hilbert space case requires a considerable effort and different techniques.

In order to give an idea of the connection between forward-backward systems and partial differential equations, that may seem rather implicit, we will give a heuristic proof of the fact that if v is a strict solution then it must coincide with $Y_0(t,x)$, according to (3.8).

If v is sufficiently regular then by the Ito formula we have

$$\begin{aligned} d\,v(t-\tau,X_\tau) &= -\frac{\partial}{\partial t}v(t-\tau,X_\tau)d\tau + \mathcal{L}v(t-\tau,X_\tau)d\tau \\ &\quad +\langle \nabla v(t-\tau,X_\tau), G(X_\tau)dW_\tau\rangle \\ &= -\psi(X_\tau, v(t-\tau,X_\tau), G(X_\tau)^*\nabla v(t-\tau,X_\tau))\,d\tau \\ &\quad +\langle G(X_\tau)^*\nabla v(t-\tau,X_\tau), dW_\tau\rangle, \end{aligned}$$

since v solves the equation $\frac{\partial}{\partial t}v = \mathcal{L}v + \psi$. Moreover,

$$v(t-\tau,X_\tau)\Big|_{\tau=t} = v(0,X_t) = \phi(X_t).$$

Setting

$$\widetilde{Y}_\tau = v(t-\tau,X_\tau),\ \widetilde{Z}_\tau = G(X_\tau)^*\nabla v(t-\tau,X_\tau),$$

we note that the pair $\widetilde{Y},\widetilde{Z}$ solves the backward equation

$$\begin{aligned} d\widetilde{Y}_\tau &= \widetilde{Z}_\tau\,dW_\tau - \psi(X_\tau,\widetilde{Y}_\tau,\widetilde{Z}_\tau)\,d\tau, \\ \widetilde{Y}_t &= \phi(X_t), \quad \tau\in[0,t]. \end{aligned}$$

By uniqueness, $Y=\widetilde{Y}, Z=\widetilde{Z}$ and in particular

$$Y_0 = \widetilde{Y}_0 = v(t,X_0) = v(t,x),$$

which shows that v is given by the formula (3.8). The proof is only heuristic, since v does not have the regularity required to apply the Ito formula, in general.

4. Applications to stochastic optimal control

Let us consider the following control system on the interval $[t, T]$:

$$\begin{aligned} dX_\tau &= AX_\tau\, d\tau + F(X_\tau)\, d\tau + G(X_\tau)\, dW_\tau + u(\tau)\, d\tau, \\ X_t &= x \in H, \quad \tau \in [t, T]. \end{aligned} \tag{4.1}$$

Here X_τ can be thought of as the state at time τ of a system that evolves in time and is subject to random perturbation (noise) represented by the term W. The evolution of the system is influenced by a control parameter u, taking values in the Hilbert space H, whose value can be chosen at every time. The aim is to choose an admissible control $u : [t, T] \times \Omega \to H$ that minimizes the cost functional

$$J(t, x) = \mathbb{E} \int_t^T [g(X_\tau) + |u(\tau)|^2]\, d\tau + \mathbb{E}\, \phi(X_T),$$

where $g : H \to \mathbb{R}$, $\phi : H \to \mathbb{R}$ are given, and t, x are the initial data in (4.1). By admissible control u we mean a square summable process adapted to the filtration $(\mathcal{F}_t)$ (compare with (2.3)) and subject to the constraint $|u(\tau)| \le R$, where $R \in (0, \infty)$ is given. The unconstrained case is also important; we will treat both cases simultaneously setting $R = \infty$ in the unconstrained case.

We mention in passing that we are not considering more complicated, yet interesting, models where an unbounded operator appears in front of the term $u(\tau)$ in (4.1). These control problems arise, for instance, in connection with boundary control of partial differential equations.

We define the *value function*

$$v(t, x) = \inf J(t, x),$$

where the infimum is taken over all admissible controls. In many cases v is the solution of the so-called Hamilton-Jacobi-Bellman equation:

$$\begin{cases} -\dfrac{\partial}{\partial t} v(t, x) = \dfrac{1}{2} \operatorname{Trace}\left(G(x)G(x)^* \nabla^2 v(t, x)\right) + \langle Ax + F(x), \nabla v(t, x)\rangle \\ \qquad\qquad\qquad + g(x) + \psi_0(\nabla v(t, x)), \\ v(T, x) = \phi(x), \qquad t \in [0, T],\ x \in H, \end{cases}$$

where

$$\psi_0(p) := \min_{u \in H, |u| \le R} [|u|^2 + \langle p, u\rangle]. \tag{4.2}$$

Notice that ψ_0 is a Lipschitz function if $R < \infty$, whereas it is not (globally) Lipschitz in the unconstrained case $R = \infty$ (indeed, in this case we have $\psi_0(p) = -|p|^2/4$): the latter case is therefore more difficult, since the nonlinear term in the Hamilton-Jacobi-Bellman equation is not Lipschitz with respect to the argument ∇v.

Note that the Hamilton-Jacobi-Bellman equation is backward in time. However, since the change of unknown function

$$\widetilde{v}(t, x) = v(T - t, x), \qquad t \in [0, T],\ x \in H,$$

changes the equation into

$$\begin{cases} \dfrac{\partial}{\partial t}\widetilde{v}(t,x) &= \mathcal{L}\widetilde{v}(t,x) + g(x) + \psi_0(\nabla\widetilde{v}(t,x)), \\ \widetilde{v}(0,x) &= \phi(x), \qquad t \in [0,T],\, x \in H, \end{cases}$$

we can use previous results to solve the Hamilton-Jacobi-Bellman equation.

It must be mentioned here that a different, appropriate notion of solution for this equation is that of viscosity solution, developed by many authors, in particular M. Crandall and P.L. Lions. As a general reference we cite [13]. By now, there is a vast literature on viscosity solutions for second order equations on infinite-dimensional spaces. We limit ourselves to referring the reader to the paper [24], which is close in spirit to the previous approach and is a good source for bibliographical references on this subject.

4.1. Optimal control for systems with additive noise

We first consider the case $G(x) = G$ constant. In this case we are considering control systems (4.1) where the noise does not depend on the state X. These are called *systems with additive noise*. The Hamilton-Jacobi-Bellman equation is

$$\begin{cases} -\dfrac{\partial}{\partial t}v(t,x) &= \mathcal{L}^0 v(t,x) + g(x) + \psi_0(\nabla v(t,x)) + \langle F(x), \nabla v(t,x)\rangle, \\ v(T,x) &= \phi(x), \qquad t \in [0,T],\, x \in H. \end{cases}$$

After changing t into $T-t$ as explained before we arrive at an equation of the form considered in Subsection 3.1.

Among the lots of available results we have chosen to report the following one, due to F. Gozzi, [21] Theorem 5.3, that deals with the constrained case $R < \infty$.

Theorem 7. *Assume $g \in UC_b^1(H;\mathbb{R})$, $\phi \in UC_b(H;\mathbb{R})$, $F \in UC_b^1(H;H)$ and there exist $C > 0$ and $\alpha < 1$ such that*

$$|Q_t^{-1/2} e^{tA}| \le C\, t^{-\alpha}, \qquad t \in (0,1]. \tag{4.3}$$

Let v be the unique mild solution of the Hamilton-Jacobi-Bellman equation given by Theorem 4. Then v coincides with the value function and there exists a unique optimal control u related to the optimal trajectory X by the feedback law

$$u(\tau) = \gamma(\nabla v(\tau, X_\tau)),$$

where $\gamma(p)$ is value of u where the minimum in (4.2) is attained (with $R < \infty$).

There is a vast literature on the control problem we are dealing with; here we report some of the references that are most closely connected with the previous approach. The reader may find additional material in the bibliography of the papers we are going to cite.

In the book [1], Chapter 4, the unconstrained case $R = \infty$ is treated, under the additional assumptions that Trace $GG^* < \infty$, $F = 0$. A basic assumption is that g and ϕ are convex functions. Under some regularity conditions, the authors are able to find a unique (classical) solution of the Hamilton-Jacobi-Bellman equation in a space of continuous functions with polynomial growth and to solve the optimal control problem. The main tools are techniques of convex analysis in connection with nonlinear semigroup theory.

In [3], [4] the authors treat the case where A is self-adjoint and G equals the identity, in the constrained and unconstrained case respectively, under various regularity conditions for F, g, ϕ. They prove existence and uniqueness of a mild solution in spaces of functions that are continuously differentiable with respect to x with a derivative that blows up at $t = 0$ like in (3.4). Notice that the assumption that G equals the identity implies uniform ellipticity for the operator $\mathcal{L}^0$.

In [21], [23], the results of [3], [4] are generalized to general (constant) G under the weaker condition (4.3). The proofs are based on the regularity properties of the semigroup (P_t^0) stated in Theorem 2.

In these papers it is also proved that the optimal control u is related to the corresponding optimal trajectory X by the feedback law $u(\tau) = \gamma(\nabla v(\tau, X_\tau))$.

4.2. Optimal control for systems with noise acting on the control

Let us now consider the control system:

$$\begin{array}{rcl} dX_\tau & = & AX_\tau\, d\tau + F(X_\tau)\, d\tau + G(X_\tau)\, dW_\tau + G(X_\tau) u(\tau)\, d\tau, \\ X_t & = & x \in H, \quad \tau \in [t, T], \end{array} \tag{4.4}$$

with cost functional

$$J(t, x) = \mathbb{E} \int_t^T [g(X_\tau) + |u(\tau)|^2]\, d\tau + \mathbb{E}\, \phi(X_T).$$

The control system (4.4) differs from (4.1) because of the occurrence of the operator $G(X_\tau)$ in front of the term $u(\tau)$: u acts on the system in the same way as the noise W; these systems may be called *systems with noise acting on the control.* The Hamilton-Jacobi-Bellman equation is

$$\left\{ \begin{array}{rcl} -\dfrac{\partial}{\partial t} v(t, x) & = & \mathcal{L} v(t, x) + g(x) + \psi_0(G(x)^* \nabla v(t, x)) \\ v(T, x) & = & \phi(x), \qquad t \in [0, T],\, x \in H, \end{array} \right.$$

where, as before, $\psi_0(p) := \min_{u \in H, |u| \le R} [|u|^2 + \langle p, u \rangle]$. Let $u = \gamma(p)$ denote the value where the minimum is attained.

After changing t into $T - t$ as explained before we arrive at an equation of the form considered in Subsection 3.2. The following result is contained in the paper [19]:

Theorem 8. *Suppose $g \in C^1(H; \mathbb{R})$ and $\phi \in C^1(H; \mathbb{R})$ have bounded derivatives and let v be the mild solution of the Hamilton-Jacobi-Bellman equation given by Theorem 5. Then, for every admissible control u, we have $J(t, x) \ge v(t, x)$ and equality holds if and only if the following feedback law holds:*

$$u(\tau) = \gamma(G(X_\tau)^* \nabla v(\tau, X_\tau)). \tag{4.5}$$

If the closed-loop equation

$$\begin{array}{rcl} dX_\tau & = & AX_\tau\,d\tau + F(X_\tau)\,d\tau + G(X_\tau)\,dW_\tau + G(X_\tau)\gamma(G(X_\tau)^*\nabla v(\tau, X_\tau))\,d\tau, \\ X_t & = & x \in H, \quad \tau \in [t, T], \end{array}$$

has a solution then X is an optimal trajectory and the control u given by (4.5) *is optimal.*

Remark 9. In this theorem the derivatives are understood in the sense of Gâteaux: see Remark 6.

We just ouline the proof, which is based on the following identity, sometimes called the *fundamental relation*:

$$\begin{array}{rcl} J(t,x) & = & v(t,x) + \mathbb{E}\int_t^T [\psi_0(G(X_\tau)^*\nabla v(\tau, X_\tau)) + |u(\tau)|^2 \\ & & +\langle G(X_\tau)^*\nabla v(\tau, X_\tau), u(\tau)\rangle]\,d\tau. \end{array} \tag{4.6}$$

This identity can be proved by an application of the Girsanov theorem. Denoting by $r(\tau)$ the integrand in the right-hand side of (4.6), it follows from the definition of the function ψ_0 that $r(\tau) \geq 0$, so that $J(t,x) \geq v(t,x)$. Equality holds if and only if $r(\tau) = 0$, which is equivalent to the feedback law (4.5).

5. The nonautonomous case

We consider the following equation

$$\left\{ \begin{array}{l} \dfrac{\partial}{\partial t} v(t,x) + \mathcal{L}_t v(t,x) = \psi(t, x, v(t,x), G(t,x)^*\nabla v(t,x)), \\ v(T,x) = \phi(x), \qquad t \in [0,T],\ x \in H, \end{array} \right. \tag{5.1}$$

with unknown function $v : [0,T] \times H \to \mathbb{R}$. Note that the equation is backward in time, with given final condition; however, as in the previous section, changing t into $T - t$ reduces the equation to a usual initial value problem (forward in time). The main difference here is that the operators $\mathcal{L}_t$ depend on time. We assume that $\mathcal{L}_t$ has the following form:

$$\mathcal{L}_t \phi(x) = \frac{1}{2}\mathrm{Trace}\left(G(t,x)G(t,x)^*\nabla^2\phi(x)\right) + \langle Ax + F(t,x), \nabla\phi(x)\rangle,$$

and $\psi : [0,T] \times H \times \mathbb{R} \times H \to \mathbb{R}$, $\phi : H \to \mathbb{R}$ are given functions. The operator G occurs in the nonlinear term as in (3.5).

In the rest of this section we assume the following:

Hypothesis 2.

- *A is the generator of a strongly continuous semigroup of linear operators e^{tA}, $t \geq 0$, in H.*
- *The mapping $F : [0,T] \times H \to H$ is measurable and satisfies, for some constant $L > 0$,*

$$|F(t,x) - F(t,y)| \leq L\,|x - y|, \qquad t \in [0,T],\ x, y \in H.$$

- *G is a strongly measurable mapping $[0,T] \times H \to L(H)$ such that $e^{sA}G(t,x)$ belongs to $L_{HS}(H,H)$ for every $s > 0$, $t \in [0,T]$ and $x \in H$, and*

$$\begin{aligned} &|G(t,x)|_{L(H)} \leq L\,(1+|x|), \\ &|e^{sA}G(t,x)|_{L_{HS}(H,H)} \leq L\, s^{-\gamma}(1+|x|), \\ &|e^{sA}G(t,x) - e^{sA}G(t,y)|_{L_{HS}(H,H)} \leq L\, s^{-\gamma}|x-y|, \end{aligned} \tag{5.2}$$

for some constants $L > 0$ and $\gamma \in [0,1/2)$ and for every $s > 0$, $t \in [0,T]$, $x, y \in H$.
- *For every $s > 0$ and $t \in [0,T]$ the mappings*

$$F(t,\cdot) : H \to H, \qquad e^{sA}G(t,\cdot) : H \to L_{HS}(H,H),$$

are continuous and Gâteaux differentiable with strongly continuous Gâteaux derivative.

For every $t \in [0,T]$, let us consider the stochastic equation on the interval $[t,T]$:

$$\begin{aligned} dX_\tau &= AX_\tau\, d\tau + F(\tau, X_\tau)\, d\tau + G(\tau, X_\tau)\, dW_\tau, \\ X_t &= x \in H, \quad \tau \in [t,T]. \end{aligned} \tag{5.3}$$

Under Hypothesis 2, there exists a unique solution X_τ, $\tau \in [t,T]$, i.e. a unique continuous, $(\mathcal{F}_\tau)$-adapted stochastic process satisfying $\mathbb{E}\,\sup_{\tau\in[0,T]} |X_\tau|^p < \infty$ for every $p \in [1,\infty)$ and such that, $\mathbb{P}$-a.s.,

$$X_\tau = e^{(\tau-t)A}x + \int_t^\tau e^{(\tau-\sigma)A}F(\sigma, X_\sigma)\, d\sigma + \int_t^\tau e^{(\tau-\sigma)A}G(\sigma, X_\sigma)\, dW_\sigma, \quad \tau \in [t,T]. \tag{5.4}$$

The σ-algebras $\mathcal{F}_\tau = \sigma\{W_r \,:\, r \in [0,\tau]\} \vee \mathcal{N}$, are defined as before (compare (2.3)). Sometimes we will denote by $X_\tau(t,x)$ the solution, to stress dependence on the initial data t, x. We denote by $(P_{t\tau})$ the two-parameters family of transition probabilities, i.e. we set, for every $\phi \in \mathcal{B}_b(H)$,

$$P_{t\tau}[\phi](x) = \mathbb{E}\,\phi(X_\tau(t,x)).$$

It is known that, under additional assumptions, the function $v(t,x) = P_{tT}[\phi](x)$ is a classical solution of the (linear) backward Kolmogorov equation

$$\begin{cases} \dfrac{\partial}{\partial t} v(t,x) + \mathcal{L}_t v(t,x) = 0, \\ v(T,x) = \phi(x), \end{cases}$$

see [14] Theorem 9.17 and [37] Theorem 6.9 for precise statements. Somewhat imprecisely, we may say that $(P_{t\tau})$ is "generated" by $(\mathcal{L}_t)$. This motivates the following definition: we say that a continuous function $v : [0,T] \times H \to \mathbb{R}$ is a mild solution of equation (5.1) if it is Gâteaux differentiable with respect to x, its Gâteaux derivative is continuous (in the strong operator topology) and uniformly bounded, and if for every $t \in [0,T]$, $x \in H$, the following equation is satisfied:

$$v(t,x) = -\int_t^T P_{t,\tau}[\psi(\tau,\cdot,v(\tau,\cdot), G(\tau,\cdot)^*\nabla v(\tau,\cdot))](x)\, d\tau + P_{t,T}[\phi](x). \tag{5.5}$$

This formula has to be compared with (3.2).

On the functions ϕ, ψ we will make the following assumptions:

Hypothesis 3.

- $\phi \in C^1(H;\mathbb{R})$ *has bounded derivative;*
- *for every* $t \in [0,T]$, *the function* $\psi(t,\cdot,\cdot,\cdot)$ *belongs to* $C^1(H \times \mathbb{R} \times H;\mathbb{R})$ *and its partial derivatives are uniformly bounded;*
- *the function* $\psi(\cdot,0,0,0)$ *is bounded on* $[0,T]$.

In the statement of this Hypothesis, derivatives are understood in the sense of Gâteaux and their continuity is understood in the strong operator topology, compare Remark 6.

Under Hypotheses 2 and 3 we will be able to show existence and uniqueness of the solution of equation (5.1): see Theorem 10 below. Moreover, it is possible to give a probabilistic representation of the solution in the way we are going to indicate.

Let us consider the backward stochastic equation on the interval $[t,T]$:

$$\begin{aligned} dY_\tau &= Z_\tau\, dW_\tau + \psi(\tau, X_\tau, Y_\tau, Z_\tau)\, d\tau, \\ Y_t &= \phi(X_t), \quad \tau \in [t,T], \end{aligned} \tag{5.6}$$

where (X_τ) is the solution of (5.3). The precise meaning of the equation is as follows: a pair of $(\mathcal{F}_\tau)$-adapted, square integrable processes Y_τ, Z_τ, $\tau \in [t,T]$, is called solution of (5.6) if, $\mathbb{P}$-a.s.,

$$Y_\tau + \int_\tau^t Z_\sigma\, dW_\sigma = -\int_\tau^T \psi(\sigma, X_\sigma, Y_\sigma, Z_\sigma)\, d\sigma + \phi(X_T), \quad \tau \in [t,T]. \tag{5.7}$$

The equations (5.3)–(5.6) form a forward-backward system. Under Hypotheses 2 and 3 one can prove that there exists a unique solution $Y_\tau(t,x), Z_\tau(t,x)$, $\tau \in [t,T]$ (the notation stresses dependence on the values of x and t, fixed in advance). Let us set

$$v(t,x) = Y_t(t,x). \tag{5.8}$$

We note that $v(t,x)$ is $\mathcal{F}_t$-measurable, by adaptedness. On the other hand, it follows from (5.4) and (5.7) that $v(t,x)$ is measurable with respect to the σ-algebra

$$\sigma\{W_r - W_t \ :\ r \in [t,T]\} \vee \mathcal{N}.$$

Since this σ-algebra is independent of $\mathcal{F}_t$, we conclude that $v(t,x)$ is deterministic. It turns out that v is the required solution, as stated in the following theorem, which is the main result of this section and whose proof can be found in [19]. We also refer the reader to this paper for applications to stochastic optimal control as well as for various extensions.

Theorem 10. *Assume that Hypotheses* 2 *and* 3 *hold. Then the function* v *given by the formula* (5.8) *is the unique mild solution of equation* (5.1).

If, in addition, $\sup_{t\in[0,T], x\in H} |\psi(t,x,0,0)| < \infty$ *and* ϕ *is bounded then* u *is also bounded.*

References

[1] V. Barbu, G. Da Prato, **Hamilton-Jacobi equations in Hilbert spaces**, Pitman Research Notes in Mathematics 86, Pitman, 1983.

[2] V.I. Bogachev, M. Röckner, B. Schmuland, *Generalized Mehler semigroups and applications,* Probab. Theory Relat. Fields **105**, 1996, 193–225.

[3] P. Cannarsa, G. Da Prato, *Second-order Hamilton-Jacobi equations in infinite dimensions,* SIAM J. Control and Optimization **29** (2), 1991, 474–492.

[4] P. Cannarsa, G. Da Prato, *Direct solution of a second-order Hamilton-Jacobi equations in Hilbert spaces,* in: **Stochastic partial differential equations and applications**, eds. G. Da Prato, L. Tubaro, 72–85, Pitman Research Notes in Mathematics 268, Pitman, 1992.

[5] S. Cerrai, *A Hille-Yosida theorem for weakly continuous semigroups,* Semigroup Forum **49**, 1994, 349–367.

[6] S. Cerrai, *Weakly continuous semigroups in the space of functions with polynomial growth,* Dynam. Systems Appl. **4** (3), 1995, 351–371.

[7] S. Cerrai, F. Gozzi, *Strong solutions of Cauchy problems associated to weakly continuous semigroups,* Differential and Integral Equations, **8**, 1995, 465–486.

[8] A. Chojnowska-Michalik, B. Gołdys, *Existence, uniqueness and invariant measures for stochastic semilinear equations on Hilbert spaces,* Probab. Theory Relat. Fields **102**, 1995, 331–356.

[9] A. Chojnowska-Michalik, B. Gołdys, *On regularity properties of nonsymmetric Ornstein-Uhlenbeck semigroup in L^p spaces,* Stochastics and Stochastics Rep. **59**, 1996, 183–209.

[10] A. Chojnowska-Michalik, B. Gołdys, *Nonsymmetric Ornstein-Uhlenbeck semigroup as second quantized operator,* J. Math. Kyoto Univ. **36**, 1996, 481–498.

[11] A. Chojnowska-Michalik, B. Gołdys, *On Ornstein-Uhlenbeck generators,* to appear.

[12] A. Chojnowska-Michalik, B. Gołdys, *Symmetric Mehler semigroups in L^p: Littlewood-Paley-Stein inequalities and domains of generators,* to appear.

[13] M.G. Crandall, H. Ishii, P.L. Lions, *User's guide to viscosity solutions of second order partial differential equations,* Bull. (new series) A.M.S. **27** (1), 1992, 1–67.

[14] G. Da Prato, J. Zabczyk, **Stochastic equations in infinite dimensions.** Encyclopedia of Mathematics and its Applications, 44. Cambridge University Press, 1992.

[15] G. Da Prato, J. Zabczyk, **Ergodicity for infinite-dimensional systems.** London Mathematical Society Lecture Notes Series, 229. Cambridge University Press, 1996.

[16] M. Fuhrman, *Analyticity of transition semigroups and closability of bilinear forms in Hilbert spaces,* Studia Mathematica, **115** (1), 1995, 53–71.

[17] M. Fuhrman, *Hypercontractivity properties of nonsymmetric Ornstein-Uhlenbeck semigroups in Hilbert spaces,* Journal of Stochastic Analysis and Applications, **16** (2), 1998, 243–263.

[18] M. Fuhrman, M. Röckner, *Generalized Mehler semigroups: the non gaussian case,* Potential Analysis, **12** (1), 2000, p. 1–47.

[19] M. Fuhrman, G. Tessitore, work in preparation.

[20] B. Gołdys, *On bilinear forms related to Ornstein-Uhlenbeck semigroup on Hilbert space,* to appear.

[21] F. Gozzi, *Regularity of solutions of second order Hamilton-Jacobi equations and application to a control problem,* Comm. Partial Differential Equations, **20**, 1995, 775–826.

[22] F. Gozzi, *Smoothing properties of nonlinear transition semigroups: case of Lipschitz nonlinearities,* to appear.

[23] F. Gozzi, *Global regular solutions of second order Hamilton-Jacobi equations in Hilbert spaces with locally Lipschitz nonlinearities,* J. Math. Anal. Appl. **198**, 1996, 399–443.

[24] F. Gozzi, E. Rouy, A. Święch, *Second order Hamilton-Jacobi equations in Hilbert spaces and stochastic boundary control*, SIAM J. Control Optim. **38** (2), 2000, 400–430.

[25] Y. Hu, S. Peng, *Adapted solution of a backward semilinear stochastic evolution equation,* Stochastic Anal. Appl., **9** (4), 1991, 445–459.

[26] Z.M. Ma, M. Röckner, **Introduction to the theory of (non symmetric) Dirichlet forms**, Springer, 1992.

[27] J. Ma, J. Yong, **Forward-backward stochastic differential equations and their applications**, Lecture Notes in Mathematics 1702, Springer, 1999.

[28] J. Ma, J. Yong *Adapted solution of a degenerate backward SPDE with applications,* Stoch. Proc. Appl. **70**, 1997, 59–84.

[29] D. Nualart, **The Malliavin calculus and related topics.** Probability and its applications. Springer-Verlag, 1995.

[30] E. Pardoux, S. Peng, *Adapted solution of a backward stochastic differential equation,* Systems and Control Lett. **14**, 1990, 55–61.

[31] E. Pardoux, S. Peng, *Backward stochastic differential equations and quasilinear parabolic partial differential equations*, in: **Stochastic partial differential equations and their applications**, eds. B.L. Rozowskii, R.B. Sowers, 200–217, Lecture Notes in Control Inf. Sci. 176, Springer, 1992.

[32] S. Peszat, J. Zabczyk, *Strong Feller property and irreducibility for diffusions on Hilbert spaces,* Ann. of Probability, **23**, 1995, 157–172.

[33] E. Priola, *The Cauchy problem for a class of Markov-type semigroups,* to appear on Comm. in Applied Anal.

[34] E. Priola, *Partial differential equations with infinitely many variables,* PhD Thesis, University of Milan, 1999.

[35] M. Röckner, *L^p-analysis of finite and infinite-dimensional diffusion operators*, in: **Stochastic PDE's and Kolmogorov Equations in Infinite Dimensions**, ed. G. Da Prato, 65–116. Lecture Notes in Mathematics 1715, Springer, 1999.

[36] G. Tessitore, *Existence, Uniqueness and space regularity of the adapted solutions of a backward SPDE,* Stochastic Anal. Appl., **14** (4), 1996, 461–486.

[37] J. Zabczyk, *Parabolic equations on Hilbert spaces*, in: **Stochastic PDE's and Kolmogorov Equations in Infinite Dimensions**, ed. G. Da Prato, 117–213. Lecture Notes in Mathematics 1715, Springer, 1999.

[38] E. Pardoux, A. Rascanu, *Backward stochastic variational inequalities*, Stochastics Rep. **67** (3–4), 1999, 159–167.

Marco Fuhrman
Dipartimento di Matematica
Politecnico di Milano
piazza Leonardo da Vinci 32
I-20133 Milano, Italy
e-mail: `marco.fuhrman@polimi.it`

Progress in Nonlinear Differential Equations
and Their Applications, Vol. 50, 155–178

Uniform Attractors of Nonautonomous Dynamical Systems with Memory

Maurizio Grasselli and Vittorino Pata

Dedicated to the memory of Brunello Terreni
"...ben tetragono ai colpi di ventura" (Par., XVII, 24)

1. Introduction

The study of nonlinear dynamical systems is of basic importance in the understanding of several natural phenomena. If a certain mathematical model is described by a nonlinear dynamical system, then it is usually difficult to predict whether or not the system will evolve towards a stationary state or it will exhibit a chaotic behavior. The sensibility to the initial conditions and to the parameters characterizing the nonlinear system show that a correct approach to study its dynamics should be more geometric. This means that the evolution system must be treated as an ordinary differential equation whose solutions (trajectories) can be viewed as curves in a suitable phase space, possibly of infinite dimension. We recall, for instance, that the chaotic behavior of some nonlinear system can be explained by the existence of a so-called *strange attractor*; that is, a set, usually of finite fractal dimension, which *attracts uniformly* any trajectory. Actually, this means that in the long run the dynamics of an infinite-dimensional system is controlled by a finite number of parameters and this can be of some help for possible numerical approximations.

In many natural phenomena various kinds of dissipation are present (e.g., viscosity, friction, heat loss). This fact commonly characterizes the so-called dissipative dynamical systems. From the mathematical viewpoint, a dynamical system may be called dissipative if there exists an absorbing set; that is, a (bounded) set in the phase space which attracts any trajectory without being necessarily compact or of finite fractal dimension. A further nice feature of a dissipative dynamical system is the existence of an attractor; that is, the minimal (closed) set that attracts uniformly any bounded set in the phase space. In general, one can prove that the compactness of the attractor (universal attractor) and, sometimes, that its fractal dimension is finite. These results give information about the asymptotic global stability of the dynamical system.

Here we concentrate ourselves on a special class of dissipative infinite-dimensional dynamical system which has been recently investigated; namely, evolution systems with memory subject to time dependent external forces. These models arise in the description of several phenomena like, e.g., heat conduction in special materials, viscoelasticity, phase transitions. More precisely, in the next section, besides the basic notation and terminology, we introduce two evolution equations which are the prototypes for our following considerations. Indeed, in the same section, we show how to deal with the past histories that characterize the presence of memory effects in order to obtain a dynamical system. In particular, we formulate an additional equation which rules the evolution of the past history. This equation is analyzed in detail in Section 3; while Section 4 is devoted to the equivalence between the usual formulation and the one which accounts for the past history evolution. Some basic notions and results on the nonautonomous dissipative dynamical systems are contained in Section 5. There we also discuss the main ideas needed to prove the existence of a compact attractor for a nonlinear nonautonomous dynamical system with memory. Finally, in Section 6 we show how the presented results apply to the models previously introduced.

We conclude this brief introduction quoting S. Antman's words (see [10], references are omitted)

Nonlinear problems of continuum physics with various kinds of dissipation have only recently been explored. During the last 50 *years, there has developed a rich theory, including semigroup theory, which treats an evolution partial differential equation as an ordinary differential equation in a function space. An exciting offshoot of this theory is the theory of infinite-dimensional dynamical systems, which brings many of the geometrical concepts from ordinary differential equations to bear on partial differential equations. But the applications of these methods to nonlinear problems of continuum physics, ..., is in its infancy.*

2. Differential systems with memory

Let $\mathcal{H}$ be a Hilbert space (or, more generally, a Banach space). We consider a differential problem of the form

$$\begin{cases} u_t = A(u(t), t) \\ u(\tau) = u_0 \in \mathcal{H} \end{cases} \tag{2.1}$$

where $A(\cdot,\cdot) : \mathcal{D} \times \mathbb{R} \to \mathcal{H}$, $\mathcal{D}$ dense in $\mathcal{H}$. For every fixed time t, $A(\cdot, t)$ is a (nonlinear) operator of domain $\mathcal{D}$.

In many cases A depends on a *functional symbol* f, which, typically, represents an external source. As an example, one may think at the semilinear reaction-diffusion equation with external source, so that

$$A(u, t) = \Delta u + g(u) + f(t).$$

We assume to have a well-posedness result; that is, given any $u_0 \in \mathcal{H}$ at any initial time $\tau \in \mathbb{R}$, there exists a unique solution $u(t)$ at any time $t > \tau$, and $u \in C([\tau, \infty), \mathcal{H})$. Moreover, for any, $t \geq \tau$, if $u_{0n} \to u_0$ in $\mathcal{H}$, then $u_n(t) \to u(t)$ in $\mathcal{H}$.

In this case it is possible to describe the solutions to (2.1) by means of a two-parameter family of continuous operators on $\mathcal{H}$, which we denote by $U_f(t,\tau)$ to highlight the dependence on f. Namely, we denote the solution at time t to (2.1) with initial datum u_0 given at time τ, by $U_f(t,\tau)u_0$.

The family $U_f(t,\tau)$, with $\tau \in \mathbb{R}$, $t \geq \tau$, fulfills the following properties:

1. $U_f(t,\tau) : \mathcal{H} \to \mathcal{H}$ for all $t \geq \tau$, $\tau \in \mathbb{R}$;
2. $U_f(\tau,\tau)$ is the identity on $\mathcal{H}$ for all $\tau \in \mathbb{R}$;
3. $U_f(t,s)U_f(s,\tau) = U_f(t,\tau)$ for all $t \geq s \geq \tau$, $\tau \in \mathbb{R}$;
4. $U_f(\cdot,\tau)z \in C([\tau,\infty),\mathcal{H})$ for all $z \in \mathcal{H}$, $\tau \in \mathbb{R}$;
5. $U_f(t,\tau) \in C(\mathcal{H},\mathcal{H})$ for all $\tau \in \mathbb{R}$, $t \geq \tau$.

Such an object is called a *strongly continuous process* of continuous (nonlinear) operators on the phase space $\mathcal{H}$, according to Haraux's definition [24]. If the system is autonomous (i.e., A is time-independent), then the evolution depends solely on the difference $t-\tau$; in particular, $U_f(t,\tau) = U_f(t-\tau,0)$. In that case, there is no loss of generality in assuming $\tau = 0$, and $U_f(t,0)$ turns out to be a strongly continuous semigroup.

We are focused on the study of differential equations in presence of convolution terms, which, from the physical viewpoint, describe those models where the dynamics is influenced by the "past history" of the system. These models are of a certain interest for instance, in studying high viscosity liquids at low temperatures and the thermomechanical behavior of polymers (see, e.g., [11, 22, 25, 32] and references therein).

To see a simple, albeit fairly significant example, we consider the constitutive equations for a homogeneous and isotropic heat conductor that occupies a three-dimensional domain. Supposing small variations of the absolute temperature and temperature gradient from equilibrium reference values, we have

$$\begin{aligned} e &= e_0 + c\vartheta \\ \mathbf{q} &= -\omega\nabla\vartheta \end{aligned}$$

where e_0, c and ω are given positive constants and ∇ is the spatial gradient. Here we recall that e is the internal energy, ϑ is the temperature, and $\mathbf{q}$ is the heat flux. These relationships, through the energy balance

$$e_t + \nabla \cdot \mathbf{q} = r \tag{2.2}$$

entail the well-known heat equation. Here $\nabla\cdot$ is the spatial divergence and $r : \Omega \times \mathbb{R} \to \mathbb{R}$ is a given heat supply. On the other hand, there are materials for which the heat conduction is better described by the following constitutive equation

$$\mathbf{q}(t) = -\omega\nabla\vartheta - \int_0^\infty k(s)\nabla\vartheta(t-s)ds \tag{2.3}$$

where k is a suitable memory or relaxation kernel and ω may be either positive (see [8]) or null (see [21]). This yields the heat equation with memory

$$c\vartheta_t - \omega\Delta\vartheta - \int_0^\infty k(s)\Delta\vartheta(t-s)ds = r \tag{2.4}$$

in $\Omega \times \mathbb{R}$, where Δ is the Laplacian with respect to the spatial variables.

A further basic example refers to the viscoelasticity of Boltzmann type (see, e.g., [11, 32]). If $u(x,t)$ denotes the (transverse) displacement of a viscoelastic filament with unit density occupying a finite interval $[0, L]$, then, according to the Boltzmann superposition principle, the stress-strain relationship that characterizes viscoelasticity is

$$\sigma(u)(t) = \int_0^\infty \psi(s)u_{xt}(t-s)ds$$

being ψ the relaxation kernel which accounts for the viscoelastic behavior. Provided that ψ is smooth enough and the past history of the strain u_x satisfies reasonable physical assumptions, the former constitutive law reduces to

$$\sigma(u)(t) = \psi(0)u_x(t) + \int_0^\infty \psi'(s)u_x(t-s)ds$$

where $\psi(0)$ is positive. One can also suppose that an internal friction may be present (cf. [16] and references therein); that is,

$$\sigma(u)(t) = a_0u_{xt}(t) + \psi(0)u_x(t) + \int_0^\infty \psi'(s)u_x(t-s)ds$$

where a_0 is a nonnegative constant. Consequently, combining the stress-strain law with the motion equation and assuming that the surrounding medium also exhibits a damping effect

$$u_{tt} + b_0u_t + \sigma_x = f$$

where b_0 is a nonnegative constant and $f : [0, L] \times \mathbb{R} \to \mathbb{R}$ is a given force acting on the filament. Summing up, u is ruled by the evolution equation

$$u_{tt} + b_0u_t - a_0u_{xxt} - \psi(0)u_{xx} - \int_0^\infty \psi'(s)u_{xx}(t-s)ds = f \tag{2.5}$$

in $(0, L) \times \mathbb{R}$.

The setting of concrete cases in the framework of processes, gives rise to some problems. In particular, the approach to parabolic or hyperbolic systems is quite different. Anyway, a first difficulty shared by all models is represented by the memory terms.

To be more specific, let H, V be real Hilbert spaces, with continuous and dense embeddings

$$V \hookrightarrow H \equiv H^* \hookrightarrow V^*$$

(V^* being the dual space of V). Let A_0 and let A_1 be (nonlinear) operators from V to V^* and B be a linear bounded operator from V to V^* which satisfies the coercivity hypothesis

$$\langle Bu, u\rangle \geq \omega \|u\|_V^2 \qquad \forall\, u \in V,\ \omega > 0 \tag{2.6}$$

where $\langle \cdot, \cdot \rangle$ denotes the duality product between V^* and V. Finally, let $f \in L^1_{\text{loc}}(\mathbb{R}, H)$. Having in mind equations (2.4) and (2.5), we concentrate ourselves on the following two prototypes.

E1. First order equation

$$\begin{aligned} &u_t(t) + A_0(u(t)) - \int_0^\infty a(s) B u(t-s)\, ds = f(t) \qquad t > \tau \\ &u(t) = u_0(t) \qquad t \leq \tau. \end{aligned}$$

E2. Second order equation

$$\begin{aligned} &u_{tt}(t) + A_0(u(t)) + A_1(u_t(t)) - \int_0^\infty b'(s) B u(t-s)\, ds = f(t) \qquad t > \tau \\ &u(t) = u_0(t) \qquad t \leq \tau \\ &u_t(\tau) = u_1. \end{aligned}$$

We assume that, under suitable hypotheses on the kernels a and b, and once u is known up to the initial time τ, the two Cauchy problems here above have solutions $u \in C([\tau, \infty), H)$ and $u \in C([\tau, \infty), V) \cap C^1([\tau, \infty), H)$ for any initial data $u(\tau) \in H$ and $(u(\tau), u_t(\tau)) \in V \times H$, respectively.

Our aim is to describe these equations in terms of processes, or, rather, in terms of families of processes depending on f. Clearly, we have to prescribe the initial data at the initial time τ. Nonetheless the initial datum alone is not enough; indeed, the *past history* of the variable u has to be known, up to time τ, since it enters in the convolution integral.

We are interested in evaluating the solutions corresponding to *different* initial times. More precisely, we want to build a machinery that associates with an initial datum given at *any* initial time τ the solution of the equation at time $t > \tau$. Thus, for instance, if the initial datum is given at time $\tau_1 > \tau$, then we need to know also the values of u from τ to τ_1.

A frequently used idea to bypass this problem is to redefine $f(t)$ as

$$f(t) + \int_{t-\tau}^\infty a(s) B u(t-s)\, ds \qquad \text{or} \qquad f(t) + \int_{t-\tau}^\infty b'(s) B u(t-s)\, ds$$

hence including the memory terms for past times into the source term. This approach is applicable if one restricts the analysis on a privileged initial time (typically $\tau = 0$). Indeed, the main problem is that now the initial time τ appears explicitly within the equation.

The only way to associate a process with such equations is then to view the past history of u as a new variable of the system, which will be ruled by a supplementary equation. As far as we know, this idea goes back to the seventies, and it was first introduced by Dafermos [9] for linear viscoelasticity.

We have to use a different approach, depending whether we are considering first or second order equations, respectively.

E1. We introduce the auxiliary variable

$$\eta^t(s) = \int_0^s u(t-y)\,dy = \int_{t-s}^t u(y)\,dy \qquad s \geq 0,\ t > \tau. \tag{2.7}$$

Suppose now that the relaxation kernel a is positive, nonincreasing, and summable along with its first derivative. Then a formal integration by part yields

$$\int_0^\infty a(s)Bu(t-s)\,ds = \int_0^\infty \mu(s)B\eta^t(s)\,ds$$

having set $\mu(s) = -a'(s)$. Hence the original equation turns into

$$u_t(t) + A_0(u(t)) - \int_0^\infty \mu(s)B\eta^t(s)\,ds = f(t). \tag{2.8}$$

Differentiating equality (2.7), we obtain the first-order linear equation

$$\eta_t^t(s) = -\eta_s^t(s) + u(t) \tag{2.9}$$

which is to be regarded as a supplementary equation to be added to (2.8). Also, from (2.7) we have

$$\eta^t(0) = 0 \qquad \forall\, t > \tau. \tag{2.10}$$

Hence we translated the original problem into the system (2.8)–(2.9), together with the boundary condition (2.10). Concerning initial conditions, they are to be rewritten as

$$\begin{cases} u(\tau) = u_0 \\ \eta^\tau(s) = \eta_0(s) \end{cases}$$

where we set

$$u_0 = u_0(\tau)$$

and

$$\eta_0(s) = \int_0^s u_0(\tau - y)\,dy. \tag{2.11}$$

E2. In this case, the auxiliary variable to be considered is

$$\eta^t(s) = u(t) - u(t-s). \tag{2.12}$$

Assuming now that the relaxation kernel b is positive, nonincreasing, and b' is summable, adding and subtracting to equation the term $(b(0) - b(\infty))Bu(t)$, we obtain

$$u_{tt}(t) + A_0(u(t)) + (b(0) - b(\infty))Bu(t) + A_1(u_t(t)) - \int_0^\infty \mu(s)B\eta^t(s)\,ds = f(t) \tag{2.13}$$

where $\mu(s) = -b'(s)$. Differentiation of (2.12) gives the first order linear equation

$$\eta_t^t(s) = -\eta_s^t(s) + u_t(t). \tag{2.14}$$

Again, this is as a supplementary equation to be added to (2.13). So that the original problem turns into the system (2.13)–(2.14), together with the boundary condition (2.10), which follows from (2.12). Concerning initial conditions, we now have

$$\begin{cases} u(\tau) = u_0 \\ u_t(\tau) = u_1 \\ \eta^\tau(s) = \eta_0(s) \end{cases}$$

where we set

$$u_0 = u_0(\tau)$$

and

$$\eta_0(s) = u_0 - u_0(\tau - s). \tag{2.15}$$

When $\mu \equiv 0$, the phase spaces for **E1** and **E2** are H and $V \times H$, respectively. If we require a certain regularity for the memory kernel, for instance, $\mu \in C^1(\mathbb{R}^+) \cap L^1(\mathbb{R}^+)$, we can introduce the Hilbert space

$$\mathcal{M} := L^2_\mu(\mathbb{R}^+, V) = \left\{ \xi : \mathbb{R}^+ \to V : \int_0^\infty \mu(s) \|\xi(s)\|_V^2 < \infty \right\}$$

endowed with the inner product

$$\langle \xi_1, \xi_2 \rangle_{\mathcal{M}} = \int_0^\infty \mu(s) \langle \xi_1(s), \xi_2(s) \rangle_V \, ds.$$

By force of the coercivity hypothesis (2.6), an equivalent inner product is given by

$$\langle \xi_1, \xi_2 \rangle_{\mathcal{M}} = \int_0^\infty \mu(s) \langle B\xi_1(s), \xi_2(s) \rangle \, ds.$$

This is the natural space where to settle the variable η. Indeed, reasonable multiplications in order to find energy estimates for (2.8)–(2.9) and for (2.13)–(2.14), are the duality products between (2.8) and u, and between (2.13) and u_t, respectively, yielding the terms

$$\int_0^\infty \mu(s) \langle B\eta^t(s), u(t) \rangle \, ds \qquad \text{and} \qquad \int_0^\infty \mu(s) \langle B\eta^t(s), u_t(t) \rangle \, ds$$

which cancel each other out when we add the (formal) products in $\mathcal{M}$ of (2.9) and η, and of (2.14) and η, respectively.

We also mention that in some models it might also appear another convolution term, which physically arises when assuming a dependence of the internal energy on the past history of the variable, that is, a term of the type

$$\int_0^\infty \nu(s) \eta^t(s) \, ds$$

to be added to the right-hand sides of (2.8) and (2.13). In this case the proper setting for η is the Hilbert space

$$L^2_\nu(\mathbb{R}^+, H) \cap L^2_\mu(\mathbb{R}^+, V).$$

This happens, for instance, in the theory of heat conduction with memory when the internal energy e may also depend on the past history of the temperature (see, e.g., [12] and references therein).

Our aim is then to write systems (2.8)–(2.9) and (2.13)–(2.14) in the form (2.1), with phase space $\mathcal{H} = H \times \mathcal{M}$ or $\mathcal{H} = V \times H \times \mathcal{M}$, respectively. In order to do that, we need to understand better the role of the supplementary equations (2.9) and (2.14), and, particularly, how to include the boundary condition (2.10) into the abstract formulation.

3. The role of the supplementary equation

Assume $\mu \in C^1(\mathbb{R}^+) \cap L^1(\mathbb{R}^+)$, $\mu(s) \geq 0$ and $\mu'(s) \leq 0$, for every $s \in \mathbb{R}^+$. Consider the differential equation on the time interval $[0, t_0]$

$$\eta_t^t(s) = -\eta_s^t(s) + \vartheta(t) \tag{3.1}$$

where $\vartheta \in L^1([0, t_0], V)$.

Our aim is to define in some way a solution $\eta \in C([0, t_0], \mathcal{M})$ of the above equation, endowed with the initial condition

$$\eta^0 = \eta_0 \in \mathcal{M}.$$

We also have to take into account the boundary condition $\eta^t(0) = 0$ for all $t \in [0, t_0]$. Let then T be the linear operator of domain

$$\mathcal{D}(T) = \Big\{\eta \in \mathcal{M} : \eta' \in \mathcal{M},\ \eta(0) = 0\Big\}.$$

acting as

$$T\eta = -\eta'.$$

Here *prime* stands for the distributional derivative with respect to the variable s. Notice that, if $\eta \in \mathcal{D}(T)$, then

$$\frac{\eta(\cdot + \sigma) - \eta(\cdot)}{\sigma} \longrightarrow \eta'(\cdot) \quad \text{in } \mathcal{M} \quad \text{as } \sigma \to 0.$$

We therefore interpret (3.1) as the ordinary differential equation on $\mathcal{M}$

$$\frac{d}{dt}\eta^t = T\eta^t + \vartheta(t).$$

Theorem 3.1. *The operator T is the infinitesimal generator of a strongly continuous semigroup $S(t)$ acting on $\mathcal{M}$.*

Proof. First we show that T is dissipative, i.e.,

$$\langle T\eta, \eta\rangle_{\mathcal{M}} \leq 0, \qquad \forall \eta \in \mathcal{D}(T). \tag{3.2}$$

Indeed, for $\eta \in \mathcal{D}(T)$, integration by parts gives

$$\begin{aligned}\langle T\eta, \eta\rangle_{\mathcal{M}} &= -\frac{1}{2}\int_0^\infty \mu(s)\frac{d}{ds}\|\eta(s)\|_V^2\, ds \qquad (3.3)\\ &= \lim_{y\downarrow 0}\frac{1}{2}\Big(- \mu(1/y)\|\eta(1/y)\|_V^2 + \mu(y)\|\eta(y)\|_V^2 + \int_y^{1/y} \mu'(s)\|\eta(s)\|_V^2\, ds\Big).\end{aligned}$$

Notice that

$$\begin{aligned}\lim_{y\downarrow 0}\mu(y)\|\eta(y)\|_V^2 &\le \limsup_{y\downarrow 0}\mu(y)\Big(\int_0^y \|\eta'(z)\|_V\,dz\Big)^2\\ &\le \limsup_{y\downarrow 0}\Big(\int_0^y \mu(z)^{1/2}\|\eta'(z)\|_V\,dz\Big)^2\\ &\le \limsup_{y\downarrow 0} y\int_0^y \mu(z)\|\eta'(z)\|_V^2\,dz = 0.\end{aligned}$$

But the left-hand side of (3.3) is bounded, and the remaining two terms of the right-hand side are negative. Hence conclude that both the integral and the limit exist and are finite. In particular, this forces the limit to equal zero. We thus obtain

$$\langle T\eta,\eta\rangle_{\mathcal{M}} = \int_0^\infty \mu'(s)\|\eta(s)\|_V^2\,ds \le 0.$$

Next, we show that

$$\operatorname{range}(\mathbb{I}-T) = \mathcal{M}. \tag{3.4}$$

Take $\hat\eta \in \mathcal{M}$, and consider the equation

$$(\mathbb{I}-T)\eta = \hat\eta.$$

An immediate integration entails

$$\eta(s) = \int_0^s e^{-(s-y)}\hat\eta(y)\,dy.$$

Clearly, $\eta(0)=0$, and by comparison, $\eta' \in \mathcal{M}$ if and only if $\eta \in \mathcal{M}$. We are thus left to show that $\eta \in \mathcal{M}$. Indeed, for $\tilde\eta \in \mathcal{M}$ with $\|\tilde\eta\|_{\mathcal{M}} = 1$, we have

$$\begin{aligned}&\Big|\int_0^\infty \mu(s)\int_0^s e^{-(s-y)}\langle\hat\eta(y),\tilde\eta(s)\rangle_V\,dy\,ds\Big|\\ &\le \int_0^\infty \mu(s)^{1/2}\|\tilde\eta(s)\|_V\int_0^s e^{-(s-y)}\mu(y)^{1/2}\|\hat\eta(y)\|_V\,dy\,ds\\ &\le \|\hat\eta\|_{\mathcal{M}}.\end{aligned}$$

Here we applied a well-known result concerning convolutions of functions on $\mathbb{R}^+$; namely, if $f,\ g \in L^2(\mathbb{R}^+)$ and $h \in L^1(\mathbb{R}^+)$, then

$$\big|\langle f, g*h\rangle\big| \le \|f\|_2\|g\|_2\|h\|_1.$$

On account of (3.2) and (3.4), by the Lumer-Phillips theorem (see [31], Theorem 4.3), T is the infinitesimal generator of a strongly continuous semigroup of contractions $S(t)$ on $\mathcal{M}$. □

The next step is to identify the semigroup $S(t)$. Introduce to this purpose the *right-translation* strongly continuous semigroup $U(t)$ on $\mathcal{M}$ defined by

$$(U(t)\eta)(s) = \begin{cases}\eta(s-t) & s \ge t\\ 0 & s < t\end{cases}$$

and denote by T' its infinitesimal generator. If $\eta \in \mathcal{D}(T)$, then

$$\lim_{t\downarrow 0} \frac{(U(t)\eta)(s)-\eta(s)}{t} = \lim_{t\downarrow 0} \begin{cases} \dfrac{\eta(s-t)-\eta(s)}{t} & s \geq t \\ -\dfrac{\eta(s)}{t} & s < t. \end{cases} \tag{3.5}$$

Observe that

$$\begin{aligned} \lim_{t\downarrow 0} \frac{1}{t^2} \int_0^t \mu(s)\|\eta(s)\|_V^2 \, ds &\leq \limsup_{t\downarrow 0} \frac{1}{t^2} \int_0^t \mu(s) \Big(\int_0^s \|\eta'(y)\|_V \, dy \Big)^2 ds \\ &\leq \limsup_{t\downarrow 0} \frac{1}{t^2} \int_0^t s\mu(s) \int_0^s \|\eta'(y)\|_V^2 \, dy \, ds \\ &= \limsup_{t\downarrow 0} \frac{1}{t^2} \int_0^t \|\eta'(y)\|_V^2 \int_y^t s\mu(s) \, ds \, dy \\ &\leq \frac{1}{2} \limsup_{t\downarrow 0} \int_0^t \mu(y)\|\eta'(y)\|_V^2 \, dy \;=\; 0. \end{aligned}$$

We conclude that the limit (3.5) equals $-\eta'$, that is, $T'_{|\mathcal{D}(T)} \equiv T$. This yields the equality $S(t) \equiv U(t)$. Hence, if $\eta_0 \in \mathcal{D}(T)$, and ϑ is, e.g., differentiable almost everywhere with $\vartheta' \in L^1([0,t_0],\mathcal{M})$, the function

$$\eta^t(s) = \begin{cases} \eta_0(s-t) + \displaystyle\int_0^t \vartheta(y)\, dy & s \geq t \\ \displaystyle\int_{t-s}^t \vartheta(t)\, dy & s < t. \end{cases} \tag{3.6}$$

belongs to $C([0,t_0],\mathcal{M})$ and fulfills equation (3.1) in the strong sense, with initial datum η_0 (cf. [31], Corollary 2.10). Nonetheless, the above expressions make sense for any given $\eta_0 \in \mathcal{M}$ and $\vartheta \in L^1([0,t_0],\mathcal{M})$, and the resulting η still belongs to $C([0,t_0],\mathcal{M})$. In this case η is said to be a *mild* solution to (3.1).

It is important to point out that $\partial_t\eta$ and $\partial_s\eta$ have a very poor regularity. On the other hand, we read from (3.1) that the sum $\partial_t\eta + \partial_s\eta$ have the same regularity of ϑ (thought as a vector of $\mathcal{M}$ constant in s). Hence it makes sense to define *variational* solution to (3.1) a function η such that

$$\langle \partial_t\eta + \partial_s\eta, \varphi\rangle_{\mathcal{M}} = \langle \vartheta, \varphi\rangle_{\mathcal{M}} \tag{3.7}$$

for every $\varphi \in \mathcal{M}$. Since we read $-\partial_s$ as the operator T (in order to include the information on the boundary condition for η), the left-hand side of the above equation has to be interpreted as the limit as $\eta_n \to \eta$, $\eta_n \in \mathcal{D}(T)$, of the sum $\partial_t\eta_n + \partial_s\eta_n$, where η_n is the solution to (3.8) with initial datum η_{0n} which tends to η_0. It is apparent that the mild solution and the variational solution to (3.1) coincide.

One might object that equation (3.7) is of little use, since all the information about η is already contained in the representation formula (3.6). Yet, even if redundant, (3.7) is crucial for finding energy estimates in a regularization scheme (recall by Section 2 that (3.7) appears within a system of equations). In particular, the term $\langle \vartheta, \varphi \rangle_{\mathcal{M}}$, with a shrewd choice of the test function φ, provides a crucial cancellation in the other equation (see for instance the Faedo-Galerkin procedure used in [12, 13, 14]).

4. Back to the original equation

As we saw in the previous sections, in order to settle **E1** and **E2** in the framework of dynamical systems, we actually study modified equations. Of course one might ask whether this procedure is consistent; namely, whether there is a link between **E1**, **E2** and (2.8)–(2.9), (2.13)–(2.14), respectively. Indeed, it turns out that they are the same thing, or, to be more precise, the modified equations are in fact a generalization of the original equations. To see that, let for simplicity $\tau = 0$. Assume first that u is a variational solution to **E1**, with $u(t) = u_0(t)$ for $t \leq 0$. Then, for every $w \in V$ and every $t > 0$, there holds

$$\langle u_t(t), w\rangle + \langle A_0(u(t)), w\rangle - \int_0^t a(s)\langle Bu(t-s), w\rangle \, ds = \langle \tilde{f}(t), w\rangle \tag{4.1}$$

where

$$\tilde{f}(t) = f(t) + \int_t^\infty a(s)\langle Bu_0(t-s), w\rangle \, ds. \tag{4.2}$$

If (u, η) is a solution to (2.8)–(2.9), with $(u(0), \eta^0) = (u_0, \eta_0)$, η_0 being given by (2.11), for every $w \in V$ and every $t > 0$, we have

$$\langle u_t(t), w\rangle + \langle A_0(u(t)), w\rangle - \int_0^\infty \mu(s)\langle B\eta^t(s), w\rangle \, ds = \langle f(t), w\rangle.$$

Let us suppose that $\eta_0 \in \mathcal{D}(T)$. Then, using the representation formula (3.6) and integration by parts, we get

$$\begin{aligned}
&- \int_0^\infty \mu(s)\langle B\eta^t(s), w\rangle \, ds \\
&= - \int_0^t \mu(s)\langle B\int_{t-s}^t u(y)\, dy, w\rangle \, ds - \int_t^\infty \mu(s)\langle B\int_0^t u(y)\, dy, w\rangle \, ds \\
&\quad - \int_t^\infty \mu(s)\langle B\eta_0(s-t), w\rangle \, ds \\
&= a(t)\langle B\int_0^t u(y)\, dy, w\rangle - \int_0^t a(s)\langle Bu(t-s), w\rangle \, ds \\
&\quad -a(t)\langle B\int_0^t u(y)\, dy, w\rangle + \int_t^\infty a(s)\langle B\eta_0'(s-t), w\rangle \, ds \\
&= - \int_0^t a(s)\langle Bu(t-s), w\rangle \, ds + \int_t^\infty a(s)\langle Bu_0(t-s), w\rangle \, ds.
\end{aligned}$$

Thus, on account of (4.2), we recover (4.1).

Arguing similarly, if u is a variational solution to **E2**, with $u(t) = u_0(t)$ for any $t \leq 0$ and $u_t(0) = u_1$. Then, for every $w \in V$, and every $t > 0$, there holds

$$\langle u_{tt}(t), w\rangle + \langle A_0(u(t)), w\rangle + \langle A_1(u_t(t)), w\rangle - \int_0^t b'(s)\langle Bu(t-s), w\rangle\, ds = \langle \tilde{f}(t), w\rangle \tag{4.3}$$

where

$$\tilde{f}(t) = f(t) + \int_t^\infty b'(s)\langle Bu_0(t-s), w\rangle\, ds.$$

If (u,η) is a solution to (2.8)–(2.9), with $(u(0), u_t(0), \eta^0) = (u_0, u_1, \eta_0)$, η_0 being given by (2.15), for every $w \in V$ and every $t > 0$, we have

$$\begin{aligned}&\langle u_{tt}(t), w\rangle + \langle A_0(u(t)), w\rangle + (b(0) - b(\infty))\langle Bu(t), w\rangle + \langle A_1(u_t(t)), w\rangle \\ &- \int_0^\infty \mu(s)\langle B\eta^t(s), w\rangle\, ds = \langle f(t), w\rangle.\end{aligned}$$

Notice that, due to (2.14), the representation formula (3.6) in this case reads

$$\eta^t(s) = \begin{cases} u(t) - u_0(t-s) & s \geq t \\ u(t) - u(t-s) & s < t. \end{cases}$$

Hence we deduce

$$\begin{aligned}&(b(0) - b(\infty))\langle Bu(t), w\rangle - \int_0^\infty \mu(s)\langle B\eta^t(s), w\rangle\, ds \\ &= (b(0) - b(\infty))\langle Bu(t), w\rangle - \int_0^t \mu(s)\langle B\eta^t(s), w\rangle\, ds \\ &\quad - \int_t^\infty \mu(s)\langle B\eta^t(s), w\rangle\, ds \\ &= (b(0) - b(\infty))\langle Bu(t), w\rangle + (b(t) - b(0))\langle Bu(t), w\rangle \\ &\quad - \int_0^t b'(s)\langle Bu(t-s), w\rangle\, ds + (b(\infty) - b(t))\langle Bu(t), w\rangle \\ &\quad - \int_t^\infty b'(s)\langle Bu_0(t-s), w\rangle\, ds \\ &= -\int_0^t b'(s)\langle Bu(t-s), w\rangle\, ds - \int_t^\infty b'(s)\langle Bu_0(t-s), w\rangle\, ds.\end{aligned}$$

Therefore u solves (4.3).

It is worth observing that, in general, the usual approach requires some additional regularity on the past history up to the initial time. For instance, to get (4.1), we must assume that $\eta_0 \in \mathcal{D}(T)$; while to get (4.3), we have to suppose $\tilde{f} \in L^1_{\text{loc}}(\mathbb{R}, H)$ to ensure the existence of a weak solution. Hence there is a regularity gap between the past history prior to the initial time and afterwards. On the contrary, in the history space setting, η^t belongs to the same space of η_0, for any $t > 0$.

5. Asymptotic behavior

The asymptotic analysis for systems with memory in the framework of dynamical systems (with particular reference to the existence of attractors) apparently has only recently been investigated. On the other hand, the study of the asymptotic behavior of processes of operators is relatively new, and goes back to the works of Haraux [24] and Chepyzhov and Vishik [5, 6, 7, 34]. The reader is also referred to the seminal books [2, 23, 33] for a background on the theory of attractors of semigroups.

The basic idea is to consider the evolution of a set of initial data, rather than focusing on the evolution of a single datum. Therefore, in order to have a satisfactory control of the system, it is necessary to have uniform estimates. More precisely, it is not enough to know that, for every fixed datum, the solution has a certain behavior, but we need to have a uniform (in time) control on the trajectories originating from a given bounded set of data.

A significant example is a linear equation of the form $u_t = Lu$, where L is the infinitesimal generator of a contraction semigroup, that is, $u(t) = S(t)u_0$, with $\|S(t)\| \leq 1$ for all $t \geq 0$. Then, either $S(t)$ has an exponential decay or $\|S(t)\| = 1$ for all t. In the latter case it is possible that $S(t)u_0 \to 0$ for every given initial datum u_0. However, it does not exist a decay pattern valid for all data taken, say, in the unit ball. In fact, for every $\varepsilon \in (0,1)$, and for every $t > 0$, there exists an element u_1 in the unit ball such that $\|S(t)u_1\| > \epsilon$.

Of course, in presence of nonlinearities the situation is much more complicated. The best one can say about nonlinear systems is that the trajectories originating from a fixed bounded set, after a certain time interval (depending on the size of the set) will be close enough to a compact set, possibly of finite fractal dimension (see Section 6).

We now formalize this argument. In the sequel, rather than taking a single process, we will be interested in considering a family of processes $\{U_f(t,\tau),\ f \in F\}$, where F is a complete metric space. This allows to pursue the analysis not just of a single differential equality, but of a family of differential equality depending on the parameter f (which usually represents the external source).

The dissipativeness of the system is commonly characterized by the existence of a bounded *uniformly absorbing set.*

Definition 5.1. A set $\mathcal{B}_0 \subset \mathcal{H}$ is said to be *uniformly absorbing* (with respect to $f \in F$) for the family $\{U_f(t,\tau),\ f \in F\}$ if for every bounded set $\mathcal{B} \subset \mathcal{H}$ there is $t_0 = t_0(\mathcal{B})$ such that

$$\bigcup_{f\in F} U_f(t,\tau)\mathcal{B} \subset \mathcal{B}_0$$

for every $\tau \in \mathbb{R}$ and every $t \geq \tau + t_0$.

The existence of a (bounded) uniformly absorbing set is derived from suitable uniform in time estimates (see, e.g., Theorem 6.2 below).

It is usually very hard for a system to have a compact uniformly absorbing set, unless there is a sufficient regularization of initial data (such as in the purely parabolic models). Therefore it is preferable to introduce another set, which ensures a weaker dissipativity than the above, but has more chances of being compact. Of course, the construction of such a set is based on the existence of a (bounded) uniformly absorbing set.

Definition 5.2. A set $\mathcal{K} \subset \mathcal{H}$ is *uniformly attracting* for $\{U_f(t,\tau),\ f \in F\}$ if for every $\tau \in \mathbb{R}$ and every bounded set $\mathcal{B} \subset \mathcal{H}$,

$$\lim_{t\to\infty}\left[\sup_{f\in F}\delta_{\mathcal{H}}(U_f(t,\tau)\mathcal{B},\mathcal{K})\right]=0$$

where $\delta_{\mathcal{H}}$ denotes the *Hausdorff semidistance* in $\mathcal{H}$, defined as

$$\delta_{\mathcal{H}}(\mathcal{B}_1,\mathcal{B}_2)=\sup_{z_1\in\mathcal{B}_1}\inf_{z_2\in\mathcal{B}_2}\|z_1-z_2\|_{\mathcal{H}}.$$

A family $\{U_f(t,\tau),\ f \in F\}$ possessing a compact uniformly attracting set is said to be *uniformly asymptotically compact.*

Definition 5.3. A closed set $\mathcal{A} \subset \mathcal{H}$ is a *uniform attractor* for $\{U_f(t,\tau),\ f \in F\}$ if it is simultaneously uniformly attracting and contained in every uniformly attracting set.

The minimality property required in the definition, which replaces the full invariance for attractors of semigroups, implies the uniqueness of the uniform attractor.

The key tool, due to Chepyzhov and Vishik [5, 6], is the following characterization theorem.

Theorem 5.4. *Let F be a compact metric space, and let $T(t)$ be a strongly continuous semigroup on F satisfying the translation equality*

$$U_f(t+s,\tau+s)=U_{T(s)f}(t,\tau)\qquad\forall\, f\in F. \tag{5.1}$$

In addition, let $U_\bullet(t,\tau):\mathcal{H}\times F\to\mathcal{H}$ be a continuous map, for every $\tau\in\mathbb{R}$ and every $t\geq\tau$. If $\{U_f(t,\tau),\ f\in F\}$ is uniformly asymptotically compact, then it has a compact uniform attractor given by

$$\mathcal{A}=\left\{\begin{array}{l} z(0)\text{ such that }z(t)\text{ is any complete bounded trajectory}\\ \text{of }U_f(t,\tau)\text{ for some }f\in\mathcal{A}(F)\end{array}\right\}$$

where $\mathcal{A}(F)$ is the global attractor of $T(t)$ on F.

The existence of $\mathcal{A}(F)$ follows from standard arguments of the theory of attractors for semigroups (see [2, 33]).

Remark 5.5. Sometimes, due to the nature of the equations, it is not possible to prove directly the uniformly asymptotic compactness property for the family of processes. In this case, the hypotheses of the above theorem can be weakened introducing the *Kuratowski measure of noncompactness* $\alpha_{\mathcal{H}}$ of a subset $\mathcal{B} \subset \mathcal{H}$ (cf. [23, 28]):

$$\alpha_{\mathcal{H}}(\mathcal{B}) = \inf \Big\{ d > 0 : \mathcal{B} \text{ has a finite cover of balls of } \mathcal{H} \text{ of diameter less than } d \Big\}.$$

and requiring that, in place of the uniformly asymptotic compactness property, the family of processes fulfills

$$\lim_{j \to \infty} \alpha_{\mathcal{H}}\big(\textstyle\bigcup_{f \in F} U_f(t_j + \tau, \tau)\mathcal{B}_0 \big) = 0 \text{ for some } t_j \geq 0, \text{ uniformly as } \tau \in \mathbb{R} \quad (5.2)$$

where $\mathcal{B}_0$ is a bounded *invariant* uniformly absorbing set, that is, $U_f(t,\tau)\mathcal{B}_0 \subset \mathcal{B}_0$ for every $f \in F$, $\tau \in \mathbb{R}$ and $t \geq \tau$.

We now apply this machinery to our systems. In correspondence of a given external source f, the solutions to (2.8)–(2.9) and (2.13)–(2.14) are described by means of a process $U_f(t,\tau)$. Since we are working in a history space setting, the solution will be of the form $z(t) = (w(t), \eta^t)$, belonging to some Hilbert space

$$\mathcal{H} = H_1 \times L^2_\mu(\mathbb{R}^+, H_2).$$

We now allow f to move within a specific functional space F, namely, the *hull* of a function g which is *translation compact* in $L^1_{\text{loc}}(\mathbb{R}, H_3)$, where H_3 is a suitable Hilbert space. We recall that a function $g \in L^1_{\text{loc}}(\mathbb{R}, H_3)$ is said to be *translation compact* if the *hull* of g, that is the closure in $L^1_{\text{loc}}(\mathbb{R}, H_3)$ of the set $\{g(\cdot + r)\}_{r \in \mathbb{R}}$ is compact (see Section 6 below for details). Then we consider the family of processes $\{U_f(t,\tau),\ f \in F\}$. It is immediate to see that if $T(t)$ is the translation semigroup on F, then equality (5.1) holds, and $\mathcal{A}(F)$ coincides with the entire space F.

We may rewrite our problem as an ordinary differential equation on $\mathcal{H}$ of the form

$$\begin{cases} z_t = Lz + N_f(z,t) \\ z(\tau) = z_0 \in \mathcal{H} \end{cases} \tag{5.3}$$

where L is a linear operator on $\mathcal{H}$, and N_f contains the nonlinearities and the source term. According to the above notation, $z(t) = U_f(t,\tau)z_0$.

As we said, a first step towards a global asymptotic analysis is to prove the existence of a bounded uniformly absorbing set. This is usually obtained by means of uniform in time energy estimates, under the hypothesis that the memory kernel has a reasonable behavior at infinity (typically, it has to be a decreasing function with exponential decay, see [27] and Section 6 below).

Once we have a bounded uniform absorbing set $\mathcal{B}_0$, in order to show the existence of a uniform attracting set, we may restrict to consider initial data $z_0 \in \mathcal{B}_0$. The standard procedure is to split the solution z to (5.3) as the sum $z = z_L + z_N$, where

$$\begin{cases} \partial_t z_L = L z_L \\ z(\tau) = z_0 \end{cases} \qquad \text{and} \qquad \begin{cases} \partial_t z_N = L z_N + N_f(z,t) \\ z_N(\tau) = 0. \end{cases}$$

First we have to show that z_L has an exponential decay; that is,

$$\|z_L(t)\|_{\mathcal{H}} \leq e^{-\varepsilon(t-\tau)} \|z_0\|_{\mathcal{H}} \qquad \forall\, z_0 \in \mathcal{H}. \tag{5.4}$$

Again, one can proceed via energy estimates, or, alternatively, by means of semigroup techniques. However, notice that it would suffice to show that $\|z_L(t)\|_{\mathcal{H}}$ goes to 0 as t tends to ∞. This fact can be useful whenever the decomposition of z contains a solution to a nonlinear problem in place of z_L.

The next step, which is usually more delicate, is to prove that the set of trajectories originating from $\mathcal{B}_0$ are contained, for every fixed time $t+\tau$ ($t \geq 0$), in a compact set $\mathcal{K} = \mathcal{K}(t) \subset \mathcal{H}$. This fact and (5.4) yield (5.2). In particular, if $\mathcal{K}$ is independent of t, it turns out to be a uniformly attracting set.

The idea is then to obtain, for every fixed $t \geq 0$, energy estimates in higher order spaces of the type

$$\|z_N(t+\tau)\|_{\mathcal{V}} \leq C(t) \qquad \forall\, \tau \in \mathbb{R}, \quad \forall\, f \in F \tag{5.5}$$

where

$$\mathcal{V} = V_1 \times L^2_\mu(\mathbb{R}^+, V_2) \qquad \text{and the inclusions} \qquad V_1 \hookrightarrow H_1 \qquad V_2 \hookrightarrow H_2$$

are compact. Here we face another technical problem, since the embedding

$$L^2_\mu(\mathbb{R}^+, V_2) \hookrightarrow L^2_\mu(\mathbb{R}^+, H_2)$$

in general is not compact. To overcome this obstacle, we consider the set $\mathcal{C}$ of the second components of z_N, that is, η_N, and exploit the following compactness result (cf. [30]).

Lemma 5.6. *Let V, H, W be three Banach spaces such that*

$$V \hookrightarrow H \hookrightarrow W$$

(the first injection being compact). Assume that $\mathcal{C} \subset L^2_\mu(\mathbb{R}^+, H)$ fulfills the assumptions

1. *$\mathcal{C}$ is bounded in $L^2_\mu(\mathbb{R}^+, V) \cap H^1_\mu(\mathbb{R}^+, W)$;*
2. *$\sup_{\eta \in \mathcal{C}} \|\eta(s)\|^2_H \leq h(s) \quad \forall\, s \in \mathbb{R}^+$ for some $h \in L^1_\mu(\mathbb{R}^+)$.*

Then $\mathcal{C}$ is relatively compact in $L^2_\mu(\mathbb{R}^+, H)$.

Therefore, in light of (5.4)–(5.5) and the above lemma, we are able to prove that the family $\{U_f(t,\tau),\ f \in F\}$ fulfills (5.2) (or, alternatively, is uniformly asymptotically compact).

Since the continuity of $U_\bullet(t,\tau)$ as a function from $\mathcal{H} \times F$ to $\mathcal{H}$ is usually quite a direct matter, by means of Theorem 5.4 and Remark 5.5, we finally obtain the existence of a uniform attractor $\mathcal{A}$ for the family $\{U_f(t,\tau),\ f \in F\}$ given by

$$\mathcal{A} = \left\{ \begin{array}{l} z(0) \text{ such that } z(t) \text{ is any complete bounded trajectory} \\ \text{of } U_f(t,\tau) \text{ for some } f \in F \end{array} \right\}.$$

6. Applications

Here we report some results regarding the existence of compact uniform attractors for certain dynamical systems with memory effects. These results can be proved by using the approach described in the previous sections. Due to the expository nature of the present paper, we concentrate on the models outlined in Section 2 only. However, more complicated systems governing phase transition dynamics in materials with memory have also been recently investigated (see [12, 13, 17, 18, 19, 20]).

Heat conduction with memory

Referring to [14, 15, 29], we consider equation (2.4) where the heat supply has a nonlinear dependence on the temperature; that is, setting $c = 1$,

$$\vartheta_t = \omega\Delta\vartheta + \int_0^\infty k(s)\Delta\vartheta(t-s)\,ds + g(\cdot,\vartheta) + f \tag{6.1}$$

in $\Omega \times (\tau, +\infty)$. Here $\Omega \subset \mathbb{R}^n$, $n \leq 3$, is a bounded domain with a smooth boundary $\partial\Omega$ and $\tau \in \mathbb{R}$ is a given initial time.

We assume that ϑ satisfies, for the sake of simplicity, a homogeneous Dirichlet boundary condition on $\partial\Omega$. Furthermore, the value of $\vartheta(t,\cdot)$ is known for $t \leq \tau$, where $\tau \in \mathbb{R}$ is the initial time. This means

$$\omega\vartheta = 0 \quad \text{on } \partial\Omega \times (\tau, +\infty) \tag{6.2}$$

$$\vartheta(\tau) = \vartheta_0 \quad \text{in } \Omega. \tag{6.3}$$

Following the approach described in Section 2, we introduce integrated past history of the temperature

$$\eta^t(\cdot,s) = \int_{t-s}^t \vartheta(\cdot,\sigma)\,d\sigma \qquad s \in \mathbb{R}^+.$$

Then, supposing $k(\infty) = 0$ and setting $\mu = -k'$, a formal integration by parts in time leads us to formulate the following initial and boundary value problem, depending on $\omega \geq 0$.

Problem $\mathbf{P}_\omega$. *Find (ϑ, η) solution to the system*

$$\vartheta_t(t) = \omega\Delta\vartheta(t) + \int_0^\infty \mu(s)\Delta\eta^t(s)\,ds + g(\cdot,\vartheta(t)) + f(t)$$

$$\eta^t_t(s) + \eta^t_s(s) = \vartheta(t)$$

in Ω, for any $t > \tau$ and any $s > 0$, which satisfies the initial and boundary conditions

$$\begin{aligned}
\omega\vartheta &= 0 \quad \text{on } \partial\Omega \times (\tau,+\infty)\\
\eta^t &= 0 \quad \text{on } \partial\Omega \times \mathbb{R}^+ \times (\tau,+\infty)\\
\eta^t(0) &= 0 \quad \text{in } \Omega, \quad \text{for any } t > \tau\\
\vartheta(\tau) &= \vartheta_0 \quad \text{in } \Omega\\
\eta^\tau &= \eta_0 \quad \text{in } \Omega \times \mathbb{R}^+.
\end{aligned}$$

Observe that, owing to Section 4, problem $\mathbf{P}_\omega$ can be shown to be equivalent to the original problem (6.1)–(6.3).

Analysis of problem $\mathbf{P}_\omega$ strongly depends on ω. Indeed, if ω is positive, then the equation for ϑ has parabolic features. In particular, the term $\omega\Delta\vartheta$ greatly contributes to the system dissipation. On the contrary, when $\omega = 0$, we are in presence of an integrodifferential equation for ϑ which is of hyperbolic type and the dissipation is contained in the convolution term solely.

As we mentioned in Section 2, the first step in order to analyze $\mathbf{P}_\omega$ from the dynamical system point of view is to show that it generates a process on a suitable phase space.

Let us set

$$H = L^2(\Omega), \qquad V = H_0^1(\Omega)$$

and introduce first all the assumptions on the nonlinearity g that will be used in this case; that is,

(g1) $|g(x,r)| \le c_1(1+|r|) \quad \forall x \in \Omega,\ \forall r \in \mathbb{R}$

(g2) $|g(x,r) - g(x,s)| \le c_2|r-s| \quad \forall x \in \Omega,\ \forall r,s \in \mathbb{R}$

(g3) $\limsup\limits_{r\to\infty} \dfrac{g(x,r)}{r} \le 0 \quad$ uniformly as $x \in \Omega$

(g4) $g(x,\cdot) \in C^1(\Omega \times \mathbb{R})$

(g5) $g(\cdot,0) \in V$

(g6) $g'(\cdot,r) \le 0 \quad \forall r \in \mathbb{R}$

(g7) $\sup\limits_{|u|\le c} |D_x g(\cdot,u)|_H < \infty \quad \forall c \ge 0.$

Here D_x and *prime* denote differentiation with respect to the first three space variables, and differentiation with respect to the fourth variable of g, respectively. Notice that (g4) together with (g1) imply (g3). It is also worth observing that, whenever $\omega > 0$, assumptions on g can be relaxed due to the parabolic character of equation (6.1) (see [14, 15]).

Regarding the memory kernel, the assumptions we need are

(m1) $\mu \in C^1(\mathbb{R}^+) \cap L^1(\mathbb{R}^+)$

(m2) $\mu(s) \ge 0$ and $\mu'(s) \le 0 \quad \forall s \in \mathbb{R}^+$

(m3) $\displaystyle\int_0^\infty \mu(s)\,ds = k_0 > 0$

(m4) $\mu'(s) + \delta\mu(s) \le 0 \quad \forall s \in \mathbb{R}^+, \qquad$ for some $\delta > 0$

(m5) $\lim\limits_{s\to 0+} \mu(s) < \infty.$

Remark 6.1. Assumption (m4) entails the exponential decay of μ. This assumption is widely used even to prove the stability of linear systems with memory (see, for instance, [27]). Assumptions (m3) and (m5) are only needed when $\omega = 0$ since in this case the dissipative feature of equation 6.1 depends entirely on the memory term.

Theorem 3.2 in [29] implies that $\mathbf{P}_\omega$ is a dynamical system. Indeed we have

Theorem 6.2. *Let* (g1)–(g2) *and* (m1)–(m3) *hold. Then, for any* $f \in L^1_{\rm loc}(\mathbb{R}, H)$, *problem* $\mathbf{P}_\omega$ *generates a process* $U_f(t,\tau)$ *on the phase space* $\mathcal{H} = H \times L^2_\mu(\mathbb{R}^+, V)$.

We can now introduce the functional

$$\mathcal{E}(t) = \frac{1}{2}\,\|U_f^\omega(t,\tau)z_0\|^2_{\mathcal{H}} = \frac{1}{2}\bigg(\|\vartheta(t)\|^2 + \int_0^\infty \mu(s)\|\nabla\eta^t(s)\|^2 ds\bigg).$$

Moreover, to show the existence of a (uniform) absorbing set, we need to introduce a more specific symbol functional space, namely the Banach space $\mathcal{T}(X)$ of $L^1_{\rm loc}(\mathbb{R}, X)$-*translation bounded* functions with values in a Banach space X; namely,

$$\mathcal{T}(X) = \bigg\{h \in L^1_{\rm loc}(\mathbb{R}, X) : \|h\|_{\mathcal{T}(X)} = \sup_{r\in\mathbb{R}} \int_r^{r+1} \|\vartheta(y)\|_X dy < \infty\bigg\}.$$

Then we have the following uniform in time estimate (see [29], Theorem 4.1)

Theorem 6.3. *Let* (g1)–(g3) *and* (m1)–(m5) *hold. Consider a bounded set* $F \subset \mathcal{T}(H)$ *be a bounded set. Then there exist positive constants* C, ε, *and* $\Lambda = \Lambda(F)$, *all independent of* ω, *such that the relation*

$$\mathcal{E}(t) \le Ce^{-\varepsilon(t-\tau)}\mathcal{E}(\tau) + \Lambda$$

holds for every $t \ge \tau$, *every* $\tau \in \mathbb{R}$, *and every* $f \in F$. *In particular, if* $g \equiv 0$ *and* F *reduces to the null function (that is, the linear homogeneous case), then* $\Lambda = 0$.

Theorem 6.3 entails

Corollary 6.4. *Let the hypotheses of Theorem* 6.2 *be fulfilled. Then there exists a bounded absorbing set* $\mathcal{B}_0$ *in* $\mathcal{H}$ *for the family* $\{U_f(t,\tau),\ f \in \mathcal{F}\}$, *which is uniform as* $\omega \ge 0$ *and* $f \in \mathcal{F}$. *Moreover,* $\mathcal{B}_0$ *is connected whenever* $\mathcal{F}$ *is connected.*

As we pointed out in Section 5, the existence of a bounded absorbing set is a preliminary step towards the existence of a universal attractor. In this case (cf. Section 5) we need F to be a compact metric space. Therefore we introduce the notion of *translation compact* function (see [7] and references therein).

Definition 6.5. A function $h \in \mathcal{T}(X)$ is *translation compact* in $L^1_{\rm loc}(\mathbb{R}, X)$ if the *hull* of h defined as

$$\mathrm{H}(h) = \overline{\{h^r\}_{r\in\mathbb{R}}}^{L^1_{\rm loc}(\mathbb{R},X)}$$

is compact in $L^1_{\rm loc}(\mathbb{R}, X)$, where $h^r(\cdot) = h(\cdot + r)$ is the translate of h by r.

Remark 6.6. Observe that a function $f : \mathbb{R} \to X$ belongs to $\mathrm{H}(h)$ if and only if for any $\varepsilon > 0$ and any $M > 0$ there is a translate h^r of h such that $\int_{-M}^M \|f(y) - h^r(y)\|_X dy \le \varepsilon$. In particular, the relation $\|f\|_{\mathcal{T}(X)} \le \|h\|_{\mathcal{T}(X)}$ holds for every $f \in \mathrm{H}(h)$. If h is constant, then $\mathrm{H}(h)$ reduces to the singleton $\{h\}$. The class of translation compact functions in $L^1_{\rm loc}(\mathbb{R}, X)$ is quite general; for instance, it contains $L^p(\mathbb{R}, X)$ for all $p \ge 1$, the constant X-valued functions, and the class of *almost periodic* functions (cf., e.g., [1, 26]).

Using the decomposition method sketched in Section 5 it can be proved (see Theorem 5.6 in [29])

Theorem 6.7. *Let* (g1)–(g5) *and* (m1)–(m5) *hold. Suppose* $F = \mathrm{H}(h)$ *where* h *is translation compact in* $L^1_{\mathrm{loc}}(\mathbb{R}, H)$. *Then, for every* $\omega \geq 0$, *the family of processes* $\{U_f^\omega(t,\tau),\ f \in \mathrm{H}(h)\}$ *associated with problem* $\mathbf{P}_\omega$ *possesses a compact and connected uniform attractor* $\mathcal{A}_\omega$ *given by*

$$\mathcal{A}_\omega = \left\{ \begin{array}{l} z(0) \text{ such that } z(t) \text{ is any complete bounded trajectory} \\ \text{of } U_f^\omega(t,\tau) \text{ for some } f \in \mathrm{H}(h) \end{array} \right\}.$$

One may wonder whether in some sense $\mathcal{A}_\omega$ stays close to $\mathcal{A}_0$ as $\omega \to 0$. This is a problem of stability for the family of attractors $\{\mathcal{A}_\omega\}$ (see Chap. IV, Sec. 10 in [33]). Along this direction, the so-called upper semicontinuity of $\mathcal{A}_0$ can be proved (see, e.g., [17] and references therein for analogous results). We have (see Theorem 6.4 in [29])

Theorem 6.8. *Let* (g1)–(g5) *and* (m1)–(m5) *hold. Suppose* f *be a fixed and constant in time function such that*

$$f \in V.$$

Consider the semigroup $S_\omega(t) = U_f^\omega(t,0)$ *for any* $\omega \geq 0$. *Then the attractor* $\mathcal{A}_0$ *of* $S_0(t)$ *is upper semicontinuous at zero with respect to the sets*

$$\{\mathcal{A}_\omega \ \ \textit{attractor of} \ \ S_\omega(t),\ \omega > 0\}.$$

This means

$$\lim_{\omega \to 0} \delta_{\mathcal{H}}(\mathcal{A}_\omega, \mathcal{A}_0) = 0.$$

Viscoelasticity

Here we consider a more general version of equation (2.5); namely,

$$u_{tt} + b_0 u_t - a_0 \Delta u_t - \psi(0)\Delta u - \int_0^\infty \psi'(s)\Delta u(t-s)ds + g(u) = f \qquad (6.4)$$

in $\Omega \times (\tau, +\infty)$, $\Omega \subset \mathbb{R}^n$, $n \leq 3$, being a bounded domain with a smooth boundary $\partial\Omega$. If, for instance, $n = 2$, one may think, at a model of the vertical displacement motion of a viscoelastic membrane subject to an external force which also depends on the displacement itself.

For the sake of simplicity, besides the past history of u up to τ, we assume that u satisfies a homogeneous Dirichlet boundary condition on $\partial\Omega$. Hence

$$u = 0 \quad \text{on } \partial\Omega \times (\tau, +\infty) \qquad (6.5)$$

$$\vartheta(\tau) = \vartheta_0 \quad \text{in } \Omega. \qquad (6.6)$$

According to Section 2 (cf. the **E2** case), we consider the additional variable

$$\eta^t(\cdot, s) = u(\cdot, t) - u(\cdot, t-s) \qquad s \in \mathbb{R}^+.$$

and we set $\mu = -\psi'$. Consequently, we obtain the (equivalent) initial and boundary value problem

Problem $\mathbf{P}_{a_0,b_0}$. *Find (u,η) solution to the system*

$$u_t(t) = v(t)$$

$$v_t(t) = -b_0 v(t) + a_0 \Delta v(t) + h(0)\Delta u(t) + \int_0^\infty \mu(s)\Delta\eta^t(s)ds - g(u(t)) + f(t)$$

$$\eta_t^t = -\eta_s^t + u(t)$$

in Ω, for any $t > \tau$ and any $s > 0$, which satisfies the initial and boundary conditions

$$\begin{aligned} u &= 0 \quad \text{on } \partial\Omega \times (\tau, +\infty), & u(\tau) &= u_0 \quad \text{in } \Omega, \\ \eta^t &= 0 \quad \text{on } \partial\Omega \times \mathbb{R}^+ \times (\tau, +\infty), & v(\tau) &= v_0 \quad \text{in } \Omega, \\ \eta^t(0) &= 0 \quad \text{in } \Omega, \quad \text{for any } t > \tau, & \eta^\tau &= \eta_0 \quad \text{in } \Omega \times \mathbb{R}^+. \end{aligned}$$

The interesting cases are clearly the following ones

WD *Weak damping:* $a_0 = 0$, $b_0 > 0$

SD *Strong damping:* $a_0 > 0$, $b_0 = 0$.

Consider the **WD** case first. Suppose from now on $g \in C^1(\mathbb{R})$ and set

$$G(s) = \int_0^s g(y)\,dy \qquad \forall\, s \in \mathbb{R}.$$

Then assume that there are C_1, $C_2 > 0$ such that

(G1) $\liminf_{|y|\to\infty} \dfrac{G(y)}{y^2} \geq 0$

(G2) $\liminf_{|y|\to\infty} \dfrac{yg(y) - C_1 G(y)}{y^2} \geq 0$

(G3) $|g'(y)| \leq C_2(1 + |y|^\gamma)$ for some $0 \leq \gamma < 2$, $\forall\, y \in \mathbb{R}$.

Using the notation introduced in the previous application and collecting Theorems 3.2 and 3.3 in [30], we deduce

Theorem 6.9. *Let* (G1)–(G3), (m1)–(m2) *hold. Then, for any $f \in L^1_{\text{loc}}(\mathbb{R}, H)$, problem $\mathbf{P}_{0,b_0}$ generates a process $U_f(t,\tau)$ on the phase space $\mathcal{H} = V \times H \times L^2_\mu(\mathbb{R}^+, V)$.*

The exponential decay of the memory kernel implies the dissipativity of the dynamical system associated with $\mathbf{P}_{0,b_0}$. In fact, there holds (cf. [30], Theorem 4.2)

Theorem 6.10. *Let the hypotheses of Theorem* 6.9 *hold. Suppose in addition that μ satisfies* (m4) *and let F be a bounded subset of $\mathcal{T}(H)$. Then there exists a bounded absorbing set $\mathcal{B}_0$ in $\mathcal{H}$ for the family $\{U_f(t,\tau),\, f \in F\}$, which is uniform with respect to $f \in F$. Moreover, $\mathcal{B}_0$ is connected whenever F is connected.*

In order to prove the existence of a uniform attractor, we need to take $F = \mathrm{H}(h)$, h being a translation compact function in $L^1_{\text{loc}}(\mathbb{R}; H)$ (cf. Definition 6.4). Adapting a decomposition of the solution similar to the one used in [17], Section 4, to the proof of Theorem 5.3 in [30], one can deduce

Theorem 6.11. *Let* (G1)–(G3) *and* (m1)–(m2), (m4) *hold. Suppose* h *is translation compact in* $L^1_{\rm loc}(\mathbb{R}, H)$. *Then the family of processes* $\{U_f(t,\tau),\ f \in \mathrm{H}(h)\}$ *associated with problem* $\mathbf{P}_{0,b_0}$ *possesses a compact and connected uniform attractor* $\mathcal{A}$ *given by*

$$\mathcal{A} = \left\{ \begin{array}{l} z(0) \text{ such that } z(t) \text{ is any complete bounded trajectory} \\ \text{of } U_f(t,\tau) \text{ for some } f \in \mathrm{H}(h) \end{array} \right\}. \tag{6.7}$$

Consider now the **SD** case. The existence of a process is ensured by (see [4], Theorem 3.2)

Theorem 6.12. *Let* (G1)–(G2), (m1)–(m2) *hold. Suppose moreover that there exists* $C_3 > 0$ *and* $\gamma \in [1,5]$ *such that*

$$\textbf{(G4)} \quad |g(y_1) - g(y_2)| \le C_3 |y_1 - y_2| (1 + |y_1|^{\gamma-1} + |y_2|^{\gamma-1})$$

for any $y_1, y_2 \in \mathbb{R}$. *Then, for any* $f \in L^1_{\rm loc}(\mathbb{R}, H)$, *problem* $\mathbf{P}_{a_0,0}$ *generates a process* $U_f(t,\tau)$ *on the phase space* $\mathcal{H} = V \times H \times L^2_\mu(\mathbb{R}^+, V)$.

Moreover, we have (see [4], Theorem 4.2)

Theorem 6.13. *Let the hypotheses of Theorem* 6.12 *hold. Suppose in addition that* g *satisfies* (G4) *and let* F *be a bounded subset of* $\mathcal{T}(H)$. *Then there exists a bounded absorbing set* $\mathcal{B}_0$ *in* $\mathcal{H}$ *for the family* $\{U_f(t,\tau),\ f \in F\}$, *which is uniform with respect to* $f \in F$. *Moreover,* $\mathcal{B}_0$ *is connected whenever* F *is connected.*

Proceeding as in the **SD** case, one can prove the existence of a uniform attractor by combining the argument used in [17], Section 4, with the proof of Theorem 5.3 in [4]. Thus there holds

Theorem 6.14. *Let* (G1)–(G2), (m1)–(m2), (m4), *and* (G4) *with* $\gamma \in [1,5)$ *hold. Suppose* $F = \mathrm{H}(h)$ *where* h *is translation compact in* $L^1_{\rm loc}(\mathbb{R}, H)$. *Then the family of processes* $\{U_f(t,\tau),\ f \in \mathrm{H}(h)\}$ *associated with problem* $\mathbf{P}_{0,b_0}$ *possesses a compact and connected uniform attractor* $\mathcal{A}$ *given by* (6.7).

Remark 6.15. The weak damping case can also be studied on the whole $\mathbb{R}^3$ (see [28]). This case is technically more difficult than the present one due to the lack of compactness. The strong damping case on $\mathbb{R}^3$ is even worse since the speed of propagation of the solution is infinite as pointed out in [3], where the model without memory effect is analyzed in detail.

Remark 6.16. The described results can be suitably adapted to the three-dimensional viscoelastic models where the Laplacian is replaced by the linear elasticity operator (see [16]).

Acknowledgments. This work has been partially supported by the Italian MURST Research Projects "Simmetrie, Strutture Geometriche, Evoluzione e Memoria in Equazioni a Derivate Parziali" and "Problemi e Metodi nella Teoria delle Equazioni Iperboliche".

References

[1] L. Amerio, G. Prouse, *Abstract Almost Periodic Functions and Functional Equations*, Van Nostrand, New York (1971)

[2] A.V. Babin, M.I. Vishik, *Attractors of evolution equations*, North-Holland, Amsterdam (1992)

[3] V. Belleri, V. Pata, *Attractors for semilinear strongly damped wave equation on R^3*, Discrete Contin. Dynam. Systems **7**, 719–735 (2001)

[4] S. Borini, V. Pata, *Uniform attractors for a strongly damped wave equation with linear memory*, Asymptot. Anal. **20**, 263–277 (1999)

[5] V.V. Chepyzhov, M.I. Vishik, *Non-autonomous evolution equations and their attractors*, Russian J. Math. Phys. **1**, 165–190 (1993)

[6] V.V. Chepyzhov, M.I. Vishik, *Attractors of non-autonomous dynamical systems and their dimension*, J. Math. Pures Appl. **73**, 279–333 (1994)

[7] V.V. Chepyzhov, M.I. Vishik, *Non-autonomous evolutionary equations with translation compact symbols and their attractor*, C.R. Acad. Sci. Paris Sér. I Math. **321**, 153–158 (1995)

[8] B.D. Coleman, M.E. Gurtin, *Equipresence and constitutive equations for rigid heat conductors*, Z. Angew. Math. Phys. **18**, 199–208 (1967)

[9] C.M. Dafermos, *Asymptotic stability in viscoelasticity*, Arch. Rational Mech. Anal. **37**, 297–308 (1970)

[10] B. Engquist, W. Schmid (Eds.), *Mathematics Unlimited – 2001 and Beyond*, Springer, Berlin Heidelberg (2001)

[11] M. Fabrizio, A. Morro, *Mathematical problems in linear viscoelasticity*, SIAM Studies Appl. Math. 12, SIAM, Philadelphia (1992)

[12] C. Giorgi, M. Grasselli, V. Pata, *Well-posedness and longtime behavior of the phase-field model with memory in a history space setting*, Quart. Appl. Math. **59**, 701–736 (2001)

[13] C. Giorgi, M. Grasselli, V. Pata, *Uniform attractors for a phase-field model with memory and quadratic nonlinearity*, Indiana Univ. Math. J. **48**, 1395–1445 (1999)

[14] C. Giorgi, A. Marzocchi, V. Pata, *Asymptotic behavior of a semilinear problem in heat conduction with memory*, NoDEA Nonlinear Differential Equations Appl. **5**, 333–354 (1998)

[15] C. Giorgi, A. Marzocchi, V. Pata, *Uniform attractors for a non-autonomous semilinear heat equation with memory*, Quart. Appl. Math. **58**, 661–683 (2000)

[16] M. Grasselli, V. Pata, *Longtime behavior of a homogenized model in viscoelastodynamics*, Discrete Contin. Dynam. Systems **4**, 338–359 (1998)

[17] M. Grasselli, V. Pata, *Upper semicontinuous attractor for a hyperbolic phase-field model with memory*, Indiana Univ. Math. J. **50**, 1281–1308 (2001)

[18] M. Grasselli, V. Pata, *On the dissipativity of a hyperbolic phase-field system with memory*, Nonlinear Anal. **47**, 3157–3169 (2001)

[19] M. Grasselli, V. Pata, *On the longterm behavior of a parabolic phase-field model with memory*, in "Differential Equations and Control Theory" (S. Aizicovici and N.H. Pavel, eds.), 147–157, Lecture Notes in Pure and Appl. Math. 225, Dekker, New York (2002)

[20] M. Grasselli, V. Pata, F.M. Vegni, *Longterm dynamics of a conserved phase-field system with memory*, submitted

[21] M.E. Gurtin, A.C. Pipkin, *A general theory of heat conduction with finite wave speeds*, Arch. Rational Mech. Anal. **31**, 113–126 (1968)

[22] Y.M. Haddad, *Viscoelasticity of engineering materials*, Chapman & Hall, London (1985)

[23] J.K. Hale, *Asymptotic behavior of dissipative systems*, Math. Surv. Monogr. n. 25, Amer. Math. Soc., Providence (1988)

[24] A. Haraux, *Systèmes dynamiques dissipatifs et applications*, Coll. RMA n. 17, Masson, Paris (1990)

[25] J. Jäckle, *Heat conduction and relaxation in liquids of high viscosity*, Phys. A **162**, 377–404 (1990)

[26] B.M. Levitan, V.V. Zhikov, *Almost periodic functions and differential equations*, Cambridge University Press, Cambridge (1982)

[27] Z. Liu, S. Zheng, *Semigroups associated with dissipative systems*, Chapman & Hall/CRC Res. Notes Math. n. 398, Boca Raton (1999)

[28] V. Pata, *Attractors for a damped wave equation on R^3 with linear memory*, Math. Methods Appl. Sci. **23**, 633–653 (2000)

[29] V. Pata, *Hyperbolic limit of parabolic semilinear heat equation with fading memory*, Z. Anal. Anwendungen **20**, 359–377 (2001)

[30] V. Pata, A. Zucchi, *Attractors for a damped hyperbolic equation with linear memory*, Adv. Math. Sci. Appl. (to appear)

[31] A. Pazy, *Semigroups of linear operators and applications to partial differential equations*, Springer-Verlag, New York (1983)

[32] M. Renardy, W.J. Hrusa, J.A. Nohel, *Mathematical problems in viscoelasticity*, Longman Scientific & Technical; Harlow John Wiley & Sons, Inc., New York (1987)

[33] R. Temam, *Infinite-dimensional dynamical systems in mechanics and physics*, Springer, New York (1988)

[34] M.I. Vishik, *Asymptotic behaviour of solutions of evolutionary equations*, Cambridge University Press, Cambridge (1992)

Maurizio Grasselli
Dipartimento di Matematica "F. Brioschi"
Politecnico di Milano, Italy
e-mail: `maugra@mate.polimi.it`

Vittorino Pata
Dipartimento di Matematica "F. Brioschi"
Politecnico di Milano, Italy
e-mail: `pata@mate.polimi.it`

Progress in Nonlinear Differential Equations
and Their Applications, Vol. 50, 179–195

On some Mathematical Aspects of the Ring Cavity Problem

Reinhard Illner, Horst Lange and Holger Teismann

1. Introduction

The 'ring cavity problem' of nonlinear laser optics describes the behavior of a laser beam which propagates through a 'bistable optical cavity'; this is a ring resonator system consisting of a nonlinear dielectric medium (of length ℓ), and several partially transmitting mirrors including an input and an output mirror. During one ring circulation, the propagating laser beam enters the nonlinear medium, emerges, is partially transmitted through the output mirror, is partially fed back around the circuit mirror system to the entry point at the input mirror, and added there to the input pump field beam. The system describes the spatial modulation and filtration of laser beams in order to achieve 'memory'. The appearance of a wide range of spontaneously emerging spatial coherent structures ('filaments') is of practical importance in laser optics where the physical control of these structures has led to new and natural methods for information processing ('pixel transfer'), as well as a theoretical model of a route to chaos and turbulence; for the physical background see references [3]–[11], [13]–[18].

The mathematical problem underlying the ring cavity model is the study of the infinite-dimensional map

$$u_{n-1}(.,\ell) \to u_n(.,\ell) \tag{1.1}$$

for a sequence (u_n) of complex wave-functions $u_n(t,x)$ obeying a series of initial value problems

$$i\frac{\partial u_n}{\partial t} = -\Delta u_n + N(|u_n|^2)u_n, \tag{1.2}$$

$$u_n(0,x) = a(x) + Re^{i\theta}u_{n-1}(\ell,x). \tag{1.3}$$

Here, N is a real-valued function depending on the nonlinear medium through which the laser beam is propagating; $a(x)$ is the given optical pump field, R denotes the mirror loss rate factor ($0 < R < 1$); θ is a phase shift coefficient, and ℓ denotes the longitudinal length of the device; some further physical parameters have been set equal to one in our presentation. Also, let us remark here that in the physical model the variable t used here does not denote time but stands for the longitudinal coordinate (usually denoted by z) whereas x denotes the rest of the spatial transversal variables; thus one has $x \in \mathbb{R}^d$ where either $d = 1$ or

$d = 2$. We have preferred to use this notation since then for fixed n, which may be seen as a discrete time parameter, (1.2)–(1.3) has the standard form of an initial value problem for a time-dependent nonstationary nonlinear Schrödinger equation. However, one should be aware that from a physical point of view (1.2)–(1.3) is a *stationary* model; the *nonstationary* equations have the form (see [8]):

$$i\frac{\partial u}{\partial t} = -\Delta u - i\frac{\partial u}{\partial z} + N(|u|^2)u,$$

where now t stands for time, z is the longitudinal coordinate in the direction of the beam propagation, and orthogonal to x, y, the transversal coordinates; Δ denotes the two-dimensional Laplacian $\partial^2/\partial x^2 + \partial^2/\partial y^2$.

The functions N which actually are used in the applications are the *Kerr nonlinearity* $N(s) = 1 - \alpha s$ or the *saturating nonlinearity* $N(s) = 1/(1+\beta s)$ (α, β positive constants). In the case of the Kerr nonlinearity the system (1.2) consists of the well-known *focusing cubic nonlinear Schrödinger equation*; mathematically this case is more interesting than the one of the saturating nonlinearity.

The physical and mathematical goal is the study of the infinite-dimensional dynamical system $u_{n-1}(\cdot, \ell) \to u_n(\cdot, \ell)$ and of the sequence of wave-functions $u_n(t, x)$, esp. the behavior of these functions as $n \to \infty$. Much of the work done on this problem is *numerical* in character, whereas mathematically rigorous results are rare. Most of the interesting effects in the dynamical behavior of the ring cavity model have been seen numerically (as sort of experimental mathematics). The system displays the appearance of stable and unstable fixed points (period-1-solutions) of the map (1.1), period-m-solutions of the system for various m's (e.g. $m = 2$, $m = 4$, $m = 6$, etc.), chaotic attractors and Cantor like structures, and period doublings; many of these coherent structures have been observed in the numerical calculation of the field intensity contour 'filaments' in cuts through the nonlinear dielectric region in transversal direction (see [14] for impressive pictures of the filaments).

In this survey paper we try to present most of the available rigorous mathematical results on the ring cavity problem (which are not too manifold). In *Section 2* we look at plane wave solutions; here also some new results on special plane wave period-m-solutions for $m = 2$ are presented. In *Section 3* we briefly summarize the results of McLaughlin et al. from [11] on various reductions and approximations of the system (1.1)–(1.3) in the solitary wave case whereas in *Section 4* we discuss the results of Merle [12] on the existence of fixed points of the map (1.1), and present some new results on the same topic and on the nonlinear 'stability' of plane wave fixed points of the map (1.1) in the transversal periodic situation. In *Section 5* we discuss further topics and present some open problems.

The dynamical operator T which belongs to the system (1.2)–(1.3) can be described as follows: Let $a(x)$, $\theta \in \mathbb{R}$, $R \in (0,1)$, $\ell > 0$ be given parameter functions values resp.; for a given function $g(x)$ let $u(t, x)$ be the solution of the initial value problem

$$i\frac{\partial u}{\partial t} = -\Delta u + N(|u|^2)u, \tag{1.4}$$

$$u(0,x) = a(x) + Re^{i\theta}\, g(x)\,; \tag{1.5}$$

then define

$$Tg := u(\ell,.)\,. \tag{1.6}$$

The n-th circuit of the system (1.2)–(1.3) is described by the iterates T^n of T where the 0-th circuit is given by a fixed function $u_o(x) = g(x)$. Thus, these iterates describe the dynamics of the system. The simplest dynamics is given by *fixed points* of T; period-m-solutions are fixed points of T^m which are not fixed for T^k with any natural number $k < m$.

2. Plane Waves

The simplest solutions to the ring cavity problem are *plane wave* solutions. Plane wave solutions of (1.2)–(1.3) are functions of type

$$u(t,x) = \alpha e^{i\left[kx - \left(N\left(|\alpha|^2\right)+k^2\right)t\right]},\ x \in \mathbb{R}^d \tag{2.1}$$

with some vector $k \in R^d$ and a complex constant α; N is the function from (1.2) (solutions of this type are discussed in [11], [5]). Functions of type (2.1) are solutions of the nonlinear Schrödinger equation (1.2). If the pump field $a(x)$ is also a planar wave with the same wave number k, $a(x) = a\ e^{ikx}$, $a \in \mathbb{C}$, then the existence of a fixed point of problem (1.2)–(1.3) is equivalent to the validity of the transcendental equation

$$\alpha\left(1 - Re^{i\left[\theta - \left(N\left(|\alpha|^2\right)+k^2\right)\ell\right]}\right) = a \tag{2.2}$$

Setting $I := |\alpha|^2$ (='intensity') one sees that (2.2) is equivalent to the solvability of the real transcendental equation

$$I\left(1 + R^2 - 2R\cos\left[\theta - \left(N(I) + k^2\right)\ell\right]\right) = |a|^2\,; \tag{2.3}$$

(2.3) may be written as

$$\cos\left[\theta - \left(N(I) + k^2\right)\ell\right] = \frac{1}{2}\left(\frac{1}{R} + R - \frac{|a|^2}{RI}\right); \tag{2.4}$$

(2.4) implies that the fixed points are intersection points of the two curves given by the left- and right-hand side of (2.4) as functions of I; one sees that there is always at least one fixed point, but depending on the parameters there may be several. For special cases of these parameters it is claimed in [11] that the intensity I as a function of $|a|^2$ has the usual form of a *hysteresis response curve* (see Fig. 4 in [11]). If (2.3) or (2.4) is solved for I then α is given by

$$\alpha = \frac{|a|^2}{1 - Re^{i[\theta - (N(I)+k^2)\ell]}};$$

we remark that this formula makes sense since $0 < R < 1$. If one writes (2.2) in the form

$$\alpha = a + Re^{i\left[\theta - \left(N\left(|\alpha|^2\right) + k^2\right)\ell\right]}\alpha =: \tilde{T}(\alpha), \tag{2.5}$$

it is readily seen that one may get the solutions of (2.2) (being the fixed points of the mapping $\tilde{T}$) by iterating this mapping which acts as a composition of a rotation (by an angle depending nonlinearly on the amplitude), a contraction, and a translation in the complex plane; this intuitively illustrates the existence of fixed points (see also Fig. 2 of [11]). For several functions N (e.g. the Kerr nonlinearity or the saturating nonlinearity) the solvability has been discussed almost completely (see [11], [5]).

By linearizing the mapping $\tilde{T}$ around a fixed point α, and computing the Jacobian determinant of $\tilde{T}$ at fixed points (which is R^2) and the eigenvalues of the Jacobian matrix $D_\alpha \tilde{T}$ (at a fixed point α), it is shown in [11] that these eigenvalues μ_1, μ_2 satisfy the relations

$$\mu_1\mu_2 = R^2 < 1,\ \mu_1 + \mu_2 = \left(1 + ilN'(|\alpha|^2)\,|\alpha|^2\right) Re^{i\Gamma} + \left(1 - ilN'(|\alpha|^2)\,|\alpha|^2\right) Re^{-i\Gamma} \tag{2.6}$$

where $\Gamma := \theta - \left(N(|\alpha|^2) + k^2\right)\ell$. Here only two cases are possible: either both μ_1 *and* μ_2 are real, or $\mu_1 = \overline{\mu}_2$. Since a fixed point of $\tilde{T}$ is linearly stable under forward iterations if and only if $|\mu_j| \le 1$ for $j = 1, 2$, it is proved in [11] that in the second case ($\mu_1 = \overline{\mu}_2$) the fixed point is always linearly stable; in the first case (μ_1 *and* μ_2 both real), either both eigenvalues satisfy $|\mu_j| < 1$ (which implies linear stability), or only one satisfies this condition and the other does not; then, as an eigenvalue crosses $+1$, the instability which occurs is a 'saddle node' bifurcation; as an eigenvalue crosses -1 the instability is a period-2-bifurcation.

The considerations of [11], [5], and the numerical experiments cited there, suggest the existence of *period-m-solutions* of plane wave type of (1.2) for any m, i.e. fixed points of T^m which are not fixed for T^k with any natural number $k < m$; esp. the existence of cascades of *period doublings* are observed numerically, but also period-6-solutions seem to exist, although period-3-solutions have never been seen numerically. Nevertheless, it seems that no rigorous analytical proof of the existence of period-m-solutions for $m > 1$ has been published. We present here a sketch of an elementary analytical proof of the existence of branches of period-2-solutions of plane wave type emerging from fixed points. We take $N(s) = -s$.

Let us consider a plane wave pump function $a(x) = \gamma e^{ikx}$, $\gamma \in \mathbb{C}$ and $k \in \mathbb{R}$; for fixed parameters θ and l we make the ansatz

$$u(t,x) = \alpha e^{i\left[kx + \left(|\alpha|^2 - k^2\right)t\right]},$$
$$v(t,x) = \beta e^{i\left[kx + \left(|\beta|^2 - k^2\right)t\right]} \quad \left(x \in \mathbb{R}^d,\quad t \in \mathbb{R},\ \alpha, \beta \in \mathbb{C}\right)$$

for two plane wave solutions of

$$i\frac{\partial u}{\partial t} = -\Delta u - |u|^2 u. \tag{2.7}$$

Then a pair of different fixed points of T^2 (which are not a fixed points of T) will be given by the plane waves $u(\ell,x) = \alpha e^{i\left[kx+\left(|\alpha|^2-k^2\right)\ell\right]}$ and $v(\ell,x) = \beta e^{i\left[kx+\left(|\beta|^2-k^2\right)\ell\right]}$ if the complex numbers α and β are different and satisfy the nonlinear system

$$\alpha = \gamma + Re^{i\left[\theta+\left(|\beta|^2-k^2\right)\ell\right]}\beta,\ \beta = \gamma + Re^{i\left[\theta+\left(|\alpha|^2-k^2\right)\ell\right]}\alpha. \tag{2.8}$$

System (2.8) can be written as

$$\alpha - Re^{i\left[\theta+\left(|\beta|^2-k^2\right)\ell\right]}\beta = \gamma,\ \beta - Re^{i\left[\theta+\left(|\alpha|^2-k^2\right)\ell\right]}\alpha = \gamma. \tag{2.9}$$

By solving (2.9) formally for α and β one sees that (2.8) is equivalent to the solvability of the equations

$$\alpha = \gamma\frac{1+R_{\theta,\ell}e^{i|\beta|^2\ell}}{1-R_{\theta,\ell}^2e^{i\left(|\alpha|^2+|\beta|^2\right)\ell}},\ \beta = \gamma\frac{1+R_{\theta,\ell}e^{i|\alpha|^2\ell}}{1-R_{\theta,\ell}^2e^{i\left(|\alpha|^2+|\beta|^2\right)\ell}}, \tag{2.10}$$

where $R_{\theta,\ell} = R\,e^{i\left[\theta-k^2l\right]}$. Setting $a := |\alpha|^2$, $b := |\beta|^2$, $c := |\gamma|^2$ we see that solutions must satisfy

$$a = c\frac{\left|1+R_{\theta,\ell}e^{ib\ell}\right|^2}{\left|1-R_{\theta,\ell}^2e^{i(a+b)\ell}\right|^2},\ b = c\frac{\left|1+R_{\theta,\ell}e^{ia\ell}\right|^2}{\left|1-R_{\theta,\ell}^2e^{i(a+b)\ell}\right|^2}. \tag{2.11}$$

On the other hand, if a and b are solutions of (2.11) then

$$\alpha := \gamma\frac{1+R_{\theta,\ell}e^{ib\ell}}{1-R_{\theta,\ell}^2e^{i(a+b)\ell}},\ \beta := \gamma\frac{1+R_{\theta,\ell}e^{ia\ell}}{1-R_{\theta,\ell}^2e^{i(a+b)\ell}} \tag{2.12}$$

are solutions of (2.10). These solutions are different if and only if the solutions a and b of (2.11) are different, namely, if α and β are defined by (2.12), and $\alpha = \beta$ we have $a = |\alpha|^2$, $b = |\beta|^2$ by (2.10), and so $a = b$. Thus, it is enough to solve the real nonlinear two-by-two system (2.11) for solutions a, b which are different.

We will show that in a neighborhood of any plane wave fixed point $u_o(t,x) = \alpha_o e^{i\left[kx+\left(|\alpha_o|^2-k^2\right)t\right]}$ there bifurcate at least two different period-2 plane wave solutions. The fixed point $u_o(t,x)$ is determined by the 'parameters' $a_o := |\alpha_o|^2$, $c_o := c(\gamma_o) = |\gamma_o|^2$ for a specific pump function $\gamma_o(x) = \gamma_o e^{ikx}$, $\gamma_o \in \mathbb{C}$; here the parameters (a_o, c_o) have to satisfy the relation (2.3) which now can be written as

$$c_0 = c(\gamma_o) = \left(R^2+1\right)a_o - 2R\cos\left(\theta_{k,\ell} + a_o\ell\right)a_o. \tag{2.13}$$

Theorem 2.1. *For any set of parameters $\theta > 0$, $\ell > 0$, $0 < R < 1$ there is a plane wave fixed point $u_o(x, c_o) = \alpha_o e^{i\left[k_ox+\left(a_o-k_0^2\right)\ell\right]}$ of the cubic Schrödinger equation (2.7) with parameters (a_o, c_o) satisfying (2.13) (and $a_o := |\alpha|^2$, $c_o := c(\gamma_o) = |\gamma_o|^2$), and at least two different plane wave period-2-solutions*

$$u(x,c) = \alpha e^{i\left[kx+\left(a(c)-k^2\right)\ell\right]},\ v(x,c) = \beta e^{i\left[kx+\left(b(c)-k^2\right)\ell\right]} \tag{2.14}$$

of (2.7) *with parameters* $(a(c), c)$, $(b(c), c)$ *resp., bifurcating from that fixed point, i.e.* $a(c)$, $b(c)$ *are continuous functions of* c *in a neighborhood of* c_o *such that* $a(c) = |\alpha|^2$, $b(c) = |\beta|^2$, *and*

$$\lim_{c \longrightarrow c_o} u(x, c) = u_o(x, c_o), \quad \lim_{c \longrightarrow c_o} v(x, c) = u_o(x, c_o), \quad \lim_{c \longrightarrow c_o} a(c) = \lim_{c \longrightarrow c_o} b(c) = a_o$$

uniformly in x; *the functions* $a(c)$ *and* $b(c)$ *satisfy equations* (2.11).

Sketch of a proof (of Theorem 2.1): From our remarks it is clear that it is enough to solve the system (2.11) for a and b locally near a_o such that a is different from b. The idea is to apply the implicit function theorem and Morse's Lemma to the functions involved in (2.11). For simplicity we allow $k = k_o$, and choose the parameters in such a way that

$$\cos(\theta_{k,\ell} + a_o \ell) = 0, \ \sin(\theta_{k,\ell} + a_o \ell) = 1, \ c_o = c(x_o) = \left(R^2 + 1\right) a_o,$$

where $\theta_{k,\ell} := \theta - k^2 \ell$; then (2.13) is fulfilled, and it will turn out later that we have to choose

$$a_o = \frac{1 + R^2}{2R\ell}. \tag{2.15}$$

We remark that (2.11) can be written in the form

$$\Phi(a, b) := (\Phi_1(a, b), \Phi_2(a, b)) := (cf(b) - ah(a + b), cf(a) - bh(a + b)) = 0 \tag{2.16}$$

where

$$f(x) := \left|1 + R_{\theta,\ell} e^{ix\ell}\right|^2, h(z) := \left|1 - R_{\theta,\ell}^2 e^{iz\ell}\right|^2,$$

i.e. we have to find solutions (a, b) of (2.16) such that $(a, b) \notin \Delta$ where Δ is the diagonal of R^2. The Jacobian of Φ at the fixed point (a_o, a_o) has the form

$$J_{(a_o,a_o)} \Phi(a, b; c) = \begin{pmatrix} A & B \\ B & A \end{pmatrix},$$

$$A := -h(2a_o) - a_o h'(2a_o), \quad B := cf'(a_0) - a_o h'(2a_o)$$

with eigenvalues $A + B$ and $A - B$. The above choices of parameters and (2.15) lead to $A = B$.

Now, we write $\varphi := \Phi_1 + \Phi_2$, $\psi := \Phi_1 - \Phi_2$, and we have to look for the zero set of the mapping $(a, b) \longmapsto (\varphi(a, b; c), \psi(a, b; c))$ in a neighborhood of $(a_o, a_o; c_o)$.

The idea is to solve the equation

$$\varphi(a, b; c) = 0 \tag{2.17}$$

for a as a function of (b, c) in a neighborhood of (a_o, c_o), i.e to find a unique smooth function $s(b, c)$ such that locally $\varphi(s(b, c), b; c) = 0$ and $(s(b, c), b) \notin \Delta$. Then, to insert this function into the equation $\psi(a, b; c) = 0$, and to find the zero set of the function

$$\tilde{\psi}(b, c) := \psi(s(b, c), b; c) \tag{2.18}$$

near (a_o, c_o). The local solvability of (2.17) is guaranteed by the fact that by our choice of parameters (and esp. by the choice of the eigenvalues of the Jacobian of Φ) one has

$$\nabla\varphi\left(a_o, a_o; c_o\right) = -2\left(1+R^2\right)^2 \begin{pmatrix}1\\1\end{pmatrix} \neq 0.$$

On the other hand, one computes $\nabla\psi\left(a_o, a_o; c_o\right) = 0$, but the Hessian of $\tilde{\psi}\left(b, c\right) = \psi\left(s\left(b, c\right)b, c\right)$ with respect to the variables (b, c) (at the point (a_o, c_o)) is given by

$$\left(H\tilde{\psi}\right)\left(a_o, c_o\right) = 4R\ell \begin{pmatrix} 0 & -1 \\ -1 & 1/\left(1+R^2\right)\end{pmatrix},$$

and has one positive and one negative eigenvalue. This means that we can find the local zero set of $\tilde{\psi}\left(b, c\right) = \psi\left(s\left(b, c\right), b; c\right)$ by applying Morse's Lemma.

Finally, the condition $\left(s\left(b, c\right), b\right) \notin \Delta$ is satisfied locally, since otherwise one would have that $s(b, c) \equiv b$ locally which would imply $\partial_b s\left(b, c\right) \equiv 1, \partial_c s\left(b, c\right) \equiv 0$; but by our choice of parameters one has $\left(\partial_b s\right)\left(a_o, c_o\right) = -1, \left(\partial_c s\right)\left(a_o, c_o\right) = \frac{1}{1+R^2}$.

An alternative way of looking for period-2-solutions is based on (2.11). Let, with $R_{\theta,\ell} := Re^{i\left(\theta - k^2\ell\right)}$,

$$f\left(x\right) = \left|1 + R_{\theta,\ell}e^{ix\ell}\right|^2 = 1 + R^2 + 2R\cos\left(\theta - k^2\ell + x\ell\right)$$

and

$$h\left(x\right) = \left|1 - R_{\theta,\ell}^2 e^{ix\ell}\right|^2 = 1 + R^4 - 2R^2\cos\left(2\left(\theta - k^2\ell\right) + x\ell\right).$$

(cf. (2.10)). We can then write (2.11) as

$$a = \frac{c}{h\left(a+b\right)} f\left(b\right), \quad b = \frac{c}{h\left(a+b\right)} f\left(a\right).$$

This suggests to study the mapping

$$H : \mathbb{R}_+^2 \longrightarrow \mathbb{R}_+^2 \quad \text{defined by} \quad \begin{pmatrix}x_1\\y_1\end{pmatrix} := H\begin{pmatrix}x\\y\end{pmatrix} = \frac{c}{h\left(x+y\right)}\begin{pmatrix}f\left(y\right)\\f\left(x\right)\end{pmatrix}.$$

Diagonal fixed points of H_1, i.e., x_o such that

$$H\begin{pmatrix}x_o\\x_o\end{pmatrix} = \begin{pmatrix}x_o\\x_o\end{pmatrix}$$

correspond to plane wave type fixed points of the mapping $\tilde{T}$ defined earlier; off-diagonal fixed points $\binom{x_o}{y_o}$ correspond to period-2 orbits of $\tilde{T}$. Theorem 2.1 asserts that within suitable parameter ranges such orbits bifurcate from diagonal fixed points.

A direct fixed point search for H can be based on the following properties of H.

Lemma 2.1. $H : \mathbb{R}_+^2 \longleftrightarrow Q$, where

$$Q := \left[\frac{c\left(1-R\right)^2}{\left(1+R^2\right)^2}, \frac{c}{\left(1-R\right)^2}\right]^2.$$

Proof. This is elementary. For example, for the upper bound one uses

$$0 \le f(x) \le (1+R)^2 \text{ and } h(x+y) \ge \left(1-R^2\right)^2,$$

hence $c\frac{f(x)}{h(x+y)} \le \frac{c}{(1-R)^2}$. Note that Q is not empty, as $\frac{(1-R)^2}{(1+R^2)^2} \le \frac{1}{(1-R)^2}$.

Corollary. $H : Q \longrightarrow Q$.

We abbreviate $R_1 = \frac{(1-R)^2}{(1+R^2)^2}$, $R_2 = \frac{1}{(1-R)^2}$.

H has a simple symmetry invariance. Let $\Delta\binom{x}{y} = \binom{y}{x}$, then we see immediately that

$$H \circ \Delta = \Delta \circ H.$$

It follows that fixed points of H will occur either on the diagonal or in symmetric pairs. Third, there is the possibility of period-2-orbits *for* H, i.e., points $\binom{x_o}{y_o}$ such that

$$H\begin{pmatrix} x_o \\ y_o \end{pmatrix} = \begin{pmatrix} y_o \\ x_o \end{pmatrix} \quad \text{and} \quad H\begin{pmatrix} y_o \\ x_o \end{pmatrix} = \begin{pmatrix} x_o \\ y_o \end{pmatrix}.$$

In explicit form, this means

$$\frac{c}{h(x_o+y_o)} f(y_o) = y_o \quad \text{and} \quad \frac{c}{h(x_o+y_o)} f(x_o) = x_o.$$

Let us investigate the linear stability properties of fixed points on the diagonal, i.e., $x_o \in (cR_1, cR_2)$ such that $x_o = c\frac{f(x_o)}{h(2x_o)}$. The point $\binom{x_o}{x_o}$ is linearly (and hence asymptotically) stable if both eigenvalues of the matrix $M := \left.\frac{\partial H}{\partial(x,y)}\right|_{\binom{x_o}{x_o}}$ satisfy the condition $|\lambda| < 1$. If both $|\lambda_1| > 1$ and $|\lambda_2| > 1$, then $\binom{x_o}{x_o}$ is ejective in the sense of Browder ([1], [2]).

We compute

$$\left.\frac{\partial H}{\partial(x,y)}\right|_{\binom{x_o}{x_o}} = \frac{c}{h^2(2x_o)}$$

$$\times \begin{pmatrix} -f(x_o)\,h'(2x_o) & f'(x_o)\,h(2x_o) - f(x_o)\,h'(2x_o) \\ f'(x_o)\,h(2x_o) - f(x_o)\,h'(2x_o) & -f(x_o)\,h'(2x_o) \end{pmatrix}.$$

This matrix is of the type $\begin{pmatrix} a & b \\ b & a \end{pmatrix}$, with characteristic polynomial $(a-\lambda)^2 - b^2$, and eigenvalues $\lambda_1 = a+b$, $\lambda_2 = a-b$, i.e.,

$$\lambda_{1,2} = \frac{c}{g^2(2x_o)}\left[-f(x_o)\,h'(2x_o) \pm \left[f'(x_o)\,h(2x_o) - f(x_o)\,h'(2x_o)\right]\right].$$

By using the definitions of f and h,

$$\lambda_{1,2} = \frac{c\cdot\ell}{\left[1+R^4-2R^2\cos\left(2\left(\theta-k^2\ell\right)+2x_o\ell\right)\right]^2}$$
$$\cdot\left[-2R^2\left(1+R^2+2R\cos\left(\theta-k^2\ell+x_o\ell\right)\right)\cdot\sin\left(2\left(\theta-k^2\ell\right)+2x_o\ell\right)\right.$$
$$\pm\left[-2R\sin\left(\theta-k^2\ell+x_o\ell\right)\cdot\left(1+R^4-2R^2\cos\left(2\left(\cdots\right)+2x_o\ell\right)\right)\right.$$
$$\left.\left.-\left(1+R^2+2R\cos\left(\theta-k^2\ell+x_o\ell\right)\right)2R^2\sin\left(2\left(\cdots\right)+2x_o\ell\right)\right]\right].$$

The factor ℓ suggests that *in general*, as $\ell \nearrow$ or $c \nearrow$, both $|\lambda_1|$ and $|\lambda_2|$ will become larger than 1, i.e., (x_o, x_o) will become ejective.

Remark. As x_o is assumed to be a fixed point, we have

$$x_o \cdot h\left(2x_o\right) = cf\left(x_o\right)$$

and this leads to the simpler representation

$$\lambda_1 = -c\frac{f'\left(x_o\right)}{h\left(2x_o\right)}, \qquad \lambda_2 = -2x_o\frac{h'\left(2x_o\right)}{h\left(2x_o\right)} + c\frac{f'\left(x_o\right)}{h\left(2x_o\right)}.$$

It will be exceptional for λ_1 or λ_2 to be in the range $|\lambda| \leq 1$, but as x_o and even the number of possible fixed points on the diagonal depend on c and ℓ, it is difficult to obtain consistent information at all diagonal fixed points.

Remark. From the symmetry properties of H and the symmetry properties of $\left.\frac{\partial H}{\partial(x,y)}\right|_{\binom{x_o}{x_o}}$ it follows that one eigenvector of H will be parallel to the diagonal, the other orthogonal to the diagonal.

A further analysis of H can be done by setting $\Delta_1 = \left\{\binom{x}{y} \in Q;\ y \leq x\right\}$. Define $\mathbb{H} : \Delta_1 \longrightarrow \Delta_1$ by

$$\mathbb{H}\binom{x}{y} = \begin{cases} H\binom{x}{y} & \text{if } \ H\binom{x}{y} \in \Delta_1 \\ \Delta \circ H\binom{x}{y} & \text{if } \ \Delta \circ H\binom{x}{y} \in \Delta_1 \end{cases}.$$

This $\mathbb{H}$ is continuous and maps Δ_1 into itself. By Schauder's Theorem it follows that $\mathbb{H}$ possesses at least one fixed point in Δ_1 (this does not convey any new information, because we already know about all the fixed points on the diagonal). If all the diagonal fixed points are ejective, it would follow from theorems due to Browder [1] and Nussbaum [2] that $\mathbb{H}$ possesses at least one non-ejective fixed point (which corresponds to either a period-2 orbit of $\tilde{T}$, or a period-2 orbit of H; in the second case the relationship to the ring cavity problem is not clear).

The mappings H (or $\mathbb{H}$) offer themselves to explicit study; for example, it is immediate that lines $x = y + \frac{2\pi}{\ell}\cdot m$, $m \in \mathbb{Z}$, are mapped onto the diagonal; similarly one can identify subsets of Δ_1 which are mapped back into Δ_1, etc. As the significance of such studies is questionable, we do not pursue them further.

For the unphysical but interesting singular case when $R = 1$ we can construct *explicit period-2-solutions* for any nonlinearity $N(s)$ of (1.2). This is done by the following procedure: We consider a plane wave pump function $a(x) = ae^{ikx}$, $a \in C$ and $k \in \mathbb{R}^d$ given; also, for fixed parameters θ and ℓ, choose numbers $m \in \mathbb{Z}$ and $\kappa \in R\backslash(2m+1)\pi\mathbb{Z}$, $m \in \mathbb{Z}$, such that one has

$$N\left(c^2\right) = \frac{\theta - (2m+1)\pi}{\ell} - k^2 \tag{2.19}$$

where $c^2 := \frac{|a|^2}{2(1+\cos\kappa)}$; we remark that it is always possible to satisfy (2.19) for specific m and κ in both cases of the Kerr and the saturating nonlinearity. Now, two plane wave solutions $u(t,x)$ and $v(t,x)$ of the nonlinear Schrödinger equation (1.4) of the form

$$u(t,x) = \alpha e^{i\left[kx - \left(N\left(|\alpha|^2\right)+k^2 t\right)\right]},\ v(t,x) = \beta e^{i\left[kx - \left(N\left(|\alpha|^2\right)+k^2 t\right)\right]}\ \left(x \in \mathbb{R}^d\right)$$

will generate a plane wave period-2-solution of (1.2)–(1.3) (or (1.4)–(1.5)) (i.e. a fixed point of the mapping T^2) if and only if the relations

$$u(0,x) = ae^{ikx} + e^{i\theta} v(\ell,x),\ v(0,x) = ae^{ikx} + e^{i\theta} u(\ell,x),\ u(\ell,.) \neq v(\ell,.) \tag{2.20}$$

hold; if (2.20) is true, then $g(x) := v(\ell,x)$ will be the period-2-solution. One can fulfill (2.20) by setting

$$\alpha := \frac{a}{1+e^{i\kappa}}, \beta := \frac{a}{1+e^{-i\kappa}}$$

and therefore

$$|\alpha|^2 = |\beta|^2 = \frac{|a|^2}{2(1+\cos\kappa)} = c^2,\ \beta = e^{i\kappa}\alpha$$

which implies $u(\ell,.) \neq v(\ell,.)$; also, by (2.19) $e^{i\theta}e^{-i\left(N\left(c^2\right)+k^2\right)\ell.} = -1$ which leads to

$$\begin{aligned} e^{-ikx}\left(u(0,x) - \left[ae^{ikx} + e^{i\theta}v(\ell,x)\right]\right) &= \alpha - a - e^{i\theta}e^{-i\left(N\left(c^2\right)+k^2\right)\ell}\beta \\ &= \alpha\left(1+e^{i\kappa}\right) - a = 0, \\ e^{-ikx}\left(v(0,x) - \left[ae^{ikx} + e^{i\theta}u(\ell,x)\right]\right) &= \beta - a - e^{i\theta}e^{-i\left(N\left(c^2\right)+k^2\right)\ell}\alpha \\ &= \alpha\left(e^{i\kappa}+1\right) - a = 0. \end{aligned}$$

Thus, (2.20) is true with this choice of parameters.

3. Solitary Wave Reduction

In this section we summarize the mathematical results of H. Adachihara, D.W. McLaughlin, J.V. Moloney, A.C. Newell from [11]. Details may be found there. Their goal is to filter out those solutions to (1.2)–(1.3) which behave approximatively like solitary waves, the next simplest case of qualitative behavior of solutions in comparison with the simplest case, the plane waves.

In the case $d = 1$, at least for the Kerr and the saturating nonlinearity, there exist special solutions to (1.2) of *solitary wave type*, namely solutions of the form

$$u_s(t,x;\lambda) = S(\lambda x;\lambda)\exp[i\mu t] \tag{3.1}$$

where S is a real, even solution of the differential equation

$$S'' - \frac{\mu}{\lambda^2} S = \frac{1}{\lambda^2} N\left(S^2\right) S, \tag{3.2}$$

vanishing at infinity. Often, solutions of type (3.1) are called *stationary solutions*. For the case of the Kerr or saturating nonlinearities, it is convenient to take $\mu = \lambda^2 - 1$; thus S should satisfy the equation

$$S'' - S = \frac{1}{\lambda^2}\left[N\left(S^2\right) - 1\right] S. \tag{3.3}$$

By using symmetries of equation (1.2) one finds a more general family of solitary wave solutions of (1.2) of the form

$$u_s\left(t, x; \lambda, \gamma\right) = S\left(\lambda\left(x - vt - a\right); \lambda\right) \exp\left[i\left(\lambda^2 - 1\right)t + vx + \gamma\right]; \tag{3.4}$$

here a, v and γ are free parameters (see [19], [20] for the most general class of solitary wave type solutions of equations like (1.2)). In [11], because of symmetry properties of the physical problem, mainly solutions of type

$$u_s\left(t, x; \lambda, \gamma\right) = S\left(\lambda x; \lambda\right) \exp\left[i\left(\lambda^2 - 1\right)t + \gamma\right] \tag{3.5}$$

are discussed. In the case of the Kerr nonlinearity (i.e., $1 - N\left(S^2\right) = 2S^2$) one has the explicit formula $S\left(z; \lambda\right) = \lambda \text{ sech } (z)$, $z = \lambda x$; in the saturating case one has $1 - N\left(S^2\right) = 2S^2(1 + 2S^2)$, and the existence of smooth, even, and decaying solutions to (3.3) can be shown, but no explicit formulas may be available (see [19], [20]).

The linearization of (1.2) around solitary waves of type (3.4) has the form

$$i\frac{\partial \tilde{u}}{\partial t} = -\tilde{u}_{xx} + \left[N\left(S^2\right) + N'\left(S^2\right)S^2\right]\tilde{u} + \left[N'\left(S^2\right)u_s^2\right]\overline{\tilde{u}}. \tag{3.6}$$

One can get a family of four linear independent solutions of (3.6) from the formulas (see [11]):

$$\tilde{u}^{(1)} = \left.\frac{\partial}{\partial a} u_s\right|_{a=0,\nu=0,\gamma=0}, \tag{3.7a}$$

$$\tilde{u}^{(2)} = \left.\frac{\partial}{\partial \gamma} u_s\right|_{a=0,\nu=0,\gamma=0}, \tag{3.7b}$$

$$\tilde{u}^{(3)} = \left.\frac{\partial}{\partial \nu} u_s\right|_{a=0,\nu=0,\gamma=0}, \tag{3.7c}$$

$$\tilde{u}^{(4)} = \left.\frac{\partial}{\partial \lambda} u_s\right|_{a=0,\nu=0,\gamma=0}, \tag{3.7d}$$

where u_s is the solution of type (3.4); these four solutions generate a 4-dimensional subspace of the solution space of the linearized equation (see [11; Appendix B]).

To obtain an approximative formal solitary wave reduction of the dynamical system given by (1.2)–(1.3) one can make the ansatz

$$u = u_s + \tilde{u} + \tilde{R} \tag{3.8}$$

for solutions; here u_s is a solitary wave, $\tilde{u}$ solves the linearized equation, and $\tilde{R}$ is a small remainder term such that u is a solution of (1.2); a function of type $u = u_s + \tilde{u}$ is in general not a solution of the nonlinear equation (1.2), and neither is the sum of two solitary waves; the term $\tilde{R}$ is omitted in [11]. More precisely, one may consider the solutions of the system (1.2)–(1.3) for any circuit to have the form

$$u(t,x) = u_s(t,x;\lambda^\varepsilon,\gamma^\varepsilon,a^\varepsilon,\nu^\varepsilon) + \varepsilon\tilde{u}(t,x) + \tilde{R}_\varepsilon(t,x), \tag{3.9}$$

$\tilde{u}(t,x)$ being a solution of the linearized system (with linearization around u_s), and a remainder term $\tilde{R}_\varepsilon$ of order $O(\varepsilon^2)$ in some metric; again, this remainder term is not considered in [11].

If one assumes that the space generated by the solutions (3.7a–d) is only 2-dimensional due to the symmetry assumptions of the physical system, and furthermore, under the assumption that the output of the n-th circuit down the nonlinear medium is (approximately) a solitary wave, i.e. of the form (3.8) with parameters λ_n, γ_n, then the solitary wave component of the solution of (1.2) of the $(n+1)$-st circuit with parameters $\lambda_{n+1}, \gamma_{n+1}$, and the projection of this solution onto the two-dimensional subspace resp., describe the solitary wave components of the wave function of the $(n+1)$-st pass through the device. These components (which are given by the parameters $\lambda_{n+1}, \gamma_{n+1}$) can formally be computed (and this is done in [11]) by the ansatz

$$i\frac{\partial u_{n+1}}{\partial t} = -u_{n+1,xx} + N(|u_{n+1}|^2)u_{n+1}, \tag{3.10a}$$

$$u_{n+1}(0,x) = a(x) + Re^{i\theta}e^{i[(\lambda_n^2-1)\ell+\gamma_n]}S(\lambda_n x;\lambda_n). \tag{3.10b}$$

The result is a set of recursion formulas for the parameter sequence (λ_n, γ_n) (see [11]) which may be seen as a 2-dimensional dynamical system, and which represents some 2-dimensional reduction of the given infinite dimensional system around solitary wave components. For the Kerr case this 2-dimensional mapping is given explicitly in [11] in terms of integrals over the given pump field $a(x)$ and hyperbolic functions.

This ansatz to get the structure of the solitary wave components is formal inasmuch as the error terms are not treated rigorously. It is not clear whether these error terms do not include further solitary wave type components of the dynamical system. Nevertheless, in the Kerr case, there is a similarity between the transcendental equations for the plane wave fixed points of *Section 2* and the recursion formulas for the parameter sequence (λ_n, γ_n) of the reduced solitary wave system (3.10a–b).

4. Existence results. Infinite-dimensional case.

Virtually all rigorous analytical results about the model deal with the "plane wave" case discussed earlier; i.e. the pump function a and thus the solutions u_n are assumed not to depend on the transverse variables $x \in \mathbb{R}^d$ $(d = 1, 2)$. In this case the dynamical system $\{T^n\}_{n\in\mathbb{N}}$ is finite-dimensional. Describing analytically transverse effects (i.e. effects with explicit dependence on the transverse variables) turns out to be a hard problem. In this section we briefly describe what is known in this infinite-dimensional case.

In [12] F. Merle proved a "pure" existence result by means of Schauder's fixed point theorem, not addressing questions of uniqueness and/or stability.

Theorem 4.1 (Merle) *Let $d = 2$ and $N(s) := -s + \varepsilon_0 s^2$, where $\varepsilon_0 > 0$ is a given number. Then, for any $R \in (0,1)$, $\theta \in \mathbb{R}$, $\ell \in (0,\infty)$ and $a \in X := \{u \in H^1(\mathbb{R}^d) \mid \int |x|^2|u(x)|^2dx < \infty\}$, there exists a fixed point $f \in X$ of the operator T. (Of course, the operator T is to be understood as the one associated with the parameter set R, ℓ, θ, a; see (1.6)).*

Note that the nonlinearity appearing in the theorem does not coincide with the Kerr nonlinearity (which can be interpreted as a first-order approximation of the saturating nonlinearity). Rather, it consists of the first *two* terms of the Taylor expansion of the saturating term. (At the end of [12] more general nonlinearities, including the saturating one, are considered.) This modification guarantees the global in-time solvability of the nonlinear Schrödinger equation involved. The crucial step of the proof consists of identifying a suitable compact and convex subset $K \subset L^2(\mathbb{R}^d)$ which can be shown to be invariant under the map T. Such a set is

$$K := \{f \in H^1(\mathbb{R}^d) \mid \|f\|_2 \le k_0, \quad E_+(f) \le k_1, \quad \int |x|^2|f(x)|^2dx \le k_2\},$$

where

$$E_+(f) := \int |\nabla f|^2 dx + \frac{\varepsilon_0}{3}\int |f|^6 dx.$$

We have checked that, under additional assumptions, Merle's result holds true in the Kerr case ($\varepsilon_0 = 0$) as well.

Theorem 4.2 *Let $N(s) = -s$ (Kerr nonlinearity). Then the assertion of Theorem* 4.1 *remains valid if*

(i) $d = 1$ or

(ii) $d = 2$ *and* $\|a\|_{L^2} \le (1-R)(1-\rho^2)^{1/2}C$ *for some $\rho \in (R,1)$. Here C denotes the "best" constant with the following property: The initial value problem to NLS with the Kerr nonlinearity is globally well-posed for initial data $\varphi \in H^1(\mathbb{R}^2)$ satisfying $\|\varphi\|_{L^2(\mathbb{R}^2)} < C$ (cf. [21], Theorem 3.17).*

The existence of fixed points when N is given by the Kerr nonlinearity can also be established under the assumption that either ℓ or $a \in H^1$ is small. These fixed points can be obtained by iterating the map T and may be interpreted as the infinite-dimensional analogues of the stable "lower-branch" fixed points of the plane wave map.

Theorem 4.3 *Let $N(s) = -s$, $d \in \{1,2\}$, $R \in (0,1)$, $\theta \in \mathbb{R}$, $\ell \in (0,\infty)$ and $a \in H^1(\mathbb{R}^d)$. Then there exist functions $c_d = c_d(s)$ with $c_d(s) \to 0$ as $s \to 0$*

($d = 1, 2$) such that the sequence $(T^n(0))$ converges to a (locally unique) fixed point of T if one of the following conditions is fulfilled.

(i) $d = 1$ *and* $c_1(\|a\|_{H^1(\mathbb{R})})\ell < 1 - R$.

(ii) $d = 2$, $\|a\|_{L^2} < (1 - R)(1 - \rho^2)^{1/2}C$ *and* $c_2(\|a\|_{H^1(\mathbb{R}^2)})\ell^{1-2/\sigma} < 1 - 2^{2/\sigma}R$ *for some* $\rho \in (R, 1)$, $\sigma \in (2, \infty)$, *where C is the constant introduced in part (ii) of the previous theorem.*

The theorem is proved by verifying that T is contractive: We consider the sequence $u_n \in X := L^\sigma(0, \ell; L^q(\mathbb{R}^d)) \cap C([0, \ell], L^2(\mathbb{R}^d))$, defined by $u_0(t) := 0$ and (1.2), (1.3); i.e., $T^n 0 = u_n(., \ell)$ and

$$u_{n+1}(t) := e^{it\Delta}(a + Re^{i\theta}u_n(\ell)) + (GN(|u_{n+1}|^2)u_{n+1})(t).$$

Here (σ, q) is an admissible pair of exponents and G is the integral operator $(Gf)(t) := -i\int_0^t e^{i(t-s)\Delta}f(s)ds$. Then, a uniform $H^1(\mathbb{R}^d)$-bound, which follows from the assumptions (i) and (ii), together with standard tools such as Strichartz estimates and Sobolev embeddings imply the convergence of the sequence $(u_n) \subset X$. Thus the sequence $u_n(\ell) = T^n(0)$ converges in $L^2(\mathbb{R}^d)$ and we get $f := \lim u_{n+1}(\ell) = \lim Tu_n(\ell) = T \lim u_n(\ell) = Tf$.

By utilizing an Implicit Function Theorem (IFT) the existence of fixed points (solutions of the *infinite*-dimensional ring cavity problem) close to plane wave fixed points (solutions of the *finite*-dimensional ring cavity problem) can be established.

We consider the case $d = 1$ and work with periodic boundary conditions, as the functional setting has to accommodate plane waves, i.e. constant functions. We want to apply the IFT to the map $\Phi : X \times X \to X$, $(a, f) \mapsto T_a f - f$, where $X := H^1_{per}(0, A)$ (here A is the "aperture" of the cavity, i.e. the width of the cross section of the cavity) and T_a denotes the "ring cavity operator" associated with the pump field a. Clearly, $f \in X$ is a fixed point of T_a iff $\Phi(a, f) = 0$. Now let g be a plane wave fixed point associated with an x-independent pump field b. Then the IFT will guarantee the existence of fixed points f_a (lying in a certain neighbourhood of g) of T_a for all pump fields a taken from a certain neighbourhood of b if the operator $\partial_f\Phi(b, g)$ can be shown to possess a bounded inverse. Moreover, if g is stable, a suitable choice of the neighbourhood of g will make sure that the fixed points f_a are stable as well. To determine $\partial_f\Phi(b, g)$ let v_z be the solution of the linearized NLS

$$iv_t + v_{xx} = -[N + N'|h|^2] \cdot v - [e^{2iNt}N'h^2] \cdot \bar{v}\,, \qquad v(0) = Re^{i\theta}z, \tag{4.1}$$

where $h := b + Re^{i\theta}g$, $N := N(|h|^2)$ and $N' := N'(|h|^2)$. Then $\partial_f\Phi(b, g) = v_z(\ell) - z$ and we have the following criterion.

Theorem 4.4

(i) *For every $k \in \mathbb{Z}$ define* $\alpha := -\frac{4\pi^2}{A^2}k^2 + N'|h|^2$, $\omega^2 := \frac{16\pi^4}{A^4}k^4 - 2N'|h|^2\frac{4\pi^2}{A^2}k^2$ *and assume that the function*

$$P(k) = R^2 + 1 - 2R\left[\cos(N\ell + \theta)\cos(\omega\ell) - \frac{\alpha}{\omega}\sin(N\ell + \theta)\sin(\omega\ell)\right]$$

has no zeros. Then the map $\partial_f\Phi(b, g)$ has a bounded inverse.

(ii) *Let the condition in (i) be satisfied. Then, for any (x-dependent) pump field which is sufficiently close to b, there exists a fixed point of the infinite-dimensional ring cavity map (with periodic boundary conditions) close to the plane wave fixed point associated with b. Since these (infinite-dimensional) fixed points have the same stability properties as the plane wave fixed points, this shows the occurrence of bistability in the infinite-dimensional case. (Recall that the plane wave map exhibits hysteresis and bistable behavior.)*

Part (ii) of the theorem complements an *in*stability result by McLaughlin et al. [4, 5, 6] who argue that the plane wave fixed points become unstable in the infinite-dimensional setting if the condition in (i) is violated.

5. Open problems

There are many open problems in the rigorous treatment of ring cavity systems. We list some of them:

1.) Rigorous analytical treatment of the questions discussed in *Section 3*, e.g. inclusion of the error terms which arise when exact solutions of the corresponding nonlinear Schrödinger equation are involved; also, this needs a nonlinear stability theory for this class of equations which seems not to have been done.

2.) Most of the mathematical results are only known for the case of dimension $d = 1$ (see *Section 1*), i.e. one longitudinal and one transversal dimension; the case $d = 2$ is usually not considered, but is the physically important one; also, the solitary wave reduction is an open question in higher dimensions since e.g. explicit formulas for the solitary waves are not available here; the same is true for an exact treatment of the linearized equations.

3.) Even for dimension $d = 1$ a rigorous theory of the existence of period-m-loops ($m > 2$) and their stability, and of the period doublings is missing; the same can be stated for the analysis of the (numerically observed) filament patterns in case $d = 2$, and for a global stability analysis of the dynamical system at all.

References

[1] F. Browder, Another generalization of the Schauder fixed point theorem, *Duke Math. J.* **32** (1965), 399–406.

[2] R.D. Nussbaum, Periodic solutions of some nonlinear autonomous functional differential equations, *Ann. Mat. Pura Appl.* **101** (1974), 263–306.

[3] A.J. Lichtenberg, M.A. Lieberman, *Regular and Stochastic Motion*, Springer-Verlag, Berlin 1983.

[4] D.W. McLaughlin, J.V. Moloney, A.C. Newell, Solitary waves as fixed points of infinite-dimensional maps for an optical bistable ring cavity. Fluid and plasmas: Geometry and dynamics, Proc. Conference Boulder, Colorado 1983; Contemp. Math. Vol. 28, 369–376, AMS, Providence, RI 1984.

[5] S.M. Hammel, C.K.R.T. Jones, J.V. Moloney, Global dynamical behavior of the optical field in a ring cavity, *J. Opt. Soc. Am. B* **2**, no.4 (1985), 552–564.

[6] D.W. McLaughlin, J.V. Moloney, A.C. Newell, New class of instabilities in passive optical cavities. *Phys. Rev. Lett.* **54** (1985), 681–684.

[7] H.M. Gibbs, *Optical Bistabilty: Controlling Light with Light*, Academic Press, London 1985.

[8] A. Aceves, H. Adachihara, C. Jones, J.C. Lerman, D.W. McLaughlin, J.V. Moloney, A.C. Newell, Chaos and coherent structures in partial differential equations, *Physica D* **18** (1986), 85–112.

[9] J.V. Moloney, Plane wave modulational instabilities in passive optical resonators. Nonlinear phenomena and chaos. Malvern Phys. Ser., Vol. 2, 214–245, Hilger, Bristol 1986.

[10] J.V. Moloney, H. Adachihara, D.W. McLaughlin, A.C. Newell, Fixed points and chaotic dynamics of an infinite-dimensional map. Proc. Conf. Chaos, Noise and Fractals, Como 1986; Malvern Phys. Series, Vol. 3, 137–186, Hilger, Bristol 1987.

[11] H. Adachihara, D.W. McLaughlin, J.V. Moloney, A.C. Newell, Solitary waves as fixed points of infinite-dimensional maps for an optical ring cavity. Analysis, *J. Math. Phys.* **19** (1986), 63–85.

[12] F. Merle, A note on the existence of a solution for a model of McLaughlin, Moloney, and Newell for optical ring cavities, *J. Math. Phys.* **30** (1989), 1643–1647.

[13] N.B. Abraham, W.J. Firth, Overview of transverse effects in nonlinear-optical systems, *J. Opt. Soc. Am. B* **2**, no.6 (1990), 951–962.

[14] J.V. Moloney, H. Adachihara, R. Indik, C. Lizarraga, R. Northcutt, D.W. McLaughlin, A.C. Newell, Modulational-induced optical pattern formulation in a passive optical-feedback system, *J. Opt. Soc. Am. B* **7**, no.6 (1990), 1039–1044.

[15] G.S. McDonald, W.J. Firth, Spatial solitary-wave optical memory, *J. Opt. Soc. Am. B* **7**, no. 7 (1990), 1328–1335.

[16] P. Mandel, *Theoretical Problems in Cavity Nonlinear Optics*, Cambridge Univ. Press, Cambridge 1997.

[17] Y. Chen, D.W. McLaughlin, Focusing-defocusing effects for diffusion-dominated bistable optical array, *J. Opt. Soc. Am. B* **16**, no.7 (1999), 1087–1098.

[18] Y. Chen, D.W. McLaughlin, Diffraction effects on diffusive optical arrays and optical memory, *Physica D* **138** (2000), 163–195.

[19] L. Brüll, H. Lange, Stationary and solitary wave type solutions of singular nonlinear Schrödinger equations, *Math. Methods Appl. Sciences* **8** (1986), 559–575.

[20] L. Brüll, H. Lange, Solitary waves for quasilinear Schrödinger equations, *Expositiones Math.* **4** (1986), 279–288.

[21] C. Sulem, P.-L. Sulem, *The Nonlinear Schrödinger Equation. Self-Focusing and Wave Collapse.* Applied Mathematical Sciences, 139. Springer-Verlag, New York 1999.

Reinhard Illner
Department of Mathematics and Statistics
University of Victoria
P.O. Box 3045
Victoria, B.C. V8W 3P4
Canada
e-mail: `rillner@math.uvic.ca`

Horst Lange
Mathematisches Institut
Universität Köln
Weyertal 86–90
D-50931 Köln
e-mail: `lange@math.uni-koeln.de`

Holger Teismann
Department of Mathematics and Statistics
University of Saskatchewan
236 McLean Hall, 106 Wiggins Road
Saskatoon, SK, Canada S7N 5E6
e-mail: `teismann@math.usask.ca`

Progress in Nonlinear Differential Equations
and Their Applications, Vol. 50, 197–216

Hadamard Well-posedness of Weak Solutions in Nonlinear Dynamic Elasticity-full von Karman Systems

Herbert Koch and Irena Lasiecka*

1. Introduction

Dynamic systems of nonlinear elasticity described by von Karman equations, one of the fundamental equations in mathematical physics, have a long tradition in the literature [11, 5, 24, 33, 16, 15, 14, 8]. Their importance stems from the fact that many physical phenomena related to oscillation theory are described by dynamic elastic models. Propagation of waves, oscillations and vibrations of membranes, plates, shells, etc. are governed by nonlinear elastic systems involving wave and plate equations or combination thereof.

From the mathematical point of view, these models are of hyperbolic type, with no inherent smoothing mechanism typically present in parabolic like dynamics. This fact, in the case of nonlinear dynamics, makes the issue of well-posedness of solutions rather subtle. Standard methods of nonlinear analysis typically fail and one has to look for more ad hoc strategies. In fact, many problems related to existence/ uniqueness/regularity of such solutions are still in a category of open problems. One of the technical difficulties is that there is an intrinsic "gap" between the existence and uniqueness for "weak" solutions. Indeed, while existence of weak solutions is often a rather easy task which can be accomplished by using Faedo-Galerkin method, the uniqueness of weak solutions is more difficult. This is due to the fact that the nonlinear terms, while having some "conservation" of energy properties (typical for Hamiltonian systems), they are not bounded in *finite energy space*, i.e: the space where weak solutions have a priori bounds and exist globally.

On the other hand, uniqueness, and also continuous dependence of solutions with respect to initial conditions, are the properties which are critical in almost all applications. Most recently, with a stimulation coming from control theory, there has been a renewed interest in the theory of well-posedness for these systems. Indeed, problems such as controllability, stabilization, theory of attractors call for a good grasp of properties of finite energy solutions.

* Research partially supported by the National Science Foundation under Grant DMS-0104305

A rigorous justification of many results/estimates obtained in these fields pend upon a good understanding of well-posedness and regularity of solutions, particularly with respect to finite energy topology. Of particular interest are models with dissipative boundary conditions. This is because boundary dissipation is an attractive tool for stabilizing the model or for affecting long time behaviour (attractors) of nonlinear dynamics. This explains why there has been recently a rather intense activity in the area of well-posedness of such systems, with several new results in place.

In what follows we shall concentrate on a fairly general (rather complex) model – referred to as *full von Karman system* – which comprises of strongly coupled wave and plate equations [11, 16, 15, 14, 8]. This model describes nonlinear oscillations of a plate with large displacements and takes into account in plane accelerations. This latter feature is responsible for a vectorial structure of the PDE model.

The goal of this paper is to present full well-posedness of weak and strong solutions for von Karman model. We shall first provide a brief discussion of the results available in the literature which will be followed by presentation of some new results.

Though the techniques presented are tailroed to the problem in hand, we believe and hope that some of the strategies and ideas can be also used in the context of different nonlinear (hyperbolic) dynamics which display similar set of difficulties.

2. Full von Karman system – PDE model

Here, the variables w and $u = (v_1, v_2)$ represent, respectively, the vertical and the in plane displacements of a thin plate occupying a two dimensional domain Ω with sufficiently smooth boundary Γ. With the above notation, the governing equations to be considered are given by [14, 16]:

$$\begin{aligned} u_{tt} - div(\mathcal{C}[\epsilon(u) + f(\nabla w)]) = 0 \text{ in } \Omega \times (0,\infty) \\ (I - \gamma\Delta)w_{tt} + D\Delta^2 w - div(\mathcal{C}[\epsilon(u) + f(\nabla w)]\nabla w) = 0 \text{ in } \Omega \times (0,\infty) \end{aligned} \tag{1}$$

with traction forces prescribed for the horizontal displacements $u = (v_1, v_2)$

$$[\mathcal{C}[\epsilon(u) + f(\nabla w)]\nu + ah(u_t) = 0; \text{ on } \Gamma \times (0,\infty) \tag{2}$$

and Dirichlet (**clamped**) boundary conditions imposed on the vertical displacement w

$$w = \nabla w = 0 \text{ on } \Gamma \times (0,\infty) \tag{3}$$

The vector ν denotes an unit outward normal vector to the boundary Γ. With (1) we associate the initial conditions

$$u(0) = u_0, u_t(0) = u_1, \;\; w(0) = w_0, w_t(0) = w_1; \text{ in } \Omega \tag{4}$$

D represents the flexural rigidity, the constant $0 < \mu < 1/2$ is Poisson's modulus and the constant $0 \leq \gamma \leq M$ is proportional to the square of the thickness of the plate. The constant $\gamma > 0$ represents the presence of rotational inertia.

The fourth order tensor $\mathcal{C}$ is defined by

$$\mathcal{C}(\epsilon) \equiv \frac{E}{(1-2\mu)(1+\mu)}[\mu \; trace \; \epsilon \; I + (1-2\mu)\epsilon]$$

with the strain tensor given by

$$\epsilon(u) \equiv 1/2(\nabla u + \nabla^T u)$$

It can be easily verified that the tensor $\mathcal{C}$ is symmetric and strictly positive. The function f is given by the formula

$$f(s) \equiv \frac{1}{2} s \otimes s; \;\; s \in R^2$$

Vector function $h : R^2 \to R^2$ appearing in the boundary conditions (2) represents potential (nonlinear) dissipation on the boundary. This dissipation is often applied [14, 26, 21] in order to control the asymptotic behaviour of the solutions. We make the following assumption imposed on h (which is typical in boundary stabilization problems).

Assumption 1. *Vector function $h : R^2 \to R^2$ is assumed continuous, monotone increasing and of at most linear growth at infinity. This is to say, we assume the following condition*

$$(h(s) - h(y), s - y)_{R^2} \geq 0; \quad \forall \; s, y \in R^2 \tag{5}$$

$$m|s| \leq |h(s)| \leq M|s|; \quad \forall s \in R^2; \quad |s| \geq R, \quad 0 < m \leq M < \infty \tag{6}$$

The constant a in (2) is nonnegative. If $a = 0$, there is no dissipation on the boundary.

The natural energy functional associated with the system (1) and representing kinetic and potential energy is given by:

$$E(t) = E_k(t) + E_p(t) \tag{7}$$

The kinetic energy and potential energy are respectively defined by:

$$E_k(t) = \int_\Omega |u_t|^2 + w_t^2 + \gamma|\nabla w_t|^2 d\Omega \tag{8}$$

$$E_p(t) = \int_\Omega (\mathcal{C}N(u,w) \cdot N(u,w)) + D|\Delta w|^2 d\Omega \tag{9}$$

where the stress resultants $N(u, w)$ is defined by

$$N(u,w) \equiv \epsilon(u) + f(\nabla w)$$

It is well known that $E_p(t)$ is topologically equivalent to $[H^1(\Omega)]^2 \times H^2(\Omega)$ topology. In particular, the following inequalities resulting from Korn's inequality and Sobolev's embeddings will be used frequently:

$$\begin{aligned} |u|_{H^1(\Omega)} &\leq C[|N(u,w)|_{L_2(\Omega)} + |\nabla w|^2_{L_4(\Omega)} + |u|_{L_2(\Omega)}] \\ &\leq C[|N(u,w)|_{L_2(\Omega)} + |w|^2_{H^2(\Omega)} + |u|_{L_2(\Omega)}] \end{aligned} \tag{10}$$

Solutions whose energy is bounded are called *finite energy solutions.* In fact, based on topological equivalence of potential energy with $[H^1(\Omega)]^2 \times H^2(\Omega)$, finite energy solutions are solutions such that

$$(u(t), w(t)) \in [H^1(\Omega)]^2 \times H_0^2(\Omega), \ (u_t(t), w_t(t)) \in [L_2(\Omega)]^2 \times H_\gamma^1(\Omega)$$

where

$$H_\gamma^1(\Omega) \equiv L_2(\Omega), \ \gamma = 0; \ H_\gamma^1(\Omega) \equiv H_0^1(\Omega); \ \gamma > 0$$

In what follows we shall adopt the following notation:

$$|u|_{s,D} \equiv |u|_{H^s(D)}, (u,v)_D \equiv \int_D uv dx$$

where for D we shall substitute either Ω or Γ. Spaces $H^s(D), s \geq 0$ are classical Sobolev's spaces [10] with $H^{-s}(D) \equiv H_0^s(D)', s > 0$.

Our main interest is the study of *weak* (finite energy) solutions corresponding to (1). Such solutions are defined, as usual, variationally. To this end this we introduce test functions $\xi \in [H^1(\Omega)]^2$ and $\psi \in H_0^2(\Omega)$, which lead to the following classical variational formulation of the original system (1) with $\gamma > 0$.

Find

$$u(t) \in [H^1(\Omega)]^2, w(t) \in H_0^2(\Omega), u_t(t) \in [L_2(\Omega)]^2, w_t(t) \in H_0^1(\Omega)$$

such that *for all test functions* $\xi \in [H^1(\Omega)]^2$ *and* $\psi \in H_0^2(\Omega)$ *the following identities are valid:*

$$\frac{d}{dt}(u_t(t), \xi)_\Omega + (\mathcal{C}[\epsilon(u(t)) + f(\nabla w(t))], \epsilon(\xi))_\Omega + a(h(u_t(t)), \xi)_\Gamma = 0$$

$$\begin{aligned} &\frac{d}{dt}[(w_t(t), \psi)_\Omega + \gamma(\nabla w_t(t), \nabla\psi)_\Omega] + D(\Delta w(t), \Delta\psi)_\Omega \\ &\qquad + (\mathcal{C}[\epsilon(u(t)) + f(\nabla w(t))], \nabla\psi \otimes \nabla w)_\Omega = 0 \end{aligned} \tag{11}$$

with the initial conditions

$$\begin{aligned} u(0) &= u_0 \in [H^1(\Omega)]^2; \quad u_t(0) = u_1 \in [L_2(\Omega)]^2; \\ w(0) &= w_0 \in H_0^2(\Omega), \quad w_t(0) = w_1 \in H_0^1(\Omega) \end{aligned} \tag{12}$$

Remark 2.1. *Other (than clamped) boundary conditions can be associated with von Karman plate system* (1) *as well. Typical choices are*

- (**Hinged** *boundary conditions)*

$$w = 0, \;\; \Delta w = h_1(\frac{\partial}{\partial \nu} w_t) \tag{13}$$

- (**Free** *boundary conditions)*

$$D[\Delta w + (1-\mu)B_1 w] = -h_1(\frac{\partial}{\partial \nu} w_t)$$

$$D[\frac{\partial}{\partial \nu}\Delta w + (1-\mu)B_2 w] - \gamma \frac{\partial}{\partial \nu} w_{tt} - [\mathcal{C}[\epsilon(u) + f(\nabla w)]\nu \cdot \nabla w]$$

$$= -\frac{\partial}{\partial \tau} h_2(\frac{\partial}{\partial \tau} w_t) \tag{14}$$

where the boundary operators B_1, B_2 *are defined by:*

$$B_1 w \;\; \equiv \;\; 2\nu_1\nu_2 w_{x,y} - \nu_1^2 w_{y,y} - \nu_2^2 w_{x,x};$$

$$B_2 w \;\; \equiv \;\; \frac{\partial}{\partial \tau}[(\nu_1^2 - \nu_2^2)w_{x,y} + \nu_1\nu_2(w_{y,y} - w_{x,x})].$$

Continuous, nondecreasing functions h_i *are subject to Assumption* 1 *and represent a potential nonlinear dissipation on the boundary*

3. Main Results

When *in plane accelerations are not accounted* in the model (1), this system decouples into a linear system of elasticity and nonlinear von Karman plate equation, where effects of nonlinearity are represented by the Airy's stress function. The latter equation is often referred to as "modified" von Karman equation, which is, in fact, a scalar equation but with a nonlocal nonlinear term (see e.g; [8, 10, 18, 14]). Vertical displacement of a plate is then described by the following scalar equation

$$(I - \gamma\Delta)w_{tt} + D\Delta^2 w = [\mathcal{F}(w), w] \text{ in } \Omega \times (0, \infty)$$

$$w = \nabla w = 0 \text{ on } \Gamma \times (0, \infty)$$

$$w(0) = w_0 \in H_0^2(\Omega), \;\; w_t(0) = w_1 \in H_\gamma^l(\Omega), \tag{15}$$

Here the von Karman bracket $[\cdot, \cdot]$ is given by

$$[w, v] \equiv w_{xx}v_{xx} - 2w_{xy}v_{xy} + w_{yy}v_{yy}$$

and the Airy's stress function defined as a solution to the following nonlinear boundary value problem:

$$\Delta^2 \mathcal{F}(w) = -[w, w] \text{ in } \Omega; \;\; \mathcal{F}(w) = \nabla \mathcal{F}(w) = 0 \text{ on } \Gamma$$

The energy associated with this model (15) is given by

$$E(t) \equiv \int_\Omega [|w_t(t)|^2 + \gamma|\nabla w_t(t)|^2 + |\Delta w(t)|^2 + 1/2|\Delta \mathcal{F}(w(t))|^2]dx$$

In addition to clamped boundary conditions imposed on w, other boundary conditions associated with (15) such as *hinged* (see (13) or *free* (see (14) are frequently considered as well.

Modified von Karman equations received considerable attention in the literature [10, 8, 7, 13, 4]. When rotational forces are accounted for in, i.e: $\gamma > 0$ in (15), well-posedness theory for these equations is well understood and established. In fact, existence, uniqueness and continuous dependence on the data are well known for both finite energy solutions and also strong solutions [10, 32]. As the result of a satisfactory well-posedness theory developed for the model, topics related to long time behaviour with either internal dissipation [3] or boundary dissipation [19] and existence and properties of attractors have been pursued by several authors (see [3, 19] and references therein). Instead, when $\gamma = 0$ in (15), the situation is more difficult. The nonlinear term $[\mathcal{F}(w), w]$ is not a priori bounded in the energy space. Indeed, following computations in [10]

$$w \in H^2(\Omega) \Rightarrow [w, w] \in L_1(\Omega) \subset H^{-1-\epsilon}(\Omega)$$

Hence

$$\mathcal{F}(w) \in H^{3-\epsilon}(\Omega) \Rightarrow [\mathcal{F}(w), w] \in H^{-\epsilon}(\Omega)$$

where $\epsilon > 0$. These arguments based on Sobolev's embeddings and classical elliptic theory lead to the "loss" of ϵ derivative with respect to finite energy space. This was the reason why the issue of well-posedness (uniqueness and continuous dependence) of solutions to equation (15) with $\gamma = 0$ has been open for a long time [10, 14]. Mathematical interest and challenge in this model is due to the he fact that nonlinear terms in the equation (with $\gamma = 0$) are not "a priori" bounded in the topology induced by natural energy function. While an existence of weak solutions can be easily shown by means of Faedo-Galerkin approximations [8, 10], the issue of *uniqueness of weak solutions* and *global existence of regular solutions* is much more subtle and difficult issue. However, by exploiting scalar nature of the equation together with the presence of Airy's stress function, both global existence of regular solutions [13, 4, 2, 32] and uniqueness of weak solutions [9, 7] has been established. In particular, it was shown in [9], by establishing an additional regularity displayed by the Airy's stress function, that the nonlinear term in the equation is in fact bounded in the energy space. Indeed, by using compensated compactness methods [6] together with properties of Hardy spaces H^1, it was shown in [9] that

$$|\mathcal{F}(w)|_{W^{2,\infty}(\Omega)} \leq C|w|^2_{2,\Omega}$$

This regularity allowed to prove not only the uniqueness of weak solutions, but also continuous dependence with respect to the initial data in the finite energy space. This is to say solutions $w(t)$ of (15) are in

$$C([0,T]; H^2(\Omega)) \cap C^1([0,T]; L_2(\Omega))$$

and depend continuously (in that norm) with respect to initial data $w_0 \in H_0^2(\Omega)$, $w_1 \in L_2(\Omega)$.

The situation for *full von Karman system* introduced in (1) is, however, more complex. There is a strong nonlinear coupling between waves and plate, and there is no decoupling via the Airy's stress function. While in the one-dimensional case full analysis of the well-posedness was carried out in [17], the lack of appropriate Sobolev's embeddings in two dimensions make the analysis much more difficult. In fact, for the two-dimensional model, even with $\gamma > 0$ (considered in this paper), the analysis is already challenging, since the nonlinear terms appearing in the equations are not known to be bounded in finite energy topology. To see this, let's take $w \in H^2(\Omega), u \in [H^1(\Omega)]^2$ and compute

$$div f(\nabla w) \in H^{-\epsilon}(\Omega);\ div \epsilon(u) \in H^{-1-\epsilon}(\Omega)$$

Thus, in the case $\gamma = 0$ the "loss" of differentiability with respect to finite energy topology is $1 + \epsilon$ derivative. But even in the case of $\gamma > 0$ there is a loss of ϵ derivative. This loss does not affect that much the proof of existence of solutions, but it has a major impact on uniqueness and continuous dependence. Indeed, while existence of *weak* solutions and global existence of regular solutions with linear boundary dissipation (i.e; $h(s)$ in (1) and h_i in (14) are linear) and $\gamma > 0$ was shown in [27, 26], the issue of uniqueness of weak solutions was left out as an open problem. In [31], the authors prove uniqueness of weak solutions for the case of $\Omega = R^2$ and some other special geometric configurations reducible, by symmetry, to the full space case. This is accomplished by using Strichratz estimates developed in [23, 25]. This method, however, is limited to a full space structure of the problem. In fact, as stated in [31], the issue of uniqueness of weak solutions for an arbitrary bounded domain *is an open problem.*

The uniqueness result for weak solutions of (1) with an arbitrary set of boundary conditions specified in previous section and defined on an arbitrary bounded (smooth) domain $\Omega \in R^2$ was settled in affirmative in [21] (see also [20] for other related results). Arguments in [21, 20] rely on a rather special technique introduced in [28] in the context of Marguerre-Vlasov equations (see also [29]. The Theorem stated below provides relevant results.

Theorem 3.1 ([21]). *With reference to full von Karman model* (1) *with* $\gamma > 0$ *and boundary conditions given by* (2), (3), *the following statements are valid:*

(1) **Weak solutions.** *There exists a global solution of finite energy. This is to say that for any initial data*

$$u_0, u_1 \in [H^1(\Omega)]^2 \times [L_2(\Omega)]^2;\ w_0, w_1 \in H_0^2(\Omega) \times H_0^1(\Omega),$$

there exists a solution of the corresponding variational form (11)

$$(u, w) \in C_w([0,T]; [H^1(\Omega)]^2 \times H_0^2(\Omega))$$

$$(u_t, w_t) \in C_w([0,T]; [L_2(\Omega)]^2 \times H_0^1(\Omega)) \tag{16}$$

where $T > 0$ *is arbitrary and* $C_w([0,T]; X)$ *denotes the space of weakly continuous functions with the values in* X. *Moreover, the following boundary regularity is available*

$$au_t|_\Gamma, ah(u_t)|_\Gamma \in [L_2(0,T;L_2(\Gamma))]^2$$

(2) **Uniqueness**. *In the case when h is linear,* **weak** *solutions referred to in part (1) are* **unique**.

(3) **Regular solutions.** *Let us assume, in addition, that h is a C^1 vector function. For any initial data subject to the regularity in part (1) and in addition*

$$u_0, u_1 \in [H^2(\Omega)]^2 \times [H^1(\Omega)]^2; \ w_0, w_1 \in H^3(\Omega) \times H_0^2(\Omega)$$

with compatibility conditions satisfied on the boundary:

$$[\epsilon(u_0) + f(\nabla w_0)]\nu + ah(u_1) = 0, \ \text{on } \Gamma$$

there exists a **unique, global** *solution*

$$(u, w) \in C([0,T]; [H^2(\Omega)]^2 \times H^3(\Omega)$$
$$(u_t, w_t) \in C([0,T]; [H^1(\Omega)]^2 \times H^2(\Omega) \tag{17}$$

where $T > 0$ is arbitrary.
Moreover, regular solutions depend continuously on the initial data in the topology of regular solutions (as above).

Remark 3.1. *Theorem* 3.1 *applies to other boundary conditions imposed on the plate component. In fact, one may consider either hinged boundary conditions* (13) *or free* (14) *with or without boundary dissipation. For all these cases the results stated in the theorem above remain valid with obvious modifications of boundary compatibility (see* [21]*).*

While Theorem 3.1 provides existence and uniqueness of *weak* solutions, there is no statement that such solution depends continuously with respect to the initial data. In fact, the method used in [21] to prove the uniqueness, allows to prove continuous dependence on initial data, but with respect to a *lower topology* than the one induced by the energy. This is a common feature, whenever the method of [28] is used (see [7, 29]). The basic feature of this method is that energy estimates are carried in the topology below the energy level, with initial data taken in *finite energy space.*

Our main goal in this paper is to fill this gap and to prove that *weak* solutions depend continuously on to the initial data, with respect to the topology induced by the energy function. By proving this we shall assert the Hadamard well-posedness for the *full von Karman* system. Our main result is formulated below:

Theorem 3.2. *We consider system given by* (1) *with $h(s)$ linear and $\gamma > 0$. Then,* weak *solutions to* (1) *depend continuously on to the initial data in finite energy* *norm. This is to say that for all* $T > 0$ *and all sequences of initial data such that*

$$u^n(t=0) \to u_0 \ in \ [H^1(\Omega)]^2;$$

$$u_t^n(t=0) \to v_0 \ in \ [L_2(\Omega)]^2;$$

$$w^n(t=0) \to w_0 \ in \ H_0^2(\Omega);$$

$$w_t^n(t=0) \to u_1 \ in \ H_0^1(\Omega); \tag{18}$$

the corresponding solutions

$$u^n(t) \in C([0,T];[H^1(\Omega)]^2), w^n(t) \in C([0,T];H_0^2(\Omega))$$

$$u(t) \in C([0,T];[H^1(\Omega)]^2)), w(t) \in C([0,T];H_0^2(\Omega)) \tag{19}$$

satisfy:

$$u^n \to u \ in \ C([0,T];[H^1(\Omega)]^2);$$

$$u_t^n \to u_t \ in \ C([0,T];[L_2(\Omega)]^2);$$

$$w^n \to w \ in \ C([0,T];H_0^2(\Omega));$$

$$w_t^n \to w_t \ in \ C([0,T];H_0^1(\Omega)). \tag{20}$$

Remark 3.2. *The proof of Theorem* 3.2 *relies on uniqueness property of weak solutions. Since such uniqueness result has been proved so far (see Theorem* 3.1*) for the case of linear dissipation* h*, we accordingly impose this assumption in the formulation of Theorem* 3.2*. However, proper arguments of Theorem* 3.2 *do not depend on the linearity of* h*. It suffices to assume that* h *is monotone and* $h(s)s$ *is convex. Therefore, once uniqueness of* weak *solution is established for nonlinear dissipation* h *(which is so far an open problem), the linearity assumption imposed in Theorem* 3.2 *can automatically be dispensed with.*

Remark 3.3. *The results of Theorem* 3.1 *and Theorem* 3.2 *are valid for all well-posed boundary conditions imposed on the waves and plate (i.e;* **clamped** (3), **hinged** (13) *or* **free** *(14)) with or without dissipation. The arguments provided below are not sensitive with respect to boundary conditions.*

The interest of considering boundary conditions with dissipation is motivated by variety of boundary stabilization problems, where such dissipation on the boundary is the main mechanism responsible for stabilizing the system. In fact, boundary stabilization of full von Karman system with linear boundary dissipation *acting via free boundary conditions (ie* (14), (2) *with* h, h_2, h_3 *linear) and additional tangential derivatives added to boundary conditions in* (2) *was studied first in* [26] *subject to star shaped geometric restrictions. The stabilization of the original model* (1), (14) *without geometric restrictions and with fully nonlinear boundary dissipation has been established in* [21]*. This was accomplished by using microlocal estimates for both waves and plates.*

We conclude this section by listing two *open problems* associated with eq. (1).

- Uniqueness of weak solutions in the case of nonlinear boundary dissipation.
- Uniqueness of weak solutions for the case $\gamma = 0$.

Regarding the second point listed above, we mention that the full well-posedness of weak and strong solutions corresponding to (1) with $\gamma = 0$ has been proved in [22] provided, however, that thermal effects are accounted for. It is known that thermal effects have a regularizing effect on the dynamics, making one component of the system (plate) analytic. This additional regularity is used in [22] in order to establish full Hadamard well-posedness for weak and strong solutions.

The remainder of this paper is devoted to the proof of Theorem 3.2. Here we shall briefly outline the strategy to be followed. One of the main technical ingredient is the establishment of energy *identity* for weak solutions. Once this is accomplished, in order to demonstrate validity of the statements in the Theorem 3.2 the arguments used in the context of fully nonlinear hyperbolic equations presented in [12] are used. It should be noted that for conservative or dissipative systems energy equalities are derived from physical principles of conservation of energy. However, arguments used in this process are often formal and justifiable with mathematical rigour for regular solutions only. Thus, in the case of weak solutions and strongly nonlinear dynamics, one typically obtains only the energy *inequality*. In fact, this is accomplished by using appropriate density argument and lower semicontinuity of the energy function. Obtaining energy *equality* for weak solutions is often, in nonlinear problems, an impossible task. However, in our case we take advantage of rather special type of nonlinearity which allows passage with the limit on suitable finite difference approximations of solutions.

4. Proof of Theorem 3.2

4.1. Energy identity

We recall that the energy $E(t)$ for the system (1)–(3) is given by (7). The first step in the proof is to establish *the energy identity* valid for all *weak* solutions. This is given in the Lemma below.

We denote by $B([0,T],X)$ the space of X-valued functions which are bounded on the interval $[0,T]$. This space is endowed with the usual norm $|x|_{B([0,T],X)} = \sup_{t\in[0,T]} |x(t)|_X$.

Lemma 4.1. *Let* $\gamma > 0$. *We consider solutions to* (1) *with the following a priori regularity*

$$u \in B([0,T];[H^1(\Omega)]^2);\;\; u_t \in B([0,T];[L_2(\Omega)]^2);$$

$$w \in B([0,T];H_0^2(\Omega)); w_t \in B([0,T];H_0^1(\Omega));$$

$$au_t|_\Gamma, ah(u_t)|_\Gamma \in L_2(0,T;[L_2(\Gamma)]^2) \tag{21}$$

The following identity takes place

$$E(t) + 2a \int_s^t \int_\Gamma (h(u_t(z)), u_t(z)) dx dz = E(s); \ 0 \le s \le t \le T \tag{22}$$

By combining the result of Theorem 3.2 and Lemma 4.1 one obtains the following corollary:

Corollary 4.2. *With reference to weak solutions asserted by Theorem* 3.2 *we obtain:*

$$u \in C([0,T]; [H^1(\Omega)]^2), u_t \in C([0,T]; [L_2(\Omega)]^2)$$

$$w \in C([0,T]; H_0^2(\Omega)), w_t \in C([0,T]; H_0^1(\Omega)), N(u,w) \in C([0,T]; L_2(\Omega)] \tag{23}$$

Remark 4.1. *In order to prove the result of Corollary* 4.2 *it is not necessary to have energy equality (see* [12]*). However, since the energy equality does hold for our problem and it will be critically used for the proof of continuous dependence of solutions with respect to the data, we obtain Corollary* 4.2 *as a direct consequence of the energy identity.*

Remark 4.2. *We note that energy* equality *established in Lemma* 4.1 *can be easily obtained* formally *from variational equality* (11) *applied with test functions* $\xi = u_t, \psi = w_t$. *However, low regularity of weak solutions prevents from making this argument rigorous. In fact, the issue of obtaining energy* identity *for weak solutions of nonlinear hyperbolic problems is often a very major problem whose solution requires various skillful devices (see* [30] *for abstract hyperbolic problems and* [12] *where rather general fully nonlinear wave equation is considered). A common route to handle this type of problems is to approximate weak solutions by regular solutions and pass through the limit. In our case, this strategy meets with serious difficulties due to the fact that the nonlinear term in equation* (1) *is not bounded on the energy space. This combined with the lack of monotonicity for the PDE system, makes the limit process – to say the least – problematic. Thus, the result which would follow from known in the literature arguments is* energy inequality, *instead of* energy equality. *In view of the above, the main contribution and interest of Lemma* 4.1 *is the* equality *in the energy relation* (22).

Proof of Lemma 4.1. First of all notice that the regularity in (21), when combined with the regularity of u_{tt}, w_{tt} (with the values in appropriate negative Sobolev's spaces) implies [30] *weak* continuity of solutions u, u_t, w, w_t with the values in the energy space i.e:

$$u \in C_w([0,T]; [H^1(\Omega)]^2), u_t \in C_w([0,T]; [L_2(\Omega)]^2), u_{tt} \in C_w([0,T]; [H^{-1-\epsilon}(\Omega)]^2)$$

$$w \in C_w([0,T]; H_0^2(\Omega)), w_t \in C_w([0,T]; H_0^1(\Omega)), w_{tt} \in C_w([0,T]; H^{-\epsilon}(\Omega)), \tag{24}$$

Moreover,

$$w \in C([0,T]; H^{2-\epsilon}(\Omega)), \ u \in C([0,T]; [H^{1-\epsilon}(\Omega)]^2); \quad \forall \epsilon > 0$$

$$w_t \in C([0,T]; H^{1-\epsilon}(\Omega)), \ u_t \in C([0,T]; [H^{-\epsilon}(\Omega)]^2); \quad \forall \epsilon > 0$$

Therefore, with ϵ suitably small we obtain

$$\nabla w \in C([0,T]; H^{1-\epsilon}(\Omega)) \subset C([0,T]; L_4(\Omega)) \Rightarrow f(\nabla w) \in C([0,T]; L_2(\Omega)) \quad (25)$$

This, together with a weak continuity (in time) of u with values in $H^1(\Omega)$ implies *weak* continuity of the tensor $N(u,w)$ with the values in $L_2(\Omega)$. This is to say

$$N(u,w) \in C_w([0,T]; L_2(\Omega)) \quad (26)$$

In order to derive the energy identity we shall use finite difference approximation of time derivatives (the same approximation was used in [1] in the context of a simpler von Karman model with zero boundary conditions).

Let $h > 0$ be a small parameter destinated to go to zero. Let $g \in B([0,T]; X)$ where X is a Hilbert space. We extend $g(t)$ to all $t \in R$ by defining $g(t) = g(0); t \leq 0$ and $g(t) = g(T); t \geq T$.

With above extensions we define three finite difference operators depending on the parameter h.

$$\begin{aligned} g_h^+(t) &\equiv g(t+h) - g(t) \\ g_h^-(t) &\equiv g(t) - g(t-h) \\ D_h g(t) &\equiv \frac{1}{2h}[g_h^+(t) + g_h^-(t)] \end{aligned} \quad (27)$$

Proposition 4.3.

- *Let g be weakly continuous with the values in X. Then*

$$\lim_{h\to 0} \int_0^T (g(t), D_h g(t))_X dt = \frac{1}{2}[|g(T)|_X^2 - |g(0)|_X^2]$$

- *Let $g \in H^1(0,T,X)$. Then the following limits are well defined in $L_2(0,T;X)$.*

$$\lim_{h\to 0} D_h g = g_t; \ \lim_{h\to 0} \frac{1}{h} g_h^+ = g_t; \ \lim_{h\to 0} \frac{1}{h} g_h^- = g_t;$$

  Moreover, if g_t is weakly continuous with the values in X, then for every $t \in (0,T)$, $D_h g(t) \to g_t(t)$; weakly in X, and

$$\frac{1}{h} g_h^-(T) \to g_t(T); \ \frac{1}{h} g_h^+(0) \to g_t(0); \ \textit{weakly in } X.$$

- *In addition, to previous assumptions, let $V \subset X \subset V'$, $g_{tt} \in L_2(0,T;V')$; $g \in L_2(0,T,V)$. Then*

$$\lim_{h\to 0} \int_0^T (g_{tt}(t), D_h g(t))_X dt = \frac{1}{2}[|g_t(T)|_X^2 - |g_t(0)|_X^2].$$

Proof. The proof of Proposition 4.3 is elementary. For the convenience of the reader we provide the main steps. In order to prove the first statement in Proposition 4.3

we use the definition of D_h and we perform elementary calculations based on changing the variables and cancelling the redundant terms. This gives:

$$\int_0^T (g(t), D_h g(t))_X dt$$

$$= -\frac{1}{2h}\int_{-h}^0 (g(t), g(t+h))_X dt + \frac{1}{2h}\int_{T-h}^T (g(t), g(t+h))_X dt \tag{28}$$

Rewriting both terms as

$$\frac{1}{2h}\int_{-h}^0 (g(t), g(t+h))_X dt = \frac{1}{2h}\int_{-h}^0 (g(0), g(t+h)-g(t))_X dt + \frac{1}{2h}\int_{-h}^0 |g(0)|_X^2 dt$$

and

$$\frac{1}{2h}\int_{T-h}^T (g(t), g(t+h))_X dt = \frac{1}{2h}\int_T^{T+h} (g(t-h)-g(t), g(T))_X dt + \frac{1}{2h}\int_T^{T+h} |g(T)|_X^2 dt$$

and applying weak continuity of g allows us to pass with the limit yielding the expression stated in the first part of the Proposition.

The result stated in the second part of Proposition is a well-known convergence property for the difference quotients.

As for the third part, we calculate directly the integrals involved.

$$\int_0^T (g_{tt}(t), D_h g(t))_X dt = (g_t, D_h g)_X|_0^T - \int_0^T (g_t(t), D_h g_t(t))_X dt \tag{29}$$

In order to evaluate the second term in (29) we split the interval of integration into three subintervals $[0,h], [h, T-h], [T-h, T]$, we make appropriate changes of variables in respective integrations and we note cancellation of terms. This yields to

$$\int_0^T (g_t(t), \frac{d}{dt}D_h g(t))_X dt = \frac{1}{2h}\Big[\int_0^h (g_t(t), g_t(t+h))_X dt$$

$$+\int_h^{T-h} (g_t(t), g_t(t+h)-g_t(t-h))_X dt - \int_{T-h}^T (g_t(t), g_t(t-h))_X dt\Big]$$

$$= \frac{1}{2h}\Big[\int_0^h (g_t(t), g_t(t+h))_X dt + \int_h^{T-h} (g_t(t), g_t(t+h))_X dt$$

$$-\int_0^{T-2h} (g_t(t), g_t(t+h))_X dt - \int_{T-h}^T (g_t(t), g_t(t-h))_X dt\Big]$$

$$= \frac{1}{2h}\Big[\int_0^h (g_t(t), g_t(t+h))_X dt + \int_{T-2h}^{T-h} (g_t(t), g_t(t+h)_X dt$$

$$- \int_0^h (g_t(t), g_t(t+h)_X dt - \int_{T-2h}^{T-h} (g_t(t), g_t(t+h)_X dt\Big] = 0 \tag{30}$$

On the other hand, for the first term in (29) we have:

$$(g_t(T), (D_h g)(T))_X = \frac{1}{2h}(g_t(T), g(T) - g(T-h))_X \to 1/2|g_t(T)|_X^2$$

$$(g_t(0), (D_h g)(0))_X = \frac{1}{2h}(g_t(0), g(h) - g(0))_X \to 1/2|g_t(0)|_X^2 \tag{31}$$

Combining (29)–(31) yields the desired result in the third part of the Proposition.
□

We return to the proof of Lemma 4.1. To accomplish this we shall use variational form in (11) with the test functions:

$$\xi = D_h u \in H^1(\Omega); \;\; \psi = D_h w \in H_0^2(\Omega);$$

To shorten the notation we denote

$$E_l(t) \equiv E_k(t) + D\int_\Omega |\Delta w|^2 dx$$

which represents the *linear* part of the energy. Hence

$$E_l(t) \equiv E(t) - \int_\Omega [\mathcal{C}N(u,w), N(u,w)] d\Omega$$

By using the first three formulas in Proposition 4.3 together with *a priori* regularity of weak solutions one easily derives the following identity:

$$1/2E_l(T) + \lim_{h\to 0}\Big[\int_0^T a(h(u_t), D_h u)_\Gamma dt + \int_0^T X_h dt\Big] = 1/2E_l(0) \tag{32}$$

where

$$X_h \equiv (\mathcal{C}N(u,w), \epsilon(D_h u))_\Omega + (\mathcal{C}N(u,w), \nabla w \otimes \nabla(D_h w))_\Omega$$

and we provisionally assume the existence of the said limit (to be shown below).

On the other hand, from

$$au_t, ah(u_t) \in L_2(0,T; [L_2(\Gamma)]^2)$$

we obtain from Proposition 4.3

$$a \lim_{h\to 0}\int_0^T (h(u_t), D_h u)_\Gamma dt = a\int_0^T (h(u_t), u_t)_\Gamma dt \tag{33}$$

which combined with (32) yields

$$1/2E_l(T) + a\int_0^T (h(u_t), u_t)_\Gamma dt + \lim_{h\to 0}\int_0^T X_h dt = 1/2E_l(0) \tag{34}$$

We rewrite X_h as follows:

$$X_h = (\mathcal{C}N(u,w), D_h\epsilon(u))_\Omega + (\mathcal{C}N(u,w), \nabla w \otimes D_h \nabla w)_\Omega$$

Hence

$$X_h = (\mathcal{C}N(u,w), D_h N(u,w))_\Omega - (\mathcal{C}N(u,w), D_h f(\nabla w) - \nabla w \otimes D_h \nabla w)_\Omega \quad (35)$$

On the other hand, by direct calculations

$$\begin{aligned} D_h f(\nabla w)(t) &= \frac{1}{4h}[\nabla w(t+h) \otimes \nabla w(t+h) - \nabla w(t-h) \otimes \nabla w(t-h)] \\ &= \frac{1}{2}[D_h \nabla w(t) \otimes \nabla w(t+h) + \nabla w(t-h) \otimes D_h \nabla w(t)] \quad (36) \end{aligned}$$

Since $\mathcal{C}N(u,w)$ is a symmetric tensor

$$\begin{aligned} &(\mathcal{C}N(u(t),w(t)), D_h f(\nabla w(t)) - \nabla w(t) \otimes D_h \nabla w(t))_\Omega \\ = \;& 1/2(\mathcal{C}N(u(t),w(t)), D_h \nabla w(t) \otimes [\nabla w(t+h) - \nabla w(t)])_\Omega \\ &+1/2(\mathcal{C}N(u(t),w(t)), D_h \nabla w(t) \otimes [\nabla w(t-h) - \nabla w(t)])_\Omega \\ = \;& 1/2(\mathcal{C}N(u(t),w(t)), D_h \nabla w(t) \otimes [\nabla w_h^+(t) - \nabla w_h^-(t)])_\Omega \quad (37) \end{aligned}$$

From (35)–(37) we infer

$$X_h = (\mathcal{C}N(u,w), D_h N(u,w))_\Omega - 1/2(\mathcal{C}N(u,w), D_h \nabla w \otimes [\nabla w_h^+ - \nabla w_h^-])_\Omega \quad (38)$$

and applying the first identity in Proposition 4.3 together with weak continuity of $N(u,w)$ we obtain

$$\int_0^T X_h dt \to 1/2|\mathcal{C}^{1/2}N(u,w)(T)|_{0,\Omega}^2 - 1/2|\mathcal{C}^{1/2}N(u,w)(0)|_{0,\Omega}^2 - 1/2 \lim_{h\to 0} \int_0^T Y_h dt \quad (39)$$

where

$$Y_h \equiv (\mathcal{C}N(u,w), D_h \nabla w \otimes [\nabla w_h^+ - \nabla w_h^-])_\Omega$$

Our goal is to show that

$$\lim_{h\to 0} \int_0^T Y_h = 0. \quad (40)$$

Indeed, once (40) is shown the desired energy identity follows by combining (32), (33), and (39). From (39) and (40) we obtain

$$\int_0^T X_h dt \to 1/2|\mathcal{C}^{1/2}N(u,w)(T)|_{0,\Omega}^2 - 1/2|\mathcal{C}^{1/2}N(u,w)(0)|_{0,\Omega}^2 \quad (41)$$

and from (34)

$$E_l(T) + 2a\int_0^T (h(u_t), u_t)_\Gamma dt + |\mathcal{C}^{1/2}N(u,w)(T)|_{0,\Omega}^2 = E_l(0) + |\mathcal{C}^{1/2}N(u,w)(0)|_{0,\Omega}^2$$

which provides the desired result in Lemma 4.1 with $s = 0, t = T$. Other points s, t are treated in the same way, due to the fact that the argument is local.

Thus, it suffices to prove the validity of (40). We notice first that a priori regularity of weak solutions gives $N(u,w) \in B([0,T];L_2(\Omega))$; and the assertion follows from

$$\int_0^T |D_h \nabla w \otimes [\nabla w_h^+ - \nabla w_h^-]|_{0,\Omega} dt \to 0 \text{ as } h \to 0, \tag{42}$$

which in turn follows from Hölder's inequality and

$$Z_h = \int_0^T h\|D_h \nabla w\|^2_{L_4(\Omega)} + h^{-1}(\|\nabla w_h^+\|^2_{L_4(\Omega)} + \|\nabla w_h^-\|^2_{L_4(\Omega)})dt \to 0. \tag{43}$$

To prove (43) by a simple density argument, it suffices to show

$$Z_h \le c(|w|_{L_2(0,T;H^2(\Omega))} + |w_t|_{L_2(0,T;H^1(\Omega))}).$$

The argument is similar for all three terms. The Gagliardo-Nirenberg inequality, or more precisely Sobolev's inequality $|g|_{L_2} \le c|g|_{W^{1,1}(\Omega)}$ (recall $\Omega \subset R^2$) applied with $g = |D_h \nabla w|^2$ implies the desired estimate

$$h\|D_h \nabla w\|^2_{L_4(\Omega)} \le ch|D_h \nabla w|_{1,\Omega}|D_h \nabla w|_{0,\Omega} \le 2c|\nabla w|_{1,\Omega}|D_h \nabla w|_{0,\Omega}.$$

4.2. Completion of the proof of Theorem 3.2

In what follows we shall use the following notation

$$H \equiv [H^1(\Omega)]^2 \times [L_2(\Omega)]^2 \times H_0^2(\Omega) \times H_0^1(\Omega)$$

$$U(t) = (u(t), u_t(t), w(t), w_t(t)),$$

where $u(t), w(t)$ is a weak solution of equation (1) at the time t due to initial data $U(0)$.

We start with initial data $U(0) \in H$ and a sequence of initial data $U^n(0) \in H$ such that

$$U^n(0) \to U(0) \text{ in } H, n \to \infty$$

Our aim is to prove that

$$U^n(t) \to U(t); \text{ in } C([0,T];H) \tag{44}$$

We shall follow some of the ideas presented in [12].

By Lemma 4.1 (in fact we use here only the inequality) and (10) which implies

$$|U|_H \le CE(U),\ E(U) \le C(|U|_H^4)$$

we obtain

$$|U^n(t)|_H + a\int_0^t \int_\Gamma |h(u_t^n)u_t^n| dx ds \le C(|U^n(0)|_H) \le C(|U(0)|_H) \tag{45}$$

where by $C(s)$ we denote a function which is bounded for bounded arguments.

Hence, on a subsequence denoted by the same index we obtain

$$U^n(t) \to U^* \text{ weakly* in } L_\infty(0,T;H)$$

$$au_t^n|_\Gamma \to au_t^*|_\Gamma \text{ weakly in } L_2(0,T;L_2(\Sigma)); \tag{46}$$

By using variational equality together with weak continuity of nonlinear terms, it is routine to show that U^* and $u_t^*|_\Gamma$ coincide with a weak solution to (1) due to initial data $U(0)$. By uniqueness of weak solution $U^*(t) = U(t)$. Hence we have

$$U^n(t) \to U(t) \text{ weakly* in } L_\infty(0,T;H)$$

$$au_t^n|_\Gamma \to au_t(t)|_\Gamma \text{ weakly in } L_2(0,T;L_2(\Sigma)); \tag{47}$$

In view of (47), to prove the Theorem 3.2 it suffices to show the norm convergence i.e;

$$|U^n(t)|_H \to |U(t)|_H \ in \ C[0,T]. \tag{48}$$

In order to achieve this we shall use the *equality* in the energy relation in Lemma 4.1. We denote by $E(U(t))$ the energy corresponding to solution $U(t)$. From Lemma 4.1

$$\begin{aligned} & E(u^n(t), u_t^n(t), w^n(t), w_t^n(t)) + 2a\int_0^t\int_\Gamma h(u_t^n)u_t^n dxds \\ = \quad & E(u^n(0), u_t^n(0), w^n(0), w_t^n(0)), \end{aligned}$$

which we write as

$$E(U^n(t)) + 2a\int_0^t\int_\Gamma h(u_t^n)u_t^n dxds = E(U^n(0)) \tag{49}$$

By the continuity of the energy with respect to the strong topology in H we obtain

$$\lim_{n\to\infty} E(U^n(0)) = E(U(0)) \tag{50}$$

On the other hand by applying Lemma 4.1 again

$$E(U(t)) + 2a\int_0^t\int_\Gamma h(u_t)u_t dxds = E(U(0)) \tag{51}$$

Hence, by (49)–(51) and the uniqueness of weak solutions

$$\lim_n [E(U^n(t)) + 2a\int_0^t\int_\Gamma h(u_t^n)u_t^n dxds] = E(U(t)) + 2a\int_0^t\int_\Gamma h(u_t)u_t dxds \tag{52}$$

where the limit is taken $C[0,T]$. By using weak convergence in (47) together with lower semicontinuity and convexity of $h(s)\cdot s$ we obtain

$$a\int_0^t\int_\Gamma h(u_t)u_t dxds \leq a \lim_{n\to\infty}\int_0^t\int_\Gamma h(u_t^n)u_t^n dxds \tag{53}$$

Hence, by (53)

$$E(U^n(t)) + 2a\int_0^t\int_\Gamma h(u_t)u_t dxds \leq E(U^n(t)) + 2a\lim_n\int_0^t\int_\Gamma h(u_t^n)u_t^n dxds$$

Taking the limit on both sides of the above inequality and applying (52) to the resulting right-hand side yields

$$\lim_n E(U^n(t)) + 2a \int_0^t \int_\Gamma h(u_t)u_t dx ds \leq E(U(t)) + 2a \int_0^t \int_\Gamma h(u_t)u_t dx ds$$

This implies that

$$\lim_n E(U^n(t)) \leq E(U(t));$$

which then combined with lower semicontinuity of the energy gives

$$lim_n E(U^n(t)) = E(U(t)); \ in \ C[0,T]$$

as desired.

Combining weak convergence with norm convergence we obtain the convergence of the corresponding solutions in $C([0,T];H)$. □

References

[1] A. Benabdallah and I. Lasiecka. "Exponential decay rates for a full von karman system of dynamic thermoelasticity". *J. Differential Equations*, 160:51–93, 2000.

[2] M. Bohm. "Global well-posedness of the dynamic von Karman Equations for generalized Solutions". *Manuscript*, 1994.

[3] I. Chueskov. "Finite dimensionality of an attractor in some problems of nonlinear shell theory". *Math. USSR Sbornik*, 61:411–420, 1988.

[4] I. Chueskov. "Strong solutions and the attractors of the von Karman equations". *Math. USSR Sbornik*, 69:25–35, 1991.

[5] Ph. Ciarlet and P. Rabier. *"Les Equations de von Karman"*. Springer Verlag, 1982.

[6] R. Coifman, P.L. Lions, Y. Meyer, and S. Semmers. "Compensated compactness and Hardy spaces". *J. Math. Pure Appliq.*, 72:247–286, 1993.

[7] A. Boutet de Monvel and I. Chueshov. "Uniqueness theorem for weak solutions of von Karman evolution equations". *Journal Mathematical Analysis and Applications*, 221:419–429, 1998.

[8] G. Duvaut and J.L.Lions. *"Les Inéquations en Mécaniques et en Physiques"*. Dunod, 1972.

[9] A. Favini, M. A. Horn, I. Lasiecka, and D. Tataru. "Global existence, uniqueness and regularity of solutions to a von Karman system with nonlinear boundary dissipation." *Differential and Integral Equations*, 9:267–294, 1996, with Addendum *Diff. Int. Equations* 10: 197–200, 1997.

[10] J. L. Lions. *"Quelques Méthods de Résolution des Problèmes aux Limites Nonlinéaires"*. Dunod, Paris, 1969.

[11] T. Von Karman. "Festigkeitprobleme in Maschinenbau". *Encyklopädie der Mathematischen Wissenschaften*, 4:314–385, 1910.

[12] H. Koch. "Mixed problems for fully nonlinear hyperbolic equations". *Math. Z.*, 214:9–42, 1993.

[13] H. Koch and A. Stachel. "Global existence of classical solutions to the dynamical von Karman equations". *Math. Methods in the Applied Sciences*, 16:581–586, 1993.

[14] J. Lagnese. *"Boundary Stabilization of Thin Plates"*. SIAM, 1989.

[15] J. Lagnese. "Modeling and stabilization of nonlinear plates". *International Ser. Num. Math.*, 100:247–264, 1991.

[16] J. Lagnese. "Uniform asymptotic energy estimates for solutions of the equations of dynamic plane elasticity with nonlinear dissipation at the boundary". *Nonlinear Analysis*, 16:35–54, 1991.

[17] J. Lagnese and G. Leugering. "Uniform stabilization of a nonlinear beam by nonlinear boundary feedback". *Journal of Differential Equations*, 91:355–388, 1991.

[18] J. Lagnese and J.L. Lions. *"Modelling analysis and control of thin plates"*. Masson, 1988.

[19] I. Lasiecka. "Finite dimensionality of attractors associated with von Karman plate equations and boundary damping". *Journal Differential Equations*, 117:357–389, 1995.

[20] I. Lasiecka. "Intermediate Solutions to Full von Karman System of Dynamic Nonlinear Elasticity". *Applicable Analysis*, 68:121–148, 1998.

[21] I. Lasiecka. "Uniform stabilizability of a full von Karman system with nonlinear boundary feedback". *SIAM J. on Control*, 36:1376–1422, 1998.

[22] I. Lasiecka. "Uniform decay rates for full von Karman system of dynamic elasticity with free boundary conditions and partial boundary dissipation". *Communications in Partial Differential Equations*, 24:1801–1849, 1999.

[23] H. Lindblad and C.D. Sogge. "On existence and scattering with minimal regularity for semilinear wave equation". *Journal of Functional Analysis*, 130:357–426, 1995.

[24] N. Morozov. "Non-linear vibrations of thin plates with allowance for rotational inertia". *Sov. Math*, 8:1137–1141, 1967.

[25] H. Pecher. "Nonlinear small data scattering for the wave and Klein-Gordon equations". *Math. Z. Journal of Functional Analysis*, 185:261–270, 1984.

[26] J. Puel and M. Tucsnak. "Boundary stabilization for the von Karman equations". *SIAM J. on Control*, 33:255–273, 1996.

[27] J. Puel and M. Tucsnak. "Global existence for the full von Karman system". *Applied Mathematics and Optimization*, 34:139–161, 1996.

[28] V. I. Sedenko. "On uniqueness of the generalized solutions of initial boundary value problem for Marguerre-Vlasov nonlinear oscillations of the shallow shells". *Russian Izvestiya, North-Caucasus Region, Ser. Natural Sciences*, 1-2, 1994.

[29] V. I. Sedenko. "On the uniqueness theorem to generalized solutions of initial boundary-value problems to the Maquerre-Vlasov vibrations of shallow shells with clamped boundary conditions". *Applied Mathematics and Optimization*, 39:309–327, 1999.

[30] W. Strauss. "On continuity of functions with values in various Banach spaces". *Pacific Journal of Mathematics*, 19:543–551, 1966.

[31] D. Tataru and M. Tucsnak. "On the Cauchy problem for the full von Karman system". *NODEA*, 4:325–340, 1997.

[32] W. van Wahl. "On nonlinear evolution equations in Banach spaces and on nonlinear vibrations of the clamped plate". *Bayreuther Mathematische Schriften*, 7:1–93, 1981.

[33] I. Vorovic. "On some direct methods in nonlinear theory of vibration of curved shells". *Izv. Akad. Nauk. SSSR. Mat.*, 21:747–784, 1957.

Herbert Koch
Fachbereich Mathematik
Universität Dortmund
D-44221 Dortmund, Germany
e-mail: `koch@mathematik.uni-dortmund.de`

Irena Lasiecka
University of Virginia
Department of Mathematics
Kerchof Hall
P.O. Box 400137
Charlottesville
VA 22904-4137, USA
e-mail: `il2v@virginia.edu`

Progress in Nonlinear Differential Equations
and Their Applications, Vol. 50, 217–236

Applications des sommes d'opérateurs dans l'étude du comportement singulier des solutions dans les problèmes elliptiques

Rabah Labbas

A la mémoire de mon ami Brunello Terreni,
avec toutes mes pensées affectueuses à sa famille.

1. Introduction

Le problème suivant

$$\begin{cases} -\Delta u = f & \text{dans } Q \\ u = 0 & \text{sur } \partial Q, \end{cases} \tag{1}$$

où Q est un ouvert de $\mathbb{R}^n$ a été étudié par beaucoup d'auteurs dans les espaces de Sobolev construits sur les $L^p(Q)$, $1 < p < \infty$, voir par exemple Agmon-Douglis-Nirenberg [1], [2] pour les ouverts réguliers, Grisvard [9], Dauge [7] et Kondratiev [11] pour les ouverts à points singuliers. On montre que la solution variationnelle (lorsqu'elle existe) s'écrit sous forme

$$u = u_r + u_s, \tag{2}$$

où u_r a la régularité optimale $W^{2,p}(Q)$ et u_s s'écrit explicitement au voisinage des points singuliers dans les cas d'ouverts à géométrie simple.

Lorsque Q est un cône, la technique utilisée dans le cas hilbertien ($p = 2$) est essentiellement basée sur la transformée de Fourier partielle et le théorème de Plancherel.

Dans le cas $p \neq 2$, la décomposition (2) est obtenue par Clément-Grisvard [5], grâce à l'utilisation des deux théories des sommes d'opérateurs linéaires de Da Prato-Grisvard [6] et de Dore-Venni [8].

Les techniques utilisées dans le cadre des espaces de Hölder, voir Labbas-Moussaoui-Najmi [15], sont basées sur l'application de la première théorie des sommes et sur les résultats de régularité optimale obtenus dans l'étude des problèmes à deux points pour une équation différentielle du second ordre à coefficients opérateurs et de type elliptique (voir Labbas [13].)

Ici on considère le problème (1) dans le cône infini

$$Q = \{\rho\sigma \ / \ \rho > 0, \ \sigma \in G\}, \tag{3}$$

où G est un ouvert régulier de S^{n-1}. Pour $k \in \mathbb{N}$, on note par $UC^k\left(\overline{Q}\right)$ l'espace des fonctions dont les dérivées sont uniformément continues et bornées sur $\overline{Q}$ jusqu'à l'ordre k et $C^\alpha\left(\overline{Q}\right)$, pour $0 < \alpha < 1$, l'espace des fonctions u définies sur $\overline{Q}$, bornées et uniformément α-höldériennes. Ce dernier est muni de la norme

$$\begin{aligned} \|u\|_{C^\alpha(\overline{Q})} &= \max\nolimits_{x\in\overline{Q}} |u(x)| + \sup_{\rho\sigma\neq\rho'\sigma'} \frac{|u(\rho\sigma) - u(\rho'\sigma')|}{\|\rho\sigma - \rho'\sigma'\|^\alpha} & (4) \\ &= \max_{x\in\overline{Q}} |u(x)| + [u]_{\alpha,\overline{Q}}\,. & (5) \end{aligned}$$

$C^{k+\alpha}\left(\overline{Q}\right)$ est le sous-espace de $UC^k\left(\overline{Q}\right)$ formé par les fonctions dont les dérivées d'ordre k appartiennent à $C^\alpha\left(\overline{Q}\right)$. D'une manière analogue, on définit les espaces à valeurs vectorielles $UC^k\left(\overline{Q}, X\right)$, $C^\alpha\left(\overline{Q}, X\right)$ et $C^{k+\alpha}\left(\overline{Q}, X\right)$ où X est un espace de Banach complexe.

On considèrera aussi les sous-espaces, dit petits Hölder, définis par

$$\begin{aligned} h^\alpha\left(\overline{Q}, X\right) &= \left\{u \in UC\left(\overline{Q}, X\right) / \lim_{\delta\to 0} \sup_{\|x-y\|\le\delta} \frac{\|u(x) - u(y)\|_X}{\|x-y\|^\alpha} = 0\right\} \\ h^\alpha\left(\overline{Q}\right) &= \left\{u \in UC\left(\overline{Q}\right) / \lim_{\delta\to 0} \sup_{\|x-y\|\le\delta} \frac{|u(x) - u(y)|}{\|x-y\|_2^\alpha} = 0\right\}, \end{aligned}$$

munis respectivement des normes de $C^\alpha\left(\overline{Q}, X\right)$ et $C^\alpha\left(\overline{Q}\right)$. L'espace $h^\alpha\left(\overline{Q}, X\right)$ est aussi caractérisé comme la fermeture de $UC^1\left(\overline{Q}, X\right)$ dans $C^\alpha\left(\overline{Q}, X\right)$ [ou celle de $C^\theta\left(\overline{Q}, X\right)$ dans $C^\alpha\left(\overline{Q}, X\right)$ pour $\theta > \alpha$], (Sinestrari [20], Lunardi [16], [17]).

On prouvera la validité de la décomposition (2) si $f \in h_0^\alpha\left(\overline{Q}\right)$. Ici, $h_0^\alpha\left(\overline{Q}\right)$ (resp. $C_0\left(\overline{G}\right)$) désigne l'espace des fonctions de $h^\alpha\left(\overline{Q}\right)$ (resp. de $C\left(\overline{G}\right)$) s'annulant sur ∂Q (resp. sur ∂G). On montrera précisément que

$$u_r \in C^{2+\alpha}\left(\overline{Q}\right),$$

et on donnera une description précise de la partie singulière u_s près du sommet O.

Au paragraphe 2, on présente les résultats essentiels de la théorie des sommes de Da Prato et Grisvard et de Dore et Venni dans le cas commutatif. Le paragraphe 3 sera consacré à l'écriture de l'équation (1) dans le cylindre $\Sigma = \mathbb{R} \times G$ grâce aux coordonnées sphériques. Au paragraphe 4, on appliquera la première stratégie des sommes à l'équation transformée. On utilisera ensuite le théorème de Dore-Venni pour la régularité optimale de la solution forte du problème transformé. Le paragraphe 5 sera consacré au retour au cône et à la décomposition cherchée. Au paragraphe 6, on rappellera les résultats essentiels de régularité de Labbas [13] et enfin, au paragraphe 7, on donne la décomposition dans le cadre höldérien.

2. Sommes d'opérateurs linéaires

On considère un espace de Banach complexe E et deux opérateurs linéaires fermés A et B de domaines $D(A)$ et $D(B)$. Leur somme est définie par

$$Sx = Ax + Bx, \qquad x \in D(S) = D(A) \cap D(B). \tag{6}$$

On suppose que ces deux opérateurs vérifient

$$(H.1)\begin{cases} \exists r,\ C_A, C_B > 0,\ \epsilon_A,\ \epsilon_B \in]0,\pi[\ : \\ i)\ \rho(-A) \supset \sum_{\epsilon_A} = \{z\ /\ |z| \geq r\ ,\ |Arg(z)| \leq \epsilon_A\} \\ \text{et } \left\|(A+zI)^{-1}\right\|_{L(E)} \leq C_A/|z|\ \forall\ z \in \sum_{\epsilon_A}; \\ ii)\ \rho(-B) \supset \sum_{\epsilon_B} = \{z\ /\ |z|\ \geq r\ ,\ |Arg(z)| \leq \epsilon_B\} \\ \text{et } \left\|(B+zI)^{-1}\right\|_{L(E)} \leq C_B/|z|\ \forall\ z \in \sum_{\epsilon_B}; \\ iii)\ \epsilon_A +\ \epsilon_B > \pi; \\ iv)\ \overline{D(A)+D(B)}\ = E, \end{cases}$$

$$(H.2)\begin{cases} (A+\xi I)^{-1}(B+\eta I)^{-1} - (B+\eta I)^{-1}(A+\xi I)^{-1} \\ = \left[(A+\xi I)^{-1};\ (B+\eta I)^{-1}\right] = 0\ \ \forall \xi \in \rho(-A), \forall \eta \in \rho(-B), \end{cases}$$

$$(H.3)\ \{\sigma(-A) \cap \sigma(B) = \emptyset,$$

où $\sigma(A)$ et $\sigma(-B)$ désignent respectivement les spectres de A et $-B$ et $\rho(A)$, $\rho(-B)$ leurs ensembles résolvants. Alors, grâce à Da Prato-Grisvard [5], la somme $S = A + B$ est fermable et l'opérateur défini par l'intégrale de Dunford

$$x \longmapsto -\frac{1}{2i\pi}\int_\Gamma (A+zI)^{-1}(B-zI)^{-1}x dz \tag{7}$$

coïncide avec $\left(\overline{S}\right)^{-1}$ où $\overline{S} = \overline{A+B}$ est la fermeture de $A+B$.

Γ est une courbe simple infinie entourant les spectres de $(-A)$ et B et demeurant dans $\rho(-A) \cap \rho(B)$. Le résultat suivant est prouvé dans [6]:

Théorème 1 *On suppose (H.1), (H.2) et (H.3). Si F est un sous espace de Banach qui s'injecte continûment dans E et s'il existe une constante K telle que pour un certain $\theta \in]0,1[$ on ait*

$$\|x\|_F \leq K\left(\|x\|_E + \|x\|_E^{1-\theta}\|Ax\|_E^\theta\right) \quad \forall x \in D(A),$$

alors $D\left(\overline{A+B}\right) \subset F$.

L' unique solution v de l'équation

$$\overline{S}v = \left(\overline{A+B}\right)v = f$$

est appelée solution forte de l'équation $Sv = f$.

Introduisons cette fois les hypothèses suivantes

$$(H.4)\quad E \text{ est espace de Banach U.M.D,}$$

$$(H.5)\begin{cases} \exists M \geq 1: \\ i)\ \rho(-A) \supset]-\infty, 0] \quad \text{et } \forall t \geq 0 \\ \left\| (A+tI)^{-1} \right\|_{L(E)} \leq M/(1+t), \\ ii)\ \rho(-B) \supset]-\infty, 0] \quad \text{et } \forall t \geq 0 \\ \left\| (B+tI)^{-1} \right\|_{L(E)} \leq M/(1+t), \\ iv)\ \overline{D(A)} = \overline{D(B)} = E, \end{cases}$$

$$(H.6)\begin{cases} i)\ A^{is} \in L(E)\ \forall s \in \mathbb{R} \text{ et } \exists K > 0,\ \theta_A : \\ \left\| A^{is} \right\| \leq K e^{|s|\theta_A} \quad \forall s \in \mathbb{R}, \\ ii)\ B^{is} \in L(E)\ \forall s \in \mathbb{R} \text{ et } \exists K > 0,\ \theta_B : \\ \left\| B^{is} \right\| \leq K e^{|s|\theta_B} \quad \forall s \in \mathbb{R}, \\ iii)\ \theta_A + \theta_B < \pi. \end{cases}$$

L'hypothèse (H.4) signifie grâce à Burckholder [3], que la transformée de Hilbert est linéaire continue de $L^p(\mathbb{R}; E)$ dans lui-même pour un $p \in]1, \infty[$ (et donc pour tout $p > 1$). Alors le résultat remarquable de Dore-Venni [8] est le suivant

Théorème 2 *Sous les hypothèses (H.2), (H.4), (H.5) et (H.6) l'opérateur S est fermé et inversible.*

Ces auteurs donnent en plus explicitement l'inverse de S sous la forme

$$(A+B)^{-1}x = \frac{1}{i\pi} \int_{\Gamma'} A^{-z} B^{z-1} x dz$$

où Γ' est une courbe verticale contenue dans la bande $\{z \,/\, 0 < \operatorname{Re} z < 1\}$ et orientée de $\infty e^{-i\frac{\pi}{2}}$ à $\infty e^{i\frac{\pi}{2}}$. Il est connu que l'hypothèse (H.5) est suffisante pour définir les puissances complexes A^z et B^z. Mais ces dernières ne sont pas nécessairement bornées pour $\operatorname{Re}(z) \geq 0$.

3. Le problème dans le cylindre

Les coordonnées sphériques permettent d'écrire l'équation (1) sous la forme

$$\begin{cases} D_\rho^2 u + \frac{2}{\rho} D_\rho u + \frac{1}{\rho^2} \Delta' u = f \\ u_{|\partial Q} = 0, \end{cases} \tag{8}$$

où Δ' désigne l'opérateur de Laplace-Beltrami sur la sphère S^{n-1}. Cette dernière équation devient dans Q

$$\begin{cases} (\rho D_\rho)^2 u + (\rho D_\rho) u + \Delta' u = \rho^2 f = g \\ u_{|\partial Q} = 0, \end{cases}$$

et le changement de variable $\rho = e^t$ donne dans $\Sigma = \mathbb{R} \times G$

$$\begin{cases} D_t^2 u + D_t u + \Delta' u = e^{2t} f = g \\ u_{|\partial\Sigma} = 0. \end{cases} \tag{9}$$

Si $f \in L^p(Q)$ alors $e^{(-2+n/p)t} g \in L^p(\Sigma)$. Le nombre $\gamma = (-2 + n/p)$ est exactement l'opposé de l'exposant de Sobolev de $W^{2,p}(Q)$. Cela suggère l'utilisation des nouvelles fonctions $v = e^{\gamma t} u$ et $h = e^{\gamma t} g$ pour éviter l'utilisation des espaces à poids. L'équation (8) devient finalement

$$\begin{cases} D_t^2 v + (n - 2 - 2\gamma)\, D_t v + \gamma\,(\gamma - n + 2)\, v + \Delta' v = h \\ v_{|\partial\Sigma} = 0. \end{cases} \tag{10}$$

4. Application des sommes

Dans l'espace de Banach $E = L^p(\mathbb{R}, L^p(G)) = L^p(\Sigma)$ on considère les trois opérateurs A, B et C définis par

$$\begin{cases} D(A) = L^p(\mathbb{R}, D(\Delta')) \\ (-Av)(t) = \Delta'(v(t,.)), \\ D(\Delta') = W^{2,p}(G) \cap W_0^{1,p}(G), \end{cases}$$

$$\begin{cases} D(B) = W^{2,p}(\mathbb{R}, L^p(G)) \\ (-Bv) = D_t^2 v + (n - 2 - 2\gamma)\, D_t v + \gamma(\gamma - n + 2) v, \end{cases}$$

$$\begin{cases} D(C) = W^{1,p}(\mathbb{R}, L^p(G)) \\ Cv = D_t v. \end{cases}$$

Le problème (9) s'écrit alors sous la forme $Av + Bv = h$.

Il est facile de voir qu'ici les domaines $D(A)$ et $D(B)$ sont denses dans E. D'autre part, il est connu que $\sigma(C) = i\mathbb{R}$ et que pour tout $\mu \neq 0$, $\left\|(C + \mu)^{-1}\right\| \leq 1/\left|\operatorname{Re}\mu\right|$. De plus, si on pose

$$P(z) = z^2 + (n - 2 - 2\gamma)\, z + \gamma(\gamma - n + 2)$$

alors grâce au théorème de l'application spectrale on a

$$\sigma(-B) = \left\{-y^2 + i\,(n - 2 - 2\gamma)\, y + \gamma(\gamma - n + 2)\right\}$$

et la factorisation $(-B - \lambda)^{-1} = (C - \alpha(\lambda))^{-1}(C - \beta(\lambda))^{-1}$, où $\alpha(\lambda)$ et $\beta(\lambda)$ sont les racines de $P(z) = \lambda$, permet d'avoir

$$\left\|(-B + \lambda)^{-1}\right\| = O\left(\frac{1}{\operatorname{Re}\sqrt{\lambda})^2}\right),$$

pour $|\lambda|$ assez grand. Cela implique que $\epsilon_B = \pi/2$.

D'autre part, on sait que $\sigma(A) = \sigma(-\Delta')$ est constitué par une suite de valeurs propres positives

$$0 < \lambda_1 \leq \lambda_2 \leq \dots .$$

et que $-\Delta'$ est générateur d'un semi-groupe analytique; donc $\epsilon_A = \pi/2 + \epsilon$ et pour $|Arg(\lambda)| \leq \epsilon_A$

$$\left\|(A+\lambda)^{-1}\right\| = O\left(\frac{1}{(|\lambda|}\right).$$

Dès lors, la condition (H.3) est vérifiée si toutefois

$$\gamma(\gamma - n + 2) = \left(\frac{n}{p} - 2\right)\left(\frac{n}{p} - n\right) \neq \lambda_j \quad \forall j \geq 1. \tag{11}$$

On supposera vérifiée cette condition dans la suite.

L'hypothèse (H.2) de commutativité est vérifiée puisqu'elle l'est entre les résolvantes de C et A. Pour cela il suffit d'utiliser les formules suivantes

$$\left[(C+\mu I)^{-1}\phi\right](t,\sigma) = \begin{cases} -\int_t^{\infty} e^{\mu(s-t)}\phi(s,\sigma)ds & \text{si } \operatorname{Re}\mu < 0, \\ \int_{-\infty}^{t} e^{-\mu(t-s)}\phi(s,\sigma)ds & \text{si } \operatorname{Re}\mu > 0, \end{cases}$$

et

$$\left[(A-\lambda I)^{-1}\varphi\right](t,\sigma) = \sum_{j\geq 1} \frac{1}{\lambda_j - \lambda}\left[\int_G \varphi(t,\xi) w_j(\xi) d\xi\right] w_j(\sigma),$$

où w_j est la fonction propre associée à λ_j; ici φ appartient à un sous-espace dense de E.

On peut donc déjà dire que si la condition (11) est vérifiée alors $(A+B)$ est fermable et sa fermeture $\overline{A+B}$ est inversible. Cela signifie, pour l'équation (10), qu'il existe une solution forte $v \in L^p(\Sigma)$ et une suite

$$v_n \in D(A) \cap D(B) = L^p\left(\mathbb{R}, W^{2,p}(G)\right) \cap W^{2,p}\left(\mathbb{R}, L^p(G)\right),$$

telle que

$$\begin{cases} v_n \xrightarrow{E} v \\ D_t^2 v_n + (1+2\beta) D_t v_n + \beta(\beta+1) v_n + \Delta' v_n \xrightarrow{E} h \\ v_{n|\partial\Sigma} = 0. \end{cases}$$

En particulier, v est solution au sens des distributions. Pour avoir plus de régularité sur v on va appliquer le théorème 1. Pour cela, posons

$$\begin{cases} F_1 = W^{1,p}\left(\mathbb{R}, L^p(G)\right) \subset E, \\ F_2 = L^p\left(\mathbb{R}, W_0^{1,p}(G)\right) \subset E, \end{cases}$$

alors il est connu qu'il existe une constante C telle que

$$\begin{cases} \|v'\|_{L^p(R,L^p(G))} \leq C \|v\|^{1/2}_{L^p(R,L^p(G))} \cdot \|v''\|^{1/2}_{L^p(R,L^p(G))} \\ \forall v \in W^{2,p}\left(\mathbb{R}, L^p(G)\right) = D(B). \end{cases}$$

et

$$\begin{cases} \|v\|_{L^p(R,W_0^{1,p}(G))} \leq C \|v\|^{1/2}_{L^p(R,L^p(G))} \|Av\|^{1/2}_{L^p(R,L^p(G))} \\ \forall v \in L^p\left(\mathbb{R}, D(\Delta')\right). \end{cases}$$

On en déduit que $D(\overline{S}) \subset F_1 \cap F_2 \subset W^{1,p}(\Sigma)$. D'où la proposition

Proposition 3 *Pour tout* $h \in L^p(\Sigma)$ *(où* p *vérifie la condition* (11)*), le problème* (10) *admet une unique solution forte* $v \in W^{1,p}(\Sigma)$.

En vue d'avoir plus de régularité sur v, on va appliquer le Théorème de Dore-Venni [8]. On écrit que

$$\begin{cases} D_t^2 v + \Delta' v - v = h - (n-2-2\gamma) D_t v - \gamma(\gamma - n + 2) v = k \\ v_{|\partial\Sigma} = 0, \end{cases} \tag{12}$$

ici $k \in L^p(\Sigma)$ grâce à la régularité précédente.

Soit l'opérateur B_1 défini par

$$\begin{cases} D(B_1) = W^{2,p}(\mathbb{R}, L^p(G)) = D(B) \\ (-B_1 v) = D_t^2 v - v. \end{cases}$$

Alors l'équation (12) est la somme $S_1 v = (-A - B_1)v = k$ dans l'espace UMD $L^p(\Sigma)$. L'opérateur de Laplace-Beltrami génère un semi-groupe analytique dans $L^p(G)$ et donc (H.5) est vérifiée pour A. En calquant la méthode utilisée pour B avec le polynôme $P_1(z) = z^2 - 1$, on montre (H.5) pour B_1. L'hypothèse difficile à vérifier dans les cas concrets est l'hypothèse (H.6). Du fait que Δ' génère un semi-groupe fortement continu de contraction qui préserve la positivité et grâce à la méthode de transfert de Coifman-Weiss, on a pour tout $q \in]1, \infty[$

$$\begin{cases} \left\| A^{is} \right\|_{L(L^q(\Sigma))} = O\left(|s| \, e^{\frac{\pi}{2}|s|} \right) \\ \left\| A^{is} \right\|_{L(L^2(\Sigma))} = O(1), \end{cases}$$

et donc par interpolation, il existe $\theta_A = \theta_A(p) < \frac{\pi}{2}$ tel que

$$\left\| A^{is} \right\|_{L(L^p(\Sigma))} = O\left(e^{\theta_A |s|} \right).$$

Pour B_1 on a

$$\left\| (B_1)^{is} \right\|_{L(L^p(\Sigma))} = O\left(e^{\delta |s|} \right)$$

pour tout $\delta > 0$, par un calcul direct en utilisant le symbole $(1+\tau^2)^{is}$ et le Théorème de Mikhlin. Finalement (H.6) est vérifiée et donc S_1 est inversible. On en déduit grâce au Théorème 2:

Proposition 4 *Pour tout* $h \in L^p(\Sigma)$ *(où* p *vérifie la condition* (11)*), le problème* (10) *admet une unique solution* $v \in W^{2,p}(\Sigma) \cap W_0^{1,p}(\Sigma)$.

En effet

$$v \in D(S_1) = W^{2,p}(\mathbb{R}, L^p(G)) \cap L^p\left(\mathbb{R}, W^{2,p}(G) \cap W_0^{1,p}(G) \right)$$

qui s'injecte par prolongement dans

$$W^{2,p}\left(\mathbb{R}, L^p\left(\mathbb{R}^{n-1}\right)\right) \cap L^p\left(\mathbb{R}, W^{2,p}\left(\mathbb{R}^{n-1}\right) \cap W_0^{1,p}\left(\mathbb{R}^{n-1}\right) \right),$$

(d'ailleurs il suffit pour cela que G soit à frontière lipschitzienne). Ce dernier espace coïncide avec $W^{2,p}(\mathbb{R}^n)$ par le Théorème de Mikhlin et par restriction, on revient à $W^{2,p}(\Sigma)$.

5. Le problème dans le cône

Posons

$$u_0 = e^{(2-n/p)t}v,$$

alors par retour aux anciennes variables, on peut voir que

$$\frac{u_0}{\rho^2},\ \frac{D_i u_0}{\rho^2},\ D_i D_j u_0 \in L^p(Q)$$

pour tout $i,j = 1,\dots,n$. D'autre part, u_0 qui est solution de (1), ne coïncide pas nécessairement avec la solution variationnelle u. Mais $\psi = \varphi(u - u_0)$ est harmonique et est dans $H_0^1(Q)$, (φ est une fonction de troncature de $D(\overline{Q})$, valant 1 sur une boule fixée $Q_R = \overline{B(0,R)}$). On peut alors développer ψ dans $L^2(Q)$ sur la base complète et orthonormée formée par les vecteurs propres $(w_j)_{j\geq 1}$ de $(-\Delta')$. On obtient l'existence de deux suites $(a_j)_{j\geq 1}$ et $(b_j)_{j\geq 1}$ telles que

$$\psi = \sum_{j\geq 1} a_j \rho^{1-\frac{n}{2}+\beta_j} w_j(\sigma) + \sum_{j\geq 1} b_j \rho^{1-\frac{1}{2}-\beta_j} w_j(\sigma)$$

avec

$$\beta_j^2 = \lambda_j + \left(\frac{n}{2} - 1\right).$$

Le fait que $\psi \in H^1_{loc}$ implique que $b_j = 0 \quad \forall j \geq 1$ car $\varphi\rho^\alpha w_j \in H_0^1(Q)$ si et seulement si $\mathrm{Re}\,(\alpha) > 1 - n/2$. Finalement l'équivalence

$$\varphi\rho^\alpha w_j \in W^{2,p}(Q) \Leftrightarrow \mathrm{Re}\,(\alpha) > 2 - n/p$$

nous conduit au résultat de décomposition citée en introduction dans le cadre L^P:

Théorème 5 *Soit $u \in H_0^1(Q)$ la solution variationnelle de $-\Delta u = f \in L^p(Q)$ où $p \geq 2$ est tel que $\left(\frac{n}{p} - 2\right)\left(\frac{n}{p} - n\right) \neq \lambda_j\ \forall j \geq 1$. Alors il existe une suite $(a_j)_{j\geq 1}$ telle que $\left(u - \sum_{\lambda_j < (\frac{n}{p}-2)(\frac{n}{p}-n)} a_j \rho^{1-\frac{n}{2}+\beta_j} w_j(\sigma)\right) \in W^{2,p}(Q_R)$ pour tout $R > 0$, avec $\beta_j^2 = \lambda_j + \left(\frac{n}{2} - 1\right)$.*

Par exemple, en dimension $n = 2$, dans un secteur plan d'ouverture ω et en coordonnées polaires (r,θ) on a

$$\begin{cases} \lambda_j = \frac{j^2\pi^2}{\omega^2}, \\ w_j(\theta) = \sqrt{\frac{2}{\pi}} \sin\left(\frac{j\pi\theta}{\omega}\right), \\ \left(\frac{n}{p} - 2\right)\left(\frac{n}{p} - n\right) = (2p')^2, \end{cases}$$

et donc

$$\left(u - \sum_{j < \frac{2\omega}{\pi p'}} a_j r^{\frac{j\pi}{\omega}} \sin\left(\frac{j\pi\theta}{\omega}\right)\right) \in W^{2,p}(Q_R).$$

6. Le cadre höldérien

Dans un premier temps, on va rappeler les résultats essentiels de régularité optimale pour une équation différentielle abstraite du second ordre de type elliptique de la forme

$$\begin{cases} y''(t) + Ly(t) = l(t) \\ y(0) = y_0 \\ y(1) = y_1, \end{cases} \tag{13}$$

où $y_0, y_1 \in X$ et L est un opérateur linéaire fermé de domaine $D(L)$ non nécessairement dense dans un espace de Banach complexe X et vérifiant l'unique hypothèse d'ellipticité au sens de Krein [12] :

$$\exists C > 0 \ \forall r \geq 0 \quad \exists (L - rI)^{-1} \ / \ \left\|(L - rI)^{-1}\right\|_{L(X)} \leq \frac{C}{1+r}. \tag{14}$$

Pour $\theta \in]0,1[$, on considère l'espace réel d'interpolation caractérisé dans Grisvard [10] par

$$D_L(\theta, +\infty) = \left\{x \in X \quad / \quad \sup_{r>0} r^\theta \left\|L(L - rI)^{-1}x\right\|_X < \infty\right\},$$

et le sous-espace fermé $D_L(\theta)$ (voir Sinestrari [20], Lunardi [17]) défini par

$$D_L(\theta) = \left\{x \in X \quad / \quad \lim_{r \to \infty} r^\theta \left\|L(L - rI)^{-1}x\right\|_X = 0\right\}.$$

Pour θ fixé dans $]0, 1/2[$, alors d'après Labbas [13], on a les résultats suivants:

Théorème 6 *Pour y_0 $y_1 \in D(L)$, $l \in C^{2\theta}([0,1], X)$ il existe une unique solution y du problème* (13) *telle que*

i) $y \in C^2([0,1], X) \cap C([0,1], D(L)) \iff l(0) - Ly_0$ *et* $l(1) - Ly_1$ *appartiennent à* $\overline{D(L)}$.
ii) $y'', Ly \in C^{2\theta}([0,1], X) \iff l(0) - Ly_0$ *et* $l(1) - Ly_1$ *appartiennent à* $D_L(\theta, +\infty)$.
iii) $y'' \in, L^\infty(0,1; D_L(\theta, +\infty)) \iff l(0) - Ly_0$ *et* $l(1) - Ly_1$ *appartiennent à* $D_L(\theta, +\infty)$.

Théorème 7 *Pour $l \in C([0,1], X) \cap L^\infty(0,1; D_L(\theta, +\infty))$ et y_0, $y_1 \in D(L)$ il existe une unique solution y de* (13) *telle que*

i) $y \in W^{2,\infty}(0,1; X) \cap L^\infty(0,1; D(L)) \iff l(0) - Ly_0$ et $l(1) - Ly_1$ *appartiennent à* $\overline{D(L)}$.
ii) $y'', Ly \in L^\infty(0,1; D_A(\theta, +\infty)) \iff l(y_0) - Ly_1$ et $l(1) - Ly_1$ *appartiennent à* $D_L(\theta, +\infty)$.
iii) $Ly \in C^{2\theta}([0,1]; X) \iff l(0) - Ly_0$ et $l(1) - Ly_1$ *appartiennent à* $D_L(\theta, +\infty)$.

Les mêmes résultats sont vrais si on remplace $C^{2\theta}([0,1], X)$ par $h^{2\theta}([0,1], X)$ et $D_L(\theta, +\infty)$ par $D_L(\theta)$. D'autre part, les mêmes techniques utilisées dans [13] pour l'équation sur la demi-droite $[0,\ +\infty[$

$$\begin{cases} y''(t) + Ly(t) = l(t) \in X \\ y(0) = y_0 \\ y \text{ borné sur } [0, +\infty[. \end{cases} \tag{15}$$

conduisent au résultat:

Théorème 8 *Pour $y_0 \in D(L)$ et $l \in C^{2\theta}([0,+\infty[, X)$ il existe une unique solution y de* (13) *telle que*

i) $y \in C^2([0,+\infty[; X) \cap C([0,+\infty[; D(L)) \Longleftrightarrow l(0) - Ly_0 \in \overline{D(L)}$.
ii) y'', $Ly \in C^{2\theta}([0,+\infty[; X) \Longleftrightarrow l(0) - Ly_0 \in D_L(\theta, +\infty)$.
iii) $y'' \in L^\infty([0,+\infty[; D_L(\theta,+\infty)) \Longleftrightarrow l(0) - Ly_0 \in D_L(\theta, +\infty)$.

On a un théorème analogue si on remplace $C^{2\theta}$ par $h^{2\theta}$ et $D_L(\theta,+\infty)$ par $D_L(\theta)$. De même le problème sur $]-\infty,\ 0]$ est similaire.

On considère maintenant le problème

$$\begin{cases} -\Delta u = f \\ u_{|\partial Q} = 0, \end{cases}$$

où cette fois $f \in h_0^\alpha(\overline{Q})$. Pour $(t,\sigma) \in \Sigma = \mathbb{R} \times G$, on pose

$$\begin{aligned} V(t,\sigma) &= e^{-(2+\alpha)t} u(e^t\sigma) \\ H(t,\sigma) &= e^{-\alpha t} f(e^t\sigma), \end{aligned}$$

et on définit les fonctions vectorielles à valeurs dans un espace de Banach X (qu'on précisera plus loin)

$$\begin{aligned} v &: \mathbb{R} \to X; t \longmapsto v(t)\ , \quad v(t)(\sigma) = V(t,\sigma), \\ h &: \mathbb{R} \to X; t \longmapsto h(t), \quad h(t)(\sigma) = H(t,\sigma). \end{aligned}$$

Alors v vérifie l'équation abstraite

$$\begin{cases} D_t^2 v + (1+2\beta)\, D_t v + \beta\,(\beta+1)\, v + \Delta' v = h \\ v(t) \in D(\Delta') \subset X, \end{cases} \tag{16}$$

où

$$\beta = 2 + \alpha.$$

Les lemmes suivants précisent les liens entre la petite höldérianité globale, partielle et vectorielle dans le cylindre Σ.

Lemme 9 *On a*

i) $h \in h^\alpha(\mathbb{R}, C_0(\overline{G}))$ *si et seulement si* $H \in UC(\mathbb{R} \times \overline{G})$ *et* $H(.,\sigma) \in h^\alpha(\mathbb{R})$ *uniformément en* $\sigma \in \overline{G}$.
ii) $h \in UC(\mathbb{R}, C_0(\overline{G})) \cap L^\infty(\mathbb{R}, h_0^\alpha(\overline{G}))$ *si et seulement si* $H \in UC(\mathbb{R} \times \overline{G})$ *et* $H(t,.) \in h_0^\alpha(\overline{G})$ *uniformément en* $t \in \mathbb{R}$.

Voir le lemme 6.2 de Sinestrari [20].

Lemme 10 *Soit $f \in h_0^\alpha(\overline{Q})$. Alors la fonction $H(t,\sigma) = e^{-\alpha t} f(e^t \sigma)$ est telle que*

i) $H \in UC(\mathbb{R} \times \overline{G})$ et $H(.,\sigma) \in h^\alpha(\mathbb{R})$ uniformément en $\sigma \in \overline{G}$.

ii) $H \in UC(\mathbb{R} \times \overline{G})$ et $H(t,.) \in h_0^\alpha(\overline{G})$ *uniformément en* $t \in \mathbb{R}$.

i) En effet si τ et t sont tels que $-\infty < \tau < t < +\infty$, alors

$$\begin{aligned} H(t,\sigma) - H(\tau,\sigma) &= \left(e^{-\alpha t} - e^{-\alpha t}\right) f(e^\tau \sigma) + e^{-\alpha t}\left(f(e^t\sigma) - f(e^\tau\sigma)\right) \\ &= \frac{-1}{\alpha} \sum_\tau^t e^{-\alpha\xi} d\xi \left(f(e^\tau\sigma) - f(0)\right) + e^{-\alpha t}\left(f(e^t\sigma) - f(e^\tau\sigma)\right) \\ &= \Delta_1 + \Delta_2, \end{aligned}$$

d'où

$$|\Delta_1| \leq \frac{1}{\alpha} |t-\tau| \, e^{-\alpha\tau} \frac{|f(e^\tau\sigma) - f(0)|}{\|e^\tau\sigma\|^\alpha} e^{\alpha\tau} \leq \frac{1}{\alpha} |t-\tau| \, \|f\|_{C^\alpha(\overline{Q})}$$

qui implique que $\Delta_1(.,\sigma) \in h^\nu(\mathbb{R})$ uniformément en σ pour tout $\nu \in]0,1[$. Pour Δ_2, on a

$$\begin{aligned} |\Delta_2| &\leq e^{-\alpha t} \left\|e^t\sigma - e^\tau\sigma\right\|_2^\alpha \frac{|f(e^t\sigma) - f(e^\tau\sigma)|}{\|e^t\sigma - e^\tau\sigma\|^\alpha} \\ &\leq e^{-\alpha t} \left|e^t - e^\tau\right|^\alpha \frac{|f(e^t\sigma) - f(e^\tau\sigma)|}{\|e^t\sigma - e^\tau\sigma\|^\alpha} \\ &\leq e^{-\alpha t} \left(\int_\tau^t e^\xi d\xi\right)^\alpha \frac{|f(e^t\sigma) - f(e^\tau\sigma)|}{\|e^t\sigma - e^\tau\sigma\|^\alpha} \\ &\leq e^{-\alpha t} e^{\alpha t} (t-\tau)^\alpha \frac{|f(e^t\sigma) - f(e^\tau\sigma)|}{\|e^t\sigma - e^\tau\sigma\|^\alpha}, \end{aligned}$$

d'où

$$\lim_{\delta \to 0} \sup_{|t-\tau| \leq \delta} \frac{|\Delta_2|}{(t-\tau)^\alpha} = 0$$

uniformément en σ, et donc $\Delta_2(.,\sigma) \in h^\alpha(\mathbb{R})$.

ii) Partant de

$$H(t,\sigma) = e^{-\alpha t} f(e^t\sigma) = 0, \quad \forall \sigma \in \partial G,$$

on a

$$|H(t,\sigma)| = e^{-\alpha t}\left|f(e^t\sigma) - f(0)\right| \leq \|f\|_{C^\alpha(\overline{Q})} \quad \forall \sigma \in \overline{G},$$

et

$$\begin{aligned} |H(t,\sigma) - H(t,\sigma')| &= e^{-\alpha t}\left|f(e^t\sigma) - f(e^\tau\sigma')\right| \\ &\leq e^{-\alpha t} \left\|e^t\sigma - e^t\sigma'\right\|_2^\alpha \frac{|f(e^t\sigma) - f(e^t\sigma')|}{\|e^t\sigma - e^t\sigma'\|^\alpha} \\ &\leq \|\sigma - \sigma'\|^\alpha \frac{|f(e^t\sigma) - f(e^t\sigma')|}{\|e^t\sigma - e^t\sigma'\|^\alpha}, \end{aligned}$$

d'où

$$\lim_{\delta\to 0}\sup_{\|\sigma-\sigma'\|\leq\delta}\frac{|H(t,\sigma)-H(t,\sigma')|}{\|\sigma-\sigma'\|^{\alpha}}=0.$$

Lemme 11 *Soit* $\phi\in L^{\infty}\left(\mathbb{R},h_0^{\alpha}\left(\overline{G}\right)\right)\cap h^{\alpha}\left(\mathbb{R},C_0\left(\overline{G}\right)\right)$ *alors la fonction* F *définie par* $F(t,\sigma)=\phi(t)(\sigma)$ *appartient à* $h_0^{\alpha}\left(\overline{\Sigma}\right)$.

Il est clair que $F=0$ sur $\partial\Sigma$. Soient (t,σ) et $(t',\sigma')\in\mathbb{R}\times\overline{G}$ $(t\neq t',\ \sigma\neq\sigma')$ tels que $\|\sigma-\sigma'\|_2\leq\delta/2$ et $|t-t'|\leq\delta/2$ pour $\delta>0$ fixé. Alors

$$\begin{aligned}
&|F(t,\sigma)-F(t',\sigma')|\\
\leq\ &|F(t,\sigma)-F(t,\sigma')|+|F(t,\sigma')-F(t',\sigma')|\\
\leq\ &\|\sigma-\sigma'\|^{\alpha}\frac{|F(t,\sigma)-F(t,\sigma')|}{\|\sigma-\sigma'\|^{\alpha}}+|t-t'|^{\alpha}\frac{|F(t,\sigma)-F(t',\sigma)|}{|t-t'|^{\alpha}}\\
\leq\ &\left(\|\sigma-\sigma'\|^{\alpha}+|t-t'|^{\alpha}\right)\left(\frac{|F(t,\sigma)-F(t,\sigma')|}{\|\sigma-\sigma'\|^{\alpha}}+\frac{|F(t,\sigma)-F(t',\sigma)|}{|t-t'|^{\alpha}}\right)\\
\leq\ &K\|(t,\sigma)-(t',\sigma')\|^{\alpha}\left(\frac{|\phi(t)(\sigma)-\phi(t)(\sigma')|}{\|\sigma-\sigma'\|^{\alpha}}+\frac{|\phi(t)(\sigma)-\phi(t')(\sigma)|}{|t-t'|^{\alpha}}\right),
\end{aligned}$$

et

$$\begin{aligned}
\frac{|F(t,\sigma)-F(t',\sigma')|}{\|(t,\sigma)-(t',\sigma')\|^{\alpha}}\ &\leq\ K\left(\frac{|\phi(t)(\sigma)-\phi(t)(\sigma')|}{\|\sigma-\sigma'\|^{\alpha}}+\frac{|\phi(t)(\sigma)-\phi(t')(\sigma)|}{|t-t'|^{\alpha}}\right)\\
&\leq\ K\left(\sup_{\|\sigma-\sigma'\|\leq\delta/2}\frac{|\phi(t)(\sigma)-\phi(t)(\sigma')|}{\|\sigma-\sigma'\|^{\alpha}}\right.\\
&\qquad\left.+\sup_{\|t-t'\|\leq\delta/2}\frac{|\phi(t)(\sigma)-\phi(t')(\sigma)|}{|t-t'|^{\alpha}}\right)
\end{aligned}$$

d'où

$$\begin{aligned}
&\sup_{\|(t,\sigma)-(t',\sigma')\|\leq\delta}\frac{|F(t,\sigma)-F(t',\sigma')|}{\|(t,\sigma)-(t',\sigma')\|^{\alpha}}\\
\leq\ &K\left(\sup_{\|\sigma-\sigma'\|\leq\delta/2}\frac{|\phi(t)(\sigma)-\phi(t)(\sigma')|}{\|\sigma-\sigma'\|^{\alpha}}+\sup_{\|t-t'\|\leq\delta/2}\frac{|\phi(t)(\sigma)-\phi(t')(\sigma)|}{|t-t'|^{\alpha}}\right).
\end{aligned}$$

Et puisque $\phi\in L^{\infty}\left(\mathbb{R},h_0^{\alpha}\left(\overline{G}\right)\right)\cap h^{\alpha}\left(\mathbb{R},C_0\left(\overline{G}\right)\right)$, alors

$$\lim_{\delta\to 0}\sup_{\|(t,\sigma)-(t',\sigma')\|\leq\delta}\frac{|F(t,\sigma)-F(t',\sigma')|}{\|(t,\sigma)-(t',\sigma')\|^{\alpha}}=0,$$

donc $F\in h^{\alpha}\left(\overline{\Sigma}\right)$.

Remarquons que grâce à l'hypothèse sur f, la fonction h définie par $h(t)(\sigma)=H(t,\sigma)=e^{-\alpha t}f(e^{t}\sigma)$ est dans l'espace $L^{\infty}\left(\mathbb{R},h_0^{\alpha}\left(\overline{G}\right)\right)\cap h^{\alpha}\left(\mathbb{R},C_0\left(\overline{G}\right)\right)$. Cela suggère d'appliquer les résultats du paragraphe 2 à l'équation (16) dans ces deux espaces.

Dans $E = L^\infty\left(\mathbb{R}, h_0^\alpha\left(\overline{G}\right)\right)$ normé par $\|f\|_E = \sup_{t\in R}\|f(t,.)\|_{C^\alpha(\overline{G})}$ on pose

$$\left\{\begin{array}{l} D(A) = L^\infty\left(\mathbb{R}, D(\Delta')\right) \\ (-Av)(t) = \Delta'\left(v(t,.)\right) \\ D(\Delta') = \left\{w \in C_0^\alpha\left(\overline{G}\right) \ / \ \Delta' w \in C_0^\alpha\left(\overline{G}\right)\right\}, \end{array}\right.$$

$$\left\{\begin{array}{l} D(-B) = W^{2,\infty}\left(\mathbb{R}, h_0^\alpha\left(\overline{G}\right)\right) \\ Bv = D_t^2 v + (1+2\beta)\, D_t v + \beta\,(\beta+1)\, v. \end{array}\right.$$

Ici on a $\overline{D(B)} \neq E$ alors que $\overline{D(A)} = E$, grâce à la caractérisation des petits espaces de Hölder. Et

$$\sigma(-B) = \left\{-\xi^2 + (1+2\beta)\xi i + \beta(\beta+1)\ ,\ \xi \in \mathbb{R}\right\}$$

donc $\rho(-B)$ contient le secteur

$$S_\beta = \{z \in C \ \ / \ |z| \ \geq \ 2\beta(\beta+1) : |Arg(z)| < \epsilon_B\}$$

avec $\epsilon_B > \pi/2$. Comme dans la première partie, on obtient

$$\left\|(-B-\lambda I)^{-1}\right\|_{L(E)} = O\left(\frac{1}{\left(\operatorname{Re}\sqrt{\lambda}\right)^2}\right) \quad \forall \lambda \in S_\beta.$$

Ici aussi les propriétés spectrales de $(-A)$ sont celles de sa réalisation Δ'. Grâce à Campanato [4], Δ' génère un semi-groupe analytique fortement continu dans $h_0^\alpha\left(\overline{G}\right)$. D'où l'existence de $\epsilon_A > \pi/2$ tel que A vérifie i) de $(H.1)$ avec $r = 0$. De même l'hypothèse $(H.3)$ est vérifiée si toutefois aucune des valeurs propres λ_j ne coïncide avec $(\alpha+2)(\alpha+3)$. On supposera

$$\beta(\beta+1) = (\alpha+2)(\alpha+3) \neq \lambda_j \quad \forall j \geq 1.$$

La commutativité $(H.2)$ se vérifie comme dans le cas L^p.

Considérons $F_{11} = W^{1,\infty}\left(\mathbb{R}, h_0^\alpha\left(\overline{G}\right)\right) \subset E$. Alors, grâce aux espaces de classe K_θ de Lions-Peetre (voir en appendice), il existe une constante $C > 0$ telle que

$$\left\{\begin{array}{l} \|v'\|_{L^\infty(R,h_0^\alpha(\overline{G}))} \leq C\,\|v\|^{1/2}_{L^\infty(R,h_0^\alpha(\overline{G}))} \cdot \|v''\|^{1/2}_{L^\infty(R,h_0^\alpha(\overline{G}))} \\ \forall v \in W^{2,\infty}\left(\mathbb{R}, h_0^\alpha\left(\overline{G}\right)\right) = D(B). \end{array}\right.$$

Du Théorème 1 on déduit la

Proposition 12 *Pour toute* $h \in L^\infty\left(\mathbb{R}, h_0^\alpha\left(\overline{G}\right)\right)$ *le problème* (16) *admet une unique solution forte* v *qui est dans* $F_{11} = W^{1,\infty}\left(\mathbb{R}, h_0^\alpha\left(\overline{G}\right)\right)$.

Prenons maintenant $F_{12} = L^\infty\left(\mathbb{R}, C^{1+\alpha}\left(\overline{G}\right) \cap h_0^\alpha\left(\overline{G}\right)\right) \subset E$. Alors il existe une constante positive C telle que

$$\|w\|_{C^{2+\alpha}(\overline{G})} \leq C\,\|\Delta' w\|_{C_0^\alpha(\overline{G})} \quad \forall w \in D(\Delta'),$$

(voir Campanato [4]). Par interpolation, on obtient

$$\|v(t,.)\|_{C^{1+\alpha}(\overline{G})} \leq C\,\|v(t,.)\|^{1/2}_{C_0^\alpha(\overline{G})}\,\|\Delta'\left(v(t,.)\right)\|^{1/2}_{C_0^\alpha(\overline{G})} \quad \forall v(t,.) \in D(\Delta'),$$

et donc

$$\|v\|_F \leq C \|v\|_E^{1/2} \|Av\|_E^{1/2} \quad \forall v \in D(A).$$

Le Théorème 1 s'applique et donne la

Proposition 13 *Pour toute* $h \in L^\infty\left(\mathbb{R}, h_0^\alpha\left(\overline{G}\right)\right)$ *le problème* (16) *admet une unique solution forte* v *qui est dans* $F_{12} = L^\infty\left(\mathbb{R}, C^{1+\alpha}\left(\overline{G}\right) \cap h_0^\alpha\left(\overline{G}\right)\right)$.

Les deux résultats précédents impliquent que

$$v \in W^{1,\infty}\left(\mathbb{R}, h_0^\alpha\left(\overline{G}\right)\right) \cap L^\infty\left(\mathbb{R}, C^{1+\alpha}\left(\overline{G}\right) \cap h_0^\alpha\left(\overline{G}\right)\right).$$

Considérons maintenant l'espace de Banach $E = h^\alpha\left(\mathbb{R}, C_0\left(\overline{G}\right)\right)$ et posons

$$\begin{cases} D(A) = h^\alpha\left(\mathbb{R}, D(\Delta')\right) \\ (-Av)(t) = \Delta'\left(v(t,.)\right) \\ D(\Delta') = \left\{w \in \bigcap_{q\geq 1} W^{2,q}(G) \cap C_0\left(\overline{G}\right),\ \Delta' w \in C_0\left(\overline{G}\right)\right\} \end{cases}$$

$$\begin{cases} D(B) = C^{2+\alpha}\left(\mathbb{R}, C_0\left(\overline{G}\right)\right) \\ (-B)v = D_t^2 v + (1+2\beta) D_t v + \beta(\beta+1) v. \end{cases}$$

Alors $\overline{D(B)} = E$ et les mêmes propriétés spectrales précédentes restent vraies. Pour A, voir Stewart [21] par exemple. Les inégalités de convexité du Théorème 1 sont vérifiées dans $F_{21} = C^{1+\alpha}\left(\mathbb{R}, C_0\left(\overline{G}\right)\right)$ et dans $F_{22} = h^\alpha\left(\mathbb{R}, W_0^{1,q}\left(\overline{G}\right)\right)$, $\forall q \geq 1$. D'où la

Proposition 14 *pour toute* $h \in h^\alpha\left(\mathbb{R}, C_0\left(\overline{G}\right)\right)$ *le problème* (16) *admet une unique solution forte* v *telle que*

$$v \in C^{1+\alpha}\left(\mathbb{R}, C_0\left(\overline{G}\right)\right) \cap h^\alpha\left(\mathbb{R}, W_0^{1,q}(G)\right).$$

La fonction v vérifie

$$v''(t) + \Delta'\left(v(t)\right) = h(t) - (1+2\beta)v'(t) - \beta(\beta+1)v(t) = k(t),$$

avec

$$k \in C^\alpha\left(\mathbb{R}, C_0\left(\overline{G}\right)\right) \cap L^\infty\left(\mathbb{R}, h_0^\alpha\left(\overline{G}\right)\right).$$

On étudie maintenant cette dernière équation sur les deux demi-droites $[t_0, +\infty[$ et $]-\infty, t_0]$ pour un $t_0 > 0$, fixé. Soit Ψ une fonction de troncature telle que

$$\begin{cases} \Psi = 1 & \text{si } t \geq t_0 \\ \Psi = 0 & \text{si } t \leq 0, \end{cases}$$

alors $w = \Psi.v$ est solution de

$$\begin{cases} w''(t) + \Delta'\left(w(t)\right) = \Psi(t)k(t) + \Psi'(t)v'(t) + \Psi''(t)v(t) = l(t) \\ w(0) = 0 \\ w \quad \text{bornée sur } [0,\infty[, \end{cases} \tag{17}$$

où

$$l \in C^\alpha\left([0,\infty[, C_0\left(\overline{G}\right)\right) \cap L^\infty\left(0,\infty; h_0^\alpha\left(\overline{G}\right)\right).$$

On fait de même sur $]-\infty, 0]$.

Dans $X = C_0\left(\overline{G}\right)$ on définit L par

$$\left\{\begin{array}{l} D(L) = \left\{w \in \bigcap_{q\geq 1} W^{2,q}(G) \cap C_0\left(\overline{G}\right),\ \Delta' w \in C_0\left(\overline{G}\right)\right\} \\ Lw = \Delta' w. \end{array}\right. \tag{18}$$

Utilisant la première régularité de l, le Théorème 8 donne

Proposition 15 *La solution forte v vérifie:*

i) $v'',\ \Delta' v \in C^{\alpha}\left([0,\infty[, C_0\left(\overline{G}\right)\right)$,
ii) $v'' \in L^{\infty}\left(0,\infty; D_{\Delta'}\left(\alpha/2,+\infty\right)\right)$.

En effet l'hypothèse (14) est vérifiée grâce à Miranda [18] et Stewart [21] tandis que la condition de compatibilité

$$l(0) \in D_{\Delta'}\left(\alpha/2,+\infty\right)$$

découle du fait que l'espace d'interpolation $D_{\Delta'}\left(\alpha/2,+\infty\right)$ coïncide avec $C_0^{\alpha}\left(\overline{G}\right)$ (Lunardi [14]), d'autre part puisque $v \in W^{1,\infty}\left(\mathbb{R}, h_0^{\alpha}\left(\overline{G}\right)\right)$, alors

$$v(t)\ ,\ v'(t) \in h_0^{\alpha}\left(\overline{G}\right) \text{ p.p } t \in \mathbb{R},$$

et donc $l(0) \in h_0^{\alpha}\left(\overline{G}\right) = D_{\Delta'}\left(\alpha/2\right) \subset D_{\Delta'}\left(\alpha/2,+\infty\right)$.

De même, de la seconde régularité l et de l'équivalent du Théorème 8, on a

Proposition 16 *La solution forte v vérifie:*

i) $v'',\ \Delta' v \in L^{\infty}\left(0,\infty; h_0^{\alpha}\left(\overline{G}\right)\right)$,
ii) $\Delta' v \in h^{\alpha}\left([0,\infty[, C_0\left(\overline{G}\right)\right)$.

Finalement, après l'étude sur $]-\infty,0]$, on obtient les régularités suivantes

$$\left\{\begin{array}{l} i)\ v \in W^{1,\infty}\left(\mathbb{R}, h_0^{\alpha}\left(\overline{G}\right)\right) \cap L^{\infty}\left(\mathbb{R}, C^{1+\alpha}\left(\overline{G}\right) \cap h_0^{\alpha}\left(\overline{G}\right)\right), \\ ii)\ v \in C^{1+\alpha}\left(\mathbb{R}, C_0\left(\overline{G}\right)\right) \cap h^{\alpha}\left(\mathbb{R}, W_0^{1,q}\left(\overline{G}\right)\right), \forall q > 3, \\ iii)\ v \in C^{2+\alpha}\left(\mathbb{R}, C_0\left(\overline{G}\right)\right) \cap C\left(\mathbb{R}, D(\Delta')\right) \cap W^{2,\infty}\left(\mathbb{R}, h_0^{\alpha}\left(\overline{G}\right)\right), \\ iv)\ \Delta' v \in L^{\infty}\left(\mathbb{R}, h_0^{\alpha}\left(\overline{G}\right)\right) \cap h^{\alpha}\left(\mathbb{R}, C_0\left(\overline{G}\right)\right). \end{array}\right. \tag{19}$$

Les points *iii)*, *iv)* et les résultats de Najmi [19] donnent

$$V(t,\sigma) = v(t)(\sigma) \in C^{2+\alpha}\left(\overline{\Sigma}\right).$$

D'où le Théorème

Théorème 17 *Soit $h \in L^{\infty}\left(\mathbb{R}, h_0^{\alpha}\left(\overline{G}\right)\right) \cap h^{\alpha}\left(\mathbb{R}, C_0\left(\overline{G}\right)\right)$ avec $\alpha \in]0,1[$ tel que*

$$(2+\alpha)(3+\alpha) \neq \lambda_j \quad \forall j \geq 1,$$

où λ_j , $j = 1, 2, \ldots.$ sont les valeurs propres de $(-\Delta')$ sur G avec condition de Dirichlet. Alors le problème

$$\left\{\begin{array}{l} D_t^2 v + (5+2\alpha) D_t v + (\alpha+2)(\alpha+3) v + \Delta' v = h \text{ dans } \Sigma \\ v_{|\partial\Sigma} = 0, \end{array}\right. \tag{20}$$

admet une unique solution v telle que $V(t,\sigma) = v(t)(\sigma) \in C^{2+\alpha}\left(\overline{\Sigma}\right) \cap C_0\left(\overline{\Sigma}\right)$.

Ce résultat implique l'existence de $u_0 = e^{(\alpha+2)t}v$, solution de (1) et vérifiant

$$\begin{cases} \rho^{-2}u_0 \in C^\alpha\left(\overline{Q \cap B_R}\right) \\ \rho^{-1}D_i u_0 \in C^\alpha\left(\overline{Q \cap B_R}\right) \\ D_{ij}u_0 \in C^\alpha\left(\overline{Q \cap B_R}\right), \end{cases}$$

où $B_R = B(O, R)$. D'où

$$u_0 \in C^{2+\alpha}\left(\overline{Q \cap B_R}\right).$$

Si u est la solution variationnelle (lorsqu'elle existe) du problème

$$\begin{cases} \Delta u = f \in C_0^\alpha\left(\overline{Q}\right) \\ u \in H_0^1(Q), \end{cases}$$

alors, dans $B_R \cap Q$, la fonction

$$Z = u - u_0$$

est harmonique et appartient à $H^1(B_R \cap Q)$. On peut alors l'écrire, près de l'origine, sur la base des fonctions propres w_j de $(-\Delta')$ dans L^2 . D'où l'existence de deux suites $(a_j)_{j\geq 1}$ et $(b_j)_{j\geq 1}$ telles que

$$Z = \sum_{j\geq 1} a_j \rho^{-\frac{1}{2}+\beta_j} w_j(\sigma) + \sum_{j\geq 1} b_j \rho^{-\frac{1}{2}-\beta_j} w_j(\sigma) \qquad \text{où} \qquad \beta_j^2 = \lambda_j + \left(\frac{n}{2} - 1\right).$$

Du fait que $Z \in H^1_{loc}(Q)$, tous les b_j sont nuls. D'autre part l'équivalence

$$\rho^\nu w_j \in C^{2+\alpha} \Longleftrightarrow \operatorname{Re}\nu \geq 2 + \alpha,$$

implique que

$$\begin{aligned} u &= u_0 + Z \\ &= \Big(u_0 + \Big(Z - \sum_{j\in I} a_j \rho^{-\frac{1}{2}+\sqrt{\lambda_j + \frac{n}{2} - 1}} w_j(\sigma)\Big)\Big) \\ &\quad + \Big(\sum_{j\in I} a_j \rho^{-\frac{1}{2}+\sqrt{\lambda_j + \frac{n}{2} - 1}} w_j(\sigma)\Big) \\ &= u_r + u_s, \end{aligned}$$

où

$$\begin{aligned} u_r &= u_0 + \Big(Z - \sum_{j\in I} a_j \rho^{-\frac{1}{2}+\sqrt{\lambda_j + \frac{n}{2} - 1}} w_j(\sigma)\Big) \in C^{2+\alpha}\Big(\overline{Q \cap B_R}\Big), \\ u_s &= \sum_{j\in I} a_j \rho^{-\frac{1}{2}+\sqrt{\lambda_j + \frac{n}{2} - 1}} w_j(\sigma) \end{aligned}$$

et

$$I = \Big\{j \geq 1 \ / \ \lambda_j < \Big(\alpha + 2\Big)\Big(\alpha + 3\Big)\Big\}.$$

D'où le résultat final.

Remarques

Lorsque f appartient seulement à $C_0^\alpha\left(\overline{Q}\right)$, la décomposition précédente reste encore vraie avec $u_r \in C^{2+\alpha'}\left(\overline{Q \cap B_R}\right)$, où $0 < \alpha' < \alpha < 1$. Il suffit d'utiliser le fait que $h_0^{\alpha'}\left(\overline{Q}\right) \supset C_0^\alpha\left(\overline{Q}\right)$.

Que se passe-t-il si le second membre est seulement dans $C^\alpha\left(\overline{Q}\right)$ (ou $h^\alpha\left(\overline{Q}\right)$)? On va amener un élément de réponse en examinant le cas simple du cône particulier en dimension $n = 2$

$$Q = \{\varrho\sigma,\ \rho > 0,\ \sigma \in]0, \pi/2[\} =]0, \infty[\times]0, \infty[.$$

En écrivant

$$\begin{cases} D(A) = \left\{u \in UC^2([0, \infty[; UC([0, \infty[) : u(0, .) = 0\right\} \\ Au = u_{xx} \end{cases}$$

et

$$\begin{cases} D(B) = \left\{u \in UC([0, \infty[; UC^2([0, \infty[) : u(., 0) = 0\right\} \\ Bu = u_{yy}, \end{cases}$$

il n'est pas difficile de voir que ces deux opérateurs vérifient i), ii), iii) de (H.1) et (H.2) mais que leurs domaines ne sont denses ni dans $C^\alpha\left(\overline{Q}\right)$, ni dans $h^\alpha\left(\overline{Q}\right)$ à cause des conditions aux limites. Les Théorèmes 1 et 2 ne permettent pas de décrire la fermeture $\overline{(A+B)}$. Cependant on peut affirmer, grâce à Labbas [14] qu'il existe une extension fermée $\widetilde{(A+B)}$ inversible telle que $D\left(\widetilde{(A+B)}\right)$ contienne strictement $D\left(\overline{(A+B)}\right)$. Pour cela il suffit d'utiliser la fonction

$$u = \frac{1}{\pi}\rho^2\left(\ln\rho\sin(2\sigma) + \sigma\cos(2\sigma)\right) + \frac{1}{2}\left(\rho\sin\sigma\right)^2$$

qui vérifie

$$\begin{cases} \Delta u = 1 \\ u_{|\sigma=0} = u_{|\sigma=\pi/2} = 0. \\ u \in C(\overline{Q}) \\ u \notin C^2(\overline{Q}) \\ u \in D\left(\widetilde{(A+B)}\right) \\ u \notin D\left(\overline{(A+B)}\right), \end{cases}$$

la dernière propriété est vérifiée car si $u \in D\left(\overline{(A+B)}\right)$ alors il existerait une suite $(u_n) \in C_x^2 \cap C_y^2$ avec $u_{n|\sigma=0} = u_{n|\sigma=\pi/2} = 0$ telle que

$$\begin{cases} u_n \longrightarrow u \\ \Delta u_n \longrightarrow \overline{(A+B)}u = 1, \end{cases}$$

ce qui n'est pas possible car on aurait alors $\overline{(A+B)}u(0,0) = 0 = 1$.

Appendice

Dans ce paragraphe on rappelle la définition des espaces de classe K'_θ et la preuve de l'inégalité de convexité donnée dans la section 5 dans le cas de l'espace $C^\alpha\left(\mathbb{R}, C_0\left(\overline{G}\right)\right)$.

Soient E_0 et E_1, deux espaces de Banach contenus dans un espace topologique séparé T. D'après Lions-Peetre, l'espace de Banach X appartient à la classe $K'_\theta\left(E_0, E_1\right)$ si

$$\begin{cases} i)\ E_0 \cap E_1 \subset X \subset E_0 + E_1 \\ ii)\ \exists C > 0\ /\ \|x\|_X \leq C\,\|x\|_{E_0}^{1-\theta}\,\|x\|_{E_1}^{\theta} \quad \forall x \in E_0 \cap E_1. \end{cases}$$

La proposition suivante donne des situations fréquentes d'espace X vérifiant i) et ii).

Proposition *Soit Λ un opérateur linéaire fermé de domaine $D(\Lambda) \subset E$ où E est espace de Banach. On suppose que $\rho(\Lambda) \supset \mathbb{R}_+$ et qu'il existe $C_\Lambda > 0$ telle que*

$$\left\|(\Lambda - \lambda I)^{-1}\right\|_{L(E)} \leq \frac{C_\Lambda}{\lambda} \quad \forall \lambda > 0,$$

alors $D(\Lambda) \in K'_{1/2}\left(D\left(\Lambda^2\right), E\right)$.

En effet pour $x \in D\left(\Lambda^2\right)$, $x \neq 0$ et $\lambda > 0$ on a

$$x = (\Lambda - \lambda I)^{-1}\Lambda x - \lambda(\Lambda - \lambda I)^{-1} x,$$

et

$$\begin{aligned} \Lambda x &= (\Lambda - \lambda I)^{-1}\Lambda^2 x - \lambda\Lambda(\Lambda - \lambda I)^{-1} x \\ \|\Lambda x\| &\leq \frac{C_\Lambda}{\lambda}\left\|\Lambda^2 x\right\| + (C_\Lambda + 1)\lambda\|x\|. \end{aligned}$$

Avec

$$\lambda = \lambda_0 = \sqrt{\frac{C_\Lambda}{C_\Lambda + 1}\frac{\|\Lambda^2 x\|}{\|x\|}},$$

on a

$$\|\Lambda x\| \leq 2\sqrt{C_\Lambda(C_\Lambda + 1)}\left\|\Lambda^2 x\right\|^{1/2}\|x\|^{1/2}.$$

Posons $E = C^\alpha\left(\mathbb{R}, C_0\left(\overline{G}\right)\right)$ et

$$\begin{cases} D(\Lambda) = \{u \in E\ /\ u' \in E\} = C^{1+\alpha}\left(\mathbb{R}, C_0\left(\overline{G}\right)\right) \\ \Lambda u = u', \end{cases}$$

alors

$$\begin{cases} D(\Lambda^2) = C^{2+\alpha}\left(\mathbb{R}, C_0\left(\overline{G}\right)\right) \\ \Lambda^2 u = u'', \end{cases}$$

et pour $\lambda > 0$ on a

$$\left[(\Lambda - \lambda I)^{-1} f\right](x) = -\int_x^\infty e^{-\lambda(s-x)} f(s)ds = -\int_0 \infty e^{-\lambda\xi} f(x+\xi)d\xi,$$

et donc

$$\left\|(\Lambda-\lambda I)^{-1} f\right\|_{C(R,C_0(\overline{G}))} \leq \frac{1}{\lambda}\|f\|_{C(R,C_0(\overline{G}))}.$$

D'autre part

$$\begin{aligned}\left|\left[(\Lambda-\lambda I)^{-1} f\right](x)-\left[(\Lambda-\lambda I)^{-1} f\right](y)\right| & \leq \int_0^{\infty} e^{-\lambda \xi}|f(x+\xi)-f(y+\xi)| d \xi \\ & \leq \frac{1}{\lambda}|x-y|^\alpha[f]_\alpha .\end{aligned}$$

d'où

$$\left\|(\Lambda-\lambda I)^{-1} f\right\|_{C^\alpha(R,C_0(\overline{G}))} \leq \frac{1}{\lambda}\|f\|_{C^\alpha(R,C_0(\overline{G}))}.$$

La proposition précédente donne l'existence d'une constante $C>0$ telle que

$$\|u'\|_{C^\alpha(R,C_0(\overline{G}))} \leq C\|u\|_{C^\alpha(R,C_0(\overline{G}))}^{1/2}\|u''\|_{C^\alpha(R,C_0(\overline{G}))}^{1/2} \quad \forall u \in C^{2+\alpha}\left(\mathbb{R}, C_0\left(\overline{G}\right)\right)$$

et donc

$$\begin{aligned}&\|u\|_{C^{1+\alpha}(R,C_0(\overline{G}))} \\ \leq\ & \sup(1, C)\left(\|u\|_{C^\alpha(R,C_0(\overline{G}))}+\|u\|_{C^\alpha(R,C_0(\overline{G}))}^{1/2}\|u''\|_{C^\alpha(R,C_0(\overline{G}))}^{1/2}\right).\end{aligned}$$

Références

[1] Agmon, S., Douglis, A. and Nirenberg, L.: *Estimates near the boundary for solutions of elliptic partial differential equations satisfying general boundary conditions* I, Comm. Pure Appl. Math. **12**, 1959), 623–727.

[2] Agmon, S., Douglis, A. and Nirenberg, L.: *Estimates near the boundary for solutions of elliptic partial differential equations satisfying general boundary conditions* II, Comm. Pure Appl. Math., **17** (1964), 35–92.

[3] Burkholder, D.L.: *A geometrical characterization of Banach spaces in which martingale difference sequences are unconditional,* Ann. Probab. **9** (1981), 997–1011.

[4] Campanato, S.: *Generation of analytic semi group by elliptic operators of second order in Hölder spaces,* Ann. Scuola Nor. Sup. Pisa **8** (1981).

[5] Clément, P., Grisvard, P: *Sommes d'opérateurs et régularité L^p dans les problèmes aux limites*, C.R.Acad. Sci. Paris, Série I **314** (1992), 821–824,

[6] Da Prato, G., Grisvard, P.: *Sommes d'opérateurs linéaires et équations différentielles opérationnelles*, J. Math. Pures Appl. IX Ser. **54**, (1975), 305–387.

[7] Dauge, M.: *Elliptic boundary value problems on corner domains*, Lecture Notes, **1341**, Springer Verlag, 1988.

[8] Dore G., Venni, A.: *On the closedness of the sum of two closed operators*, Mathematische Zeitschrift **196** (1987), 270–286.

[9] Grisvard, P.: Elliptic problems in non smooth domains, Monographs and studies in Mathematics, **24**, Pitman publishing compagny, 1985.

[10] Grisvard, P.: *Commutativité de deux foncteurs d' interpolation et applications*, J. Math. pures et appli. **45**, (1966), 143–290.

[11] Kondratiev, V.A.: *Boundary value problems for elliptic equations in domains with conical or angular points.* Transactions of the Moscow Mathematical Society **16**, (1967), 227–313,

[12] Krein, S.G.: *Linear differential equations in Banach spaces*, Moscou, 1967.

[13] Labbas, R.: *Equation elliptique abstraite du second ordre et équation parabolique pour le problème de Cauchy abstrait*, C.R.Acd. Sci. Paris, Série I **305**,(1987), 785–788.

[14] Labbas, R.: *Some results on the sum of linear operator with nondense domaines*, Annali di Matematica pura ed applicata (IV), **CLIV**, (1989), 91–97.

[15] Labbas, R., Moussaoui, M. and Najmi, M.: *Singular behavior of the Dirichl'et problem in Hölder spaces of the solutions to the Dirichlet problem in a cone.* Rend. Istit. Mat. Univ. Trieste, **XXX**, (1998), 155–179.

[16] Lunardi, A.: *Interpolation spaces between domains of elliptic operators and spaces of continuous functions*, Math. Nachr. **121**. (1985), 295–318.

[17] Lunardi, A.: *Analytic Semigroups and Optimal Regularity in Parabolic Problems*, Birkäuser Verlag, Boston, 1995.

[18] Miranda, C.: *Partial differential equations of elliptic type*, Springer Verlag, 1970.

[19] Najmi, M.: *Problème de régularité-singularité dans les espaces de Hölder pour un opérateur elliptique*, B.U.M.I. **10B** (1996), no. 7, 513–547.

[20] Sinestrari, E.: *Continuous interpolation spaces and spatial regularity in nonlinear Volterra integrodifferential equations*, J. Integral Equations **5** (1983), 287–308.

[21] Stewart, H.B.: *Generation of analytic semigroup by strongly elliptic operators under general boundary conditions*, Trans. Amer. Math. Soc. **259** (1980), 299–310.

Rabah Labbas
Faculté des Sciences et Techniques
25, rue Philippe Lebon
B.P 540
F-76058 Le Havre Cedex, France
e-mail: `labbas@univ-lehavre.fr`

Progress in Nonlinear Differential Equations
and Their Applications, Vol. 50, 237–262

Some Identification Problems Related to Thermal Materials with Loss of Memory

Alfredo Lorenzi

Dedicated to the memory of Brunello Terreni and to his courageous wife Raimonda and daughters Ester, Maria Pia and Noemi.

1. A first abstract integrodifferential identification problem

In this paper we are concerned with some new identification problems related to isotropic thermal bodies with a *loss* of memory. We assume that they are governed by the following state equation

$$u'(t)+D_t\int_{\alpha(t)}^{t} h_0(t,s)u(s)\,ds-Au(t)-\int_{\alpha(t)}^{t} h_1(t,s)Bu(s)\,ds=f(t),\qquad \forall t\in[0,T], \tag{1.1}$$

where A and B are given linear (differential) operators, while u and h_0, h_1 stand for the temperature and the memory functions. Moreover, function α – *the loss of memory* – satisfies $0 \le \alpha(t) < t$ for any $t \in (0,T]$.

For a physical justification of Equation (1.1) when $\alpha \equiv 0$, see, e.g., [1, 2, 6, 7, 18, 19], under the assumption that the history of the material is known in the time interval $(-\infty, 0]$.

As far as the identification of kernels h_0 and/or h_1, when $\alpha \equiv 0$, is concerned, we limit ourselves to recalling that so a large amount of work has been done in these last fifteen years that we like better to omit any quotation of papers, since it should be impossible to give here a complete list of the contributions in this field. So we confine ourselves to quoting the papers by M. Grasselli [8], J. Janno and L. v. Wolfersdorf [9, 10, 11, 12, 13, 14], Sinestrari [15, 16], since these mathematicians are the ones nearer to our research.

We recall that Equation (1.1) with $\alpha \not\equiv 0$ can be derived adapting to the thermal case the model proposed by Fichera in [5] for *viscoelastic* materials with fading memory. The basic assumption made by Fichera is expressed by the following relationship relating stresses $\sigma(u)$ and strains $\varepsilon(u)$ in the case of fading memory:

$$\sigma(u)(x,t)=\varepsilon(u)(x,t)+\int_{\alpha(t)}^{t} k(t,s)\varepsilon(u)(x,s)\,ds. \tag{1.2}$$

A somewhat different approach of heat diffusion in isotropic thermal bodies with linear memory is proposed and developed in [6, 18, 19]. There the authors prove, under suitable hypotheses including $\alpha(t) \equiv -\infty$, that the equations

$$h_j(t,s) = h_j(t-s) \qquad 0 \le s \le t \le T,\ j = 0,1, \tag{1.3}$$

must hold. However, in this case the thermal body has a *everlasting memory*, i.e. it does not forget *anything* of its past history. Further, we note that Fichera's definition of "fading memory" strongly differs from the one proposed in [4] (cf. also [3]) where the memory effects of a far past are weak but do *always* exist.

On the contrary, Fichera's model assumes that the far past simply can be erased from memory, function α pointing out what part of the past is still affecting the present state.

In this paper we are interested in the case where the general memory kernels h_j ($j = 0,1$) admit the simple representations (1.3), Moreover, $h_0, h_1 : [0,T] \to \mathbf{R}$ are *unknown* functions, while $\alpha : [0,T] \to \mathbf{R}$ may be both known and unknown.

Under Assumption (1.3), Equation (1.1) simplifies to

$$u'(t)+D_t \int_{\alpha(t)}^{t} h_0(t-s)u(s)\,ds - Au(t) - \int_{\alpha(t)}^{t} h_1(t-s)Bu(s)\,ds = f(t), \quad \forall t \in [0,T]. \tag{1.4}$$

We notice that when $\alpha \equiv 0$ we have the usual model [19] for materials with memory, while, when α is positive, the memory, at any instant positive t, takes account of the past, but limitedly to the interval $[\alpha(t), t]$ (cf. (1.1)). In other words, at any positive instant t the body has already "forgotten" what has happened in the time interval $[0, \alpha(t)]$.

As a consequence, α and h_j ($j = 0,1$) describe, respectively, the *memory loss* and the *memory intensities.*

In the first three sections of this paper we will assume that function α is *known* and enjoys the following properties for some $\beta \in (0,1)$:

$$\alpha \in C^{2+\beta}([0,T];\mathbf{R}), \quad \alpha(0) = 0, \qquad 0 \le \alpha'(t) < 1, \quad \forall t \in [0,T]. \tag{1.5}$$

Moreover, let (X,Y) be a pair of Banach spaces such that $Y \hookrightarrow X$ and denote by $(X,Y)_{\delta,\infty}$ the intermediate space of order $\delta \in (0,1)$ between X and Y. Let A, $B : Y \subset X \to X$ be closed linear operators, A being in addition sectorial [17], i.e. the resolvent set of A contains the open sector $\lambda_1 + \Sigma_\varphi$, where $\Sigma_\varphi = \{\lambda \in \mathbf{C} : |\arg \lambda| < \varphi\}$ for some $\lambda_1 \in \mathbf{R}$ and $\varphi \in (\pi/2, \pi)$, and the following estimate holds:

$$\|(\lambda - A)^{-1}\|_{\mathcal{L}(X)} \le C|\lambda - \lambda_1|^{-1}, \qquad \forall \lambda \in \lambda_1 + \Sigma_\varphi. \tag{1.6}$$

Further, let Φ_0, $\Phi_1 \in X^*$, X^* being the space dual to X.

We now consider the following abstract identification problem: *search for three functions*

$$u \in C^{2+\beta}([0,T];X) \cap C^{1+\beta}([0,T];Y)$$

and

$$h_0 \in C^{1+\beta}([0,T-\alpha(T)];\mathbf{R}), \quad h_1 \in C^{\beta}([0,T-\alpha(T)];\mathbf{R}) \quad (\beta \in (0,1))$$

such that

$$u'(t) + h_0(0)u(t) - \alpha'(t)h_0(t-\alpha(t))u(\alpha(t)) - Au(t)$$
$$+ \int_{\alpha(t)}^{t} [h_0'(t-s)u(s) - h_1(t-s)Bu(s)]\, ds = f(t), \quad \forall t \in [0,T], \quad (1.7)$$

$$u(0) = u_0, \quad (1.8)$$

$$\Phi_j[u(t)] = g_j(t), \qquad \forall t \in [0,T],\ j = 0,1. \quad (1.9)$$

We note that Equation (1.7) can be easily derived from (1.4) by simply differentiating the first integral with respect to t.

Finally we assume the data (u_0, f, g_0, g_1) enjoy the following properties for some $\delta \in (\beta, 1)$:

$$f \in C^{1+\beta}([0,T];X),\ g_0,\ g_1 \in C^{2+\beta}([0,T];\mathbf{R}),\ u_0 \in Y,\ Au_0 + f(0) \in Y, \quad (1.10)$$

$$Au_0,\ A[Au_0 + f(0)] + f'(0) \in (X,Y)_{\beta,\infty},\ Bu_0 \in (X,Y)_{\delta,\infty}, \quad (1.11)$$

$$g_0(0) \neq 0, \quad \Phi_j[u_0] = g_j(0), \qquad j = 0,1. \quad (1.12)$$

2. Solving the first identification problem

First we introduce the new unknown function

$$v(t) = u'(t) \quad \Longleftrightarrow \quad u(t) = u_0 + \int_0^t v(s)\, ds. \quad (2.1)$$

To derive the system of equations satisfied by the triplet (v, h_0, h_1) we need the following formulae

$$D_t \int_{\alpha(t)}^{t} h(t-s)f(s)\, ds = D_t \int_0^{t-\alpha(t)} h(s)f(t-s)\, ds$$
$$= \quad [1-\alpha'(t)]h(t-\alpha(t))f(\alpha(t)) + \int_{\alpha(t)}^{t} h(t-s)f'(s)\, ds, \qquad \forall t \in [0,T]. \quad (2.2)$$

Differentiating (1.7) and using (2.2), we find that (v, h_0, h_1) solves the identification problem

$$v'(t) - Av(t) + h_0(0)v(t) = N_1(v, h_0, h_1)(t)$$
$$-[1-\alpha'(t)]^2 h_0'(t-\alpha(t))\Big[u_0 + \int_0^{\alpha(t)} v(s)\, ds\Big]$$
$$+[1-\alpha'(t)]h_1(t-\alpha(t))\Big[Bu_0 + \int_0^{\alpha(t)} Bv(s)\, ds\Big], \qquad \forall t \in [0,T], \quad (2.3)$$

$$v(0) = -[1-\alpha'(0)]h_0(0)u_0 + Au_0 + f(0) := v_0, \tag{2.4}$$

$$\Phi_j[v(t)] = g_j'(t), \qquad \forall t \in [0,T],\ j=0,1, \tag{2.5}$$

where

$$N_1(v,h_0,h_1)(t) = [\alpha'(t)]^2\Big[h_0(0) + \int_0^{t-\alpha(t)} h_0'(s)\,ds\Big]\Big[v_0 + \int_0^{\alpha(t)} v'(s)\,ds\Big]$$
$$+\alpha''(t)h_0(t-\alpha(t))\Big[u_0 + \int_0^{\alpha(t)} v(s)\,ds\Big]$$
$$-\int_{\alpha(t)}^t [h_0'(t-s)v(s) - h_1(t-s)Bv(s)]\,ds + f'(t), \qquad \forall t \in [0,T]. \tag{2.6}$$

Remark 2.1. Observe that

$$\begin{aligned} N_1(v,h_0,h_1)(0) &= [\alpha'(0)]^2 h_0(0)v_0 + \alpha''(0)h_0(0)u_0 + f'(0) \\ &= \widetilde{N}_1(u_0,f,\alpha,h_0(0)) + f'(0). \end{aligned} \tag{2.7}$$

To determine the initial values of $h_0(0)$ first we apply functionals Φ_0 and Φ_1 to both sides in (1.7) and then we compute them at $t=0$. Taking Equations (1.8), (1.9) and (1.12) into account, we find the algebraic system

$$[1-\alpha'(0)]h_0(0)g_j(0) = -g_j'(0) + \Phi_j[Au_0 + f(0)], \qquad j=0,1. \tag{2.8}$$

For $h_0(0)$ to be uniquely defined we need the consistency condition

$$g_1(0)\{-g_0'(0) + \Phi_0[Au_0+f(0)]\} = g_0(0)\{-g_1'(0) + \Phi_1[Au_0+f(0)]\}. \tag{2.9}$$

By virtue of Assumptions (1.5), (1.9) we get

$$h_0(0) = \frac{1}{g_0(0)[1-\alpha'(0)]}\{-g_0'(0) + \Phi_0[Au_0+f(0)]\} := \chi_{0,0}(f,u_0,g_0,\alpha). \tag{2.10}$$

Applying then the linear functionals Φ_j $(j=0,1)$ to both sides in Equation (2.3) and taking advantage of consistency equations in (1.12), we find the following system, where $t\in[0,T]$ and $j=0,1$:

$$[1-\alpha'(t)]g_j(\alpha(t))h_0'(t-\alpha(t)) - \Big[\Phi_j[Bu_0] + \int_0^{\alpha(t)} \Phi_j[Bv(s)]\,ds\Big]h_1(t-\alpha(t))$$
$$= \frac{1}{1-\alpha'(t)}\Big\{-g_j''(t) + \Phi_j[Av(t)] - h_0(0)g_j'(t) + \Phi_j[N_1(v,h_0,h_1)(t)]\Big\}. \tag{2.11}$$

Remark 2.2. When $\alpha \equiv 0$ the coefficients of h_0' and h_1 in the left-hand side in (2.11) reduce simply to constants so that we can (globally) solve for (h_0', h_1) under the assumption

$$g_1(0)\Phi_0[Bu_0] \neq g_0(0)\Phi_1[Bu_0]. \tag{2.12}$$

On the contrary, when $\alpha \not\equiv 0$ we can write the previous system as a fixed-point one for (h_0', h_1) if we assume, e.g.,

$$\begin{aligned} &\left| g_1(\tau) \int_0^\tau \Phi_0[Bv(s)]\, ds - g_0(\tau) \int_0^\tau \Phi_1[Bv(s)]\, ds \right| \\ < \quad & |g_1(\tau)\Phi_0[Bu_0] - g_0(\tau)\Phi_1[Bu_0]|, \qquad \forall \tau \in [0, \alpha(T)]. \qquad (2.13) \end{aligned}$$

As a consequence, the present problem with a loss of memory turns out to be more complicated than the usual identification problem with $\alpha \equiv 0$. Moreover, condition (2.13), involving the unknown v, forces us to search for a local in time solution to our identification problem.

We can now state the main result of this section.

Theorem 2.1. *Under the basic Assumptions (1.10)–(1.12), (2.9), (2.12) the identification problem (1.7)–(1.9) admits a unique local in time solution* $(u, h_0, h_1) \in [C^{2+\beta}([0,T];\, X) \cap C^{1+\beta}([0,T]; Y)] \times C^{1+\beta}([0,T]) \times C^{\beta}([0,T])$.

We now introduce the function $\gamma : [0, T - \alpha(T)] \to [0, T]$ inverse to $t \to t - \alpha(t)$, i.e.

$$\gamma(\tau) - \alpha(\gamma(\tau)) = \tau, \qquad \forall \tau \in [0, T - \alpha(T)]. \qquad (2.14)$$

We need now the formulae

$$\alpha(\gamma(\tau)) = -\tau + \gamma(\tau), \quad \alpha'(\gamma(\tau)) = 1 - \frac{1}{\gamma'(\tau)}, \quad \forall \tau \in [0, T - \alpha(T)]. \qquad (2.15)$$

Making the change of variable $\tau = t - \alpha(t)$, from (2.11) we obtain the following system, where $\tau \in [0, T - \alpha(T)]$, $j = 0, 1$:

$$\begin{aligned} & [1 - \alpha'(\gamma(\tau))] g_j(\alpha(\gamma(\tau))) h_0'(\tau) - \left[\Phi_j[Bu_0] + \int_0^{\alpha(\gamma(\tau))} \Phi_j[Bv(s)]\, ds \right] h_1(\tau) \\ = \quad & \frac{1}{1 - \alpha'(\gamma(\tau))} \Big\{ - g_j''(\gamma(\tau)) + \Phi_j[Av(\gamma(\tau))] - h_0(0) g_j'(\gamma(\tau)) \\ & \qquad + \Phi_j[N_1(v, h_0, h_1)(\gamma(\tau))] \Big\} \\ =: \quad & N_{2+j}(v, h_0, h_1)(\tau) + [1 - \alpha'(\gamma(\tau))]^{-1} \Phi_j[Av(\gamma(\tau))]. \qquad (2.16) \end{aligned}$$

Remark 2.3. Observe that

$$\begin{aligned} & N_{2+j}(v, h_0, h_1)(0) = \frac{1}{1 - \alpha'(0)} \Big\{ - g_j''(0) - h_0(0) g_j'(0) \\ & + [\alpha'(0)]^2 h_0(0) \Phi_j[v_0] + \alpha''(0) h_0(0) \Phi_j[u_0] + \Phi_j[f'(0)] \Big\} \\ =: \quad & \widetilde{N}_{2+j}(u_0, f, g_0, g_1, \alpha, h_0(0)), \qquad j = 0, 1. \qquad (2.17) \end{aligned}$$

From Assumptions (2.12), (2.13) and Formula (2.16) we deduce the following system, where $\tau \in [0, T-\alpha(T)]$:

$$\begin{aligned} h_0'(\tau) &= \Big\{g_0(\alpha(\gamma(\tau)))\Big[\Phi_1[Bu_0] + \int_0^{\alpha(\gamma(\tau))} \Phi_1[Bv(s)]\,ds\Big] \\ &\quad -g_1(\alpha(\gamma(\tau)))\Big[\Phi_0[Bu_0] + \int_0^{\alpha(\gamma(\tau))} \Phi_0[Bv(s)]\,ds\Big]\Big\}^{-1} \\ &\quad \times[1-\alpha'(\gamma(\tau))]^{-1}\Big\{\Big[\Phi_0[Bu_0] + \int_0^{\alpha(\gamma(\tau))} \Phi_0[Bv(s)]\,ds\Big] \\ &\quad \times\Big[N_3(v,h_0,h_1)(\tau) + [1-\alpha'(\gamma(\tau))]^{-1}\Phi_1[Av(\gamma(\tau))]\Big] \\ &\quad -\Big[\Phi_1[Bu_0] + \int_0^{\alpha(\gamma(\tau))} \Phi_1[Bv(s)]\,ds\Big] \\ &\quad \times\Big[N_2(v,h_0,h_1)(\tau) + [1-\alpha'(\gamma(\tau))]^{-1}\Phi_0[Av(\gamma(\tau))]\Big]\Big\} \\ &=: N_4(v,h_0,h_1)(\tau) + J_{0,0}(v)(\tau)\Phi_0[Av(\gamma(\tau))] \\ &\quad +J_{0,1}(v)(\tau)\Phi_1[Av(\gamma(\tau))], \end{aligned} \tag{2.18}$$

$$\begin{aligned} h_1(\tau) &= \Big\{g_0(\alpha(\gamma(\tau)))\Big[\Phi_1[Bu_0] + \int_0^{\alpha(\gamma(\tau))} \Phi_1[Bv(s)]\,ds\Big] \\ &\quad -g_1(\alpha(\gamma(\tau)))\Big[\Phi_0[Bu_0] + \int_0^{\alpha(\gamma(\tau))} \Phi_0[Bv(s)]\,ds\Big]\Big\}^{-1} \\ &\quad \times\Big\{g_0(\gamma(\tau))\Big[N_3(v,h_0,h_1)(\tau) - [1-\alpha'(\gamma(\tau))]^{-1}\Phi_1[Av(\gamma(\tau))]\Big] \\ &\quad -g_1(\gamma(\tau))\Big[N_2(v,h_0,h_1)(\tau) - [1-\alpha'(\gamma(\tau))]^{-1}\Phi_0[Av(\gamma(\tau))]\Big]\Big\} \\ &=: N_5(v,h_0,h_1)(\tau) + J_{1,0}(v)(\tau)\Phi_0[Av(\gamma(\tau))] \\ &\quad +J_{1,1}(v)(\tau)\Phi_1[Av(\gamma(\tau))], \end{aligned} \tag{2.19}$$

$$h_0(0) = \chi_{0,0}(f, u_0, g_0, \alpha). \tag{2.20}$$

Remark 2.4. From (2.18), (2.19) we can compute the initial values $h_0'(0)$ and $h_1(0)$ in terms of the data. By virtue of the consistency conditions in (1.12) we easily compute

$$\begin{aligned} h_0'(0) &= [1-\alpha'(0)]^{-2}\{g_0(0)\Phi_1[Bu_0] - g_1(0)\Phi_0[Bu_0]\}^{-1} \\ &\quad \times\Big\{\Phi_0[Bu_0]\Big[[1-\alpha'(0)]\widetilde{N}_3(v,h_0,h_1)(0) + \Phi_1[Av_0]\Big] \\ &\quad -\Phi_1[Bu_0]\Big[[1-\alpha'(0)]\widetilde{N}_2(v,h_0,h_1)(0) + \Phi_0[Av_0]\Big]\Big\} \\ &=: \chi_{0,1}(u_0, f, g_0, g_1, \alpha), \end{aligned} \tag{2.21}$$

$$\begin{aligned} h_1(0) &= [1-\alpha'(0)]^{-1}\{g_0(0)\Phi_1[Bu_0]-g_1(0)\Phi_0[Bu_0]\}^{-1} \\ &\quad \times\Big\{g_0(0)\Big[[1-\alpha'(0)]\widetilde{N}_3(v,h_0,h_1)(0)+\Phi_1[Av_0]\Big] \\ &\quad -g_1(0)\Big[[1-\alpha'(0)]\widetilde{N}_2(v,h_0,h_1)(0)+\Phi_0[Av_0]\Big]\Big\} \\ &=: \chi_{1,0}(u_0,f,g_0,g_1,\alpha). \end{aligned} \tag{2.22}$$

Moreover, the initial values of $N_{4+j}(v,h_0,h_1)$ at $t=0$ depend on the data only and are given (cf. (1.5), (2.12)) by

$$\begin{aligned} N_4(v,h_0,h_1)(0) &= [1-\alpha'(0)]^{-1}\{g_0(0)\Phi_1[Bu_0]-g_1(0)\Phi_0[Bu_0]\}^{-1} \\ &\quad \times\{\Phi_0[Bu_0]\widetilde{N}_3(u_0,f,g_0,g_1,\alpha,\chi_{1,0}(u_0,f,g_0,g_1,\alpha)) \\ &\quad -\Phi_1[Bu_0]\widetilde{N}_2(u_0,f,g_0,g_1,\alpha,\chi_{1,0}(u_0,f,g_0,g_1,\alpha))\}, \end{aligned} \tag{2.23}$$

$$\begin{aligned} N_5(v,h_0,h_1)(0) &= \{g_0(0)\Phi_1[Bu_0]-g_1(0)\Phi_0[Bu_0]\}^{-1} \\ &\quad \times\{g_0(0)\widetilde{N}_3(u_0,f,g_0,g_1,\alpha,\chi_{1,0}(u_0,f,g_0,g_1,\alpha)) \\ &\quad -g_1(0)\widetilde{N}_2(u_0,f,g_0,g_1,\alpha,\chi_{1,0}(u_0,f,g_0,g_1,\alpha))\}, \end{aligned} \tag{2.24}$$

where $\widetilde{N}_{2+j}\,(u_0,f,g_0,g_1,\alpha,h_0(0))$ $(j=0,1)$ are defined by (2.17).

Then we replace Equation (2.18) with the equivalent one

$$\begin{aligned} h_0(\tau) &= h_0(0)+\int_0^\tau h_0'(s)\,ds \\ &= \chi_{0,0}(f,u_0,g_0,\alpha)+\int_0^\tau N_4(v,h_0,h_1)(s)\,ds \\ &\quad +\int_0^\tau \{J_{0,0}(v)(s)\Phi_0[Av(\gamma(s))]+J_{0,1}(v)(s)\Phi_1[Av(\gamma(s))]\}\,ds. \end{aligned} \tag{2.25}$$

Finally, we substitute in (2.3) the right-hand sides in (2.18) and (2.19) for h_0' and h_1. We get the Cauchy problem

$$\begin{aligned} & v'(t)-Av(t)+\chi_{0,0}(f,u_0,g_0,\alpha)v(t) \\ =\; & \Big\{N_1(v,h_0,h_1)(t)-[1-\alpha'(t)]^2N_4(v,h_0,h_1)(t-\alpha(t))\Big[u_0+\int_0^{\alpha(t)} v(s)\,ds\Big] \\ & +[1-\alpha'(t)]N_5(v,h_0,h_1)(t-\alpha(t))\Big[Bu_0+\int_0^{\alpha(t)} Bv(s)\,ds\Big] \\ & +\Phi_0[Av(t)]Q_{0,1}(v)(t)+\Phi_1[Av(t)]Q_{1,1}(v)(t)\Big\} \\ & +\Big\{\Phi_0[Av(t)]Q_{0,0}(v)(t)+\Phi_1[Av(t)]Q_{1,0}(v)(t)\Big\} \\ =:\; & N_6(v,h_0,h_1)(t)+N_7(v)(t), \qquad \forall t\in[0,T], \end{aligned} \tag{2.26}$$

$$v(0)=v_0, \tag{2.27}$$

where we have set (cf. (2.18), (2.19))

$$\begin{aligned} Q_{0,0}(v)(t) &= [1-\alpha'(t)]\{-[1-\alpha'(t)]J_{0,0}(v)(t-\alpha(t))u_0 \\ &\quad +J_{1,0}(v)(t-\alpha(t))Bu_0\}, \end{aligned} \tag{2.28}$$

$$\begin{aligned} Q_{0,1}(v)(t) &= -[1-\alpha'(t)]^2 J_{0,0}(v)(t-\alpha(t))\int_0^{\alpha(t)} v(s)\,ds \\ &\quad +[1-\alpha'(t)]J_{1,0}(v)(t-\alpha(t))\int_0^{\alpha(t)} Bv(s)\,ds, \end{aligned} \tag{2.29}$$

$$\begin{aligned} Q_{1,0}(v)(t) &= [1-\alpha'(t)]\{-[1-\alpha'(t)]J_{0,1}(v)(t-\alpha(t))u_0 \\ &\quad +J_{1,1}(v)(t-\alpha(t))Bu_0\}, \end{aligned} \tag{2.30}$$

$$\begin{aligned} Q_{1,1}(v)(t) &= -[1-\alpha'(t)]^2 J_{0,1}(v)(t-\alpha(t))\int_0^{\alpha(t)} v(s)\,ds \\ &\quad +[1-\alpha'(t)]J_{1,1}(v)(t-\alpha(t))\int_0^{\alpha(t)} Bv(s)\,ds. \end{aligned} \tag{2.31}$$

Remark 2.5. Observe that $Q_{j,1}(v)(0)=0$, $j=0,1$, while $Q_{j,0}(v)(t)\in(X,Y)_\delta$ for any $t\in[0,T]$ according to the assumption for Bu_0 in (1.11).

We now denote by $\{\mathrm{e}^{tA}\}_{t\geq 0}$ the analytic semigroup generated by the sectorial operator A. Hence, the analytic semigroup generated by $A-\chi_{0,0}(f,u_0,g_0,\alpha)I$ is given by

$$S(t)=\mathrm{e}^{-t\chi_{0,0}(f,u_0,g_0,\alpha)}\mathrm{e}^{tA}, \qquad \forall t\in\mathbf{R}_+. \tag{2.32}$$

To solve problem (2.26), (2.27) we need to solve the following auxiliary Cauchy problem, where $\varepsilon\in\mathbf{R}$:

$$v'(t)-Av(t)+\varepsilon v(t)=f_1(t)+f_2(t), \tag{2.33}$$

$$v(0)=v_0, \tag{2.34}$$

where f_1 and f_2 belong to *different* function spaces.

Theorem 2.2. *Let $v_0\in Y$, $f_1\in C^\beta([0,T];X)$, $f_2\in C([0,T];(X,Y)_{\delta,\infty})$, $\delta\in(\beta,1)$, $Av_0+f_1(0)\in(X,Y)_{\beta,\infty}$ and let α satisfy properties (1.5). Then problem (2.33), (2.34) has a unique solution $v\in C^{1+\beta}([0,T];X)\cap C^\beta([0,T];Y)$. Such a solution can be represented by*

$$v=R_0(v_0,f_1(0))(t)+R_1(f_1-f_1(0))(t)+R_2(f_2)(t), \tag{2.35}$$

where the linear operators R_0, R_1 and R_2 satisfy the following estimates for any $\tau\in(0,T]$:

$$\begin{aligned} &\|R_0(v_0,f_1(0))\|_{C^{1+\beta}([0,\tau];X)\cap C^\beta([0,\tau];Y)} \\ \leq\ & C(T)\big(\|v_0\|_X+\|f_1(0)\|_X+\|Av_0+f_1(0)\|_{(X,Y)_{\beta,\infty}}\big), \end{aligned} \tag{2.36}$$

$$\begin{aligned} & \|R_1(f_1 - f_1(0))\|_{C^{1+\beta}([0,\tau];X)\cap C^\beta([0,\tau];Y)} \\ \le\ & C(T)\|f_1 - f_1(0)\|_{C^\beta([0,\tau];X)}, \end{aligned} \tag{2.37}$$

$$\begin{aligned} & \|R_2(f_2)\|_{C^{1+\beta}([0,\tau];X)\cap C^\beta([0,\tau];Y)} \\ \le\ & T^{\delta-\beta}C(T)\|f_2\|_{C([0,\tau];(X,Y)_{\delta,\infty})}, \end{aligned} \tag{2.38}$$

$C(T)$ being a continuous T-function.

Proof. It is well known [17] that the solution to problem (2.33), (2.34) is represented by

$$\begin{aligned} v(t) &= S(t)v_0 + \int_0^t S(s)f_1(0)\,ds + \int_0^t S(t-s)[f_1(s) - f_1(0)]\,ds \\ &\quad + \int_0^t S(t-s)f_2(s)\,ds, \qquad \forall t \in [0,T], \end{aligned} \tag{2.39}$$

where in the definition of S we have replaced $\chi_{0,0}(f, u_0, g_0, \alpha)$ with ε. □

Assume for the time being that the solution (v, h_0, h_1) satisfies, in addition to (2.13), also the properties

H1 $Av_0 + N_6(v, h_0, h_1)(0) \in (X,Y)_{\beta,\infty}$;
H2 $N_6(v, h_0, h_1) \in C^\beta([0,T]; X)$;
H3 $N_7(v) \in C([0,T]; (X,Y)_{\delta,\infty})$ for some $\delta \in (\beta, 1)$.

Introduce now the complete metric spaces depending on the triplet $(\beta, r, T) \in (0,1) \times \mathbf{R}_+ \times \mathbf{R}_+$:

$$\begin{aligned} \mathcal{Z}^\beta(r,T) = \{(v,h_0,h_1) &\in [C^{1+\beta}([0,T];X) \cap C^\beta([0,T];Y)] \\ &\times C^{1+\beta}([0,T-\alpha(T)]) \times C^\beta([0,T-\alpha(T)]) : \\ \|v\|_{C^{1+\beta}([0,T];X)\cap C^\beta([0,T];Y)} &+ \|h_0\|_{C^{1+\beta}([0,T-\alpha(T)])} + \|h_1\|_{C^\beta([0,T-\alpha(T)])} \le r\} \end{aligned} \tag{2.40}$$

Then, by virtue of Theorem 2.1, our identification problem, in a *local* form, is equivalent to the following fixed-point problem: *find a pair (r,T) and a triplet $(v, h_0, h_1) \in \mathcal{Z}^\beta(r,T)$ solving the system*

$$\begin{aligned} v(t) &= R_0(v_0, N_6(v,h_0,h_1)(0)) + R_1(N_6(v,h_0,h_1) - N_6(v,h_0,h_1)(0))(t) \\ &\quad + R_2(N_7(v(t))) =: N_8(v,h_0,h_1)(t), \qquad \forall t \in [0,T], \end{aligned} \tag{2.41}$$

$$\begin{aligned} h_0(\tau) &= \chi_{0,0}(f,u_0,g_0,\alpha) + \int_0^\tau N_4(v,h_0,h_1)(s)\,ds \\ &\quad + \int_0^\tau \{J_{0,0}(v)(s)\Phi_0[AN_8(v,h_0,h_1)(\gamma(s))] \\ &\qquad\qquad + J_{0,1}(v)(s)\Phi_1[AN_8(v,h_0,h_1)(\gamma(s))]\}\,ds \\ &=: N_9(v,h_0,h_1)(\tau), \qquad \forall \tau \in [0, T-\alpha(T)], \end{aligned} \tag{2.42}$$

$$\begin{aligned} h_1(\tau) &= N_5(v,h_0,h_1)(\tau) + J_{1,0}(v)(\tau)\Phi_0[AN_8(v,h_0,h_1)(\gamma(\tau))] \\ &\quad + J_{1,1}(v)(\tau)\Phi_1[AN_8(v,h_0,h_1)(\gamma(\tau))] \\ &=: N_{10}(v,h_0,h_1)(\tau), \qquad \forall\tau \in [0, T-\alpha(T)]. \end{aligned} \tag{2.43}$$

Remark 2.6. In Formulae (2.19), (2.25) we have replaced v with $N_8(v, h_0, h_1)$ into the two terms $\Phi_j[Av(\gamma(\tau))]$ $(j = 0, 1)$.

Observe first that Assumption (2.13) is satisfied provided the pair (r, T) is chosen to be a solution to the following inequality for some (fixed) $\varepsilon \in (0,1)$:

$$\begin{aligned} &(1+\lambda)r\alpha(T)\|B(A-\lambda I)^{-1}\|_{\mathcal{L}(X)}\big(\|g_1\|_{C([0,T])}\|\Phi_0\|_{X^*} + \|g_0\|_{C([0,T])}\|\Phi_1\|_{X^*}\big) \\ \leq\ & \varepsilon|g_1(\alpha(T))\Phi_0[Bu_0] - g_0(\alpha(T))\Phi_1[Bu_0]|, \end{aligned} \tag{2.44}$$

where λ is any *real positive* element in the resolvent set of A. Indeed we have

$$\begin{aligned} &\|Bv(s)\| \leq \|B(A-\lambda I)^{-1}\|_{\mathcal{L}(X)}\|(A-\lambda I)v(s)\| \\ \leq\ & \|B(A-\lambda I)^{-1}\|_{\mathcal{L}(X)}(1+\lambda)r, \end{aligned} \tag{2.45}$$

$$\begin{aligned} &|g_1(\tau)|\int_0^\tau |\Phi_0[Bv(s)]|\,ds + |g_0(\tau)|\int_0^\tau |\Phi_1[Bv(s)]|\,ds \\ \leq\ & (1+\lambda)r\tau\|B(A-\lambda I)^{-1}\|_{\mathcal{L}(X)}\big(|g_1(\tau)|\|\Phi_0\|_{X^*} + |g_0(\tau)|\|\Phi_1\|_{X^*}\big) \\ \leq\ & \varepsilon|g_1(\alpha(T))\Phi_0[Bu_0] - g_0(\alpha(T))\Phi_1[Bu_0]|, \qquad \forall\tau \in [0,\alpha(T)]. \end{aligned} \tag{2.46}$$

Then we observe that, owing to (2.26), $Av_0 + N_6(v, h_0, h_1)(0)$ is *independent* of the triplet $(v, h_0, h_1) \in \mathcal{Z}^\beta(r, T)$ and belongs to $(X, Y)_{\beta,\infty}$ if and only if our data $(f, u_0, g_0, g_1, \alpha)$ satisfy

$$\begin{aligned} &Av_0 + N_6(v,h_0,h_1)(0) \\ =\ & Av_0 + f'(0) + [\alpha''(0)\chi_{0,0}(f,g_0,g_1,\alpha)) - \widetilde{N}_4(u_0,f,g_0,g_1,\alpha)]u_0 \\ & + [\alpha'(0)]^2\chi_{0,0}(f,g_0,g_1,\alpha))v_0 + \widetilde{N}_5(u_0,f,g_0,g_1,\alpha)Bu_0 \in (X,Y)_{\beta,\infty}. \end{aligned} \tag{2.47}$$

Remark 2.7. Since $u_0 \in Y$ (cf. (1.10)), a sufficient, but simpler, condition for (2.47) to hold is the following

$$v_0,\ Av_0 + f'(0),\ Bu_0 \in (X,Y)_{\beta,\infty}. \tag{2.48}$$

Consequently, from the Definition (2.4) of v_0 and property (1.11) for Bu_0 we easily conclude that conditions (2.48) are equivalent to the following

$$Au_0 + f(0),\ A(Au_0 + f(0)) + f'(0) \in (X,Y)_{\beta,\infty}. \tag{2.49}$$

These latter conditions are nothing but the first two conditions in (1.11).

In order to show that $N_6(v, h_0, h_1) \in C^\beta([0,T]; X)$ we have to prove (cf. (2.6)) that $N_1(v, h_0, h_1) \in C^\beta([0,T]; X)$ and $N_j(v, h_0, h_1) \in C^\beta([0,T])$ $(j = 2,3,4,5)$. For this purpose we need the following well-known properties that we collect in Lemma 2.1.

Lemma 2.1. *Let*

$$f \in C^\beta([0,T];X), \quad g \in C^\beta([0,T]), \quad \delta \in C^1([0,T]), \quad \ell \in C^\beta([0,\delta(T)];X).$$

Then gf, $\ell \circ \delta \in C^\beta([0,T];X)$ *and the following estimates hold:*

$$\|gf\|_{C^\beta([0,T];X)} \le \|g\|_{C^\beta([0,T])}\|f\|_{C^\beta([0,T];X)}, \tag{2.50}$$

$$\|\ell \circ \delta\|_{C([0,T];X)} \le \|\ell\|_{C(\delta([0,T]);X)}, \tag{2.51}$$

$$[\ell \circ \delta]_{C^\beta([0,T];X)} \le [\ell]_{C^\beta(\delta([0,T]);X)}\|\delta'\|^\beta_{C([0,T])}, \tag{2.52}$$

where $[\ell]_{C^\beta(\delta([0,T]);X)}$ *denotes the Hölder coefficient of* ℓ *of order* β.

We need also the following Lemma 2.2.

Lemma 2.2. *Let* $k \in C^\beta([0,T])$ *and let* γ, $\delta \in C^1([0,T])$ *be a pair of scalar functions such that* $\gamma(0) = \delta(0)$ *and* $0 \le \gamma(t) \le \delta(t)$ *for any* $t \in [0,T]$. *Then the linear operator*

$$L_0 f(t) = \int_{\gamma(t)}^{\delta(t)} k(t-s)f(s)\,ds, \qquad \forall t \in [0,T], \tag{2.53}$$

maps $C([0,T];X)$ *continuously into* $C^\beta([0,T];X)$ *and satisfies the estimate*

$$\begin{aligned}\|L_0 f\|_{C^\beta([0,T];X)} &\le \big(\|\gamma'\|_{C([0,T])} + \|\delta'\|_{C([0,T])}\big)\max(2T, T^{1-\beta})\\ &\quad \times \|k\|_{C^\beta([0,T])}\|f\|_{C([0,T];X)}, \quad \forall f \in C([0,T];X).\end{aligned} \tag{2.54}$$

Proof. Our assertion easily follows from Definition (2.53) and the following formula, where (t_1,t_2) is any pair such that $0 \le t_1 \le t_2 \le T$:

$$\begin{aligned} L_0 f(t_2) - L_0 f(t_1) &= \int_{\delta(t_1)}^{\delta(t_2)} k(t_2-s)f(s)\,ds - \int_{\gamma(t_1)}^{\gamma(t_2)} k(t_2-s)f(s)\,ds\\ &\quad + \int_{\gamma(t_1)}^{\delta(t_1)} [k(t_2-s)-k(t_1-s)]f(s)\,ds. \end{aligned} \tag{2.55}$$

□

From Formula (2.6), Definition (2.40) and Lemmata 2.1 and 2.2 we easily derive that N_1 satisfies the following estimates with $j=1$ and $U=X$:

$$\begin{aligned}\|N_j(v,h_0,h_1)\|_{C^\beta([0,T];U)} &\le \phi_{j,0}(u_0,f,g_0,g_1,\alpha) + (T+T^{1-\beta})\\ &\quad \times\big[r\phi_{j,1}(u_0,f,g_0,g_1,\alpha) + r^2\phi_{j,2}(u_0,f,g_0,g_1,\alpha)\big],\\ &\quad \forall (v,h_0,h_1) \in \mathcal{Z}^\beta(r,T), \end{aligned} \tag{2.56}$$

where the functionals $\phi_{j,i}$ $(i=0,1,2)$ can be explicitly computed in terms of the data. Likewise, we can show that the increments of N_1 satisfy the following estimate with $j = 1$ and $U = X$:

$$\|N_j(v,h_0,h_1) - N_j(\widetilde{v},\widetilde{h}_0,\widetilde{h}_1)\|_{C^\beta([0,T];U)} \le (T + T^{1-\beta})(1+r)\phi_{j,3}(u_0,f,g_0,g_1,\alpha)$$
$$\times\Big\{\|v-\widetilde{v}\|_{C^{1+\beta}([0,T];X)\cap C^\beta([0,T];Y)} + \|h_0 - \widetilde{h}_0\|_{C^{1+\beta}([0,T-\alpha(T)])}$$
$$+\|h_1-\widetilde{h}_1\|_{C^\beta([0,T-\alpha(T)])}\Big\}, \qquad \forall (v,h_0,h_1),\ (\widetilde{v},\widetilde{h}_0,\widetilde{h}_1) \in \mathcal{Z}^\beta(r,T). \tag{2.57}$$

Then from Formulae (2.16), (2.18), (2.19), (2.26), (2.29) (2.31) and inequality (2.45) we easily derive that operators N_j $(j = 2,3,4,5)$ and N_6 satisfy, respectively, estimates (2.56) and (2.57) with $j = 2,3,4,5$, $U = \mathbf{R}$ and $j = 6$ and $U = X$. When $j = 4,5$ the interval $[0,T]$ in the left-hand sides in (2.56), (2.57) has to be replaced with $[0, T-\alpha(T)]$.

Finally, from Formulae (2.26), (2.28), (2.30) we deduce the estimates

$$\begin{aligned}&\|N_7(v)\|_{C([0,T];(X;Y)_{\delta,\infty})}\\ \le\ &(1+r^2)\phi_{7,1}(u_0,f,g_0,g_1,\alpha), \qquad \forall (v,h_0,h_1) \in \mathcal{Z}^\beta(r,T),\end{aligned} \tag{2.58}$$

$$\begin{aligned}&\|N_7(v) - N_7(\widetilde{v})\|_{C([0,T];(X;Y)_{\delta,\infty})}\\ \le\ &(1+r)\phi_{7,2}(u_0,f,g_0,g_1,\alpha)\|v-\widetilde{v}\|_{C^{1+\beta}([0,T];X)\cap C^\beta([0,T];Y)},\\ &\qquad\qquad \forall (v,h_0,h_1),\ (\widetilde{v},\widetilde{h}_0,\widetilde{h}_1) \in \mathcal{Z}^\beta(r,T).\end{aligned} \tag{2.59}$$

Then from Theorem 2.2 and the definition of N_8 in (2.41) we obtain the estimates

$$\begin{aligned}&\|N_8(v,h_0,h_1)\|_{C^{1+\beta}([0,T];X)\cap C^\beta([0,T];Y)}\\ \le\ &\phi_{8,1}(u_0,f,g_0,g_1,\alpha) + \max(T^{1-\beta},T^{\delta-\beta})(1+r^2)\phi_{8,2}(u_0,f,g_0,g_1,\alpha),\\ &\qquad\qquad \forall (v,h_0,h_1) \in \mathcal{Z}^\beta(r,T),\end{aligned} \tag{2.60}$$

$$\begin{aligned}&\|N_8(v,h_0,h_1) - N_8(\widetilde{v},\widetilde{h}_0,\widetilde{h}_1)\|_{C^{1+\beta}([0,T];X)\cap C^\beta([0,T];Y)}\\ \le\ &\max(T^{1-\beta},T^{\delta-\beta})(1+r)\phi_{8,3}(u_0,f,g_0,g_1,\alpha)\\ &\times\{\|v-\widetilde{v}\|_{C^{1+\beta}([0,T];X)\cap C^\beta([0,T];Y)} + \|h_0-\widetilde{h}_0\|_{C^{1+\beta}([0,T-\alpha(T)])}\\ &+\|h_1-\widetilde{h}_1\|_{C^\beta([0,T-\alpha(T)])}\}, \qquad \forall (v,h_0,h_1),\ (\widetilde{v},\widetilde{h}_0,\widetilde{h}_1) \in \mathcal{Z}^\beta(r,T).\end{aligned} \tag{2.61}$$

Finally, from (2.42) and (2.43) we deduce the following estimates with $j = 0,1$:

$$\begin{aligned}&\|N_{9+j}(v,h_0,h_1)\|_{C^{1-j+\beta}([0,T-\alpha(T)])}\\ \le\ &\phi_{9+j,1}(u_0,f,g_0,g_1,\alpha) + \max(T^{1-\beta},T^{\delta-\beta})\\ &\times r(1+r)\phi_{9+j,2}(u_0,f,g_0,g_1,\alpha), \qquad \forall (v,h_0,h_1) \in \mathcal{Z}^\beta(r,T),\end{aligned} \tag{2.62}$$

$$\begin{aligned}&\|N_{9+j}(v,h_0,h_1) - N_{9+j}(\widetilde{v},\widetilde{h}_0,\widetilde{h}_1)\|_{C^{1-j+\beta}([0,T-\alpha(T)])}\\ \le\ &\max(T^{1-\beta},T^{\delta-\beta})(1+r)\phi_{9+j,3}(u_0,f,g_0,g_1,\alpha)\\ &\times\Big\{\|v-\widetilde{v}\|_{C^{1+\beta}([0,T];X)\cap C^\beta([0,T];Y)} + \|h_0-\widetilde{h}_0\|_{C^{1+\beta}([0,T-\alpha(T)])}\\ &\quad+\|h_1-\widetilde{h}_1\|_{C^\beta([0,T-\alpha(T)])}\Big\}, \quad \forall (v,h_0,h_1),\ (\widetilde{v},\widetilde{h}_0,\widetilde{h}_1) \in \mathcal{Z}^\beta(r,T).\end{aligned} \tag{2.63}$$

We are now in a position to apply the fixed-point theorem in the *complete* metric space $\mathcal{Z}^\beta(r,T)$ to system (2.41)–(2.43) provided the pair (r,T) satisfies the following system of inequalities for some *real positive* element λ in the resolvent set of A (cf. (2.44), (2.60)–(2.63)) and some $\varepsilon \in (0,1)$:

$$(1+\lambda)r\alpha(T)\|B(A-\lambda I)^{-1}\|_{\mathcal{L}(X)}\big(\|g_1\|_{C([0,T])}\|\Phi_0\|_{X^*} + \|g_0\|_{C([0,T])}\|\Phi_1\|_{X^*}\big)$$
$$\le \varepsilon|g_1(\alpha(T))\Phi_0[Bu_0] - g_0(\alpha(T))\Phi_1[Bu_0]|, \tag{2.64}$$

$$\sum_{j=8}^{10}\Big\{\phi_{j,1}(u_0,f,g_0,g_1,\alpha) + \max(T^{1-\beta},T^{\delta-\beta})(1+r^2)$$
$$\times\phi_{j,2}(u_0,f,g_0,g_1,\alpha)\Big\} \le r, \tag{2.65}$$

$$\max(T^{1-\beta},T^{\delta-\beta})(1+r)\sum_{j=8}^{10}\phi_{j,3}(u_0,f,g_0,g_1,\alpha) < 1. \tag{2.66}$$

To show that system (2.64)–(2.66) is solvable we choose

$$r = 2\sum_{j=8}^{10}\phi_{j,1}(u_0,f,g_0,g_1,\alpha) := r_0 \tag{2.67}$$

and we note that such a system with $r = r_0$ is solvable for small enough T.

In conclusion, we have proved that system (2.41)–(2.43) admits a unique *local* in time solution $(v,h_0,h_1) \in \mathcal{Z}^\beta(r,T)$. As a consequence, also our identification problem (1.7)–(1.9) admits a unique *local* in time solution $(u,h_0,h_1) \in [C^{2+\beta}([0,T];X) \cap C^{1+\beta}([0,T];Y)] \times C^{1+\beta}([0,T-\alpha(T)]) \times C^\beta([0,T-\alpha(T)])$ under the basic Assumptions (1.10)–(1.12), (2.9), (2.12). We have thus proved Theorem 2.1.

3. An application to the first identification problem

Consider the following identification integro-differential problem related to a bounded region Ω with a boundary $\partial\Omega$ of class C^2:

$$D_t u(t,x) + D_t\int_{\alpha(t)}^t h_0(t-s)u(s,x)\,ds - A(x,D_x)u(t,x)$$
$$-\int_{\alpha(t)}^t h_1(t-s)B(x,D_x)u(s,x)\,ds = f(t,x), \qquad \forall (x,t)\in\Omega\times[0,\tau], \tag{3.1}$$

$$u(t,x) = 0, \qquad \forall (x,t)\in\partial\Omega\times[0,\tau], \tag{3.2}$$

$$u(0,x) = u_0(x), \qquad \forall x\in\Omega, \tag{3.3}$$

$$\Phi_j[u(t,\cdot)] = g_j(t), \qquad \forall t\in[0,\tau],\ j=0,1. \tag{3.4}$$

The linear second-order differential operators $A(x, D_x)$ and $B(x, D_x)$ are defined by

$$A(x, D_x) = \sum_{i,j=1}^{n} a_{i,j}(x) D_{x_i} D_{x_j} + \sum_{j=1}^{n} a_{0,j}(x) D_{x_j} + a_{0,0}(x), \tag{3.5}$$

$$B(x, D_x) = \sum_{i,j=1}^{n} b_{i,j}(x) D_{x_i} D_{x_j} + \sum_{j=1}^{n} b_{0,j}(x) D_{x_j} + b_{0,0}(x). \tag{3.6}$$

We assume that the coefficients $a_{i,j}$ and $b_{i,j}$ enjoy the properties:

$$a_{i,j},\, b_{i,j} \in C(\overline{\Omega}), \qquad a_{i,j} = a_{j,i}, \quad b_{i,j} = b_{j,i}, \quad i, j = 0, 1, \ldots, n, \tag{3.7}$$

$$\sum_{i,j=1}^{n} a_{i,j}(x) \xi_i \xi_j \geq c_0 |\xi|^2, \qquad \forall x \in \overline{\Omega},\ \forall \xi \in \mathbf{R}^n, \tag{3.8}$$

c_0 being a positive constant.

For any $p \in (1, +\infty)$ introduce then the linear operators

$$\mathcal{D}(A) = W^{2,p}(\Omega) \cap W_0^{1,p}(\Omega), \qquad Au = A(x, D_x)u, \tag{3.9}$$

$$\mathcal{D}(B) = W^{2,p}(\Omega), \qquad Bu = B(x, D_x)u. \tag{3.10}$$

Then condition (1.6) holds for any $\varphi \in (\pi/2, \pi)$ [17].

We assume now that function α enjoys properties (1.5), while the linear functionals Φ_0 and Φ_1 belong to $L^p(\Omega)^*$, i.e. they are continuous on $L^p(\Omega)$.

As far as the data (u_0, f, g_0, g_1) are concerned, we make the following assumptions, where $\beta \in (0, 1)$ and $\delta \in (\beta, 1) \setminus \{1/(2p)\}$:

$$u_0,\ Au_0 + f(0, \cdot) \in W^{2,p}(\Omega) \cap W_0^{1,p}(\Omega), \quad f \in C^{1+\beta}([0, \tau]; L^p(\Omega)), \tag{3.11}$$

$$Bu_0 \in \mathcal{W}^{2\delta,p}(\Omega), \quad A(Au_0 + f(0, \cdot)) \in \mathcal{W}^{2\beta,p}(\Omega), \tag{3.12}$$

$$g_0, g_1 \in C^{2+\beta}([0, T]), \quad g_0(0) \neq 0, \quad \Phi_j[u_0] = g_j(0), \quad j = 0, 1, \tag{3.13}$$

$$g_1(0)\{-g_0'(0) + \Phi_0[Au_0 + f(0)]\} = g_0(0)\{-g_1'(0) + \Phi_1[Au_0 + f(0)]\}, \tag{3.14}$$

$$g_1(0)\Phi_0[Bu_0] \neq g_0(0)\Phi_1[Bu_0]. \tag{3.15}$$

We have set

$$\mathcal{W}^{2\varepsilon,p}(\Omega) = \begin{cases} W^{2\varepsilon,p}(\Omega), & \text{if } \varepsilon \in (0, 1/(2p)), \\ \{u \in W^{2\varepsilon,p}(\Omega) : u = 0 \quad \text{on } \partial\Omega\}, & \text{if } \varepsilon \in (1/(2p), 1). \end{cases} \tag{3.16}$$

According to Sections 4.3.3 and 4.3.1 in [21] the interpolation space $(X; Y)_{\varepsilon,\infty}$ contains $\mathcal{W}^{2\varepsilon,p}(\Omega)$ for any $\varepsilon \in (0, 1) \setminus \{1/(2p)\}$. Hence, in our application we shall replace the assumptions concerning $(X; Y)_{\varepsilon,\infty}$ with similar assumptions concerning $\mathcal{W}^{2\varepsilon,p}(\Omega)$.

Consequently, our abstract Theorem 2.2 applies. Therefore the identification problem (3.1)–(3.4) admits a unique local in time solution $(u, h_0, h_1) \in [C^{2+\beta}([0,T];$

$L^p(\Omega)) \cap [C^{1+\beta}([0,T];\cap W^{2,p}(\Omega) \cap W_0^{1,p}(\Omega))] \times C^{1+\beta}([0,T-\alpha(T)]) \times C^{\beta}([0,T-\alpha(T)])$ for some $T \in (0,\tau]$.

Remark 3.1. Under the same assumptions as above we can replace the reference space $L^p(\Omega)$ with $C(\overline{\Omega})$ provided we restrict $B(x,D_x)$ to the specific form $B(x,D_x) = b(x)A(x,D_x)$, where $b \in C(\overline{\Omega})$. In this case the realization of $A(x,D_x)$ has to be defined by

$$\mathcal{D}(A) = \{u \in \bigcap_{p>n} W^{2,p}(\Omega) : u = 0 \text{ on } \partial\Omega,\ A(x,D_x)u \in C(\overline{\Omega})\}, \quad Au = A(x,D_x)u.$$

Finally, the corresponding interpolation space is $\mathcal{W}^{2\varepsilon,\infty}(\Omega) = \{u \in C^{2\varepsilon}(\overline{\Omega}) : u = 0 \text{ on } \partial\Omega\}$, $\varepsilon \in (0,1) \setminus \{1/2\}$, while the functionals Φ_0 and Φ_1 can be chosen in $C(\overline{\Omega})^*$.

Remark 3.2. Of course our results, with $p \in (1,+\infty)$ and $p = +\infty$, apply also when the Dirichlet condition (3.2) is replaced with the first-order *regular* oblique derivative boundary condition (for the details cf. [17, Theorems 3.1.2 and 3.1.22]).

4. A second abstract integrodifferential identification problem

In this section we are assuming that the memory loss α is *itself* unknown and satisfies (1.5). Because of the computational complexity of this problem, we limit ourselves to investigating the simplified case $h_0 = 0$. Therefore we are going to consider the following abstract identification problem: *search for three functions $u \in C^{4+\beta}([0,T];X) \cap C^{3+\beta}([0,T];Y)$, $\alpha \in C^{2+\beta}([0,T];\mathbf{R})$ and $h \in C^{1+\beta}([0,T-\alpha(T)];\mathbf{R})$ such that*

$$u'(t) - Au(t) - \int_0^{t-\alpha(t)} h(s)Bu(t-s)\,ds = f(t), \qquad \forall t \in [0,T], \tag{4.1}$$

$$u(0) = u_0, \tag{4.2}$$

$$\Phi_1[u(t)] = g_1(t), \qquad \forall t \in [0,T], \tag{4.3}$$

$$\Phi_2[u(t)] = g_2(t), \qquad \forall t \in [0,T]. \tag{4.4}$$

Here operators A, B and functionals Φ_1, Φ_2 enjoy the same properties as in Section 1, while u_0, f, g_1, g_2 satisfy the following relationships for some $\beta \in (0,1)$ and $\delta \in (\beta,1)$:

$$f \in C^{3+\beta}([0,T];X), \quad g_1, g_2 \in C^{4+\beta}([0,T];\mathbf{R}), \quad u_0 \in Y, \tag{4.5}$$

$$B^2u_0,\ ABu_0,\ AB[Au_0 + f(0)],\ B[A(Au_0 + f(0)) + f'(0)] \in (X,Y)_{\beta,\infty}, \tag{4.6}$$

$$A\{A[A(Au_0 + f(0)) + f'(0)] + f''(0)\} + f'''(0) \in (X,Y)_{\beta,\infty}, \tag{4.7}$$

$$Bu_0 \in (X,Y)_{\delta,\infty}, \quad B[Au_0 + f(0)] \in (X,Y)_{\delta,\infty}. \tag{4.8}$$

Moreover, we assume that the data satisfy the consistency conditions

$$\Phi_j[u_0] = g_j(0), \quad \Phi_j[Au_0 + f(0)] = g_j'(0), \qquad j = 1, 2. \tag{4.9}$$

Then we introduce the following new unknowns $k : [0,T] \to \mathbf{R}$ and $v : [0,T] \to X$, where function $\gamma : [0, T - \alpha(T)] \to [0,T]$ is defined in (2.14):

$$k(t) = [1 - \alpha'(t)]h(t - \alpha(t)) \iff$$

$$\iff \quad h(\tau) = \frac{k(\gamma(\tau))}{1 - \alpha'(\gamma(\tau))}, \quad \tau \in [0, T - \alpha(T)], \tag{4.10}$$

$$v(t) = u'''(t) \quad \iff \quad u(t) = u_0 + t[Au_0 + f(0)]$$

$$+ \frac{1}{2}t^2\{A[Au_0 + f(0)] + k(0)Bu_0 + f'(0)\} + \frac{1}{2}\int_0^t (t-s)^2 v(s)\, ds. \tag{4.11}$$

Remark 4.1. Simple computations, which make use of Formula (4.10) and Equations (4.1), (4.2), show that the initial values of u' and u'' are given by

$$u'(0) = Au_0 + f(0), \qquad u''(0) = A[Au_0 + f(0)] + k(0)Bu_0 + f'(0). \tag{4.12}$$

Observe now that v solves the Cauchy problem

$$v'(t) - Av(t) = k''(t)Bu_0 + k(0)\alpha''(t)B(Au_0 + f(0)) + N_0(Bv, \alpha, k)(t), \tag{4.13}$$

$$v(0) = A[A(Au_0 + f(0)) + k(0)Bu_0 + f'(0)] + k'(0)Bu_0$$
$$+[1 + \alpha'(0)]k(0)B[Au_0 + f(0)] + f''(0) := v_0, \tag{4.14}$$

$$\Phi_1[v(t)] = g_1'''(t), \qquad \forall t \in [0,T], \tag{4.15}$$

$$\Phi_2[v(t)] = g_2'''(t), \qquad \forall t \in [0,T], \tag{4.16}$$

where, for any $t \in [0,T]$, we have set (cf. (4.12))

$$N_0(\zeta, \alpha, k)(t) = f'''(t) + k''(t)\Big[\alpha(t)B(Au_0 + f(0))$$
$$+ \frac{1}{2}[\alpha(t)]^2 B(Au'(0) + k(0)Bu_0 + f'(0)) + \frac{1}{2}\int_0^{\alpha(t)} [\alpha(t) - s]^2 \zeta(s)\, ds\Big]$$
$$+ \Big\{[1 + 2\alpha'(t)]k'(t) + \alpha''(t)\int_0^t k'(s)\, ds\Big\}\Big[B(Au_0 + f(0))$$
$$+ \alpha(t)B[A(Au_0 + f(0)) + k(0)Bu_0 + f'(0)] + \int_0^{\alpha(t)} [\alpha(t) - s]\zeta(s)\, ds\Big]$$
$$+ k(t)\{1 + \alpha'(t) + [\alpha'(t)]^2\}\Big[B[Au'(0) + k(0)Bu_0 + f'(0)] + \int_0^{\alpha(t)} \zeta(s)\, ds\Big]$$

$$-\frac{k(t)}{1-\alpha'(t)}\Big[Bu''(0)+\int_0^{\alpha(t)}\zeta(s)\,ds\Big]+\frac{k(0)}{[1-\alpha'(0)]^2}\Big[Bu''(0)+\int_0^t\zeta(s)\,ds\Big]$$
$$+\int_0^t\Big\{\frac{k'(\sigma)}{1-\alpha'(\sigma)}+\frac{\alpha''(\sigma)k'(\sigma)}{[1-\alpha'(\sigma)]^2}\Big\}\Big[Bu''(0)+\int_0^{t-\sigma+\alpha(s)}\zeta(\rho)\,d\rho\Big]d\sigma. \quad (4.17)$$

To compute the right-hand side in (4.13) we have used the following basic formulae (cf. (4.10)), where $t\in[0,T]$:

$$D_t^3\int_0^{t-\alpha(t)}h(s)Bu(t-s)\,ds$$
$$= D_t^2\Big\{k(t)Bu(\alpha(t))+\int_0^{t-\alpha(t)}h(s)Bu'(t-s)\,ds\Big\}$$
$$= k''(t)Bu(\alpha(t))+\{[1+2\alpha'(t)]k'(t)+k(t)\alpha''(t)]\}Bu'(\alpha(t))$$
$$+k(t)\{1+\alpha'(t)+[\alpha'(t)]^2\}Bu''(\alpha(t))+\int_0^{t-\alpha(t)}h(s)Bu'''(t-s)\,ds. \quad (4.18)$$

Moreover, we have taken advantage of the following identities, where we make use of (4.10) and the change of variable $s=\sigma-\alpha(\sigma)$ as well as of an integration by parts:

$$\int_0^{t-\alpha(t)}h(s)Bu'''(t-s)\,ds=\int_0^t[1-\alpha'(\sigma)]h(\sigma-\alpha(\sigma))Bu'''(t-\sigma+\alpha(\sigma))\,d\sigma$$
$$=-\int_0^t\frac{k(\sigma)}{1-\alpha'(\sigma)}D_\sigma\{Bu''(t-\sigma+\alpha(\sigma))\}\,d\sigma=-\frac{k(t)}{1-\alpha'(t)}Bu''(\alpha(t))$$
$$+\frac{k(0)}{1-\alpha'(0)}Bu''(t)+\int_0^t\Big[\frac{k'(\sigma)}{1-\alpha'(\sigma)}+\frac{\alpha''(\sigma)k'(\sigma)}{[1-\alpha'(\sigma)]^2}\Big]Bu''(t-\sigma+\alpha(s))\,d\sigma$$
$$=-\frac{k(t)}{1-\alpha'(t)}\Big[Bu''(0)+\int_0^{\alpha(t)}Bu'''(s)\,ds\Big]+\frac{k(0)}{1-\alpha'(0)}\Big[Bu''(0)+\int_0^t Bu'''(s)\,ds\Big]$$
$$+\int_0^t\Big\{\frac{k'(\sigma)}{1-\alpha'(\sigma)}+\frac{\alpha''(\sigma)k'(\sigma)}{[1-\alpha'(\sigma)]^2}\Big\}\Big[Bu''(0)+\int_0^{t-\sigma+\alpha(\sigma)}Bu'''(\rho)\,d\rho\Big]d\sigma. \quad (4.19)$$

We need now to compute $k(0)$. Applying the linear functionals Φ_1 and Φ_2 to both sides in the latter equation in (4.12) and assuming

$$\Phi_1[Bu_0]\neq 0, \quad (4.20)$$

we find

$$k(0)=\{\Phi_1[Bu_0]\}^{-1}\{g_1''(0)-\Phi_1[A(Au_0+f(0))+f'(0)]\}=:k_0(u_0,f,g_1)\equiv k_0 \quad (4.21)$$

as well as the *consistency condition*

$$\begin{aligned}&\Phi_2[Bu_0]\{g_1''(0)-\Phi_1[A(Au_0+f(0))+f'(0)]\}\\ =\;&\Phi_1[Bu_0]\{g_2''(0)-\Phi_2[A(Au_0+f(0))+f'(0)]\}. \qquad (4.22)\end{aligned}$$

In order to compute the pair $(k'(0),\alpha'(0))$ from (4.1)–(4.4) and (4.10), (4.12) we easily deduce that $z=u''$ satisfies

$$\begin{aligned}&z'(t)-Az(t)=k'(t)Bu(\alpha(t))+[1+\alpha'(t)]k(t)Bu'(\alpha(t))\\ &+\int_0^{t-\alpha(t)}h(s)Bv(t-s)\,ds+f''(t),\qquad \forall t\in[0,T], \qquad (4.23)\end{aligned}$$

$$z(0)=A[Au_0+f(0)]+f'(0)+k_0Bu_0, \qquad (4.24)$$

$$\Phi_j[z(t)]=g_j''(t),\qquad \forall t\in[0,T],\ j=1,2. \qquad (4.25)$$

Applying then the linear functionals Φ_j $(j=1,2)$ to both sides in (4.23) and setting $t=0$, we find the following system for the pair $(k'(0),\alpha'(0))$:

$$k'(0)\Phi_1[Bu_0]+k_0(u_0,f,g_1)\Phi_1[B(Au_0+f(0))]\alpha'(0)=\Gamma_1(u_0,f,g_1,g_2), \qquad (4.26)$$

$$k'(0)\Phi_2[Bu_0]+k_0(u_0,f,g_1)\Phi_2[B(Au_0+f(0))]\alpha'(0)=\Gamma_2(u_0,f,g_1,g_2), \qquad (4.27)$$

where, for $j=1,2$, we have set

$$\begin{aligned}\Gamma_j(u_0,f,g_1,g_2)\;=\;&g_j'''(0)-\Phi_j[A\{A(Au_0+f(0)+k_0(u_0,f,g_1)Bu_0+f'(0))\}]\\ &-k_0(u_0,f,g_1)\Phi_j[B(Au_0+f(0))]-\Phi_j[f''(0)]. \qquad (4.28)\end{aligned}$$

Assuming that

$$\begin{aligned}k_0(u_0,f,g_1)\neq 0,\quad \rho(u_0,f)\;=:\;&\Phi_1[B(Au_0+f(0))]\Phi_2[Bu_0]\\ &-\Phi_1[Bu_0]\Phi_2[B(Au_0+f(0))]\neq 0, \qquad (4.29)\end{aligned}$$

from (4.26), (4.27) we can easily express $(k'(0),\alpha'(0))$ in terms of the data:

$$\begin{aligned}k'(0)\;=\;&\rho(u_0,f)^{-1}\{\Phi_2[B(Au_0+f(0))]\Gamma_1(u_0,f,g_1,g_2)\\ &-\Phi_1[B(Au_0+f(0))]\Gamma_2(u_0,f,g_1,g_2)\}=:k_1 \qquad (4.30)\end{aligned}$$

$$\begin{aligned}\alpha'(0)\;=\;&k_0(u_0,f,g_1)^{-1}\rho(u_0,f)^{-1}\\ &\times\{-\Phi_2[Bu_0]\Gamma_1(u_0,f,g_1,g_2)+\Phi_1[Bu_0]\Gamma_2(u_0,f,g_1,g_2)\}=:\alpha_1. \qquad (4.31)\end{aligned}$$

Since $\alpha'(0)$ must belong to $[0,1)$, the data (u_0,f,g_1,g_2) must satisfy the bounds

$$\begin{aligned}0\;\leq\;&k_0(u_0,f,g_1)^{-1}\rho(u_0,f)^{-1}\\ &\times\{-\Phi_2[Bu_0]\Gamma_1(u_0,f,g_1,g_2)+\Phi_1[Bu_0]\Gamma_2(u_0,f,g_1,g_2)\}<1. \qquad (4.32)\end{aligned}$$

Finally, observe that (cf. (4.12) and (4.17))

$$\begin{aligned} N_0(Bv,\alpha,k)(0) &= f'''(0)+(1+2\alpha_1)k_1B(Au_0+f(0)) \\ &\quad +k_0(1+\alpha_1+\alpha_1^2)B[(A(Au_0+f(0))+k_0Bu_0+f'(0)] \\ &\quad +\alpha_1k_0(1-\alpha_1)^{-2}Bu''(0)=:\widetilde{N}_0(u_0,f,g_1,g_2), \end{aligned} \tag{4.33}$$

and, according to (4.6)–(4.8), that

$$Av_0+\widetilde{N}_0(u_0,f,g_1,g_2)\in (X,Y)_{\beta,\infty}, \tag{4.34}$$

(k_0,k_1,α_1) being defined by (4.21), (4.30), (4.31).

Then from (4.32)–(4.34), Assumptions (4.8) and well-known results [20] we conclude that any solution to problem (4.13), (4.14) belongs to the space $C^{1+\beta}([0,T];X)\cap C^{\beta}([0,T];Y)$ and solves the operator equation (cf. (4.21))

$$\begin{aligned} v(t) &= S(t)v_0+\int_0^t S(t-s)[k''(s)Bu_0+k_0\alpha''(s)B(Au_0+f(0))]\,ds \\ &\quad +\int_0^t S(t-\tau)N_0(Bv,\alpha,k)(s)\,ds, \end{aligned} \tag{4.35}$$

where $\{S(t)\}_{t>0}$ denotes the semigroup generated by A.

Then, we introduce the three auxiliary unknowns defined on $[0,T]$:

$$w(t)=(A-\lambda_0I)v(t) \iff v(t)=(A-\lambda_0I)^{-1}w(t), \tag{4.36}$$

$$m(t)=\alpha''(t) \iff \alpha(t)=\alpha_1t+\int_0^t(t-s)m(s)\,ds=:\alpha(m)(t), \tag{4.37}$$

$$q(t)=k''(t) \iff k(t)=k_0+k_1t+\int_0^t(t-s)q(s)\,ds=:k(q)(t), \tag{4.38}$$

where λ_0 denotes (any fixed) real in the resolvent set of A.

Our next task consists in deriving a fixed-point system for the triplet (w,k,m).

First, applying operator $A-\lambda_0I$ to both sides in (4.35) we derive the following fixed-point equation for $w=(A-\lambda_0I)v$ and any $t\in[0,T]$:

$$\begin{aligned} w(t) &= S(t)(A-\lambda_0I)v_0 \\ &\quad +\int_0^t(A-\lambda_0I)S(t-s)[q(s)Bu_0+k_0m(s)B(Au_0+f(0))]\,ds \\ &\quad +(A-\lambda_0I)\int_0^t S(t-s)[N_1(w,m,q)(s)-N_1(w,m,q)(0)]\,ds \\ &\quad +[S(t)-(1+\lambda_0)I]N_1(w,m,q)(0)=:N_2(w,m,q)(t), \end{aligned} \tag{4.39}$$

where

$$N_1(w,m,q)(t)=N_0(Ew,\alpha(m),k(q))(t), \qquad \forall t\in[0,T], \tag{4.40}$$

and the bounded operator $E \in \mathcal{L}(X)$ is defined by

$$E = B(A - \lambda_0 I)^{-1}. \tag{4.41}$$

Then we apply functionals Φ_1 and Φ_2 to both sides in (4.13) and obtain the system

$$\begin{aligned} & q(t)\Phi_1[Bu_0] + m(t)k_0\Phi_1[B(Au_0 + f(0))] \\ = \quad & g_1^{(4)}(t) - \Phi_1[A(A - \lambda_0 I)^{-1}w(t)] - \Phi_1[N_1(w,m,q)(t)], \qquad \forall t \in [0,T], \quad (4.42) \\ & q(t)\Phi_2[Bu_0] + m(t)k_0\Phi_2[B(Au_0 + f(0))] \\ = \quad & g_2^{(4)}(t) - \Phi_2[A(A - \lambda_0 I)^{-1}w(t)] - \Phi_2[N_1(w,m,q)(t)], \qquad \forall t \in [0,T]. \quad (4.43) \end{aligned}$$

Observe now that system (4.39), (4.42), (4.43) is equivalent to the following

$$w(t) = N_2(w,m,q)(t), \tag{4.44}$$

$$q(t)\Phi_1[Bu_0] + m(t)k_0\Phi_1[B(Au_0 + f(0))] = N_3(w,m,q)(t), \tag{4.45}$$

$$q(t)\Phi_2[Bu_0] + m(t)k_0\Phi_2[B(Au_0 + f(0))] = N_4(w,m,q)(t). \tag{4.46}$$

The nonlinear operators N_{2+j} $(j = 1,2)$ are defined by

$$N_{2+j}(w,m,q)(t) = D_t^4 g_j(t) - \Phi_j[(A - \lambda_0 I)^{-1}N_2(w,m,q)(t) + N_1(w,m,q)(t)]. \tag{4.47}$$

Assume now

$$\chi(u_0, f(0)) := \Phi_1[B(Au_0+f(0))]\Phi_2[Bu_0] - \Phi_1[Bu_0]\Phi_2[B(Au_0+f(0))] \neq 0. \tag{4.48}$$

Then from (4.44)–(4.46) we derive the fixed-point system

$$w = N_2(w,m,q), \tag{4.49}$$

$$m = N_5(w,m,q), \tag{4.50}$$

$$q = N_6(w,m,q), \tag{4.51}$$

where

$$N_5(w,m,q) = \chi_{1,1}N_3(w,m,q) + \chi_{1,2}N_4(w,m,q), \tag{4.52}$$

$$N_6(w,m,q) = \chi_{2,1}N_2(w,m,q) + \chi_{2,2}N_4(w,m,q), \tag{4.53}$$

and

$$\begin{aligned} & \begin{pmatrix} \chi_{1,1} & \chi_{1,2} \\ \chi_{2,1} & \chi_{2,2} \end{pmatrix} \\ = \quad & \chi(u_0, f(0))^{-1} \begin{pmatrix} k_0^{-1}\Phi_2[Bu_0] & -k_0^{-1}\Phi_1[Bu_0] \\ -\Phi_2[B(Au_0 + f(0))] & \Phi_2[B(Au_0 + f(0))] \end{pmatrix}. \quad (4.54) \end{aligned}$$

The Banach space where we can show that system (4.49)–(4.51) admits a solution is

$$\begin{aligned}\mathcal{Z}^\beta(r,T) &= \Big\{(w,m,q)\in C^\beta([0,T];X)\cap C^\beta([0,T];\mathbf{R})\times C^\beta([0,T];\mathbf{R}):\\ &\qquad \|w\|_{C^\beta([0,T];X)}+\|m\|_{C^\beta([0,T];\mathbf{R})}+\|q\|_{C^\beta([0,T];\mathbf{R})}\le r,\\ &\qquad \int_0^T |m(s)|\,ds<\min(\alpha_1,1-\alpha_1)\Big\}. \qquad (4.55)\end{aligned}$$

Remark 4.1. Observe that the integral condition in (4.55) concerning m is needed to ensure that $\alpha(m)$ (cf. (4.37)) satisfies the basic inequalities in (1.5).

We can now state the main result of this section.

Theorem 4.1. *Under the basic Assumptions (4.5)–(4.9), (4.20), (4.22), (4.29), (4.32), (4.48), the identification problem (4.1)–(4.4) admits a unique local in time solution* $(u,\alpha,h)\in[C^{4+\beta}([0,T];X)\cap C^{3+\beta}([0,T];Y)]\times C^{2+\beta}([0,T])\times C^{2+\beta}([0,T-\alpha(T)])$.

First we need the following Lemma 4.1.

Lemma 4.1. *For any* $z\in(X;Y)_{\delta,\infty}$ ($\delta\in(\beta,1)$) *the linear operator*

$$Lg(t)=\int_0^t g(s)AS(t-s)z\,ds,\qquad \forall t\in[0,T], \tag{4.56}$$

maps $C([0,T];\mathbf{R})$ *into* $C^\beta([0,T];X)$ *and satisfies the estimates*

$$\|Lg\|_{C^\beta([0,T];\mathbf{R})}\le C(T^\delta+T^{\delta-\beta})\|g\|_{C([0,T];\mathbf{R})}\|z\|_\delta, \tag{4.57}$$

where $\|\cdot\|_\delta$ *denotes the norm in* $(X;Y)_{\delta,\infty}$.

Proof. The assertion easily follows from (4.56) and the following identity where $0<t_1\le t_2\le T$:

$$Lg(t_2)-Lg(t_1)=\int_{t_1}^{t_2} g(s)AS(t-s)z\,ds+\int_0^{t_1} g(s)\,ds\int_{t_1-s}^{t_2-s} A^2S(t)z\,dt, \tag{4.58}$$

as well as from the estimates

$$\|A^jS(t)z\|\le Ct^{-j+\delta}\|z\|_\delta,\qquad \forall t\in(0,T],\ j=0,1, \tag{4.59}$$

$$\begin{aligned}\int_0^{t_1} ds\int_{t_1-s}^{t_2-s} t^{-2+\delta}\,dt &\le \int_0^{t_1}(t_1-s)^{-1+\delta-\beta}\,ds\int_{t_1-s}^{t_2-s} t^{-1+\beta}\,dt\\ &\le \beta^{-1}(t_2-t_1)^\beta\int_0^{t_1}(t_1-s)^{-1+\delta-\beta}\,ds\\ &\le \beta^{-1}(\delta-\beta)^{-1}(t_2-t_1)^\beta T^{\delta-\beta},\qquad 0<t_1\le t_2\le T. \qquad (4.60)\end{aligned}$$

□

Observe then that from (4.33) and (4.40) we easily deduce the following formula showing that $N_1(w,m,q)(0)$ depends on data, only (cf. (4.12)):

$$N_1(w,m,q)(0) = f'''(0) + (1+2\alpha_1)k_1 B(Au_0 + f(0))$$
$$+k_0(1+\alpha_1+\alpha_1^2)B[A(Au_0+f(0)) + f'(0) + k(0)Bu_0]$$
$$+\frac{\alpha_1 k_0}{(1-\alpha_1^2)^2}Bu''(0) =: \Lambda(f,g_1,g_2,u_0). \tag{4.61}$$

According to Lemma 4.1 and a well-known result (cf. [20]), from (4.39) we easily deduce the estimate

$$\|N_2(w,m,q)\|_{\beta,\infty} \le C\Big\{rT^\delta + \|N_1(w,m,q)(0)\| + \|Av_0 + N_1(w,m,q)(0)\|_\delta$$
$$+\|N_1(w,m,q) - N_1(w,m,q)(0)\|_{\beta,\infty}\Big\}, \tag{4.62}$$

$$\|N_2(w_2,m_2,q_2) - N_2(w_1,m_1,q_1)\|_{\beta,\infty} \le C\Big\{T^\delta(\|m_2-m_1\|_{0,\infty} + \|q_2-q_1\|_{0,\infty})$$
$$+\|N_1(w_2,m_2,q_2) - N_1(w_1,m_1,q_1)\|_{\beta,\infty}\Big\}. \tag{4.63}$$

In estimate (4.62) we have taken into account that $N_1(w,m,q)(0)$ depends on the data, only.

From Definitions (4.52), (4.53) and (4.47) we easily derive the estimates

$$\sum_{j=1}^2 \|N_{4+j}(w,m,q)\|_{\beta,\infty} \le C_1\sum_{j=1}^2 \|N_{2+j}(w,m,q)\|_{\beta,\infty}$$
$$\le \quad C_1\sum_{j=1}^2 \|g_j^{(4)}\|_{\beta,\infty} + C_2\sum_{j=1}^2 \|N_j(w,m,q)\|_{\beta,\infty}, \tag{4.64}$$

$$\sum_{j=1}^2 \|N_{4+j}(w_2,m_2,q_2) - N_{4+j}(w_1,m_1,q_1)\|_{\beta,\infty}$$
$$\le \quad C_1\sum_{j=1}^2 \|N_{2+j}(w_2,m_2,q_2) - N_{2+j}(w_1,m_1,q_1)\|_{\beta,\infty}$$
$$\le \quad C_2\sum_{j=1}^2 \|N_j(w_2,m_2,q_2) - N_j(w_1,m_1,q_1)\|_{\beta,\infty}. \tag{4.65}$$

It remains to estimate $N_1(w,m,q)$ and its increments. For this purpose first we observe that the following estimates are easy consequences of Definitions (4.37) and (4.38), where C_1 and C_2 denote continuous functions of their arguments and

$[x]$ stands for the largest integer not exceeding x:

$$\sum_{j=1}^{2}(T^{(\beta-1)[1-j/2]}\|D_t^j\alpha(m)\|_{\beta,\infty}+\|D_t^jk(q)\|_{\beta,\infty})\le C_1(r,T), \tag{4.66}$$

$$\sum_{j=1}^{2}(T^{(\beta-1)[1-j/2]}\|D_t^j\alpha(m_2)-D_t^j\alpha(m_1)\|_{\beta,\infty}+\|D_t^jk(q_2)-D_t^jk(q_1)\|_{\beta,\infty}) \tag{4.67}$$

$$\le C_2(r,T)(\|m_2-m_1\|_{\beta,\infty}+\|q_2-q_1\|_{\beta,\infty}). \tag{4.68}$$

Now a lengthy, but simple, analysis of operator N_0 defined by (4.17) yields the following estimates, where we make use of Definition (4.40) and functions C_3 and C_4 enjoy the same properties as functions C_1 and C_2 :

$$\|N_1(w,m,q)\|_{\beta,\infty}\le\|N_1(0,0,0)\|_{\beta,\infty}+T^{1-\beta}C_3(r,T), \tag{4.69}$$

$$\|N_1(w_2,m_2,q_2)-N_1(w_1,m_1,q_1)\|_{\beta,\infty}$$

$$\le\ T^{1-\beta}C_4(r,T)(\|w_2-w_1\|_{\beta,\infty}+\|m_2-m_1\|_{\beta,\infty}+\|q_2-q_1\|_{\beta,\infty}). \tag{4.70}$$

Observe now that estimates (4.62)–(4.65), (4.68), (4.69) yield the inequality

$$\int_0^T|N_5(w,m,q)(t)|\,dt\le TC(r,T). \tag{4.71}$$

Moreover, according to (4.21), (4.28), (4.29), (4.31) it immediately follows that α_1 (depending only on the data) is *independent* of T. Finally, reasoning as at the end of Section 2, from estimates (4.62)–(4.65), (4.68)–(4.70) we easily conclude that the nonlinear vector operator $N=(N_2,N_5,N_6)$ is a contraction mapping from $\mathcal{Z}^\beta(r,T)$ into itself for large enough r and small enough T. Therefore the fixed-point system (4.49)–(4.51) admits a unique solution $(w,m,q)\in C^\beta([0,T];X)\times C^\beta([0,T];\mathbf{R})\times C^\beta([0,T-\alpha(T)];\mathbf{R})$ for small enough T. As a consequence, also our identification problem (4.1)–(4.4) admits a unique *local* in time solution $(u,\alpha,h)\in[C^{4+\beta}([0,T];X)\cap C^{3+\beta}([0,T];Y)]\times C^{2+\beta}([0,T])\times C^{2+\beta}([0,T-\alpha(T)])$ (cf. (4.10)) under the assumptions listed in Theorem 4.1.

5. An application to the second identification problem

Consider the following identification integro-differential problem related to a bounded region Ω with a boundary $\partial\Omega$ of class C^2:

$$D_tu(t,x)-A(x,D_x)u(t,x)-\int_{\alpha(t)}^{t}h(t-s)B(x,D_x)u(s,x)\,ds$$

$$=f(t,x),\qquad\forall(x,t)\in\Omega\times[0,\tau], \tag{5.1}$$

$$u(t,x)=0\qquad\forall(x,t)\in\partial\Omega\times[0,\tau], \tag{5.2}$$

$$u(0,x) = u_0(x), \qquad \forall x \in \Omega, \tag{5.3}$$

$$\Phi_j[u(t,\cdot)] = g_j(t), \qquad \forall t \in [0,\tau],\ j = 1,2. \tag{5.4}$$

The linear second-order differential operators $A(x, D_x)$ and $B(x, D_x)$ are defined by (3.5) and (3.6), while their coefficients $a_{i,j}$ and $b_{i,j}$ enjoy properties (3.7) and (3.8).

For any $p \in (1, +\infty)$ we define the realizations A and B of $A(x, D_x)$ and $B(x, D_x)$, respectively, by (3.9) and (3.10). Then condition (1.6) holds for any $\varphi \in (\pi/2, \pi)$ [17].

We assume now that the linear functionals Φ_1 and Φ_2 belong to $L^p(\Omega)^*$, i.e. they are continuous on $L^p(\Omega)$.

As far as the data (u_0, f, g_1, g_2) are concerned, we make the following assumptions, where $\beta \in (0,1)$, $\delta \in (\beta, 1)$ and $\mathcal{W}^{2\varepsilon,p}(\Omega)$ is defined by (3.16):

$$u_0 \in W^{2,p}(\Omega) \cap W_0^{1,p}(\Omega), \quad f \in C^{3+\beta}([0,\tau]; L^p(\Omega)), \tag{5.5}$$

$$Bu_0,\ A[Au_0 + f(0,\cdot)] \in \mathcal{W}^{2\delta,p}(\Omega), \quad B^2u_0,\ A^2Bu_0 \in \mathcal{W}^{2\beta,p}(\Omega), \tag{5.6}$$

$$AB[Au_0 + f(0,\cdot)],\ B[A(Au_0 + f(0,\cdot)) + D_tf(0,\cdot)] \in \mathcal{W}^{2\beta,p}(\Omega), \tag{5.7}$$

$$A\{A[A(Au_0 + f(0)) + D_tf(0,\cdot)] + D_t^2f(0,\cdot)\} + D_t^3f(0,\cdot) \in \mathcal{W}^{2\beta,p}(\Omega), \tag{5.8}$$

$$g_1, g_2 \in C^{4+\beta}([0,T]),\ \Phi_j[u_0] = g_j(0)\ \Phi_j[Au_0 + f(0,\cdot)] = g_j'(0),\ j = 1,2, \tag{5.9}$$

$$\Phi_1[B(Au_0 + f(0,\cdot))]\Phi_2[Bu_0] - \Phi_1[Bu_0]\Phi_2[B(Au_0 + f(0,\cdot))] \neq 0. \tag{5.10}$$

$$\begin{aligned} &0 \neq \Phi_2[Bu_0]\{g_1''(0) - \Phi_1[A(Au_0 + f(0,\cdot)) + f'(0,\cdot)]\} \\ &= \Phi_1[Bu_0]\{g_2''(0) - \Phi_2[A(Au_0 + f(0,\cdot)) + f'(0,\cdot)]\}. \end{aligned} \tag{5.11}$$

$$\begin{aligned} &0 \leq -\Phi_2[Bu_0]\Gamma_1(u_0, f, g_1, g_2) + \Phi_1[Bu_0]\Gamma_2(u_0, f, g_1, g_2) \\ &< k_0(u_0, f, g_1)\rho(u_0, f), \end{aligned} \tag{5.12}$$

where

$$\begin{aligned} &k_0(u_0, f, g_1) \\ &= \{\Phi_1[Bu_0]\}^{-1}\{g_1''(0) - \Phi_1[A(Au_0 + f(0,\cdot)) + D_tf(0,\cdot)]\}, \end{aligned} \tag{5.13}$$

$$\begin{aligned} &\Gamma_1(u_0, f, g_1, g_2) \\ &= k_1((u_0, f, g_1)\Phi_1[Bu_0] + k_0(u_0, f, g_1)\Phi_1[B(Au_0 + f(0,\cdot))]\alpha_1(u_0, f, g_1, g_2), \end{aligned} \tag{5.14}$$

$$\begin{aligned} &\Gamma_2(u_0, f, g_1, g_2) \\ &= k_1(u_0, f, g_1)\Phi_2[Bu_0] + k_0(u_0, f, g_1)\Phi_2[B(Au_0 + f(0,\cdot))]\alpha_1(u_0, f, g_1, g_2), \end{aligned} \tag{5.15}$$

$k_1(u_0, f, g_1) = k_1$ and $\alpha_1(u_0, f, g_1, g_2) = \alpha_1$ being defined by (4.30) and (4.31).

In the previous Formulae (5.6)–(5.8) we have chosen not to write down the explicit membership of elements u_0, $Au_0 + f(0)$, etc. Therefore, e.g., $B^2 u_0 \in \mathcal{W}^{2\delta,p}(\Omega)$ *implies* that u_0 is an element in the domain of B^2.

By virtue of Assumptions (5.5)–(5.15) we can apply our abstract Theorem 4.1. Hence the identification problem (5.1)–(5.4) admits a unique local in time solution $(u, \alpha, h) \in [C^{4+\beta}([0,T]; L^p(\Omega)) \cap C^{3+\beta}([0,T]; W^{2,p}(\Omega) \cap W_0^{1,p}(\Omega))] \times C^{2+\beta}([0,T]) \times C^{2+\beta}([0, T - \alpha(T)])$.

Remark 5.1. Remarks 3.1 and 3.2 apply also in the present case.

References

[1] J. Baumeister: *Boundary control of an integrodifferential equation,* J. Math. Anal. Appl. 93 (1983), 550–570.

[2] B.D. Coleman, M.E. Gurtin: *Equipresence and constitutive equations for rigid heat conductors,* Z. Angew. Math. Phys. 18 (1967), 199–208.

[3] B.D. Coleman, V.J. Mizel: *Norms and semigroups in the theory of fading memory,* Arch. Rational Mech. Anal. 23 (1966), 87–123;

[4] B.D. Coleman, W. Noll: *Foundations of linear viscoelasticity,* Rev. of Modern Physics 33 (1961), 239–249;

[5] G. Fichera: *Sui materiali elastici con memoria,* (in Italian), Atti Acc. Lincei Rend. fis., 82 (1998), pp. 473–478;

[6] M. Gurtin, A. C. Pipkin: *A general theory of heat conduction with finite wave speed,* Arch. Rational Mech. Anal. 31 (1968), 113–126;

[7] H. Grambüller: *On linear theory of heat conduction in materials with memory. Existence and uniqueness theorems for the final value problem,* Proc. Royal Soc. Edinburgh 76A (1976), 119–137.

[8] M. Grasselli: *An identification problem for a linear integro-differential equation occurring in heat flow,* Math. Meth. in Appl. Sci. 15 (1992), 167–186.

[9] J. Janno, L. v. Wolfersdorf: *On identification of memory kernels in linear theory of heat conduction*, Math. Meth. Appl. Sci. 17 (1994), 919–932.

[10] J. Janno, L. v. Wolfersdorf: *Identification of weakly singular memory kernels in heat conduction*, Z. Angew. Math. Mech. 77 (1997), 243–257.

[11] J. Janno, L. v. Wolfersdorf: *Identification of memory kernels in general linear flow,* J. Inv. Ill-Posed Problems 6 (1998), 141–164.

[12] J. Janno, L. von Wolfersdorf: *An inverse problem for identification of a time- and space-dependent memory kernel of a special kind in heat conduction*, Inverse Problems 15 (1999), 1455–1467.

[13] J. Janno, L. von Wolfersdorf: *Identification of memory kernels in heat conduction and viscoelasticity*, Optimal control of partial differential equations (Chemnitz, 1998), 301–308, Internat. Ser. Numer. Math., 133, Birkhäuser, Basel, 1999.

[14] J. Janno, L. von Wolfersdorf: *Inverse problems for memory kernel by Laplace transform method*, Z.A.A 19 (2000), 489–510.

[15] A. Lorenzi, E. Sinestrari: *Stability results for a partial integro-differential inverse problem,* Pitman Research Notes in Math. vol. 190, (1989), 271–294 (Proceedings of the Meeting on "Volterra Integro-differential Equations in Banach Spaces and Applications", Trento (1987));

[16] A. Lorenzi, E. Sinestrari: *An inverse problem in the theory of materials with memory I,* Nonlinear Anal. 12 (1988), 1317–1335;

[17] A. Lunardi: *Analytic semigroups and optimal regularity in parabolic problems,* Birkhäuser, 1995;

[18] J. Meixner: *On the linear theory of heat conduction,* Arch. Rational Mech. Anal. 39 (1970), 108–130;

[19] J. W. Nunziato: *On heat conduction in materials with memory,* Quart. Appl. Math. 29 (1971), 187–204.

[20] E. Sinestrari: *On the abstract Cauchy problem of parabolic type in spaces of continuous functions,* J. Math. Anal. Appl. 107 (1985), 16–66.

[21] H. Triebel: Interpolation Theory, Function Spaces, Differential Operators, North Holland, Amsterdam (1978).

Alfredo Lorenzi
Departimento di Matematica
Università di Milano
Milano, Italia
email: `Alfredo.Lorenzi@mat.unimi.it`

Progress in Nonlinear Differential Equations
and Their Applications, Vol. 50, 263–277

On Generators of Noncommuting Semigroups: Sums, Interpolation, Regularity

Alessandra Lunardi

1. Introduction

Let $A : D(A) \mapsto X$, $B : D(B) \mapsto X$ be densely defined linear operators in general Banach space X, satisfying

$$\begin{cases} \rho(A) \supset (\omega_A, +\infty), \ \ \sup_{\lambda>\omega_A}(\lambda - \omega_A)\|R(\lambda, A)\| < +\infty, \\ \rho(B) \supset (\omega_B, +\infty), \ \ \sup_{\lambda>\omega_B}(\lambda - \omega_B)\|R(\lambda, B)\| < +\infty. \end{cases} \tag{1.1}$$

Many results about the interactions between A and B are known in the commutative case. For instance, as far as real interpolation is concerned, it is easy to see that if $R(\lambda, A)$ and $R(\mu, B)$ commute for some λ, μ, then

$$(X, D(A) \cap D(B))_{\theta,p} = (X, D(A))_{\theta,p} \cap (X, D(B))_{\theta,p}, \ \ 0 < \theta < 1, \ \ 1 \le p \le \infty. \tag{1.2}$$

Deeper results, concerning several properties of the operator $A + B$, may be found in [2, 4].

Not much is known in the noncommutative case, except of course when one of the operators A, B is a small perturbation (in a suitable sense) of the other one. Here we shall consider the case where the commutator $[A, B] = AB - BA$ has a closed extension Z, which satisfies

$$\rho(Z) \supset (\omega_Z, +\infty), \ \ \sup_{\lambda>\omega_Z}(\lambda - \omega_Z)\|R(\lambda, Z)\| < +\infty, \tag{1.3}$$

and Z commutes with both A and B, in the resolvent sense:

$$[R(\lambda, A), R(\mu, Z)] = [R(\lambda, B), R(\mu, Z)] = 0, \ \ \lambda > \max\{\omega_A, \omega_B\}, \ \mu > \omega_Z. \tag{1.4}$$

The simplest significant examples are the realizations of suitable vector fields $\mathcal{A}$, $\mathcal{B}$ in L^p spaces or in spaces of uniformly continuous and bounded functions, such as for instance

$$\mathcal{A} = \frac{\partial}{\partial x} + 2y\frac{\partial}{\partial z}, \ \ \mathcal{B} = \frac{\partial}{\partial y} - 2x\frac{\partial}{\partial z},$$

in $X = L^p(\mathbb{R}^3)$, or

$$\mathcal{A} = \frac{\partial}{\partial x}, \ \ \mathcal{B} = x\frac{\partial}{\partial y},$$

in $X = L^p(\mathbb{R}^2)$.

Coming back to the general case, it was proved in [8] that formula (1.2) still holds, as well as other interpolation results relevant to the commutative case. In this paper we extend the interpolation results obtained in [8] (Section 2), and we use such extensions together with interpolation techniques to study further optimal regularity for the problem

$$\lambda v - (\mathcal{A}^2 + \mathcal{B}^2)v = f,$$

with $\lambda > 0$, in suitable nonisotropic Hölder and fractional Sobolev spaces. Here $\mathcal{A}$ and $\mathcal{B}$ are the above operators, so that $\mathcal{A}^2 + \mathcal{B}^2$ is the Heisenberg Laplacian in the first example, and the Grushin operator in the second example. See Section 3.

Sufficient conditions in order that (1.3) and (1.4) hold were recently given in [3], in the case where A and B generate strongly continuous groups. In this case, also Z and the closure $\overline{A+B}$ of $A+B$ generate strongly continuous groups, and simple representation formulae hold both for e^{tZ} and for $e^{t\overline{A+B}}$. The results of [3] will be summarized in Section 4.

2. Interpolation

We recall the definition of real interpolation spaces. Let X, Y be Banach spaces, with $Y \subset X$. For $0 < \theta < 1$, $1 \le p \le \infty$, we have

$$\begin{cases} (X,Y)_{\theta,p} = \{x \in X : t \mapsto t^{-\theta}K(t,x,X,Y) \in L^p_*(0,+\infty)\}, \\ \|x\|_{(X,Y)_{\theta,p}} = \|t^{-\theta}K(t,x,X,Y)\|_{L^p_*(0,\infty)}; \end{cases}$$

where $K(t,x,X,Y) = \inf_{x=a+b,\ a\in X,\ b\in Y} (\|a\|_X + t\|b\|_Y)$. If I is any interval contained in $(0,+\infty)$, $L^p_*(I)$ is the Lebesgue space L^p with respect to the measure dt/t in I. In particular, $L^\infty_*(I) = L^\infty(I)$.

If $Y = D(L)$, where $L : D(L) \mapsto X$ is a linear operator such that $\rho(L) \supset (\omega,+\infty)$, and $\sup_{\lambda>\omega}(\lambda-\omega)\|R(\lambda,L)\| < +\infty$, we shall use two well-known characterizations of $(X,D(L))_{\theta,p}$ whose proof may be found in [10, §1.14.3].

Proposition 2.1. *For $0 < \theta < 1$, $1 \le p \le \infty$ we have*

$$(X,D(L))_{\theta,p} = \{x \in X : \ \varphi(\lambda) = \lambda^\theta\|LR(\lambda,L)x\| \in L^p_*(a,+\infty)\}$$

for some (and hence for all) $a > \omega$, and the norm

$$\|x\| + \|\varphi\|_{L^p_*(a,+\infty)}$$

is equivalent to the norm of $(X,D(L))_{\theta,p}$. Moreover,

$$(X,D(L))_{\theta,p} = \{x \in X : \ \psi(\lambda) = \lambda^\theta\|L^2R(\lambda,L)^2x\| \in L^p_*(a,+\infty)\}$$

for some (and hence for all) $a > \omega$, and the norm

$$\|x\| + \|\psi\|_{L^p_*(a,+\infty)}$$

is equivalent to the norm of $(X,D(L))_{\theta,p}$.

The following notation will be used. For each $m \in \mathbb{N}$ we denote by $\mathcal{L}_m$ the set of all operators of the type

$$A^{n_1} B^{m_1} A^{n_2} B^{m_2} \cdot \ldots \cdot A^{n_r} B^{m_r},$$

with $r \in \mathbb{N}$, $n_i, m_i \in \mathbb{N} \cup \{0\}$ for $i = 1, \ldots, r$, $\sum_{i=1}^{r}(n_i + m_i) = m$.

For any integer $m \geq 1$ the space K^m is defined by

$$K^m = \bigcap_{L \in \mathcal{L}_m} D(L).$$

K^m is endowed with the intersection norm

$$\|x\|_{K^m} = \|x\| + \sum_{L \in \mathcal{L}_m} \|Lx\|.$$

Therefore, $K^1 = D(A) \cap D(B)$, $K^2 = D(A^2) \cap D(AB) \cap D(BA) \cap D(B^2)$.

Fix any $a > \max\{\omega_A, \omega_B, \omega_Z\}$. Then there are $M_A, M_B, M_Z > 0$ such that for each $\lambda \geq a$

$$\|\lambda R(\lambda, A)\| \leq M_A, \quad \|\lambda R(\lambda, B)\| \leq M_B, \quad \|\lambda R(\lambda, Z)\| \leq M_Z, \tag{2.1}$$

and hence

$$\|AR(\lambda, A)\| \leq (M_A + 1), \quad \|BR(\lambda, B)\| \leq (M_B + 1), \quad \|ZR(\lambda, Z)\| \leq (M_Z + 1). \tag{2.2}$$

Estimates (2.1) and (2.2) will be used throughout this section.

We need a further assumption, that there is a dense subspace D such that for $\lambda, \mu > \max\{\omega_A, \omega_B\}$ we have

$$R(\mu, B)R(\lambda, A)(D) \subset D(BA), \quad R(\mu, A)R(\lambda, B)(D) \subset D(AB). \tag{2.3}$$

Now we recall some results proved in [8] which will be used in this paper. To be precise, in [8] we assumed $\omega_A = \omega_B = \omega_Z = 0$, $M_A = M_B = M_Z = 1$. However, these restrictions were made only for notational simplicity, and it is easy to verify that the results hold in the general case, under the assumptions (1.1), (1.3), (1.4), (2.3).

Lemma 2.2. *For every $\lambda > \max\{\omega_A, \omega_B\}$, $R(\lambda, B)$ maps $D(A) \cap D(Z)$ into $D(A)$, $R(\lambda, A)$ maps $D(B) \cap D(Z)$ into $D(B)$, and*

$$\begin{aligned} &\text{(i)} \qquad [A, R(\lambda, B)] = R^2(\lambda, B)Z, \quad \text{on } D(A) \cap D(Z), \\ &\text{(ii)} \qquad [B, R(\lambda, A)] = -R^2(\lambda, A)Z, \quad \text{on } D(A) \cap D(Z). \end{aligned} \tag{2.4}$$

Lemma 2.3. *Let $1 \leq p \leq \infty$, $0 < \sigma < 1$. Then*

$$\text{(i)} \qquad D(A) \cap D(B) \subset D_Z(1/2, \infty),$$

$$\text{(ii)} \qquad D_A(\sigma, p) \cap D_B(\sigma, p) \subset D_Z(\sigma/2, p),$$

with continuous embeddings.

Theorem 2.4. *Let $1 \le p \le \infty$, $0 < \theta < 1$. Then*

$$(X, K^1)_{\theta,p} = (X, D(A))_{\theta,p} \cap (X, D(B))_{\theta,p},$$

$$\begin{cases} (X, K^2)_{\theta,p} = (X, D(A))_{2\theta,p} \cap (X, D(B))_{2\theta,p}, & \theta < 1/2, \\ (X, K^2)_{\theta,p} = \{x \in K^1 \cap (X, D(Z))_{\theta,p} : \\ \qquad Ax,\ Bx \in (X, D(A))_{2\theta-1,p} \cap (X, D(B))_{2\theta-1,p}\}, & \theta > 1/2, \end{cases}$$

with equivalence of the respective norms.

For further use we need to improve the last result. To be more precise, we have to show that one of the conditions in the right-hand side of the last formula is superfluous.

Lemma 2.5. *For each $\theta \in (1/2, 1)$, $p \in [1, +\infty]$ we have*

$$\{x \in K^1 :\ Ax,\ Bx \in (X, D(A))_{2\theta-1,p} \cap (X, D(B))_{2\theta-1,p}\} \subset (X, D(Z))_{\theta,p}.$$

Proof. For every $x \in K^1$ with $Ax,\ Bx \in (X, D(A))_{2\theta-1,p} \cap (X, D(B))_{2\theta-1,p}$ we shall estimate the norm of the function

$$u(\lambda) = \lambda^{2\theta}(ZR(\lambda^2, Z))^2 x$$

in $L^p_*(a, +\infty; X)$, with $a > \max\{\omega_A,\ \omega_B,\ \omega_Z\}$; the statement will follow from proposition 2.1. We write $u(\lambda)$ as

$$\begin{aligned} u(\lambda) &= \lambda^{2\theta+1}(ZR(\lambda^2, Z))^2 R(\lambda, B)x + \lambda^{2\theta}(ZR(\lambda^2, Z))^2 R(\lambda, B)Bx \\ &= \lambda^{2\theta+1}(ZR(\lambda^2, Z))^2 R(\lambda, B)\lambda^2 R(\lambda, A)^2 x \\ &\quad -\lambda^{2\theta+1}(ZR(\lambda^2, Z))^2 R(\lambda, B)2\lambda A R(\lambda, A)^2 x \\ &\quad +\lambda^{2\theta+1}(ZR(\lambda^2, Z))^2 R(\lambda, B)A^2 R(\lambda, A)^2 x \\ &\quad +\lambda^{2\theta} ZR(\lambda^2, Z)R(\lambda, B)ZR(\lambda^2, Z)Bx \\ &= \sum_{i=1}^{4} u_i(\lambda). \end{aligned}$$

Using the commutation formula (2.4)(ii), we write $u_1(\lambda)$ as

$$\begin{aligned} u_1(\lambda) &= \lambda^{2\theta+3} ZR(\lambda^2, Z)^2 R(\lambda, B)(R(\lambda, A)Bx - BR(\lambda, A)x) \\ &= \lambda^{2\theta+3} R(\lambda^2, Z)R(\lambda, B)R(\lambda, A)ZR(\lambda^2, Z)Bx \\ &\quad -\lambda^{2\theta+2} R(\lambda^2, Z)BR(\lambda, B)R(\lambda, A)ZR(\lambda^2, Z)Ax \\ &\quad -\lambda^{2\theta+2} R(\lambda^2, Z)R(\lambda, B)ZR(\lambda^2, Z)Bx \end{aligned}$$

so that

$$\begin{aligned} &\|u_1(\lambda)\| \le M_Z M_B M_A \|\lambda^{2\theta-1} ZR(\lambda^2, Z)Bx\| \\ &+M_Z(M_B+1)M_A\|\lambda^{2\theta-1} ZR(\lambda^2, Z)Ax\| + M_Z M_B \|\lambda^{2\theta-1} ZR(\lambda^2, Z)Bx\|, \end{aligned}$$

whereas

$$\|u_2(\lambda)\| \le 2(1+M_Z)M_B M_A^2 \|\lambda^{2\theta-1} ZR(\lambda^2, Z)Ax\|,$$

$$\|u_3(\lambda)\| \le (1+M_Z)^2 M_B M_A \|\lambda^{2\theta-1} AR(\lambda, A)Ax\|,$$

$$\|u_4(\lambda)\| \le (1+M_Z)M_B \|\lambda^{2\theta-1} ZR(\lambda^2, Z)Bx\|$$

The function $\lambda \mapsto \|\lambda^{2\theta-1}AR(\lambda, A)Ax\|$ is in $L_*^p(a, +\infty)$ because Ax belongs to $(X, D(A))_{2\theta-1,p}$, and the functions $\lambda \mapsto \|\lambda^{2\theta-1}ZR(\lambda^2, Z)Ax\|$ and $\lambda \mapsto \|\lambda^{2\theta-1}Z$ $R(\lambda^2, Z)Bx\|$ are in $L_*^p(a, +\infty)$ because Ax and Bx belong to $(X, D(Z))_{\theta-1/2,\infty}$ due to Lemma 2.2(ii). Therefore, $u \in L_*^p(a, +\infty; X)$, and the statement follows. □

We need the following commutation formula, which generalizes (2.4).

Lemma 2.6. *Let $L = A^{n_1}B^{m_1}A^{n_2}B^{m_2} \cdot \dots \cdot A^{n_r}B^{m_r} \in \mathcal{L}_n$. For every $x \in D(Z^n)$ such that $Z^k x \in K^n$ for $k = 0, \dots, n$ and for every $\lambda > \omega_B$ we have*

$$LR(\lambda, B)x$$
$$= \sum_{i=1}^{r} \sum_{h_i=0}^{n_i} c_{n,h_1,\dots,h_r} R(\lambda, B)^{1+\Sigma_i h_i} A^{n_1-h_1}B^{m_1} \cdot \dots \cdot A^{n_r-h_r}B^{m_r} Z^{\Sigma_i h_i} x,$$

while for every $\lambda > \omega_A$ we have

$$LR(\lambda, A)x$$
$$= \sum_{i=1}^{r} \sum_{k_i=0}^{m_i} c_{n,k_1,\dots,k_r} R(\lambda, B)^{1+\Sigma_i k_i} A^{n_1}B^{m_1-k_1}A^{n_2} \cdot \dots \cdot A^{n_r}B^{m_r-k_r}(-Z)^{\Sigma_i k_i} x,$$

with suitable integers $c_{n,h_1,\dots,h_r}$. The coefficient $c_{n,0,\dots,0}$ is 1.

Proof. We argue by recurrence. The statement is true for $n = 1$ thanks to Lemma 2.2. Assume that the formulae hold for an integer n, and fix $L \in \mathcal{L}_{n+1}$. Then we have either $L = A\widetilde{L}$, or $L = B\widetilde{L}$, with $\widetilde{L} = A^{n_1}B^{m_1}A^{n_2} \cdot \dots \cdot A^{n_r}B^{m_r} \in \mathcal{L}_n$.

Using formula (i) of Lemma 2.2 we get, for every $k \in \mathbb{N}$ and $y \in D(A) \cap D(Z)$,

$$[A, R(\lambda, B)^k]y = kR(\lambda, B)^{k+1}Zy,$$

and since

$$\widetilde{L}R(\lambda,B)x = \sum_{i=1}^{r} \sum_{h_i=0}^{n_i} c_{n,h_1,\dots,h_r} R(\lambda,B)^{1+\Sigma_i h_i} A^{n_1-h_1}B^{m_1} \cdot \dots \cdot A^{n_r-h_r}B^{m_r} Z^{\Sigma_i h_i} x,$$

we get

$$A\widetilde{L}R(\lambda, B)x$$
$$= \sum_{i=1}^{r} \sum_{h_i=0}^{n_i} c_{n,h_1,\dots,h_r} R(\lambda, B)^{1+\Sigma_i h_i} A^{n_1+1-h_1}B^{m_1} \cdot \dots \cdot A^{n_r-h_r}B^{m_r} Z^{\Sigma_i h_i} x$$
$$+ \sum_{i=1}^{r} \sum_{h_i=0}^{n_i} c_{n,h_1,\dots,h_r} (1 + \Sigma_i h_i) R(\lambda, B)^{2+\Sigma_i h_i} \cdot$$
$$\cdot A^{n_1-h_1}B^{m_1}A^{n_2-h_2}B^{m_2} \cdot \dots \cdot A^{n_r-h_r}B^{m_r} Z^{1+\Sigma_i h_i} x$$

while

$$
\begin{aligned}
&B\widetilde{L}R(\lambda, B)x \\
&= \sum_{i=1}^{r} \sum_{h_i=0}^{n_i} c_{n,h_1,\dots,h_r} R(\lambda, B)^{1+\Sigma_i h_i} B A^{n_1-h_1} B^{m_1} \cdot \dots \cdot A^{n_r-h_r} B^{m_r} Z^{\Sigma_i h_i} x,
\end{aligned}
$$

and the statement follows. □

Theorem 2.7. *For each $m \in \mathbb{N}$, $\theta \in (0,1)$, $p \in [1, +\infty]$ we have*

$$(K^m, K^{m+1})_{\theta,p} = \{x \in K^m : \ Lx \in (X, D(A))_{\theta,p} \cap (X, D(B))_{\theta,p}, \ \forall L \in \mathcal{L}_m\}$$

and the norm

$$|x| = \|x\| + \sum_{L \in \mathcal{L}_m} (\|Lx\|_{(X,D(A))_{\theta,p}} + \|Lx\|_{(X,D(B))_{\theta,p}})$$

is equivalent to the norm of $(K^m, K^{m+1})_{\theta,p}$.

Proof. The embedding $\subset$ comes easily from the definition of real interpolation spaces. To prove the other embedding we fix any $a > \max\{\omega_A, \ \omega_B, \ \omega_Z\}$ and for each $x \in (K^m, K^{m+1})_{\theta,p}$ we set

$$u(\lambda) = \lambda^{2m+4} R(\lambda^2, Z)^{m+1} R(\lambda, A) R(\lambda, B) x, \quad \lambda \geq a.$$

We shall show that $\lambda \mapsto \lambda^\theta \|u(\lambda) - x\|_{K^m}$ and $\lambda \mapsto \lambda^{\theta-1} \|u(\lambda)\|_{K^{m+1}}$ are in $L^p_*(a, +\infty)$, with norms not exceeding $C|x|$. It will follow that the function $t \mapsto t^{-\theta}(\|u(1/t) - x\|_{K^m} + t\|u(1/t)\|_{K^{m+1}})$ belongs to $L^p_*(0, a)$, and therefore $t \mapsto t^{-\theta} K(t, x, \ K^m, K^{m+1})$ belongs to $L^p_*(0, a)$ (and hence it belongs to $L^p_*(0, +\infty)$) with norm not exceeding $C|x|$. This will prove the other embedding.

As a first step we estimate $\|u(\lambda)\|_{K^{m+1}}$.

We have of course $\|u(\lambda)\| \leq M_Z^{m+1} M_A M_B \|x\|$.

Let $L = A^{n_1} B^{m_1} A^{n_2} B^{m_2} \cdot \dots \cdot A^{n_r} B^{m_r} \in \mathcal{L}_{m+1}$. Let us assume that $n_1 \geq 1$. By Lemma 2.6, $Lu(\lambda)$ is a linear combination of elements of the type

$$
\begin{aligned}
v(\lambda) = \lambda^{2m+4} A R(\lambda, A)^{1+\Sigma_j h_j} R(\lambda, B)^{1+\Sigma_i k_i} Z^{\Sigma_j h_j + \Sigma_i k_i} R(\lambda^2, Z)^{1+m} \\
\cdot A^{n_1-1-h_1} B^{m_1-k_1} A^{n_2} B^{m_2-k_2} \cdot \dots \cdot A^{n_r-h_r} B^{m_r-k_r} x,
\end{aligned}
$$

where k_i runs between 0 and m_i, for $i = 1, \dots, r$; h_1 runs between 0 and $n_1 - 1$, and h_j runs between 0 and n_j for $j = 2, \dots, r$. We have to distinguish between the case in which $\sum_{j=1}^r h_j + \sum_{i=1}^r k_i$ is even and the case in which it is odd. In the first case we rewrite $v(\lambda)$ as

$$\lambda^{2m+4} A R(\lambda, A)^{1+\Sigma_j h_j} R(\lambda, B)^{1+\Sigma_i k_i} Z^{(\Sigma_j h_j + \Sigma_i k_i)/2} R(\lambda^2, Z)^{1+m} y,$$

where

$$
\begin{aligned}
y = Z^{(\Sigma_j h_j + \Sigma_i k_i)/2} A^{n_1-1-h_1} B^{m_1-k_1} A^{n_2} B^{m_2-k_2} \cdot \dots \cdot A^{n_r-h_r} B^{m_r-k_r} x \\
\in (X, D(A))_{\theta,p} \cap (X, D(B))_{\theta,p},
\end{aligned}
$$

so that, if $\sum_{j=1}^r h_j + \sum_{i=1}^r k_i > 0$,

$$\|v(\lambda)\| \le \lambda(M_A+1)M_A^{\Sigma_j h_j} M_B^{1+\Sigma_i k_i}(M_Z+1)^{(\Sigma_j h_j+\Sigma_i k_i)/2-1} \cdot M_Z^{m-(\Sigma_j h_j+\Sigma_i k_i)/2+1}\|ZR(\lambda^2,Z)y\|.$$

If each h_j and k_i vanish we have

$$v(\lambda) = \lambda^{2m+4}R(\lambda^2,Z)^{m+1}AR(\lambda,A)R(\lambda,B)\widetilde{L}x,$$

with $\widetilde{L} = A^{n_1-1}B^{m_1}A^{n_2}B^{m_2}\cdot \dots \cdot A^{n_r}B^{m_r}$, and we rewrite it as

$$\lambda^{2m+3}R(\lambda^2,Z)^{m+1}AR(\lambda,A)(\lambda R(\lambda,B)\widetilde{L}x - \widetilde{L}x) + \lambda^{2m+3}R(\lambda^2,Z)^{m+1}AR(\lambda,A)\widetilde{L}x$$

so that

$$\|v(\lambda)\| \le \lambda M_Z^{m+1}((M_A+1)\|BR(\lambda,B)\widetilde{L}x\| + \|AR(\lambda,A)\widetilde{L}x\|).$$

If $\sum_{j=1}^r h_j + \sum_{i=1}^r k_i$ is odd, we rewrite $v(\lambda)$ as

$$\lambda^{2m+4}AR(\lambda,A)^{1+\Sigma_j h_j}R(\lambda,B)^{1+\Sigma_i k_i}Z^{(\Sigma_j h_j+\Sigma_i k_i)/2+1/2}R(\lambda^2,Z)^{1+m}Mx,$$

where

$$M = Z^{(\Sigma_j h_j+\Sigma_i k_i)/2-1/2}A^{n_1-1-h_1}B^{m_1-k_1}A^{n_2}B^{m_2-k_2}\cdot \dots \cdot A^{n_r-h_r}B^{m_r-k_r}$$

is in $\mathcal{L}_{m-1}$, so that $Mx \in K_1$ and AMx, $BMx \in (X,D(A))_{\theta,p} \cap (X,D(B))_{\theta,p}$. By Lemma 2.5, $Mx \in (X,D(Z))_{(\theta+1)/2,p}$, so that $\lambda \mapsto \lambda^{(\theta+1)/2}\|ZR(\lambda,Z)Mx\|$ is in $L_*^p(a,+\infty)$, and therefore $\lambda \mapsto \lambda^{\theta+1}\|ZR(\lambda^2,Z)Mx\|$ belongs to $L_*^p(a,+\infty)$. We get

$$\|v(\lambda)\| \le \lambda^2(M_A+1)M_A^{\Sigma_j h_j}M_B^{1+\Sigma_i k_i}(M_Z+1)^{(\Sigma_j h_j+\Sigma_i k_i)/2-1/2}\|ZR(\lambda^2,Z)Mx\|.$$

Therefore, in any case $\lambda \mapsto \lambda^{\theta-1}\|v(\lambda)\|$ is in $L_*^p(a,+\infty)$, with norm not exceeding $C|x|$.

In the case where $n_1 = 0$, $m_1 \ne 0$ we argue similarly, taking into account that $Lu(\lambda)$ is now a linear combination of elements of the type

$$w(\lambda) = \lambda^{2m+4}R(\lambda,A)^{1+\Sigma_j h_j}BR(\lambda,B)^{1+\Sigma_i k_i}Z^{\Sigma_j h_j+\Sigma_i k_i}R(\lambda^2,Z)^{1+m} \cdot B^{m_1-1-k_1}A^{n_2-h_2}B^{m_2-k_2}A^{n_3-h_3}\cdot \dots \cdot A^{n_r-h_r}B^{m_r-k_r}x,$$

where k_1 runs between 0 and m_1-1, k_i runs between 0 and m_i, for $i=2,\dots,r$, h_j runs between 0 and n_j for $j=2,\dots,r$.

Now we estimate $\|u(\lambda)-x\|_{K^m}$. Writing

$$\begin{aligned} u(\lambda)-x &= \lambda^{2m+3}R(\lambda^2,Z)^{1+m}R(\lambda,A)(\lambda R(\lambda,B)x-x) \\ &\quad +\lambda^{2m+2}R(\lambda^2,Z)^{1+m}(\lambda R(\lambda,A)x-x) \\ &\quad +\sum_{k=0}^{m}\lambda^{2k}R(\lambda^2,Z)^k(\lambda^2R(\lambda^2,Z)x-x) \end{aligned}$$

we get easily

$$\|u(\lambda)-x\| \le C\lambda^{-1}(\|Ax\|+\|Bx\|) + C\lambda^{-2}\|Zx\|.$$

Moreover for every $L = A^{n_1}B^{m_1}A^{n_2}B^{m_2}\cdot \ldots \cdot A^{n_r}B^{m_r} \in \mathcal{L}_m$, using again Lemma 2.6 we write $Lu(\lambda) - Lx$ as $\lambda^{2m+4}R(\lambda^2, Z)^{1+m}R(\lambda, A)R(\lambda, B)Lx - Lx$ plus a linear combination of elements of the type

$$z(\lambda) = \lambda^{2m+4}R(\lambda, A)^{1+\Sigma_j h_j}R(\lambda, B)^{1+\Sigma_i k_i}Z^{\Sigma_j h_j + \Sigma_i k_i}R(\lambda^2, Z)^{1+m} \cdot A^{n_1-h_1}B^{m_1-k_1}A^{n_2}B^{m_2-k_2}\cdot \ldots \cdot A^{n_r-h_r}B^{m_r-k_r}x,$$

where k_i runs between 0 and m_i, h_j runs between 0 and n_j for $j = 1, \ldots, r$, and $h_i + k_j > 0$ for each i, j. The term $\lambda^{2m+4}R(\lambda^2, Z)^{1+m}R(\lambda, A)R(\lambda, B)Lx - Lx$ may be estimated writing it as

$$\lambda^{2m+3}R(\lambda^2, Z)^{1+m}R(\lambda, A)(\lambda R(\lambda, B)Lx - Lx) + \lambda^{2m+2}R(\lambda^2, Z)^{1+m}(\lambda R(\lambda, A)Lx - Lx) + \sum_{k=0}^{m} \lambda^{2k}R(\lambda^2, Z)^k(\lambda^2 R(\lambda^2, Z)Lx - Lx)$$

so that

$$\|\lambda^{2m+4}R(\lambda^2, Z)^{1+m}R(\lambda, A)R(\lambda, B)Lx - Lx\| \leq M_Z^{1+m}M_A\|BR(\lambda, B)Lx\| + M_Z^{1+m}\|AR(\lambda, A)Lx\| + \sum_{k=0}^{m} M_Z^k\|ZR(\lambda^2, Z)Lx\|.$$

To estimate the other terms we have to distinguish again between the case where $\Sigma_j h_j + \Sigma_i k_i$ is even and the case where it is odd, and argue as before; we get estimates quite similar to the previous ones divided by λ. This is because the factor $AR(\lambda, A)$ for $n_1 > 0$ (respectively, $BR(\lambda, B)$ if $n_1 = 0$) in the expression of $v(\lambda)$ is now replaced by $R(\lambda, A)$ (respectively, by $R(\lambda, B)$).

Eventually we get that $\lambda \mapsto \lambda^\theta\|u(\lambda) - x\|_{K^m}$ is in $\in L^p_*(a, +\infty)$, with norm not exceeding $C|x|$, and the statement follows. □

This theorem extends Theorem 2.3 of [8], where the cases $m = 0$ and $m = 1$ were considered.

3. Further regularity for the Heisenberg Laplacian and the Grushin operator

In this section we use the results of Section 2 to study regularity for the problem

$$\lambda v - \mathcal{L}v = f, \tag{3.1}$$

with $\lambda > 0$, where $\mathcal{L}$ is either the Heisenberg Laplacian in $\mathbb{R}^3$,

$$\mathcal{L}f(x, y, z) = f_{xx} + f_{yy} + 4yf_{xz} - 4xf_{yz} + 4(x^2 + y^2)f_{zz},$$

or the Grushin operator in $\mathbb{R}^2$,

$$\mathcal{L}f(x, y) = f_{xx} + x^2 f_{yy}.$$

We shall set these problems in spaces of bounded continuous functions, and in L^p spaces. Accordingly, we shall obtain optimal regularity results by interpolation in Hölder and in fractional Sobolev spaces.

We shall use the characterization of the spaces $(K^m, K^{m+1})_{\theta,p}$ of the previous section, and the next Theorem 3.1 taken from [8].

We recall that if $D \subset E \subset X$ are Banach spaces and $\beta \in (0,1)$, E is said to belong to $J_\beta(X, D)$ if there is $C > 0$ such that $\|x\|_E \leq C\|x\|_X^{1-\beta}\|x\|_D^\beta$, for all $x \in D$.

Theorem 3.1. *Let $T(t)$ be a semigroup in a general Banach space X with generator $L : D(L) \mapsto X$. Assume that there exists a Banach space $E \subset X$ and $m \in \mathbb{N}$, $0 < \beta < 1$, $\omega \in \mathbb{R}$, $C > 0$ such that*

$$\|T(t)\|_{L(X,E)} \leq \frac{Ce^{\omega t}}{t^{m\beta}}, \quad t > 0, \tag{3.2}$$

and for every $x \in X$, $t \mapsto T(t)x$ is measurable with values in E.

Then $E \in J_\beta(X, D(L^m))$.

As usual, if $L : D(L) \mapsto X$ is any closed operator and $k \in \mathbb{N} \cup \{0\}$, $\theta \in (0,1)$, $p \in [1, +\infty]$, we shall denote by $D_L(k + \theta, p)$ the set of all $x \in D(L^k)$ such that $Lx \in (X, D(L))_{\theta,p}$. If L is the generator of a semigroup, then $(X, D(L^m))_{\theta,p} = D_L(m\theta, p)$ if $m\theta$ is not integer. See e.g. [10, §1.13.5].

3.1. The Heisenberg Laplacian

We introduce the functional spaces associated to this operator.

First of all, we recall that the Heisenberg group is R^3 endowed with the sum

$$(x, y, z) \oplus (x', y', z') = (x + x', y + y', z + z' + 2(yx' - xy')).$$

We shall use the distance $d(p, \widetilde{p}) = [(-\widetilde{p}) \oplus p]$, where $[\cdot]$ is the pseudonorm defined by $[(x, y, z)] = ((x^2 + y^2)^2 + z^2)^{1/4}$.

We shall consider the usual spaces $L^p(\mathbb{R}^3)$, $1 \leq p \leq \infty$, and other functional spaces defined through the distance d and the sum $\oplus$.

The space of the uniformly continuous functions with respect to the distance d is defined by

$$BUC_H(\mathbb{R}^3) = \{f \in L^\infty(\mathbb{R}^3) : \lim_{p\to 0} \|f(\cdot \oplus p) - f\|_\infty = 0\}.$$

The derivatives D_j, $j = 1, 2, 3$ are defined by

$$D_j f(p) = \lim_{h\to 0} \frac{f(p \oplus he_j) - f(p)}{h}, \tag{3.3}$$

for all functions $f : \mathbb{R}^3 \mapsto \mathbb{R}$ such that the right-hand side exists. In particular, if f is differentiable in the usual sense,

$$D_1 f = f_x + 2yf_z, \quad D_2 f = f_y - 2xf_z, \quad D_3 f = f_z.$$

Let $k \in \mathbb{N}$, let $\alpha = (\alpha_1, \ldots, \alpha_k) \in \mathbb{N}^k$ be a multi-index with $\alpha_i \in \{1, 2\}$, and set $|\alpha| = k$, $D_\alpha = D_{\alpha_1} D_{\alpha_2} \cdots D_{\alpha_k}$.

The spaces $BUC^k_H(\mathbb{R}^3)$ are defined by

$$\begin{cases} BUC^k_H(\mathbb{R}^3) = \{f \in BUC_H(\mathbb{R}^3) : \exists D_\alpha f \in BUC_H(\mathbb{R}^3),\ |\alpha| \le k\}, \\ \|f\|_{BUC^k_H} = \|f\|_\infty + \sum_{|\alpha|=k} \|D_\alpha f\|_\infty. \end{cases} \tag{3.4}$$

For $p \in [1,\infty)$ the Sobolev spaces $W^{k,p}_H(\mathbb{R}^3)$ are defined by

$$\begin{cases} W^{k,p}_H(\mathbb{R}^3) = \{f \in L^p(\mathbb{R}^3) : \exists D_\alpha f \in L^p(\mathbb{R}^3),\ |\alpha| \le k\}, \\ \|f\|_{W^{k,p}_H} = \|f\|_{L^p} + \sum_{|\alpha|=k} \|D_\alpha f\|_{L^p}. \end{cases} \tag{3.5}$$

For $0 < \theta < 1$ the Hölder space with respect to the distance d and the sum $\oplus$ is

$$\begin{cases} C^\theta_H(\mathbb{R}^3) = \left\{ f \in L^\infty(\mathbb{R}^3) : [f]_{C^\theta_H} = \sup_{p\in\mathbb{R}^3,\, h\neq 0} \dfrac{|f(p\oplus h) - f(p)|}{[h]^\theta} < +\infty \right\}, \\ \|f\|_{C^\theta_H} = \|f\|_\infty + [f]_{C^\theta_H}. \end{cases} \tag{3.6}$$

For $0 < \theta < 1$, $k \in \mathbb{N}$ we set

$$\begin{cases} C^{\theta+k}_H(\mathbb{R}^3) = \{f \in BUC^k_H(\mathbb{R}^3) : D_\alpha f \in C^\theta_H(\mathbb{R}^3),\ |\alpha| \le k\}, \\ \|f\|_{C^{\theta+k}_H} = \|f\|_{BUC^k_H} + [f]_{C^{\theta+k}_H} = \|f\|_{BUC^k_H} + \sum_{|\alpha|=k} [D_\alpha f]_{C^\theta_H}. \end{cases} \tag{3.7}$$

The fractional Sobolev space $W^{\theta+k,p}_H(\mathbb{R}^3)$ is defined by

$$W^{\theta+k,p}_H(\mathbb{R}^3) = \left\{ f \in W^{k,p}_H(\mathbb{R}^3) : \int_{\mathbb{R}^3}\int_{\mathbb{R}^3} \frac{|D_\alpha f(\xi\oplus h) - D_\alpha f(\xi)|^p}{[h]^{4+\theta p}} d\xi\, dh < \infty,\ |\alpha| = k \right\},$$

and it is endowed with the norm

$$\|f\|_{W^{\theta+k,p}_H} = \|f\|_{L^p} + \sum_{|\alpha|=k} \|D_\alpha f\|_{L^p} + \sum_{|\alpha|=k} \left(\int_{\mathbb{R}^3}\int_{\mathbb{R}^3} \frac{|D_\alpha f(\xi\oplus h) - D_\alpha f(\xi)|^p}{[h]^{4+\theta p}} d\xi\, dh \right)^{1/p}.$$

For a systematic treatment of such spaces see [5], [7], [9].

Let L be the realization of any derivative D_j, $j = 1,2,3$, in $X = L^p(\mathbb{R}^3)$, $1 \le p < \infty$, or in $X = BUC_H(\mathbb{R}^3)$. It generates the contraction semigroup

$$(T_j(t)f)(p) = f(p \oplus te_j), \quad f \in X,\ p \in \mathbb{R}^3. \tag{3.8}$$

Let us denote respectively by A, B the realizations of D_1, D_2, in X. Then A and B satisfy assumptions (1.1), with $\Omega_A = \omega_B = 0$, $M_A = M_B = 1$. The closed extension Z of $[A,B]$ is the realization of $-4D_3$ in X, which satisfies (1.3) with $\omega_Z = 0$, $M_Z = 1$, and (1.4).

If $X = L^p(\mathbb{R}^3)$, we have

$$K^m = W^{m,p}_H(\mathbb{R}^3),$$

and (2.3) is satified with $D = C_0^\infty(\mathbb{R}^3)$ (the space of the smooth functions with compact support); whereas if $X = BUC_H(\mathbb{R}^3)$ we have

$$K^m = BUC_H^m(\mathbb{R}^3),$$

and (2.3) is satified with $D = \cap_{k\in\mathbb{N}} BUC_H^k(\mathbb{R}^3)$.

From [8] we already know that

$$(X, D(A))_{\theta,p} \cap (X, D(B))_{\theta,p} = W_H^{\theta,p}(\mathbb{R}^3), \tag{3.9}$$

if $X = L^p(\mathbb{R}^3)$, and

$$(X, D(A))_{\theta,\infty} \cap (X, D(B))_{\theta,\infty} = C_H^\theta(\mathbb{R}^3), \tag{3.10}$$

if $X = BUC_H(\mathbb{R}^3)$.

Then we apply Theorem 2.7 to get

Corollary 3.2. *For $0 < \theta < 1$, $1 \le p < \infty$, we have*

$$(W_H^{m,p}(\mathbb{R}^3), W_H^{m+1,p}(\mathbb{R}^3))_{\theta,p} = W_H^{m+\theta,p}(\mathbb{R}^3),$$

$$(BUC_H^m(\mathbb{R}^3), BUC_H^{m+1}(\mathbb{R}^3))_{\theta,\infty} = C_H^{m+\theta}(\mathbb{R}^3),$$

with equivalence of the respective norms.

The nonisotropic fractional Sobolev and Hölder spaces may be characterized as real interpolation spaces between X and the domains of the powers of the Heisenberg Laplacian.

Theorem 3.3. *Let $k \in \mathbb{N}$, $0 < \theta < 1$, $1 \le p < \infty$. If L is the realization of the Heisenberg Laplacian in $L^p(\mathbb{R}^3)$, we have*

$$W^{k+\theta,p}(\mathbb{R}^3) = D_L((k+\theta)/2, p), \tag{3.11}$$

with equivalence of the respective norms. If L is the realization of the Heisenberg Laplacian in $BUC_H(\mathbb{R}^3)$, we have

$$C_H^{k+\theta}(\mathbb{R}^3) = D_L((k+\theta)/2, \infty) \tag{3.12}$$

with equivalence of the respective norms.

Proof. Let us prove (3.11). The realization of the Heisenberg Laplacian in $L^p(\mathbb{R}^3)$, $1 \le p \le \infty$, generates an analytic semigroup $T(t)$, thanks to [5]. Using homogeneity arguments (see e.g. [8]) it is easy to prove that for each $f \in L^p(\mathbb{R}^3)$ and for every multi-index α we have

$$\|D_\alpha T(t) f\|_{L^p} \le C t^{-|\alpha|/2} \|f\|_{L^p}, \quad t > 0. \tag{3.13}$$

It follows that for every $\omega > 0$, $k \in \mathbb{N}$ there is C such that

$$\|T(t)\|_{L(L^p, W_H^{k,p})} \le \frac{Ce^{\omega t}}{t^{k/2}}, \quad t > 0,$$

$$\|T(t)\|_{L(BUC_H, BUC_H^k)} \le \frac{Ce^{\omega t}}{t^{k/2}}, \quad t > 0.$$

Applying Theorem 3.1 with $X = L^p(\mathbb{R}^3)$, $1 \le p < \infty$, $Y = W^{k,p}_H(\mathbb{R}^3)$, we obtain that $W^{k,p}_H(\mathbb{R}^3)$ belongs to the class $J_{k/2m}$ between X and $D(L^m)$, for every integer m such that $2m > k$. By the Reiteration Theorem [10, §1.10.2], for every $\theta \in (0,1)$ we have

$$(X, D(L^m))_{(k+\theta)/2m,p} \subset (W^{k,p}_H(\mathbb{R}^3), W^{k+1,p}_H(\mathbb{R}^3))_{\theta,p}$$

with continuous embedding. On the other hand, $(X, D(L^m))_{(k+\theta)/2m,p} = D_L((k+\theta)/2, p)$, and by Corollary 3.2 we have $(W^{k,p}_H(\mathbb{R}^3), W^{k+1,p}_H(\mathbb{R}^3))_{\theta,p} = W^{k+\theta,p}_H(\mathbb{R}^3)$. It follows

$$D_L((k+\theta)/2, p) \subset W^{k+\theta,p}_H(\mathbb{R}^3).$$

This embedding is the key of our proof. The other embedding is recovered in a standard way from the results of [8] and general interpolation theory. Indeed, let $f \in W^{k+\theta,p}_H(\mathbb{R}^3)$. Write $k+\theta = 2l+\sigma$, l integer, $\sigma \in (0,1) \cup (1,2)$. Then $L^l f \in W^{\sigma,p}_H(\mathbb{R}^3)$, which coincides with $(L^p(\mathbb{R}^3), D(L))_{\sigma/2,p}$ thanks to [8]. Therefore, $f \in D_L(l+\sigma/2, p) = D_L((k+\theta)/2, p)$. The proof of (3.12) is similar, and it is left to the reader. □

We are able to study regularity of the solutions to (3.1).

Theorem 3.4. *Let $\lambda > 0$, $f \in W^{k+\theta,p}_H(\mathbb{R}^3)$ with $0 < \theta < 1$, $1 \le p < \infty$. Then* (3.1) *has a unique solution $u \in W^{k+2+\theta,p}_H(\mathbb{R}^3)$, and there is $C > 0$, independent of f, such that*

$$\|u\|_{W^{k+2+\theta,p}_H(\mathbb{R}^3)} \le C\|f\|_{W^{k+\theta,p}_H(\mathbb{R}^3)}.$$

If $f \in C^{k+\theta}_H(\mathbb{R}^3)$ then (3.1) *has a unique solution $u \in C^{k+2+\theta}_H(\mathbb{R}^3)$, and there is $C > 0$, independent of f, such that*

$$\|u\|_{C^{k+\theta+2}_H(\mathbb{R}^3)} \le C\|f\|_{C^{k+\theta}_H(\mathbb{R}^3)}.$$

Proof. Let either $X = L^p(\mathbb{R}^3)$ or $X = BUC_H(\mathbb{R}^3)$, and let L be the realization of the Heisenberg Laplacian in X, let $T(t)$ be the semigroup generated by L. Since $\|T(t)\|$ is bounded with respect to $t \in (0,+\infty)$, then for $\lambda > 0$ equation (3.1) has a unique solution $u \in D(L)$. By Theorem 3.3, $f \in D_L((k+\theta)/2, p)$, so that $u \in D_L((k+\theta)/2+1, p) = W^{k+2+\theta,p}_H(\mathbb{R}^3)$ again by Theorem 3.3. □

3.2. The Grushin operator

In this subsection we set $X = L^p(\mathbb{R}^2)$, $1 \le p < \infty$. The realization of the Grushin operator $\mathcal{L}u(x,y) = u_{xx}(x,y) + x^2 u_{yy}(x,y)$ in X is a symmetric elliptic operator with regular coefficients in $\mathbb{R}^n$, and hence it generates a strongly continuous contraction semigroup. See [1, Ex. 3.2.11, Thm. 1.4.1, Thm. 1.4.2].

We write $\mathcal{L}$ as $\mathcal{A}^2 + \mathcal{B}^2$, where $\mathcal{A} = \partial/\partial x$, $\mathcal{B} = x\partial/\partial y$. The realizations of $\mathcal{A}$ and $\mathcal{B}$ in X are defined by

$$D(A) = \{f \in L^p(\mathbb{R}^2) : f_x \in L^p(\mathbb{R}^2)\}, \quad Af = f_x,$$

$$D(B) = \{f \in L^p(\mathbb{R}^2) : xf_y \in L^p(\mathbb{R}^2)\}, \quad Bf = xf_y.$$

They generate the strongly continuous contraction groups

$$e^{tA}f(x,y) = f(x+t,y), \quad t \in \mathbb{R},$$

$$e^{tB}f(x,y) = f(x,tx+y), \quad t \in \mathbb{R},$$

so that (1.1) is satisfied with $\omega_A = \omega_B = 0$, $M_A = M_B = 1$. The closed extension of their commutator is the realization Z of $\partial/\partial y$ in X, which generates the translation group

$$e^{tZ}f(x,y) = f(x,y+t), \quad t \in \mathbb{R},$$

so that (1.3) is satisfied with $\omega_Z = 0$, $M_Z = 1$. Assumption (1.4) obviously holds.

As in Subsection 3.1, it is convenient to define, for $m \in \mathbb{N}$

$$W_G^{m,p}(\mathbb{R}^2) = K^m.$$

Therefore, taking $D = C_0^\infty(\mathbb{R}^2)$, assumption (2.3) is satisfied.

The semigroup characterization ([10, §1.13.2]) yields that the interpolation spaces $(X, D(A))_{\theta,p}$, $(X, D(B))_{\theta,p}$, are given respectively by

$$(X, D(A))_{\theta,p} = \left\{ f \in L^p(\mathbb{R}^2) : [f]_{\theta,p}^p = \int_{\mathbb{R}^3} \frac{|f(x+h,y) - f(x,y)|^p}{|h|^{1+\theta p}} dx\,dy\,dh < \infty \right\}$$

with norm equivalent to

$$\|f\|_{L^p} + [f]_{\theta,p},$$

$$(X, D(B))_{\theta,p} =$$
$$= \left\{ f \in L^p(\mathbb{R}^2) : [[f]]_{\theta,p}^p = \int_{\mathbb{R}^3} \frac{(|x|^\theta |f(x,y+h) - f(x,y)|)^p}{|h|^{1+\theta p}} dx\,dy\,dh < \infty \right\},$$

with norm equivalent to

$$\|f\|_{L^p} + [[f]]_{\theta,p}.$$

So we define

$$W_G^{\theta,p}(\mathbb{R}^2) = D_A(\theta,p) \cap D_B(\theta,p),$$

and for $k \in \mathbb{N}$

$$\begin{aligned} W_G^{k+\theta,p}(\mathbb{R}^2) \quad &= \quad \{ f \in W_G^{k,p}(\mathbb{R}^2) : \mathcal{A}^{\alpha_1}\mathcal{B}^{\alpha_2} \cdot \;\cdots\; \cdot \mathcal{A}^{\alpha_{r-1}}\mathcal{B}^{\alpha_r} f \\ &\in D_A(\theta,p) \cap D_B(\theta,p),\ \alpha = (\alpha_1, \dots, \alpha_r),\ |\alpha| = k \}, \end{aligned}$$

endowing it with the intersection norm. Arguments similar to the ones used in the proof of Theorem 3.3 give

Theorem 3.5. *Let* $0 < \theta < 1$, $k \in \mathbb{N}$. *Then*

$$(W_G^{k,p}, W_G^{k+1,p}(\mathbb{R}^2))_{\theta,p} = W_G^{k+\theta,p}(\mathbb{R}^2),$$

with equivalence of the respective norms.

The above characterization gives optimal regularity results for the hypoelliptic equation (3.1), where $\mathcal{L}$ is the Grushin operator.

Theorem 3.6. *Let $\lambda > 0$, $f \in W_G^{k+\alpha,p}(\mathbb{R}^2)$ with $0 < \alpha < 1$, $k \in \mathbb{N}$, $1 \le p < \infty$. Then* (3.1) *has a unique solution $u \in W_G^{k+\alpha+2,p}(\mathbb{R}^2)$, and there is $C > 0$, independent of f, such that*

$$\|u\|_{W_G^{k+\alpha+2,p}(\mathbb{R}^2)} \le C\|f\|_{W_G^{k+\alpha,p}(\mathbb{R}^2)}.$$

Sketch of the proof. The operators A and B are homogeneous with respect to the family of dilations $\delta_r(x,y) = (rx, ry)$. This implies that for every $k \in \mathbb{N}$

$$\|T(t)\|_{L(L^p, W_G^{k,p})} \le \frac{Ce^{\omega t}}{t^{k/2}}, \quad t > 0,$$

and these estimates, used together with Theorem 3.5 as in the proof of Theorem 3.4, give the equivalence

$$(X, D(L^m))_{(k+\theta)/2m,p} = W_G^{k+\theta,p}(\mathbb{R}^2),$$

for $m > k$, which yields the statement. □

4. Representation formulae

In this section we report some results from [3], where the author proved representation formulae for e^{tZ} and $e^{t\overline{A+B}}$ in the case where A and B generate strongly continuous groups. More precisely, he assumed either that there exists a dense subspace $D \subset X$ such that

$$\begin{cases} D \subset D(AB) \cap D(BA), \\ e^{tA}(D) \subset D, \ e^{tB}(D) \subset D, \quad t \in \mathbb{R}, \\ [e^{tA}, [A,B]] = [e^{tB}, [A,B]] = 0 \ \text{ on } D, \quad t \in \mathbb{R}, \end{cases} \tag{4.1}$$

or else that there exists a dense subspace $D \subset X$ such that

$$\begin{cases} D \subset D(AB) \cap D(BA), \\ R(\lambda, A)(D) \subset D, \ R(\mu, B)(D) \subset D, \quad \lambda \in \rho(A), \ \mu \in \rho(B), \\ [R(\lambda, A), [A,B]] = [R(\mu, B), [A,B]] = 0 \ \text{ on } D, \quad \lambda \in \rho(A), \ \mu \in \rho(B). \end{cases} \tag{4.2}$$

It is possible to see that (4.1) implies (4.2), and under suitable additional density assumptions (4.2) implies (4.1). So, (4.1) and (4.2) are almost equivalent, but it may be useful to know both for ready use in the applications.

The representation formulae are the following.

Theorem 4.1. *Let either* (4.1) *or* (4.2) *hold. Then $[A,B] : D \mapsto X$ is closable, and its closure Z generates a group e^{tZ} given by*

$$\begin{cases} e^{tZ} = e^{\sqrt{t}A} e^{\sqrt{t}B} e^{-\sqrt{t}A} e^{-\sqrt{t}B}, \quad t \ge 0, \\ e^{tZ} = e^{\sqrt{-t}B} e^{\sqrt{-t}A} e^{-\sqrt{-t}B} e^{-\sqrt{-t}A}, \quad t \le 0. \end{cases} \tag{4.3}$$

Moreover $A + B : D \mapsto X$ is closable, and it closure generates the group

$$e^{t\overline{A+B}} = e^{-t^2 Z} e^{tA} e^{tB}, \quad t \in \mathbb{R}. \tag{4.4}$$

In the case where $X = \mathbb{R}^n$, (4.3) and (4.4) come immediately from the Campbell-Hausdorff formula. But in general Banach spaces they are not trivial. In the examples of Section 3 they are of not much help, since we know already an explicit representation of e^{tA}, e^{tB}, e^{tZ}. In [3] other nontrivial examples were given.

References

[1] E.B. Davies: *Heat kernels ad spectral theory*, Cambridge Univ. Press (1990).

[2] G. Da Prato, P. Grisvard: *Sommes d'opérateurs linéaires et équations différentielles opérationelles*, J. Maths. Pures Appliquées 54 (1975), 305–387.

[3] D. Di Giorgio: *Sums and Commutators of Noncommuting Operators*, preprint Dipart. Mat. Univ. Pisa (2001).

[4] G. Dore, A. Venni: *On the Closedness of the Sum of two closed Operators*, Math. Z. **196** (1987), 189–201.

[5] G.B. Folland: *Subelliptic estimates and function spaces on nilpotent Lie groups*, Ark. Mat. **13** (1975), 161–207.

[6] R. Howe: *On the role of the Heisenberg group in harmonic analysis*, Bull. Amer. Math. Soc. **32** (1980), 821–843.

[7] S.G. Krantz: *Lipschitz spaces on stratified Lie groups*, Trans. Amer. Math. Soc. **269** (1982), 29–66.

[8] A. Lunardi: *Regularity for a class of sums of noncommuting operators*, in: Topics in Nonlinear Analysis, The Herbert Amann Anniversary volume, J. Escher, G. Simonett Eds., Birkhäuser Verlag, Basel (1999), 517–533.

[9] K. Saka: *Besov spaces and Sobolev spaces on a nilpotent Lie group*, Tohoku Math. J. **31** (1979), 383–437.

[10] H. Triebel: *Interpolation theory, function spaces, differential operators*, North-Holland, Amsterdam (1978).

Alessandra Lunardi
Dipartimento di Matematica
Università di Parma
Via D'Azeglio 85/A
I-43100 Parma, Italy
e-mail: alessandra.lunardi@unipr.it

Progress in Nonlinear Differential Equations
and Their Applications, Vol. 50, 279–293

Well-posedness for Nonautonomous Abstract Cauchy Problems

Rainer Nagel and Gregor Nickel

To the Memory of Brunello Terreni

1. Wellposedness in the autonomous case

We first summarize some well-known, however instructive facts from the theory of autonomous abstract Cauchy problems

$$(ACP)\begin{cases} \dot{u}(t) = Au(t), & t \geq 0, \\ u(0) = x \end{cases} \tag{1.1}$$

for a closed operator $(A, D(A))$ on some Banach space X (compare [5], Chapter II.6).

Definition 1.1.

(1) *A function* $u : \mathbb{R}_+ \to X$ *is called a* **(classical) solution** *of* (ACP) *if* u *is continuously differentiable with respect to* X, $u(t) \in D(A)$ *for all* $t \geq 0$, *and* (ACP) *holds.*

(2) *A continuous function* $u : \mathbb{R}_+ \to X$ *is called a* **mild solution** *of* (ACP) *if* $\int_0^t u(r)\,dr \in D(A)$ *for all* $t \geq 0$ *and* $u(t) = A\int_0^t u(r)\,dr + x$.

(3) *The abstract Cauchy problem* (ACP) *is called* **well-posed** *if*

 (E) *the space* $D(A)$ *is dense in* X *and for every* $x \in D(A)$ *there exists a classical solution* $u(\cdot, x)$ *of* (ACP),

 (U) *this solution is unique, and*

 (CD) *for every sequence* $D(A) \ni x_n \to 0$ *one has* $u(t, x_n) \to 0$ *uniformly on compact intervals* $[0, T]$.

The existence of solutions and the well-posedness of (ACP) can be characterized completely in terms of strongly continuous semigroups (see [5], Cor. II.6.9).

Theorem 1.2.

(1) *The abstract Cauchy problem* (ACP) *is well posed if and only if* $(A, D(A))$ *is the generator of a strongly continuous semigroup* $(T(t))_{t \geq 0}$. *In this case, classical solutions of* (ACP) *are given by* $t \mapsto T(t)x$ *for every* $x \in D(A)$, *while mild solutions are given by* $t \mapsto T(t)x$ *for every* $x \in X$. *All solutions are exponentially bounded.*

(2) *On the other hand, every strongly continuous semigroup $(T(t))_{t\geq 0}$ is exponentially bounded and solves, in the sense of (1), an abstract Cauchy problem.*

Remark 1.3. *Weaker concepts of well-posedness and more general types of semigroups can be found, e.g., in* [1], *Chapter* 3.

2. Classical well-posedness of nonautonomous Cauchy problems

In this section we introduce the basic concepts for well-posedness of nonautonomous Cauchy problems

$$(NCP)_{s,x}\begin{cases} \dot{u}(t) = A(t)u(t), & t \geq s \in \mathbb{R}, \\ u(s) = x, \end{cases} \tag{2.2}$$

where $(A(t), D(A(t)))_{t\in\mathbb{R}}$ is a family of linear operators on some Banach space X. If the Cauchy problem is considered for all initial times $s \in \mathbb{R}$ and arbitrary initial values $x \in X$, we denote it by (NCP). We note that (NCP) can also be considered on compact time intervals.

Definition 2.1. *A continuous function $u : [s, \infty) \to X$ is called a* **classical solution** *of $(NCP)_{s,x}$ if $u \in C^1([s, \infty), X)$, $u(t) \in D(A(t))$ for all $t \geq s$, and $(NCP)_{s,x}$ holds.*

The definition of well-posedness is not so straightforward as in the autonomous case. However, the following slight modification of Kellermann's definition in [12] seems to be appropriate (and generalizes Fattorini's definition [7], p. 382).

Definition 2.2 (Wellposedness). *The nonautonomous Cauchy problem (NCP) for a family $(A(t), D(A(t)))_{t\in\mathbb{R}}$ of linear operators on the Banach space X is called* **(classically) well-posed** *with* **regularity subspaces** $(Y_s)_{s\in\mathbb{R}}$ *if the following holds.*

(i) **(Existence)** *For each $s \in \mathbb{R}$, the subspace*

$$Y_s := \{y \in X \ : \ \text{there exists a classical solution for } (NCP)_{s,y}\} \subset D(A(s))$$

is dense in X.

(ii) **(Uniqueness)** *For every $s \in \mathbb{R}$, $y \in Y_s$ the solution $u_y(\cdot, s)$ is unique.*

(iii) **(Continuous dependence)** *The solution depends continuously on the initial data s and y, i. e., if $s_n \to s \in \mathbb{R}$ and $Y_{s_n} \ni y_n \to y \in Y_s$, then we have*

$$\|\hat{u}_{y_n}(t, s_n) - \hat{u}_y(t, s)\| \to 0$$

uniformly for t in compact subsets of $\mathbb{R}$, where

$$\hat{u}_r(t, y) := \begin{cases} u_y(t, r) & \text{if } \ r \leq t \\ y & \text{if } \ r > t. \end{cases}$$

If, in addition, there exist constants $M \geq 1$ and $\omega \in \mathbb{R}$ such that

$$\|u_y(t, s)\| \leq Me^{\omega(t-s)}\|y\|$$

for all $y \in Y_s$ and $t \geq s$, then the nonautonomous Cauchy problem (NCP) is called **well-posed with exponentially bounded solutions.**

At this moment, there still are no necessary and sufficient conditions for (NCP) to be well posed. However, it were Acquistapace and Terreni [2], extending previous work by Kato, Tanabe, Sobolevsky, and others, who obtained one of the best available results (for a more complete survey with many references see [23]).

3. Evolution families

If (NCP) has sufficiently many solutions, these satisfy an algebraic property which we describe now.

Definition 3.1 (Evolution family). *A family $(U(t,s))_{t\geq s}$ of linear, bounded operators on a Banach space X is called a* **strongly continuous, exponentially bounded evolution family** *if*

(i) $U(t,r)U(r,s) = U(t,s), \qquad U(t,t) = Id \qquad$ *for all* $t \geq r \geq s \in \mathbb{R}$,

(ii) *the mapping* $\Delta := \{(t,s) \in \mathbb{R}^2 : s \leq t\} \ni (t,s) \mapsto U(t,s)$ *is strongly continuous,*

(iii) $\|U(t,s)\| \leq Me^{\omega(t-s)}$ *for some* $M \geq 1$, $\omega \in \mathbb{R}$ *and all* $t \geq s \in \mathbb{R}$.

If $(T(t))_{t\geq 0}$ is a strongly continuous semigroup, the definition $U(t,s) := T(t-s)$ yields a strongly continuous evolution family with very special properties (e.g. all operators commute). In the general case, many new phenomena appear (see below), but at least exponential boundedness can be characterized as for strongly continuous semigroups, while continuity is less easy to characterize (compare [5], Proposition I.5.5 and Proposition I.5.3).

Proposition 3.2.

(1) *The strongly continuous evolution family $(U(t,s))_{t\geq s}$ is exponentially bounded if and only if it is uniformly bounded on the set $\{(t,s) \in \Delta : s \leq t \leq s+\delta\}$ for some $\delta > 0$.*

(2) *For an evolution family $(U(t,s))_{t\geq s}$ and a dense subspace $D \subseteq X$ the following statements are equivalent.*

  (i) *The mapping $\Delta \ni (t,s) \mapsto U(t,s)x$ is continuous for all $x \in X$,*

  (ii) *The mapping $\Delta \ni (t,s) \mapsto U(t,s)y$ is continuous for all $y \in D$ and $\|U(t,s)\|$ is uniformly bounded on compact subsets of Δ,*

  (iii) *Uniformly for (t,s) in compact subsets of Δ we have*

    (a) $\lim_{s\nearrow t} U(t,s)x = x$,

    (b) *the mappings $[s,\infty) \ni t \mapsto U(t,s)x$ are continuous,*

    (c) $\|U(t,s)\|$ *is bounded.*

Proof. (1) This can be proved as Proposition I.5.5 in [5].

(2) The implication $(ii) \Rightarrow (i)$ is trivial, while $(i) \Rightarrow (ii)$ and $(i) \Rightarrow (iii)$ follow from the principle of uniform boundedness. It remains to show that $(iii) \Rightarrow (i)$. For that purpose, we prove uniform continuity of $(t,s) \mapsto U(t,s)x$ on subsets of the form $\Delta_S^T := \{(t,s) : S \leq s < t \leq T\} \subset \Delta$ for fixed $S < T \in \mathbb{R}$.

(a) Take $x \in X$, $(t,s) \in \Delta_S^T$, and $h,k > 0$ such that $t > s+k$ and consider

$$\begin{aligned} & U(t+h, s+k)x - U(t,s)x \\ &= [U(t+h,t) - Id]U(t,s+k)x - U(t,s+k)[U(s+k,s) - Id]x. \end{aligned}$$

For $h,k \searrow 0$ the second term converges to 0 since $\|U(t,s+k)\| \le M(S,T)$ is uniformly bounded and $U(s+k,s)x \to x$. Since, by the same argument,

$$U(t,s+k)x - U(t,s)x = U(t,s+k)[Id - U(s+k,s)]x \to 0,$$

we see that for every sequence $1 > k_n \to 0$ the set

$$\{U(t,s+k_n)x \,:\, n \in \mathbb{N}\} \cup \{U(t,s)x\}$$

is compact which, together with the pointwise convergence of $U(t+h,t)y \to y$, implies the convergence of the first term.

(b) Take $x \in X$, $(t,s) \in \Delta_S^T$, and $h,k > 0$ and consider

$$\begin{aligned} & U(t+h, s-k)x - U(t,s)x \\ &= [U(t+h,t) - Id]U(t,s-k)x + U(t,s)[U(s,s-k) - Id]x. \end{aligned}$$

For $h,k \searrow 0$ the second term converges to 0 since $\|U(t,s)\| \le M(S,T)$ is bounded and $U(s,s-k)x \to x$. By the same argument we see $U(t,s-k)x \to U(t,s)x$ and can argue as in (a).

(c) Take $x \in X$, $(t,s) \in \Delta_S^T$, and $h,k > 0$ such that $t-h \ge s+k$ and consider

$$\begin{aligned} & U(t-h, s+k)x - U(t,s)x \\ &= [U(t-h,s) - U(t,s)]x + U(t-h,s+k)[Id - U(s+k,s)]x. \end{aligned}$$

For $h,k \searrow 0$ the second term converges to 0 since $\|U(t-h,s+k)\| \le M(S,T)$ is uniformly bounded and $U(s+k,s)x \to x$. The continuity of U in the first variable implies that the first term converges to 0.

(d) Take $x \in X$, $(t,s) \in \Delta_S^T$, and $h,k > 0$ such that $t-h \ge s$ and consider

$$\begin{aligned} & U(t-h, s-k)x - U(t,s)x \\ &= [U(t-h,s) - U(t,s)]x + U(t-h,s)[U(s,s-k) - Id]x. \end{aligned}$$

For $h,k \searrow 0$ the second term converges to 0 since $\|U(t-h,s)\| \le M(S,T)$ is uniformly bounded and $U(s,s-k)x \to x$. The continuity of U in the first variable implies that the first term converges to 0.

Since the convergence of the above expressions is uniform on the all sets of the form Δ_S^T, we obtain strong continuity of $(U(t,s))$ on Δ. □

The connection between well-posed NCPs and strongly continuous evolution families is the following.

Proposition 3.3. *Let (NCP) be well posed and define*

$$U(t,s)x := u_x(t,s) \text{ for } t \ge s \text{ and } x \in Y_s.$$

Each $U(t,s)$ can be extended by continuity to a bounded operator $U(t,s) \in \mathcal{L}(X)$. Moreover, $(U(t,s))_{t \ge s}$ is a strongly continuous evolution family.

Definition 3.4 (EVF solving (NCP)). *A strongly continuous evolution family* $(U(t,s))_{t\geq s}$ *is called* **evolution family solving** (NCP) *if for every* $s \in \mathbb{R}$ *the regularity subspace*

$$Y_s := \{y \in X \ : \ [s, \infty) \ni t \mapsto U(t,s)y \textit{ solves } (NCP)_{s,y}\}$$

is dense in X.

In this definition as well as in Definition 2.2, it is important to distinguish between the domain $D(A(t))$ of the operator $A(t)$ and the space Y_t of possible initial values of (classical) solutions. This is in contrast to the autonomous situation where the space $D(A)$ is "uniquely" given either as space of possible initial values of solutions or as the (natural) domain of the operator in the abstract Cauchy problem. The following two examples illustrate this phenomenon.

Example 3.5. *Consider the space* $X := C_0(\mathbb{R}^2)$ *and the operators*

$$A(0) := \frac{\partial}{\partial x} \quad \textit{and} \quad A(1) := \frac{\partial}{\partial y},$$

defined on their maximal domains $D(A(0))$ *and* $D(A(1))$. *On the space*

$$Y := C_0^1(\mathbb{R}^2) := \{f \in C^1(\mathbb{R}^2) \ : \ f, f' \in C_0(\mathbb{R}^2)\} \subset D(A(0)) \cap D(A(1))$$

we can define the convex combinations $A(t) := (1-t)A(0) + tA(1)$ *for* $t \in [0,1]$. *The closure of these operators are generators of* C_0*-semigroups with domain*

$$D(A(t)) = C_t^1 := \{f \in C_0(\mathbb{R}^2) \ : \ \lim_{h \searrow 0} \frac{1}{h}[f(\cdot + (1-t)h, \cdot + ht) - f(\cdot,\cdot)] \in C_0(\mathbb{R}^2)\}.$$

It is easy to see that

$$[e^{\tau A(t)} f](x,y) = f(x + \tau(1-t), y + \tau t)$$

for $f \in C_0(\mathbb{R}^2)$, $t \in [0,1]$ *and* $\tau \geq 0$.
Consider now the nonautonomous Cauchy problem $(NCP)_{s,f}$ *for the family* $(A(t), D(A(t)))$. *It has a (classical) solution only for*

$$f \in Y_s := \cap_{r \in [s,1]} C_r^1 \supseteq C_0^1(\mathbb{R}^2)$$

given by the evolution family

$$\begin{aligned} [U(t,s)f](x,y) &= \left[e^{\int_s^t A(\tau)\, d\tau} f\right](x,y) \\ &= \left[e^{\frac{1}{2}[(1-s)^2 - (1-t)^2]A(0)} e^{\frac{1}{2}(t-s)^2 A(1)} f\right](x,y) \\ &= f\left(x + \frac{1}{2}[(1-s)^2 - (1-t)^2], y + \frac{1}{2}(t-s)^2\right), \quad t \geq s \in [0,1]. \end{aligned}$$

It follows that $U(t,s)Y_s \subseteq Y_t$, *but* $U(t,s)D(A(s)) = U(t,s)C_s^1 = C_s^1 \not\subseteq C_t^1 = D(A(t))$ *for* $s \neq t$.

Example 3.6. *Consider a Banach space X, a bounded operator $B \in \mathcal{L}(X)$, an unbounded generator $(A, D(A))$, and a continuous function $a : \mathbb{R} \to \mathbb{R}_+$ with $a(t) = 0$ for $t \leq 0$ and $a(t) > 0$ for $t > 0$. The nonautonomous Cauchy problem for the family $(a(t)A + B)_{t\in\mathbb{R}}$ has classical solutions on the regularity subspaces $Y_t = D(A)$ for $t \geq 0$ and $Y_t = \{x \in X : e^{-tB}x \in D(A)\}$ for $t \leq 0$, which, in general, can be a quite arbitrary subspace of X.*

Moreover, for a well-posed (NCP) the intersection of the domains of the corresponding operators can be trivial. Examples for this phenomenon are given in [9], but the following is even simpler.

Example 3.7 ([7], Ex. 7.3.2.). *Consider the space $X := L^p(\mathbb{R})$ and the family of pairwise commuting multiplication operators $A(t)f(\cdot) := \frac{-f(\cdot)}{(\cdot - t)^2}$ on their natural domain. The intersection of the domains is trivial $\bigcap_{t\in\mathbb{R}} D(A(t)) = \{0\}$.*

While these examples already show some of the complicated behaviour of the solutions to (NCP), our definitions allow the characterization of well-posedness by the existence of a strongly continuous evolution family (see [12], Prop. 1.4).

Proposition 3.8. *Let $(A(t), D(A(t)))_{t\in\mathbb{R}}$ a family of linear operators on a Banach space X and consider the corresponding nonautonomous Cauchy problem (NCP). The following assertions are equivalent.*

(i) *(NCP) is well posed (with exponentially bounded solutions).*
(ii) *There exists a unique strongly continuous (exponentially bounded) evolution family $(U(t,s))_{t\geq s}$ solving (NCP).*

4. A simple class of "strange" evolution families

While the semigroup law "$T(t)T(s) = T(t+s)$" implies exponential boundedness and differentiability on a dense subspace, see [5], I.5. and II.1., the weaker algebraic relation in Definition 3.1 (i) has no such consequences.

Example 4.1. *Consider the evolution family $u(t,s) := \exp(t^2 - s^2)$ on $X := \mathbb{C}$. This family satisfies (i) and (ii), but not (iii) in Definition* 3.1.

Similarly, differentiability already fails in the one-dimensional case.

Example 4.2. *Take a continuous function $p : \mathbb{R} \to [1,2]$ and define $u(t,s) := \frac{p(t)}{p(s)}$ for $s \leq t$. Then $(u(t,s))_{t\geq s}$ is a (uniformly) continuous evolution family on $\mathbb{C}$ which is not differentiable if we choose the function p to be nowhere differentiable.*

Clearly, the idea of this example can be extended to infinite dimensional spaces and yields a class of evolution families which can be characterized by invertibility. An easy calculation shows the following algebraic property of invertible evolution families.

Lemma 4.3. *An evolution family $(U(t,s))_{t\geq s}$ consists of invertible operators if and only if there exists a family of invertible bounded operators $(Q(t))_{t\in\mathbb{R}}$ such that*

$$U(t,s) = Q(t)Q(s)^{-1} \text{ for } t,s \in \mathbb{R}. \tag{4.3}$$

Moreover, by setting $U(s,t) := U(t,s)^{-1}$ for $s < t$, the evolution family can be extended to an evolution family $(U(t,s))_{(t,s)\in\mathbb{R}^2}$.

Proof. It is easy to see that the definition $U(s,t) := U(t,s)^{-1}$ for $s < t$ extends $(U(t,s))_{t\geq s}$ to an evolution family on $\mathbb{R}^2$ and $U(t,s) = U(t,0)U(0,s)$ for all $t,s \in \mathbb{R}$. Thus $Q(t) := U(t,0)$ and $Q^{-1}(t) = U(0,t)$ fulfill equation (4.3). □

For the continuity this implies a surprisingly easy characterization compared with Proposition 3.2.

Proposition 4.4. *For an evolution family $(U(t,s))_{(t,s)\in\mathbb{R}^2}$ satisfying the condition* (i) *of Definition* 3.1 *and a dense subspace $D \subset X$ the following are equivalent.*

(i) $\mathbb{R}^2 \ni (t,s) \mapsto U(t,s)x$ *is continuous for all* $x \in X$,

(ii) (a) $\lim_{s\to t} U(t,s)x = x$ *and* (b) $\lim_{t\to s} U(t,s)x = x$ *for all* $x \in X$,

(iii) (a) $\lim_{s\to t} U(t,s)x = x$ *and* (b) $\lim_{t\to s} U(t,s)x = x$ *for all* $x \in D$ *and* $U(t,s)$ *is uniformly bounded on compact subsets of* $\mathbb{R}^2$.

Proof. The implications $(i) \Rightarrow (iii) \Rightarrow (ii)$ are clear, thus, it remains to show $(ii) \Rightarrow (i)$. By Lemma 4.3 we have $U(t,s) = Q(t)Q^{-1}(s) := U(t,0)U(0,s)$. Consider thus for $t \in \mathbb{R}$, $h \in \mathbb{R}$, and $x \in X$

$$Q(t+h)x - Q(t)x = U(t+h,0)x - U(t,0)x = [U(t+h,t) - Id]U(t,0)x,$$

which converges to 0 for $h \to 0$ by property (b),

$$Q^{-1}(t+h)x - Q^{-1}(t)x = U(0,t+h)x - U(0,t)x = U(0,t)[U(t,t+h) - Id]x$$

which converges to 0 for $h \to 0$ by property (a).

The family $(U(t,s))_{(t,s)\in\mathbb{R}^2}$ is thus strongly continuous as a product of two strongly continuous families. □

Putting these two together we obtain a characterization of strongly continuous, invertible evolution families.

Proposition 4.5. *The strongly continuous evolution family $(U(t,s))_{t\geq s}$ consists of invertible operators on a Banach space X satisfying* (a) $\lim_{s\nearrow t} U(t,s)^{-1}x = x$ *and* (b) $\lim_{t\searrow s} U(t,s)^{-1}x = x$ *for every $x \in X$ and $t \leq s$ if and only if there exists a strongly continuous family of invertible bounded operators $(Q(t))_{t\in\mathbb{R}}$ with strongly continuous inverse such that*

$$U(t,s) = Q(t)Q(s)^{-1} \text{ for } t,s \in \mathbb{R}. \tag{4.4}$$

Moreover, the evolution family can be extended to a strongly continuous evolution family $(U(t,s))_{(t,s)\in\mathbb{R}^2}$ by setting $U(s,t) := U(t,s)^{-1}$ for $s < t$. The family Q can then be given by $Q(t) := U(t,0)$.

This can be applied to norm continuous evolution families.

Theorem 4.6 (Norm continuous EVF). *The family $(U(t,s))_{t\geq s}$ is a (uniformly) norm continuous evolution family if and only if there exists a (uniformly) norm continuous, invertible family $Q(t)$ with $U(t,s) = Q(t)Q(s)^{-1}$.*

Proof. It sufficies to show that $(U(t,s))_{t\geq s}$ consists of invertible operators. Consider $U(T,S)$ with $T \geq S \in \mathbb{R}$. On the compact set

$$D := \{(t,s) \,:\, S \leq s \leq t \leq T\} \subset \Delta$$

the evolution family is uniformly norm continuous. Therefore $\|U(t,s) - Id\| < 1$ for all $(t,s) \in D$ with $(t-s) < \delta$ and some sufficiently small $\delta > 0$. We thus infer that $U(t,s)$ is invertible for all $(t,s) \in D$ with $(t-s) < \delta$. The difference $(T-S)$ can now be decomposed into pieces of lenght $< \delta$, i.e., we can find a sequence $S = t_0 \leq t_1 \leq t_2 \leq \cdots \leq t_n = T$ such that $(t_{i+1} - t_i) < \delta$. We then have

$$U(T,S) = U(T,t_{n-1})U(t_{n-1},t_{n-2}) \cdots U(t_2,t_1)U(t_1,S),$$

and all the operators on the right-hand side are invertible. □

Remark 4.7. *In* [17], *Corollary* 3.4 *and Theorem* 4.2, *a similar conclusion has been obtained using evolution semigroup techniques.*

Example 4.8. *The same idea allows to construct a weakly continuous, but not strongly continuous evolution family, which is, again, in contrast to the autonomous situation (see* [5], *Thm.* I.5.8*).*

5. Strange nonautonomous ACPs

In this section we show that the concept of well-posedness, as defined in Definition 2.2, is too narrow. To that purpose we present first a paradigmatic (NCP) having "generalized" solutions, while not being well posed.

Example 5.1. *Consider the family $(B(t), D(B(t)))_{t\in\mathbb{R}}$ defined as*

$$B(t) := a(t)A + b(t)B,$$

where $(A, D(A))$ and $(B, D(B))$ are generators of semigroups $(e^{tA})_{t\geq 0}$ and $(e^{tB})_{t\geq 0}$ with $D(A) \cap D(B) = \{0\}$, $a \geq 0$ and $b \geq 0$ are continuous functions with $a(t) = 0$ for $t \geq 0$ and $b(t) = 0$ for $t \leq 0$. As domain of the operator $B(t)$ we take

$$D(B(t)) := \begin{cases} D(A) & \text{for } t < 0, \\ X & \text{for } t = 0, \\ D(B) & \text{for } t > 0. \end{cases}$$

For all $x \in X$, the expression

$$U(t,s)x := e^{\int_s^t b(r)\,dr B} e^{\int_s^t a(r)\,dr A} x$$

is a classical solution of (NCP) as long as $s \leq t \leq 0$ and $x \in D(A)$ or $t \geq s \geq 0$ and $x \in D(B)$. It should be regarded as a mild solution for every $s \leq t \in \mathbb{R}$, although for $s \leq 0$ there is no $x \neq 0$ such that $t \mapsto U(t,s)x$ is classical in the sense of Definition 2.1 *for $t > 0$.*

Moreover, the above well-posedness is not stable under bounded perturbations. This follows from Example 3.6 and can be restated as follows.

Example 5.2. *If* (NCP) *for* $(A(t))$ *is well posed and* $(B(t)) \subset \mathcal{L}(X)$ *is uniformly bounded, then* (NCP) *for* $(A(t) + B(t))$ *is, in general, not well posed.*

If $(T(t))_{t\geq 0}$ solves an (ACP) for an operator $(A, D(A))$, then $(T(\alpha t))_{t\geq 0}$ solves the (ACP) for $(\alpha A, D(A))$ for every $\alpha > 0$. This fails in the nonautonomous situation. This phenomenon is connected to "blow up", which can happen already if all $A(t)$ are generators of pairwise commuting semigroups.

Example 5.3. *There exists a family* $(A(s), D(A(s)))_{0\leq s\leq 1}$ *of generators of commuting* $\mathcal{C}_0$*-semigroups* $((e^{\tau A(s)})_{\tau\geq 0})_{0\leq s\leq 1}$ *with* $\|e^{\tau A(s)}\| \leq 2$ *on a Banach space* X *such that* (NCP) *corresponding to* $(A(s))_{0\leq s\leq 1}$ *is not well posed. More precisely, there exists an evolution family* $(U(t,s))_{0\leq s\leq t<1}$ *solving the Cauchy problem* (NCP) *corresponding to* $(A(s))_{0\leq s<1}$ *, however,* $\|U(s_m, 0)\| = 2^m$ *for a sequence* $s_m \to 1$. *However, for the operators*

$$C(s) := 2A(s) \text{ with } D(C(s)) = D(A(s))$$

(NCP) *becomes well posed (see* [20]*).*

Finally, we present an example of a well-posed (NCP), where the right-hand side operators are even not closable (compare [10]).

Example 5.4. *On the space* $X := C_b(\mathbb{R})$ *consider the* (NCP) *given by the family of operators* $(A(t), D(A(t)))_{t\in\mathbb{R}}$ *defined by* $A(t)f := f'(t)\mathbf{1}$ *for* $f \in D(A(t)) := \{f \in X : f'(t) \text{ exists}\}$. *It is well posed with regularity subspace* $Y_t := \{f \in C_b(\mathbb{R}) : f|_{[t,\infty)} \in C^1\}$, *but* $A(t)$ *is not closable on* X. *The solutions of* (NCP) *are given by* $U(t,s)f = (f(t) - f(s))\mathbf{1}$, $t \geq s$.

Remark 5.5. *The evolution family* $(U(t,s))_{t\geq s}$ *from Example* 5.4 *above consists of invertible operators. However, if we suppose* $U(t,s)f = (f(t) - f(s))\mathbf{1}$ *to be a classical solution of* (NCP) *for all* $t, s \in \mathbb{R}$, *we have to take* $f \in Y_s := C_b^1(\mathbb{R})$. *We thus loose regularity.*

These examples, among many others, indicate the need for a concept of **mild solutions** for (NCP). The following construction of an associated "evolution semigroup" may lead to this goal.

6. Evolution semigroups

The point of departure in this section is an exponentially bounded, strongly continuous evolution family $(U(t,s))_{t\geq s}$, to which we can associate a C_0-semigroup called **evolution semigroup** (first introduced in 1974 by Howland [11], see also the monography [3] and [5], Chapter VI.9). To that purpose, we choose the Banach space

$$\mathcal{C}_0 := C_0(\mathbb{R}, X) = \{f : \mathbb{R} \to X \ : \ f \text{ is continuous and } \lim_{|t|\to\infty} f(t) = 0\}$$

normed by

$$\|f\| := \sup_{t\in\mathbb{R}} \|f(t)\|, \quad f \in \mathcal{C}_0.$$

Definition 6.1 (Evolution semigroup). *For every strongly continuous, exponentially bounded evolution family $(U(t,s))_{t\geq s}$ we define the corresponding (strongly continuous)* **evolution semigroup** $(\mathcal{T}(t))_{t\geq 0}$ *on the space $\mathcal{C}_0$ by*

$$(\mathcal{T}(t)f)(s) := U(s,s-t)f(s-t) = (U(\cdot,\cdot-t)\mathcal{T}r(t)f)(s) \tag{6.5}$$

for $f \in \mathcal{C}_0$, $s \in \mathbb{R}$ and $t \geq 0$ (here $(\mathcal{T}r(t))$ denotes the left translation semigroup on $\mathcal{C}_0$). We denote its generator by $(\mathcal{G}, D(\mathcal{G}))$.

From an evolution semigroup $(\mathcal{T}(t))_{t\geq 0}$ we can recover the corresponding evolution family by $U(t,s)x = (\mathcal{T}(t-s)f)(t)$ for any $f \in \mathcal{C}_0$ with $f(s) = x$.

As a first example how evolution semigroups, hence semigroup techniques can be used for evolution families, we present a characterization of invertibility for evolution families.

Proposition 6.2. *For a strongly continuous, exponentially bounded evolution family $(U(t,s))_{t\geq s}$ the following statements are equivalent:*

(i) *The evolution family $(U(t,s))$ can be extended to a strongly continuous evolution family $(U(t,s))_{(t,s)\in\mathbb{R}^2}$.*

(ii) *The operators $U(s,s-c)$ are invertible for all $s \in \mathbb{R}$ and some fixed $c > 0$.*

Proof. Consider the corresponding evolution semigroup $(\mathcal{T}(t))_{t\geq 0}$ and observe that *(ii)* implies that the operator $\mathcal{T}(c) = U(\cdot,\cdot-c)\mathcal{T}r(c)$ is invertible. This, however, implies that $(\mathcal{T}(t))_{t\geq 0}$ can be extended to a strongly continuous group $(\mathcal{T}(t))_{t\in\mathbb{R}}$ (see [5], Exercise I.5.9.6). By the characterization of evolution semigroups, Theorem 4.2. in [23] we infer that also $(\mathcal{T}(t))_{t\leq 0}$ is an evolution semigroup. Again, by an easy calculation, we infer that the corresponding evolution family extends $(U(t,s))$ to $\mathbb{R}^2$. □

In general, no differentiability property holds for a strongly continuous evolution family, while the evolution semigroup clearly is differentiable on the domain of its generator. Differentiability of the evolution family, thus well-posedness of a corresponding (NCP), will now be characterized by properties of the generator $(\mathcal{G}, D(\mathcal{G}))$ of $(\mathcal{T}(t))_{t\geq 0}$ (see [19]). To that purpose, we define the operator $(\mathcal{A}, D(\mathcal{A}))$ on the space $\mathcal{C}_0$ by

$$\begin{aligned} D(\mathcal{A}) &:= \{f \in \mathcal{C}_0 : \lim_{t\searrow 0} \frac{[U(\cdot,\cdot-t) - Id]}{t} f(\cdot) \in \mathcal{C}_0\}, \\ \mathcal{A}f &:= s \mapsto \lim_{t\searrow 0} \frac{[U(s,s-t) - Id]}{t} f(s). \end{aligned}$$

Roughly speaking, (NCP) is well posed if there are sufficiently many orbits of $(\mathcal{T}(t))_{t\geq 0}$ staying in the intersection of $D(\mathcal{G}))$ with the space of differentiable functions $\mathcal{C}^1 := \{f \in \mathcal{C}_0 : f' \in \mathcal{C}_0\}$.

Theorem 6.3 ([19], Theorem 2.9). *Let X be a Banach space and let $(A(t), D(A(t)))_{t\in\mathbb{R}}$ be a family of linear operators on X. The following assertions are equivalent.*

(i) *The nonautonomous Cauchy problem (NCP) for the family $(A(t))_{t\in\mathbb{R}}$ is well posed (with exponentially bounded solutions).*

(ii) *There exists a unique evolution semigroup $(\mathcal{T}(t))_{t\geq 0}$ with generator $(\mathcal{G}, D(\mathcal{G}))$ and an invariant core $\mathcal{D} \subseteq \mathcal{C}^1 \cap D(\mathcal{G})$ such that*

$$\mathcal{G}f + f' = \mathcal{A}(\cdot)f = A(\cdot)f$$

for $f \in \mathcal{D}$.

It is possible to obtain well-posedness using the above characterization and semigroup methods. In the "parabolic" case (by perturbation methods) or the "hyperbolic" case (by Trotter-Kato approximation methods) case this is done in [19] or [18], respectively. The above theorem shows the need for a precise description of the domains of the generator of evolution semigroups. Since this is still not completely understood, we discuss some special cases. For the definition of the extrapolation space X_{-1}^A see [5], Section II.5.

Proposition 6.4 (The autonomous situation, [16]). *Let $(A, D(A))$ be the generator of a C_0-semigroup $(T(t))_{t\geq 0}$ on the Banach space X and define*

$$\mathcal{A}f := Af(\cdot) \text{ for } f \in D(\mathcal{A}) := \{f \in \mathcal{C}_0 \,:\, f(s) \in D(A) \text{ and } s \mapsto Af(s) \in \mathcal{C}_0\}.$$

The generator $\mathcal{G}$ of the evolution semigroup

$$(\mathcal{T}(t)f)(s) := e^{tA}f(s-t), \qquad s \in \mathbb{R}, \quad t \geq 0,$$

on the space $C_0(\mathbb{R}, X)$ is given by

$$(\mathcal{G}f)(s) = -f'(s) + A_{-1}f(s), \quad s \in \mathbb{R},$$

for all f in the domain

$$D(\mathcal{G}) = \{f \in C_0(\mathbb{R}, X) \cap C_0^1(\mathbb{R}, X_{-1}^A) : -f' + \mathcal{A}_{-1}f \in C_0(\mathbb{R}, X)\}.$$

Remark 6.5. *An invariant core for the generator $\mathcal{G}$, as required in Theorem 6.3, is given by the space $\mathcal{C}^1 \cap D(\mathcal{A})$. This simply reflects the fact that the Cauchy problem for the generator $(A, D(A))$ has classical solutions for initial values $x \in D(A)$, see* [16].

Under strong assumptions on the generators $A(t)$ an invariant core in the domain $D(\mathcal{G})$ has been found in [19]. The well-posedness result can be found in [25] and [24].

Proposition 6.6 (Parabolic case). *Assume the following conditions on the operators $(A(t))_{t\in\mathbb{R}}$.*

(P1) *The domain* $D := D(A(t))$ *is dense in* X *and independent of* $t \in \mathbb{R}$.

(P2) *Each operator* $A(t)$ *is the generator of an analytic semigroup with resolvent* $R(\lambda, A(t))$ *satisfying*

$$\|R(\lambda, A(t))\| \leq \frac{M}{|\lambda| + 1}$$

for all $Re\,\lambda \geq 0$, $t \in \mathbb{R}$ *and some constant* M.

(P3) *There exist constants* $L \geq 0$ *and* $0 < \alpha \leq 1$ *such that*

$$\|(A(t) - A(s))A(0)^{-1}\| \leq L|t - s|^{\alpha} \text{ for all } t, s \in \mathbb{R}. \tag{6.6}$$

Then the corresponding (NCP) *is well posed and yields an evolution semigroup* $(\mathcal{T}(t))_{t\geq 0}$ *on* $\mathcal{C}_0$ *(see [19]). Moreover, the Banach space* $\mathcal{X} := \mathcal{C}^1 \cap \mathcal{C}_0(\mathbb{R}, D)$ *normed by* $\|f\|_{\mathcal{X}} := \|f'\| + \sup_{s\in\mathbb{R}} \|A(s)f(s)\|$ *is invariant under* $(\mathcal{T}(t))$ *and the restriction* $(\mathcal{T}_|(t))_{t\geq 0}$ *is strongly continuous. So, for the generator* $(\mathcal{G}_|, D(\mathcal{G}_|))$ *of* $(\mathcal{T}_|(t))_{t\geq 0}$ *we obtain*

$$\begin{aligned} D(\mathcal{G}_|) &= \{f \in \mathcal{X} : A(\cdot)f - f' \in \mathcal{X}\}, \\ \mathcal{G}_| f &= A(\cdot)f - f'. \end{aligned}$$

In a parabolic L^p-setting we can use maximal regularity results in order to obtain the domain of the generator simply as the intersection of the domains of the first derivative and the multiplication operator $A(\cdot)$. As an example we quote the following result from [15].

Theorem 6.7 ([15], Theorem 2). *Let* Y *be a Banach space of class* $\mathcal{HT}$, $(A(t))_{t\geq 0}$ *a family of closed, densely defined operators in* Y *fulfilling the assumptions below.*

(i) *Each operator* $A(t)$ *admits bounded imaginary powers on* Y, *and there exist constants* $M_A, K_A > 0$, *and* $\phi_A \in (0, \pi)$ *such that*

$$\|R(\lambda, -A(t))\| \leq \frac{M_A}{1 + \lambda}$$

for all $t \geq 0$, $\lambda \in \Sigma_{\pi - \phi_A}$ *and*

$$\|A(t)^{is}\| \leq K_A e^{|s|\phi_A}$$

for all $t \geq 0$, $s \in \mathbb{R}$.

(ii) *There exist constants* $0 \leq \alpha < \delta \leq 1$ *and* $M_1 > 0$ *such that*

$$\|A(t)R(\lambda, -A(t))[A(t)^{-1} - A(s)^{-1}]\| \leq \frac{M_1|t - s|^{\delta}}{1 + |\lambda|^{1-\alpha}}$$

for all $t, s \geq 0$, *and* $\lambda \in \Sigma_{\pi - \phi_A}$.

If $M_1 > 0$ *is sufficiently small, then the operator* $(\mathcal{G}, D(\mathcal{G}))$ *on the space* $L^p(\mathbb{R}_+, Y)$ *defined by*

$$\begin{aligned} D(\mathcal{G}) &:= H_0^{1,p}(\mathbb{R}_+, Y) \cap D(A(\cdot)), \\ \mathcal{G}f &:= A(\cdot)f - f' \end{aligned}$$

is the generator of the corresponding evolution semigroup on the space $L^p(\mathbb{R}_+, Y)$.

We can now characterize invertible evolution families by properties of the generator of the corresponding evolution semigroup.

Proposition 6.8. *For a strongly continuous evolution family $(U(t,s))_{t\geq s}$ and the corresponding evolution semigroup $(\mathcal{T}(t))_{t\geq 0}$ with generator $(\mathcal{G}, D(\mathcal{G}))$ the following statements are equivalent.*

(a) *The evolution family $(U(t,s))_{t\geq s}$ consists of invertible operators and $\|U(t,s)\| \leq Me^{\omega(t-s)}$ for some $\omega \in \mathbb{R}$ and all $t, s \in \mathbb{R}$.*

(b) *The evolution semigroup $(\mathcal{T}(t))_{t\geq 0}$ is similar to the left translation semigroup with a similarity transform $\mathcal{G} = Q(\cdot)$ given by a (strongly continuous) multiplication operator on $\mathcal{C}_0$ fulfilling $\|Q(t)\| \leq Me^{\omega t}$, $\|Q(t)^{-1}\| \leq Me^{-\omega t}$.*

(c) *The domain of the generator $(\mathcal{G}, D(\mathcal{G}))$ is given by*

$$D(\mathcal{G}) = \{f \in \mathcal{C}_0 : Q(\cdot)f \in \mathcal{C}^1\}.$$

for a (strongly continuous) multiplication operator on $\mathcal{C}_0$ fulfilling $\|Q(t)\| \leq Me^{\omega t}$, $\|Q(t)^{-1}\| \leq Me^{-\omega t}$.

Proof. $(a) \Rightarrow (b), (c)$ From the invertibility we infer, by Lemma 4.3, that $U(t,s) = U(t,0)U(0,s)$ for all $t, s \in \mathbb{R}$. Thus we can define $\mathcal{G} = Q(\cdot) := e^{-\omega \cdot}U(\cdot, 0)$. By the exponential boundedness of $U(t,s)$ we obtain $\|Q(t)\| \leq M$ and $\|Q^{-1}(t)\| \leq M$ for every $t \in \mathbb{R}$, thus $\mathcal{G}, \mathcal{G}^{-1} \in \mathcal{L}(\mathcal{C}_0)$. A straight forward calculation shows $\mathcal{G}^{-1}\mathcal{T}(t)\mathcal{G} = \mathcal{T}r(t)$, and the assertion for the domain of the generator is standard semigroup theory.

The converse implications are equally easy to show. □

This can be applied to differentiable evolution families (see [17], Theorem 3.3 and Corollary 3.6; the statement (c) follows from the bounded perturbation theorem for semigroups).

Corollary 6.9 (C^1 case). *For a strongly continuous evolution family $(U(t,s))_{t\geq s}$ and the corresponding evolution semigroup $(\mathcal{T}(t))_{t\geq 0}$ with generator $(\mathcal{G}, D(\mathcal{G}))$ the following statements are equivalent.*

(a) *The evolution family $(U(t,s))_{(t,s)\in\mathbb{R}^2}$ is strongly continuously differentiable and $\|U(t,s)\| \leq Me^{\omega(t-s)}$ for some $\omega \in \mathbb{R}$ and all $t, s \in \mathbb{R}$.*

(b) *The domain of the generator of the corresponding evolution semigroup is $\mathcal{C}^1$.*

(c) *The evolution family $(U(t,s))_{t\geq s}$ solves (NCP) for a strongly continuous, uniformly bounded family operators $(A(t) \in \mathcal{L}(X)$.*

Another application is given for Lipschitz continuous evolution families, see [17], Corollary 4.4. Then, the similarity transformation becomes almost everywhere differentiable, if X has the Radon-Nikkodym property and we obtain an a.e.-solution of (NCP).

Corollary 6.10 (Lipschitz case). *Let X be a separable Banach space having the Radon Nikodym property, and let $(U(t,s))_{t\geq s}$ be a Lipschitz continuous evolution family on X. Then there exists an operator valued function $A(\cdot) \in L^\infty(\mathbb{R}, \mathcal{L}_s(X))$*

such that the generator $(\mathcal{G}, D(\mathcal{G}))$ *of the corresponding evolution semigroup* $(\mathcal{T}(t))_{t\geq 0}$ *is given by*

$$\begin{aligned} D(\mathcal{G}) &= \{f \in \mathcal{C}_0 : f \text{ is a.e. differentiable and } -f' + A(\cdot)f \in \mathcal{C}_0\} \\ \mathcal{G}f &= -f' + A(\cdot)f \end{aligned}$$

for every $f \in D(\mathcal{G})$.

References

[1] W. Arendt, C.J.K. Batty, M. Hieber, F. Neubrander: *Vector-valued Laplace Transforms and Cauchy Problems.* Birkhäuser Verlag, 2001.

[2] P. Acquistapace, B. Terreni: *A unified approach to abstract linear nonautonomous parabolic equations.* Rend. Sem. Univ. Padova **78** (1987), 47–107.

[3] C. Chicone, Y. Latushkin: Evolution Semigroups in Dynamical Systems and Differential Equations. Amer. Math. Soc., 1999.

[4] G. DaPrato, P. Grisvard: *Sommes d'opérateurs linéaires et équations différentielles opérationelles.* J. math. pures appl. **54** (1975), 305–387.

[5] K.-J. Engel, R. Nagel: One-Parameter Semigroups for Linear Evolution Equations. Springer-Verlag, 2000.

[6] D.E. Evans: *Time dependent perturbations and scattering of strongly continuous groups on Banach space.* Math. Ann. **221** (1976), 275–290.

[7] H. O. Fattorini: "The Cauchy Problem." Addison Wesley, 1983.

[8] J. Goldstein: *Abstract evolution equations.* Trans. Amer. Math. Soc. **141** (1969), 159–184.

[9] J. Goldstein: *On the absence of necessary conditions for linear evolution operators.* Amer. Math. Soc. **64** (1977), 77–80.

[10] J. Hahn: "Nichtautonome hyperbolische Cauchyprobleme." Diplomarbeit, Tübingen 1995.

[11] J.S. Howland: *Stationary scattering theory for time-dependent Hamiltonians.* Math. Ann. **207** (1974), 315–335.

[12] H. Kellermann: *Linear evolution equations with time-dependent domain.* Semesterbericht Funktionalanalysis, 1985.

[13] Y. Latushkin, S. Montgomery-Smith, and T. Randolph: *Evolutionary semigroups and dichotomies of linear skew-product flows on locally compact spaces with Banach fibers.* J. Diff. Equations **125** (1996), 73–116.

[14] Y. Latushkin, T. W. Randolph: *Dichotomy of differential equations on Banach spaces and an algebra of weighted translation operators.* Integr. Eqns. Oper. Th. **23** (1995), 472–500.

[15] S Moniaux, J. Prüss: *A theorem of the Dore-Venni type for non-commuting operators.* Tran. Amer. Math. Soc. **349** (1997), 4787–4814.

[16] R. Nagel, G. Nickel, S. Romanelli: *Identification of extrapolation spaces for unbounded operators.* Quaestiones Math. **19** (1996), 83–100.

[17] R. Nagel, A. Rhandi: *A characterization of Lipschitz continuous evolution families on Banach spaces.* Operator Theory: Advances and Application **75**, Birkhäuser Verlag, Basel 1995, 275–288.

[18] G. Nickel: "On Evolution Semigroups and Wellposedness of Nonautonomous Cauchy Problems." PhD Thesis, Tübingen, 1996.

[19] G. Nickel: *Evolution semigroups for nonautonomous Cauchy problems.* Abstract Appl. Analysis **2** (1997), 73–95.

[20] G. Nickel, R. Schnaubelt: *An extension of Kato's stability condition for nonautonomous Cauchy problems.* Taiwanese J. Math. **2** (1998), 483–496.

[21] F. Räbiger, A. Rhandi, R. Schnaubelt, J. Voigt: *Non-autonomous Miyadera perturbations.* Differential Integral Equations **13** (1999), 341–368.

[22] F. Räbiger, R. Schnaubelt: *The spectral mapping theorem for evolution semigroups on spaces of vector valued functions.* Semigroup Forum **52** (1996), 225–239.

[23] R. Schnaubelt: *Well-posedness and asymptotic behaviour of non-autonomous evolution equations.* 2001.

[24] P. E. Sobolevskii: *Equations of parabolic type in a Banach space.* Amer. Math. Soc. Transl. **49** (1966), 1–62.

[25] H. Tanabe: *Remarks on the equations of evolution in a Banach space.* Osaka Math. J. **112** (1960), 145–166.

Rainer Nagel, Gregor Nickel
Arbeitsbereich Funktionalanalysis
Mathematisches Institut
Universität Tübingen
Auf der Morgenstelle 10
D-72076 Tübingen
Germany
e-mail: `rana@michelangelo.mathematik.uni-tuebingen.de`
e-mail: `grni@michelangelo.mathematik.uni-tuebingen.de`

Progress in Nonlinear Differential Equations
and Their Applications, Vol. 50, 295–309

The Cauchy Problem for a Semilinear Heat Equation with Singular Initial Data

Bernhard Ruf and Elide Terraneo

In memory of Brunello Terreni

We consider the Cauchy problem for the semilinear heat equation

$$\begin{cases} \partial_t u = \Delta u + f(u) \\ u(0) = u_0 \end{cases}$$

where $u(t,x) : \mathbb{R}^+ \times \mathbb{R}^n \to \mathbb{R}$, and $f \in \mathcal{C}^1(\mathbb{R},\mathbb{R})$ is a given function with $f(0) = 0$. It is well known that if the initial data u_0 belong to $\mathrm{L}^\infty(\mathbb{R}^n)$ then there exist $T(u_0) > 0$ and a unique solution $u \in C([0,T[, \mathrm{L}^\infty(\mathbb{R}^n))$. In this paper we will consider initial data u_0 which do not belong to $\mathrm{L}^\infty(\mathbb{R}^n)$. At first we will review the known results about this problem; then we will present some new results concerning nonlinearities of exponential growth.

The first works in this direction are due to Weissler [24],[25], who mainly considers the case of polynomial nonlinearities $f(u) = |u|^\alpha u$, $\alpha > 0$. He studies the Cauchy problem

$$\begin{cases} \partial_t u = \Delta u + |u|^\alpha u \\ u(0) = u_0 \end{cases} \tag{1}$$

with initial data $u_0 \in \mathrm{L}^p(\mathbb{R}^n)$, with $1 \le p < +\infty$.

An easy computation shows that if the function $u(t,x)$ satisfies equation (1), then also $\lambda^{2/\alpha} u(\lambda^2 t, \lambda x)$, $\lambda > 0$ does. The only Lebesgue space invariant under this scaling (with respect to the space variables) is $\mathrm{L}^{p_0}(\mathbb{R}^n)$, with $p_0 = \frac{n\alpha}{2}$. As Weissler observed, the Lebesgue space $\mathrm{L}^{p_0}(\mathbb{R}^n)$ plays the role of "critical space" with respect to the well-posedness of the Cauchy problem (1).

Indeed, for $p \ge p_0$ the problem (1) is locally well posed, namely for every $u_0 \in \mathrm{L}^p(\mathbb{R}^n)$ there exists $T = T(u_0) > 0$ and a unique solution $u \in C([0,T[, \mathrm{L}^p(\mathbb{R}^n))$. On the other hand, there are indications that for $p < p_0$ there exists no solution in any reasonable weak sense [3], [24]. Moreover, it is known that in this case uniqueness is lost for the initial data $u_0 = 0$, see [10]. In order to establish existence, Weissler rewrites, as it is now standard to do, the differential equation (1) in the integral formulation

$$u(t) = \mathrm{e}^{t\Delta} u_0 + \int_0^t \mathrm{e}^{(t-s)\Delta} |u|^\alpha u(s) ds \tag{2}$$

where $e^{t\Delta}u_0(x) = \frac{1}{(4\pi t)^{\frac{n}{2}}} e^{-|\cdot|/4t} * u_0(x)$. Noting that the solutions of (2) are the fixed points of the operator

$$F : u(t) \to e^{t\Delta}u_0 + \int_0^t e^{(t-s)\Delta}|u|^\alpha u(s)ds \ ,$$

he shows that F is a contraction in $C([0,T], L^p(\mathbb{R}^n))$ for a suitable $T = T(u_0) > 0$. At the end of the paper we will clarify the context which ensures the equivalence between the equations (1) and (2).

Weissler's result can also be stated in the following way: for every $1 < p < +\infty$ there exists a critical power $\alpha_0 = \frac{2p}{n}$ such that for every $0 < \alpha \leq \alpha_0$ the problem is well posed. In particular, in the critical case $\alpha = \alpha_0$ (or $p = p_0$) it is possible to establish via a fixed point argument existence of global solutions in time for initial data with sufficiently small L^p norm. If we consider bounded initial data, then the restrictions on the growth behavior of the nonlinearity disappear: in fact, for initial data $u_0 \in L^\infty(\mathbb{R}^n)$ and for a generic nonlinearity $f(u)$, where $f \in \mathcal{C}^1(\mathbb{R}, \mathbb{R})$ with $f(0) = 0$, local existence can be proven.

The natural question which arises is the following: if we consider nonlinearities with exponential growth are there spaces of unbounded initial data in which a theory of existence still holds?

Before coming to the statement of our result let us make some comments on the relationship between the nonlinear heat equation and the nonlinear Schrödinger equation with polynomial nonlinearity $|u|^\alpha u$. Let us consider

$$\begin{cases} i\partial_t u + \Delta u = |u|^\alpha u \\ u(0) = u_0. \end{cases} \tag{3}$$

where $u(t,x) : \mathbb{R}^+ \times \mathbb{R}^n \to \mathbb{C}$ and u_0 belongs to the Sobolev space $H^s(\mathbb{R}^n)$.

In [5] Cazenave and Weissler established that for $0 \leq s < \frac{n}{2}$ there exists a critical power $\alpha_c(s) = \frac{4}{n-2s}$ such that for every $0 < \alpha \leq \alpha_c$ the Cauchy problem (3) is well posed in $H^s(\mathbb{R}^n)$. Moreover, in the critical case $\alpha = \alpha_c$ global existence for initial data with small H^s norm can be established by a fixed point algorithm. In the same paper they also remark that there is a striking relationship between the L^p-theory for the nonlinear heat equation and the H^s-theory for the nonlinear Schrödinger equation. More precisely, let $0 \leq s < \frac{n}{2}$ and let p such that $H^s(\mathbb{R}^n) \hookrightarrow L^p(\mathbb{R}^n)$ by the Sobolev embedding, i.e. $\frac{1}{p} = \frac{1}{2} - \frac{s}{n}$. Then one notes that the critical power α_0 associated with initial data in $L^p(\mathbb{R}^n)$ for the Cauchy problem (1) coincides with the critical power α_c for the Cauchy problem (3) for initial data in $H^s(\mathbb{R}^n)$: $\alpha_0 = \frac{2p}{n} = \frac{4}{n-2s} = \alpha_c$. On the other hand, if $s > \frac{n}{2}$, then $H^s(\mathbb{R}^n) \hookrightarrow L^\infty(\mathbb{R}^n)$, and local existence can be established without growth assumption on the nonlinearity. In fact for $f \in \mathcal{C}^k(\mathbb{C}, \mathbb{C})$ with $f(0) = 0$ and k the smallest integer larger or equal to s (differentiability refers to the real sense) one still has local well-posedness.

If we consider now the Schrödinger equation (3) in the space $H^{\frac{n}{2}}(\mathbb{R}^n)$, a local existence theory is avalaible for every polynomial nonlinearity $\alpha > 0$, so there is no critical power. This suggests the idea that in this case the critical nonlinearity should have a growth larger than every polynomial. In fact in a recent paper Nakamura and Ozawa [16] prove that the critical nonlinearity associated with the Sobolev space $\mathrm{H}^{\frac{n}{2}}(\mathbb{R}^n)$ is of exponential type, more precisely,

Theorem 1. *Let $n \geq 4$. Then there exists an $\varepsilon > 0$ such that for all $u_0 \in \mathrm{H}^{\frac{n}{2}}(\mathbb{R}^n)$, with $\|u_0\|_{\mathrm{H}^{\frac{n}{2}}} < \varepsilon$ there exists a global solution $u \in C\left([0,+\infty[, \mathrm{H}^{\frac{n}{2}}(\mathbb{R}^n)\right)$ of the equation*

$$i\partial_t u + \Delta u = \pm(\mathrm{e}^{\lambda|u|^2} - 1)\ , \quad \lambda > 0.$$

Remark. In fact Nakamura and Ozawa's result is more general. For even integers $n \geq 4$ it ensures existence for every nonlinearity f such that $f \in C^{\frac{n}{2}}(\mathbb{C},\mathbb{C})$, $f(0) = 0$ and there exists $\lambda > 0$ such that $|f'(z)| \leq C\mathrm{e}^{\lambda|z|^2}|z|$ and $\left|f^{(k)}(z)\right| \leq C\mathrm{e}^{\lambda|z|^2}$ for every $2 \leq k \leq \frac{n}{2}$ and every $z \in \mathbb{C}$. For example it applies to polynomial nonlinearity $|u|^\alpha u$, with $\alpha + 1 > \frac{n}{2}$ and $\alpha \geq 1$. However, it turns out that the exponential growth of type $\mathrm{e}^{\lambda u^2}$ at infinity is critical, in the sense that the same technique does not apply to larger nonlinearities. In dimensions $n = 2$ or $n \geq 1$ and odd some other technical hypothesis are needed; we will not discuss this situation for the sake of simplicity.

As it is well known, the Sobolev embeddings provide no best space in the range of Lebesgue spaces into which $\mathrm{H}^{\frac{n}{2}}(\mathbb{R}^n)$ can be embedded. In fact

$$\mathrm{H}^{\frac{n}{2}}(\mathbb{R}^n) \hookrightarrow \mathrm{L}^p(\mathbb{R}^n) \quad \text{for every } p \geq 2.$$

The best space which ensures an embedding is given by an Orlicz space [1]:

$$\mathrm{H}^{\frac{n}{2}}(\mathbb{R}^n) \hookrightarrow \mathrm{L}_{\mathrm{e}^{u^2}-1}(\mathbb{R}^n).$$

Thus, the mentioned analogy between the equations (1) and (3) suggests the following result:

Theorem 2. *There exists $\varepsilon > 0$ such that for $u_0 \in \mathrm{L}_{\mathrm{e}^{u^2}-1}$ with $\|u_0\|_{\mathrm{L}_{\mathrm{e}^{u^2}-1}} < \varepsilon$ there exist a positive constant $T = T(u_0)$ and a solution $u \in \mathrm{L}^\infty([0,T], \mathrm{L}_{\mathrm{e}^{u^2}-1})$ of the equation*

$$u(t) = \mathrm{e}^{t\Delta}u_0 + \int_0^t \mathrm{e}^{(t-s)\Delta}(\mathrm{e}^{u^2(s)} - 1)ds.$$

Remark 1. We would like to stress that here $u(t,x) : \mathbb{R}^+ \times \mathbb{R}^n \to \mathbb{R}$.

Remark 2. Uniqueness holds in the space where the fixed point argument is performed.

Remark 3. The same result can be stated for more general nonlinearities, namely for $f \in \mathcal{C}^1(\mathbb{R},\mathbb{R})$ such that $f(0) = 0$ and $|f'(u)| \leq C\mathrm{e}^{\lambda u^2}|u|$, $\lambda > 0$. So, in particular, the same result holds for $f(u) = |u|^\alpha u$, with $\alpha \geq 1$, or for $f(u) = \mathrm{e}^u - u - 1$. On the other hand, for nonlinearities of type $\mathrm{e}^{|u|^\alpha} - 1$, with $\alpha > 2$, this proof fails. The relationship with the nonlinear Schrödinger equation and the previous considerations suggest the idea that for initial data in the Orlicz space $\mathrm{L}_{\mathrm{e}^{u^2}-1}$ the critical growth is given by $e^{u^2} - 1$. We note however that, contrary to the corresponding case for the Schrödinger equation, we have not been able to prove a result of global existence for initial data with small $L_{e^{u^2}-1}$ norm.

The second author would like to thank Y. Martel for fruitful discussions on the subject and she would like to point out that Martel [14] has obtained by completely different methods existence results for the equation

$$\partial_t u = \Delta u + e^u, \tag{4}$$

with initial data u_0 such that e^{u_0} belongs to $L^{\frac{n}{2}}(\mathbb{R}^n)$.

Finally, we would like to mention that the nonlinear heat equation (4) has been widely studied from different points of view by various authors [4], [8], [18], [13], [19], [23].

1. The polynomial nonlinearity

In this section we recall the main results of existence and uniqueness of solutions u which are continuous on $[0,T]$ with values in $L^q(\mathbb{R}^n)$, $q \geq 1$. These results are mainly due to Weissler [24], [25].

Theorem 3 (Weissler). *Let $p_0 = \frac{n\alpha}{2}$ and $u_0 \in L^q(\mathbb{R}^n)$ for a fixed q such that $q > p_0$, $q \geq 1$, or $q = p_0$, $q > 1$. Then there exist $T = T(u_0) > 0$ and a unique solution u of (2) in the class of functions $C([0,T], L^q)$ satisfying*

i) $\sup_{0<t<T} t^\sigma \|u(t)\|_{L^p} < +\infty$,

ii) $\lim_{t\to 0} t^\sigma \|u(t)\|_{L^p} = 0$,

for some p, σ verifying $q(\alpha+1) \geq p \geq \alpha+1$, $p > q$ and $\sigma = \frac{n}{2}(\frac{1}{q} - \frac{1}{p})$, $0 < \sigma < \frac{1}{\alpha+1}$. Moreover, for $q \geq \alpha + 1$, $q > p_0$, uniqueness holds in the larger class of functions $C([0,T], L^q)$.

Idea of the proof. At first we observe that under the hypothesis $q > p_0$, $q \geq 1$, or $q = p_0$, $q > 1$, there exist p, σ verifying $q(\alpha + 1) \geq p \geq \alpha + 1$, $p > q$ and $\sigma = \frac{n}{2}(\frac{1}{q} - \frac{1}{p})$, $0 < \sigma < \frac{1}{\alpha+1}$. We then apply a fixed point argument. To this end we consider the space

$$X_{M,T} = \left\{ u :]0,T[\to L^p \text{ such that } \sup_{0<t<T} t^\sigma \|u(t)\|_{L^p} \leq M, \lim_{t\to 0} t^\sigma \|u(t)\|_{L^p} = 0 \right\} .$$

$X_{M,T}$ endowed with the metric $d(u,v) = \sup_{0<t<T} t^\sigma \|u(t) - v(t)\|_{L^p}$ is a complete metric space (it is a closed ball of a Banach space). We remark that the linear term $e^{t\Delta}u_0$ belongs to $X_{M,T}$. In fact via the Young inequality we have that $\forall p \geq q \geq 1$ and $\sigma = \frac{n}{2}(\frac{1}{q} - \frac{1}{p})$ there exists a constant $C > 0$ such that

$$\sup_{0<t<T} t^\sigma \left\|e^{t\Delta}u_0\right\|_{L^p} \leq C \|u_0\|_{L^q} \quad \text{for every } u_0 \in L^q(\mathbb{R}^n) \text{ and } T \in]0, +\infty].$$

Moreover, if $p > q$ then $\lim_{t\to 0} \left\|e^{t\Delta}u_0\right\|_{L^p} = 0$ (see for instance [3]), so for T small enough the linear term $e^{t\Delta}u_0$ belongs to $X_{M,T}$.

We now prove that F maps $X_{M,T}$ into itself and that it is a strict contraction. For every $u, v \in X_{M,T}$ we have by applying the Hölder and Young inequalities

$$\|F(u)(t)\|_{\mathrm{L}^p} \leq \sup_{0<t<T} t^\sigma \left\|\mathrm{e}^{t\Delta} u_0\right\|_{\mathrm{L}^p} + C \sup_{0<t<T} t^\sigma \left\| \int_0^t \mathrm{e}^{(t-s)\Delta} |u|^\alpha u(s) ds \right\|_{\mathrm{L}^p}$$
$$\leq \sup_{0<t<T} t^\sigma \left\|\mathrm{e}^{t\Delta} u_0\right\|_{\mathrm{L}^p} + C \sup_{0<t<T} t^\sigma \int_0^t (t-s)^{-\frac{n\alpha}{2p}} s^{-\sigma(\alpha+1)} (s^\sigma \|u(s)\|_{\mathrm{L}^p})^{\alpha+1} ds$$
$$\leq \sup_{0<t<T} t^\sigma \left\|\mathrm{e}^{t\Delta} u_0\right\|_{\mathrm{L}^p} + CT^{1-\frac{n\alpha}{2q}} M^{\alpha+1}$$

and

$$\|F(u)(t) - F(v)(t)\|_{\mathrm{L}^p} \leq C \sup_{0<t<T} t^\sigma \left\| \int_0^t \mathrm{e}^{(t-s)\Delta} \left(|u|^\alpha u(s) - |v|^\alpha v(s)\right) ds \right\|_{\mathrm{L}^p}$$
$$\leq C \sup_{0<t<T} \int_0^t (t-s)^{-\frac{n\alpha}{2p}} \|u(s) - v(s)\|_{\mathrm{L}^p} \left(\|u(s)\|^\alpha_{\mathrm{L}^p} + \|v(s)\|^\alpha_{\mathrm{L}^p}\right) ds$$
$$\leq C2T^{1-\frac{n\alpha}{2q}} M^\alpha \sup_{0<t<T} s^\sigma \|u(t) - v(t)\|_{\mathrm{L}^p}.$$

By choosing M and T such that $C2T^{1-\frac{n\alpha}{2q}} M^\alpha < \frac{1}{2}$ and $\sup_{0<t<T} t^\sigma \left\|\mathrm{e}^{t\Delta} u_0\right\|_{\mathrm{L}^p} < \frac{M}{2}$ we obtain that F maps $X_{M,T}$ into itself and is a strict contraction. Since $X_{M,T}$ is a complete metric space we can conclude that there exists a unique fixed point in $X_{M,T}$.

We now prove that this solution belongs also to $\mathrm{L}^\infty(]0,T[,\mathrm{L}^q)$. As already observed the linear term belongs to $\mathrm{L}^\infty(]0,T[,\mathrm{L}^q)$. For the nonlinear part by applying the Hölder and Young inequalities we have

$$\left\| \int_0^t \mathrm{e}^{(t-s)\Delta} |u|^\alpha u(s) ds \right\|_{\mathrm{L}^q}$$
$$\leq C \sup_{0<s<t} (s^\sigma \|u(s)\|_{\mathrm{L}^p})^{\alpha+1} \int_0^t (t-s)^{-\frac{\alpha n}{2p}+\sigma} s^{-\sigma(\alpha+1)} ds \tag{5}$$
$$\leq Ct^{1-\frac{n\alpha}{2q}} \sup_{0<s<t} (s^\sigma \|u(s)\|_{\mathrm{L}^p})^{\alpha+1}$$

and the integral in the last line converges because $\frac{n\alpha}{2p} - \sigma < 1$, thanks to the hypothesis $p > q$ and $q \geq p_0$. The proof of the continuity is then standard and can be found in [21]. Finally the uniqueness result in $C([0,T],\mathrm{L}^q)$ for $q \geq \alpha + 1$, $q > p_0$ comes from the continuity properties of the integral operator. If u and v are two solutions of the equation (2) belonging to $C([0,T],\mathrm{L}^q)$ then

$$\sup_{0<t<T} \|u(t) - v(t)\|_{\mathrm{L}^q} \leq \sup_{0<t<T} \left\| \int_0^t \mathrm{e}^{(t-s)\Delta} (|u|^\alpha u(s) - |v|^\alpha v(s)) ds \right\|_{\mathrm{L}^q}$$
$$\leq C \sup_{0<t<T} \int_0^t (t-s)^{-\frac{n\alpha}{2q}} \|u(s) - v(s)\|_{\mathrm{L}^q} \left(\|u(s)\|^\alpha_{\mathrm{L}^q} + \|v(s)\|^\alpha_{\mathrm{L}^q}\right) ds$$
$$\leq CT^{1-\frac{n\alpha}{2q}} \sup_{0<s<T} \|u(s) - v(s)\|_{\mathrm{L}^q} \sup_{0<s<T} \left(\|u(s)\|^\alpha_{\mathrm{L}^q} + \|v(s)\|^\alpha_{\mathrm{L}^q}\right)$$

and for T small enough

$$CT^{1-\frac{n\alpha}{2q}} \sup_{0<s<T} \left(\|u(s)\|_{\mathrm{L}^q}^{\alpha} + \|v(s)\|_{\mathrm{L}^q}^{\alpha}\right) < \frac{1}{2}$$

and so $u(t) = v(t)$ at least for $0 \leq t \leq T$. Then the uniqueness on the entire interval can be recovered by usual considerations (see [24], [25]). □

We emphasize that if the initial data u_0 belongs to the Lebesgue space L^{p_0} ($p_0 > 1$) which is invariant under the scaling of equation (1), and if $\|u_0\|_{\mathrm{L}^{p_0}}$ is small enough, then Weissler [25] established global existence of the solution:

Theorem 4. *Let $p_0 = \frac{n\alpha}{2} > 1$. Then for a suitable $\varepsilon > 0$ and for every $u_0 \in \mathrm{L}^{p_0}(\mathbb{R}^n)$ such that $\|u_0\|_{\mathrm{L}^{p_0}} < \varepsilon$ there exists a global solution $u \in C\left([0, +\infty[, \mathrm{L}^{p_0}(\mathbb{R}^n)\right)$ of (2).*

More recently, this result of global existence was sharpened by Ribaud (see [20], [6]): he requires the initial data $u_0 \in \mathrm{L}^{p_0}(\mathbb{R}^n)$ to have a small norm in a certain singular Besov space into which $\mathrm{L}^{p_0}(\mathbb{R}^n)$ is continuously embedded, without assuming any condition on the size of $\|u_0\|_{\mathrm{L}^{p_0}}$.

Finally, we recall that there are solutions which blow up at finite time: in fact, for $p_0 = 2$ and for initial data of negative energy, namely

$$\frac{1}{2}\left\|\vec{\nabla} u_0\right\|_{\mathrm{L}^2}^2 - \frac{1}{\alpha+2}\|u_0\|_{\alpha+2}^{\alpha+2} < 0,$$

Ball [2] and Payne [10] proved that

$$\|u(t)\|_{\mathrm{L}^2} \to +\infty \quad \text{if} \quad t \to T_{\max} < +\infty.$$

The solution constructed by Weissler exhibits some interesting properties. At first it satisfies $\sup_{0<t<T} t^{\sigma} \|u(t)\|_{\mathrm{L}^p} < +\infty$ and $\lim_{t\to 0} t^{\sigma} \|u(t)\|_{\mathrm{L}^p} = 0$ for every p, σ verifying $q < p \leq +\infty$ and $\sigma = \frac{n}{2}(\frac{1}{q} - \frac{1}{p})$. In particular, it belongs to $\mathrm{L}^{\infty}_{\mathrm{loc}}(]0, T], \mathrm{L}^{\infty})$ and it is a classical solution [3]. Moreover, Giga [9] pointed out that this solution also belongs to every space $\mathrm{L}^r(]0, T[, \mathrm{L}^s)$ for every $r, s > q$ and $r > \alpha + 1$ with $\frac{1}{r} = \frac{n}{2}(\frac{1}{q} - \frac{1}{s})$.

We end this section by some remarks about uniqueness. The result of Weissler left an open question: Weissler established uniqueness in $C\left([0, T], \mathrm{L}^q\right)$ only under the additional condition that the solution satisfies $\sup_{0<t<T} t^{\sigma} \|u(t)\|_{\mathrm{L}^p} < +\infty$ and $\lim_{t\to 0} t^{\sigma} \|u(t)\|_{\mathrm{L}^p} = 0$ for some p and σ as in the theorem or for $q > p_0$, $q \geq d+1$. For $q = p_0$, $q \geq \alpha+1$ the equation (2) is well defined for a generic function in $C\left([0, T], \mathrm{L}^q\right)$ in the distributional sense (in fact $|u|^{\alpha} u \in \mathrm{L}^{\infty}(]0, T[, \mathrm{L}^1_{\mathrm{loc}})$). Therefore it is meaningful to ask whether uniqueness still holds in the larger class $C\left([0, T], \mathrm{L}^q\right)$. This question is now completely solved.

Theorem 5 (Brézis-Cazenave). *Let $p_0 > \alpha+1$ and $u_0 \in \mathrm{L}^{p_0}$. If $u, v \in C\left([0, T], \mathrm{L}^{p_0}\right)$ are two solutions of* (2) *associated with the same initial data u_0 then $u(t) = v(t)$ for every $t \in [0, T]$.*

The proof of this result can be found in [3] and it uses in an essential way the fact that $|u|^\alpha u \in C([0,T],\mathrm{L}^r)$ with $r = \frac{p_0}{\alpha+1} > 1$ (remember that $u \in C([0,T],\mathrm{L}^{p_0})$, with $p_0 > \alpha+1$).

Here we present another proof of this result. It is well known that the integral operator in the equation (2) is not continuous on $\mathrm{L}^\infty(]0,T[,\mathrm{L}^{p_0})$ (see for instance [21]). In spite of this it can be proven that it is continuous on $\mathrm{L}^\infty(]0,T[,\mathrm{L}^{p_0,\infty})$, where $\mathrm{L}^{p_0,\infty}$ is the weak-L^{p_0} space defined by

$$\text{weak-}\mathrm{L}^{p_0} = \{f \text{ meas.} : \exists\, C>0 \text{ s. th. } |\{x\in\mathbb{R}^n : |f(x)|>\lambda\}| < \frac{C}{\lambda^{p_0}},\ \forall\,\lambda>0\}.$$

In fact, by following the same lines as in the proof of the continuity of the integral operator associated with the Navier-Stokes equations in $\mathrm{L}^\infty(]0,T[,\mathrm{L}^{3,\infty})$ of Meyer [15] one can prove that

$$\sup_{0<t<T}\left\|\int_0^t \mathrm{e}^{(t-s)\Delta}|u|^\alpha u(s)ds\right\|_{\mathrm{L}^{p_0,\infty}} \le C \sup_{0<t<T}\|u(t)\|^{\alpha+1}_{\mathrm{L}^{p_0,\infty}}.$$

Then in a similar way as in Theorem 3 plus some other technical considerations one can obtain uniqueness.

In the remaining case $p_0=\alpha+1$ the nonlinear term belongs to $C\left([0,T],\mathrm{L}^1\right)$ and the aforementioned proofs do not apply. In spite of that Brézis and Cazenave [3] proved by using a fine argument of compactness that the additional conditions in Theorem 3 which ensures uniqueness can be weakened.

Theorem 6 (Brézis-Cazenave). *Let $p_0=\alpha+1$, that is $p_0=\frac{n}{n-2}$ and $\alpha=\frac{2}{n-2}$, $n\ge 3$ and let $u_0\in\mathrm{L}^{\frac{n}{n-2}}$. Let $u,v\in C\left([0,T],\mathrm{L}^{\frac{n}{n-2}}\right)$ be two solutions of (2) associated with the same initial data u_0. Then, if u,v belong to $\mathrm{L}^\infty_{\mathrm{loc}}(]0,T],\mathrm{L}^p)$ for some $p\in(\frac{n}{n-2},\frac{n^2}{(n-2)^2})$ (as in Theorem 3), $u(t)=v(t)$ for every $t\in[0,T]$.*

We give a sketch of the proof which can be found in [3]. The central remark is that with every compact K in L^{p_0} and every $\varepsilon>0$ is associated a positive constant $T^*=T^*(K,\varepsilon)$ such that for every $u_0\in K$ Weissler's solution exists at least on $[0,T^*]$ and $s^\sigma\|u(s)\|_{\mathrm{L}^p}<\varepsilon$ for $s\in[0,T^*]$. Now let $u\in C([0,T],\mathrm{L}^{p_0})$ be a solution of (2) belonging to $\mathrm{L}^\infty_{\mathrm{loc}}(]0,T],\mathrm{L}^p)$ for some p belonging to $(\frac{n}{n-2},\frac{n^2}{(n-2)^2})$, as in Theorem 3 and let $K=\{u(t),\text{with } t\in[0,T]\}$. We will prove that $u(t)$ also verifies property $i)$ of Theorem 3 and so it coincides with Weissler's solution. Let $t\in(0,T]$ and $w\in C([0,T^*],\mathrm{L}^q)$, $T^*=T^*(K,\varepsilon)$ be Weissler's solution to (2) with initial data $u(t)$ and such that $s^\sigma\|w(s)\|_{\mathrm{L}^p}<\varepsilon$ for $s\in[0,T^*]$. Then $u(t+s)=w(s)$ on $[0,\min(T^*,T-t)]$ since $u(t+s)$ and $w(s)$ satisfies $i)$ and $ii)$ in Theorem 3. Therefore

$$s^\sigma\|u(t+s)\|_{\mathrm{L}^p}<\varepsilon\ ,$$

for every $s\in[0,\min(T^*,T-t)]$, and by letting $t\to0$ one obtains

$$s^\sigma\|u(s)\|_{\mathrm{L}^p}<\varepsilon\ ,$$

for every $s\in[0,\min(T^*,T)]$. Since $\varepsilon>0$ was arbitrary chosen we conclude that u verifies $i)$ and so we obtain uniqueness.

The surprising fact is that in Theorem 6 uniqueness is lost if the condition $u \in L^\infty_{\mathrm{loc}}(]0,T[, L^p)$ is dropped [21]:

Theorem 7 (Terraneo). *Let (as in Theorem 6) $p_0 = \frac{n}{n-2}$ and $\alpha = \frac{2}{n-2}$, $n \geq 3$. There exist a function $u_0 \in \mathrm{L}^{\frac{n}{n-2}}$ and a constant $T(u_0) > 0$ such that the equation (2) admits at least two solutions in $C\left([0,T], \mathrm{L}^{\frac{n}{n-2}}\right)$ with the same initial data u_0.*

This result extends to $\mathbb{R}^n$ a theorem of Ni and Sacks [17] in the case where the underlying space is the ball $B(0,1)$ of center 0 and radius 1. Ni and Sacks consider the elliptic equation

$$\begin{cases} \Delta U + U^{\frac{n}{n-2}} = 0 & in\ B \\ U(x) = 0 & on\ \ \partial B \\ U(x) > 0 & on\ \ B \end{cases} \tag{6}$$

and they prove that the equation (6) admits a solution $U_0 \in C^2(B(0,1) \setminus \{0\}) \cap \mathrm{L}^{\frac{n}{n-2}}(B(0,1))$, which is unbounded at the origin. This function U_0 is also a stationary solution of the parabolic equation if the underlying space is the ball $B(0,1)$. Since the theory of Weissler is still valid in this case, Ni and Sacks observed that two solutions of equation (1) on $B(0,1)$ are associated with the same initial data U_0: the stationary unbounded solution and Weissler's solution which is bounded for positive time. A similar result does not extend directly to $\mathbb{R}^n$, since the elliptic equation does not have unbounded positive solutions on $\mathbb{R}^n$ [7]. In spite of that, we consider as initial data a cut-off u_0 of U_0 which preserves the singularity at the origin and we look for solutions of type $u = u_0 + v$. It is then possible to find a bounded solution v of the perturbed equation so that $u = u_0 + v$ is an unbounded solution of (2) with initial data u_0 and thus it is different from Weissler's solution. We use some deep regularity properties of the integral operator associated to the perturbed equation in v in some Lorentz spaces. Afterwards another counterexample to uniqueness has been proposed by F. Pacard [21].

Remark. We already observed that the range of L^p spaces which ensures existence of a solution to (2) can be justified by scaling considerations. In spite of that the results of Weissler and Brézis-Cazenave remain valid for a generic nonlinerity $G(u)$ such that $G: \mathbb{R} \to \mathbb{R}$, $G(0) = 0$ and

$$|G(u) - G(v)| \leq C|u - v|(|u|^\alpha + |v|^\alpha).$$

2. The exponential nonlinearity

Before coming to the proof of our Theorem 2 we recall the definition and some useful properties of the Orlicz space $\mathrm{L}_{e^{u^2}-1}$. We refer to [1] and [12] for a wide and complete presentation of the subject.

2.1. Orlicz spaces

Let $\varphi : [0,+\infty[\to [0,+\infty[$ be a convex, strictly increasing function such that $\lim_{s\to 0^+}\varphi(s)=\varphi(0)=0$ and $\lim_{s\to+\infty}\varphi(s)=+\infty$.

Definition 8. *The Orlicz space* L_φ *is defined by*

$$\mathrm{L}_\varphi = \left\{ u \ \textit{measurable on } \mathbb{R}^n \ \textit{s. t.} \ \exists\, K>0 \ \textit{with} \int_{\mathbb{R}^n} \varphi\left(\left|\frac{u(x)}{K}\right|\right) dx < +\infty \right\}.$$

It can be proven that L_φ is a vector space (see [12]). For every $u\in \mathrm{L}_\varphi$, let us define the functional

$$\|u\|_\varphi = \inf\left\{ K>0 \text{ such that} \int_{\mathbb{R}^n} \varphi\left(\left|\frac{u(x)}{K}\right|\right) dx \le 1 \right\}$$

We remark that the definition of $\|u\|_\varphi$ is meaningful since the set on the right side is not empty. In fact $\varphi(|\frac{u(x)}{K}|)$ is decreasing with respect to K, so by the dominated convergence theorem $\lim_{K\to+\infty}\int_{\mathbb{R}^n}\varphi(|\frac{u(x)}{K}|)dx = 0$. Moreover, one can prove that

Proposition 9 ([12]). $(\mathrm{L}_\varphi, \|\cdot\|_\varphi)$ *is a Banach space.*

We now collect some properties of Orlicz space which will be useful in the following.

i) $\mathrm{L}^1\cap\mathrm{L}^\infty \subset \mathrm{L}_\varphi \subset \mathrm{L}^1+\mathrm{L}^\infty$.

ii) *(Interpolation property) Let* $T : \mathrm{L}^1+\mathrm{L}^\infty \to \mathrm{L}^1+\mathrm{L}^\infty$ *be a linear operator from* L^1 *into* L^1 *with norm* M_1 *and from* L^∞ *to* L^∞ *with norm* M_2. *Then* T *is bounded from* L_φ *into* L_φ *with norm* $M \le C\max(M_1,M_2)$ *and the constant* C *depends only on* φ.

With each space L_φ we can associate another Orlicz space. Let

$$\psi(s) = \max_{\{t\ge 0\}}\{st-\varphi(t)\},\ s\ge 0.$$

Then ψ generates an Orlicz space L_ψ, and for the two Orlicz spaces defined by φ and ψ a generalized version of the Hölder inequality holds.

Proposition 10 ([1]).

$$\left|\int_{\mathbb{R}^n} u(x)v(x)dx\right| \le 2\,\|u\|_{\mathrm{L}_\varphi}\,\|v\|_{\mathrm{L}_\psi}$$

It is easy to verify that the function $\varphi(u)=|u|^p$, $1<p<+\infty$ defines the L^p space. In the following we will consider the Orlicz space defined by $\varphi(u)=\mathrm{e}^{u^2}-1$. For this particular space we also have:

Lemma 11. *For every* $p\ge 2$ *we have* $\mathrm{L}_{\mathrm{e}^{u^2}-1}\hookrightarrow \mathrm{L}^p$ *and*

$$\|u\|_{\mathrm{L}^p} \le \left[\left(\Gamma(\frac{p}{2}+1)\right)\right]^{\frac{1}{p}} \|u\|_{\mathrm{L}_{\mathrm{e}^{u^2}-1}}. \tag{7}$$

where $\Gamma(\lambda)=\int_0^{+\infty}\mathrm{e}^{-x}x^{\lambda-1}dx$.

Proof. Let $K = \|u\|_{\mathrm{L}_{e^{u^2}-1}} > 0$, otherwise inequality (7) is obvious. We will use the property that

$$e^x - 1 \geq \frac{x^\rho}{\Gamma(\rho+1)}, \tag{8}$$

for every $\rho \geq 1$ and $x \geq 0$. This is easy proven when ρ is an integer. For $x \in (0,1]$ and for every $\rho > 1$ the inequality follows by using the fact that $\Gamma(\rho+1) \geq 1$ if $\rho \geq 1$, and $e^x - 1 \geq x^\rho$. For $x \in [1,+\infty)$ and $1 < \rho < 2$ the inequality (8) is implied by the fact that $e^x - 1 - x^2 \geq 0$ and the property $\Gamma(\rho+1) \geq 1$. The general inequality for $x \in [1,+\infty)$ and $2 < \rho$ follows from the case $1 < \rho < 2$ by an easy computation, since $\Gamma(\rho+1) = \rho\Gamma(\rho)$. By the theorem of monotone convergence

$$1 \geq \int_{\mathbb{R}^n} e^{(\frac{|u(x)|}{K})^2} - 1\,dx \geq \int_{\mathbb{R}^n} \frac{\left(\frac{|u|}{K}\right)^{2\rho}}{\Gamma(\rho+1)} dx$$

for every $\rho \geq 1$. Thus we obtain $1 \geq \frac{\|u\|_{2\rho}^{2\rho}}{K^{2\rho}\Gamma(\rho+1)}$ and we conclude that $K \geq \frac{\|u\|_{2\rho}}{(\Gamma(\rho+1))^{\frac{1}{2\rho}}}$. □

2.2. Proof of Theorem 2

The proof of Theorem 2 relies on the following lemmas.

Lemma 12 (linear term). *Let $f \in \mathrm{L}_{e^{u^2}-1}$. Then there exists $C > 0$ such that*

$$\sup_{0<t<T} \|e^{t\Delta} f\|_{\mathrm{L}_{e^{u^2}-1}} \leq C \|f\|_{\mathrm{L}_{e^{u^2}-1}}. \tag{9}$$

for every $T \in]0,+\infty]$.

Proof. The operator $L(f) = e^{t\Delta} f$ is linear from $\mathrm{L}^1 + \mathrm{L}^\infty$ into $\mathrm{L}^1 + \mathrm{L}^\infty$ and it is bounded from L^1 into L^1 with norm 1 and from L^∞ into L^∞ with norm 1, for every $t > 0$. Thus interpolation property ii) ensures (9). □

Lemma 13 (nonlinear term). *Let ρ, T be two positive real numbers and*

$$X_{\rho,T} = \left\{ u \in \mathrm{L}^\infty(]0,T[, \mathrm{L}_{e^{u^2}-1}) \text{ such that } \sup_{0<t<T} \|u(t)\|_{\mathrm{L}_{e^{u^2}-1}} \leq \rho \right\}.$$

If $\rho < \min(\sqrt{\frac{2}{n}}, \frac{1}{\sqrt{2}})$, then for every $u, v \in X_{\rho,T}$ we have

$$\left\| \int_0^t e^{(t-s)\Delta}(e^{u^2(s)} - 1) ds \right\|_{\mathrm{L}_{e^{u^2}-1}} \leq H(\rho) \max(t, t^{1-\frac{n}{2p}}) \sup_{0<s<t} \|u(s)\|_{\mathrm{L}_{e^{u^2}-1}} \tag{10}$$

and

$$\left\| \int_0^t e^{(t-s)\Delta}(e^{u^2(s)} - e^{v^2(s)}) ds \right\|_{\mathrm{L}_{e^{u^2}-1}} \leq H(\rho) \max(t, t^{1-\frac{n}{2p}}) \sup_{0<s<t} \|u(s) - v(s)\|_{\mathrm{L}_{e^{u^2}-1}} \tag{11}$$

where $H(\rho) = O(\rho)$ for $\rho \to 0$ and p is such that $\frac{1}{\rho^2} > p > \frac{n}{2}$.

Proof. Since (10) follows from (11) by choosing $v = 0$ it will suffice to prove (11). Let $u, v \in \mathrm{L}^\infty(]0,T[, \mathrm{L}_{e^{u^2}-1})$. We will prove that we can bound the L^1-norm and the L^∞-norm of the integral in (11) and then we will be able to conclude thanks to the interpolation property *ii)* of Orlicz spaces stated above.

We begin by writing

$$
\begin{aligned}
&\int_0^t \mathrm{e}^{(t-s)\Delta}(\mathrm{e}^{u^2(s)} - \mathrm{e}^{v^2(s)})ds \\
&= \int_0^t \mathrm{e}^{(t-s)\Delta}[(\mathrm{e}^{u^2(s)} - 1 - u^2(s) + u^2(s)) - (\mathrm{e}^{v^2(s)} - 1 - v^2(s) + v^2(s))]ds \\
&= \int_0^t \mathrm{e}^{(t-s)\Delta}[(\mathrm{e}^{u^2(s)} - 1 - u^2(s)) - (\mathrm{e}^{v^2(s)} - 1 - v^2(s))]ds + \\
&\quad + \int_0^t \mathrm{e}^{(t-s)\Delta}(u^2(s) - v^2(s))ds \\
&= I + II.
\end{aligned}
\tag{12}
$$

Since for every $\beta > 1$ there exists $C > 0$ such that $|(\mathrm{e}^{u^2} - 1 - u^2) - (\mathrm{e}^{v^2} - 1 - v^2)| \leq C(\mathrm{e}^{\beta^2 u^2} - 1 + \mathrm{e}^{\beta^2 v^2} - 1)|u - v|$, then for the first term we have

$$
\begin{aligned}
\|I\|_{\mathrm{L}^1} &\leq C \int_0^t \left\| \mathrm{e}^{(t-s)\Delta} |u(s) - v(s)| \, (\mathrm{e}^{\beta^2 u^2(s)} - 1 + \mathrm{e}^{\beta^2 v^2(s)} - 1) \right\|_{\mathrm{L}^1} ds \\
&\leq C \int_0^t \|u(s) - v(s)\|_{\mathrm{L}^2} \left(\left\| \mathrm{e}^{\beta^2 u^2(s)} - 1 \right\|_{\mathrm{L}^2} + \left\| \mathrm{e}^{\beta^2 v^2(s)} - 1 \right\|_{\mathrm{L}^2} \right) ds.
\end{aligned}
$$

In order to estimate $\left\| \mathrm{e}^{\beta^2 u^2} - 1 \right\|_{\mathrm{L}^2}$ we write

$$
\left\| \mathrm{e}^{\beta^2 u^2} - 1 \right\|_{\mathrm{L}^2} \leq \left\| \sum_{j=1}^{+\infty} \frac{(\beta u)^{2j}}{j!} \right\|_{\mathrm{L}^2} \leq \sum_{j=1}^{+\infty} \frac{\|\beta u\|_{\mathrm{L}^{4j}}^{2j}}{j!}.
$$

Finally, by using inequality (7) we obtain

$$
\left\| \mathrm{e}^{\beta^2 u^2} - 1 \right\|_{\mathrm{L}^2} \leq \sum_{j=1}^{+\infty} \frac{[(2j)!]^{\frac{1}{2}}}{j!} (\beta \|u\|_{\mathrm{L}_{e^{u^2}-1}})^{2j}.
$$

and since $2\beta^2\rho^2 < 1$ by hypothesis, for β close enough to 1, and $\|u\|_{\mathrm{L}_{e^{u^2}-1}} \leq \rho$ the series converges. Now since $\|u - v\|_{\mathrm{L}^2} \leq C \|u - v\|_{\mathrm{L}_{e^{u^2}-1}}$ we have

$$
\|I\|_{\mathrm{L}^1} \leq Ct \sup_{0<s<t} \|u(s) - v(s)\|_{\mathrm{L}_{e^{u^2}-1}} \sum_{j=1}^{+\infty} (\beta\rho)^{2j} \frac{[(2j)!]^{\frac{1}{2}}}{j!}.
$$

For the second term, we obtain directly

$$\|II\|_{L^1} \le C \int_0^t \left\| e^{(t-s)\Delta} |u(s) - v(s)|(|u(s)| + |v(s)|) \right\|_1 ds$$
$$\le C \int_0^t \|u(s) - v(s)\|_{L^2} (\|u(s)\|_{L^2} + \|v(s)\|_{L^2}) ds$$
$$\le C\rho t \sup_{0<s<t} \|u(s) - v(s)\|_{L_{e^{u^2}-1}} .$$

Similar estimates hold for $L^\infty(\mathbb{R}^n)$. Indeed by Young and Hölder inequalities we have

$$\|I\|_{L^\infty} \le C \left(\int_0^t \left\| e^{(t-s)\Delta} |u(s) - v(s)| \, |e^{\beta^2 u^2(s)} - 1 + e^{\beta^2 v^2(s)} - 1| \right\|_{L^\infty} ds \right)$$
$$\le C \int_0^t (t-s)^{-\frac{n}{2p}} \|u(s) - v(s)\|_{L^r} \left(\left\| e^{\beta^2 u^2(s)} - 1 \right\|_{L^q} + \left\| e^{\beta^2 v^2(s)} - 1 \right\|_{L^q} \right) ds$$

where p is chosen such that $p > \frac{n}{2}$ and $\frac{1}{r} + \frac{1}{q} = \frac{1}{p}$, $1 \le q, r < +\infty$. In the same way as before, for $1 \le q < +\infty$

$$\left\| e^{\beta^2 u^2} - 1 \right\|_{L^q} \le \sum_{j=1}^{+\infty} \frac{[\Gamma(qj+1)]^{\frac{1}{q}}}{j!} \|\beta u\|^{2j}_{L_{e^{u^2}-1}} .$$

By choosing $q > p$ close enough to p and β close enough to 1, by hypothesis $q\beta^2\rho^2 < 1$ and since $\|u\|_{L_{e^{u^2}-1}} \le \rho$ the series converges. For the second term the Hölder and Young inequalities give directly

$$\|II\|_{L^\infty} \le C\rho t^{1-\frac{n}{2p}} \sup_{0<s<t} \|u(s) - v(s)\|_{L_{e^{u^2}-1}} .$$

□

We are now in position to prove the theorem. Let

$$G : u(t) \to e^{t\Delta} u_0 + \int_0^t e^{(t-s)\Delta} (e^{u^2(s)} - 1) ds$$

we will prove that G has a fixed point in $X_{\rho,T}$ for small data and for ρ and T well chosen. For u in $X_{\rho,T}$ by using the two lemmas we have

$$\sup_{0<t<T} \|G(u)(t)\|_{L_{e^{u^2}-1}}$$
$$\le C \|u_0\|_{L_{e^{u^2}-1}} + H(\rho) \max(T, T^{1-\frac{n}{2p}}) \sup_{0<t<T} \|u(t)\|_{L_{e^{u^2}-1}}$$

and

$$\sup_{0<t<T} \|G(u)(t) - G(v)(t)\|_{L_{e^{u^2}-1}}$$
$$\le H(\rho) \max(T, T^{1-\frac{n}{2p}}) \sup_{0<t<T} \|u(t) - v(t)\|_{L_{e^{u^2}-1}}$$

and if $\|u_0\|_{L_{e^{u^2}-1}} \leq \frac{\rho}{2}$ and $H(\rho)\max(T, T^{1-\frac{n}{2p}}) \leq \frac{1}{2}$ then G maps $X_{\rho,T}$ into itself and it is a strict contraction.

Remark 2. In the case when the underlying space is the ball $B(0,1)$ the problem of existence of solution for the nonlinear heat equation

$$\partial_t u = \Delta u + \mu e^u, \qquad \mu > 0 \tag{13}$$

has already been studied by different authors (see for instance [4], [18], [19], [13]). More precisely, Péral and Vasquez [19] showed that if μ is large enough and the initial data $u_0(x) \geq -\log|x|^2$, $u_0(x) \neq -\log|x|^2$ then (13) has no solution. This result was then extended to a large class of nonlinearities by Martel [13].

It is quite easy to see that in the case of $B(0,1)$ the same result as in Theorem 2 still holds: local existence can be proven under a smallness condition on the $L_{e^{u^2}-1}$ norm of the initial data. In fact, since we are working on the bounded domain $B(0,1)$, then $L^\infty \hookrightarrow L_{e^{u^2}-1}$ and so we have only to control the L^∞ norm. Moreover we do not need vanishing properties at 0 for the nonlinearity.

This result can also be improved by considering initial data in the Orlicz space L_{e^u-u-1} with small norm. Lemma 13 and Theorem 2 can be stated in this framework and it is possible to obtain an estimate for the smallness of the norm of the initial data which ensures existence, namely $\|u_0\|_{L_{e^u-u-1}} < \frac{2}{n}$. We remark that $-\log|x|^2$ belongs L_{e^u-u-1} but its norm does not verify this smallness condition.

3. Equivalence between the differential and integral formulation

We say that $f \in L^{\alpha+1}_{\mathrm{uloc}}$ if and only if $\sup_{x_0\in\mathbb{R}^n} \int_{B(x_0,1)} |f(x)|^{\alpha+1}dx < +\infty$. For the nonlinear heat equation with polynomial nonlinearity the equivalence between the differential and integral formulation holds in the following framework:

Proposition 14. *Let $u \in L^{\alpha+1}([0,T], L^{\alpha+1}_{\mathrm{uloc}})$. Then the following statements are equivalent*

i) *u satisfies $\partial_t u = \Delta u + |u|^\alpha u$ in the sense of distributions;*

ii) *there exists $u_0 \in \mathcal{S}'$ such that $u(t) = e^{t\Delta}u_0 + \int_0^t e^{(t-s)\Delta}|u|^\alpha u(s)ds$.*

The key tool is the fact that $|u|^\alpha u$ belongs to $L^1([0,T], L^1{}_{\mathrm{uloc}})$. For a proof see for example [22] (or [11] where the same problem is analyzed for the Navier-Stokes system). In the case of exponential nonlinearity a similar proposition still holds since $e^{u^2} - 1 \in L^\infty(]0,T[, L^p)$ for some integer $p \geq 1$ whenever $u \in L^\infty(]0,T[, L_{e^{u^2}-1})$ and it has small norm in this space.

References

[1] R. Adams, *Sobolev spaces*, Acad. Press, (1975).

[2] M. Ball, *Remarks on blowup and nonexistence theorems for nonlinear evolution equations*, Quart. J. Math., 28 (1977), pp. 473–486.

[3] H. Brézis and T. Cazenave, *A nonlinear heat equation with singular initial data*, Journ. d'anal. math., 68 (1996), pp. 186–212.

[4] H. Brézis, T Cazenave, Y. Martel and A. Ramiandrisoa, *Blow-up for $u_t - \Delta u = g(u)$ revisited*, Adv. Diff. Eq., 1 (1996) pp. 73–90

[5] T. Cazenave and F. B. Weissler, *The Cauchy problem for the critical nonlinear Schrödinger equation in H^s*, Nonlinear Anal. T.M.A., 14 (1990), pp. 807–836.

[6] T. Cazenave and F. B. Weissler, *Asymptotically self-similar global solutions of the nonlinear Schrödinger and heat equations*, Math. Zeit., 228 (1998), pp. 83–120.

[7] B. Gidas and J. Spruck, *Global and local behavior of positive solutions of nonlinear elliptic equations*, Comm. Pure Appl. Math., 34 (1981), pp. 525–598.

[8] V. A. Galaktionov and J. L. Vasquez, *Continuation of blow-up solutions of nonlinear heat equations in several space dimensions*, Comm. on Pure and Applied Math., 50 (1997), no. 1, pp. 1–67.

[9] Y. Giga, *Solutions for semilinear parabolic equations in L^p and regularity of weak solutions of the Navier-Stokes system.*, J. Differential Equations, 62 (1986), pp. 186–212.

[10] A. Haraux and F. B. Weissler, *Non uniqueness for a semilinear initial value problem*, Indiana Univ. Math. J., 31 (1982), pp. 167–189.

[11] P. G. Lemarié-Rieusset, *Recent developments in the Navier-Stokes problem.* Book in preparation.

[12] L. Maligranda, *Orlicz spaces and interpolation*, IMECC, 1989.

[13] Y. Martel, *Dynamical instability of singular extremal solutions of nonlinear elliptic problems*, Adv. Math. Sci. Appl., 9 (1999), pp. 163–181.

[14] Y. Martel, *Personal communication*, (2001).

[15] Y. Meyer, *Wavelets, paraproducts and Navier-Stokes equations*, Current developements in mathematics, International Press, Cambridge, MA 02238-2872, 1996.

[16] M. Nakamura and T. Ozawa, *Nonlinear Schrödinger equations in the Sobolev space of critical order*, Journ. Funct. Anal., 155 (1998), pp. 364–380.

[17] W.-M. Ni and P. Sacks, *Singular behavior in nonlinear parabolic equations*, Trans. of the Amer. Math. Soc., 287 (1985), pp. 657–671.

[18] I. Peral and J. L. Vázquez, *Blow up solutions of some nonlinear elliptic problems*, Rev. Mat. Univ. Complut. Madrid, 10 (1997), pp. 443–469.

[19] I. Peral and J. L. Vázquez, *On the stability or instability of the singular solution of the semilinear heat equation with exponential reaction term*, Arch. Rational Mech.Anal., 129 (1995), pp. 201–224.

[20] F. Ribaud, *Cauchy problem for semilinear parabolic equations with initial data in* $H^s_p(\mathbb{R}^n)$, Rev. Mat. Iberoamericana, 14 (1998), pp. 1–46.

[21] E. Terraneo, *Application de certains espaces de l'analyse harmonique aux equations de Navier-Stokes et de la chaleur non-lineaire.* Université Evry Val d'Essonne, 1999.

[22] E. TERRANEO, *Sur la non-unicité des solutions faibles de l'équation de la chaleur non linéaire avec non-linéarité* u^3, C. R. Acad. Sci. Paris, 328 (1999), pp. 759–762.

[23] J. L. VÁZQUEZ, *Domain of existence and blow-up for the exponential reaction-diffusion equation*, Indiana Univ. Math. J., 48 (1999), 677–709.

[24] F. B. WEISSLER, *Local existence and nonexistence for semilinear parabolic equations in* L^p, Indiana Univ. Math. J., 29 (1980), pp. 79–102.

[25] F. B. WEISSLER, *Existence and nonexistence of global solutions for a semilinear heat equation*, Israel J. Math., 38 (1981), pp. 29–40.

Bernhard Ruf, Elide Terraneo
Dipartimento di Matematica "F. Enriques"
Università di Milano
via Saldini, 50
I–20133 Milano
e-mail: ruf@mat.unimi.it
e-mail: terraneo@mat.unimi.it

Progress in Nonlinear Differential Equations
and Their Applications, Vol. 50, 311–338

Well-posedness and Asymptotic Behaviour of Non-autonomous Linear Evolution Equations

Roland Schnaubelt

To the memory of Brunello Terreni

1. Introduction

There is a striking difference between autonomous and non-autonomous linear evolution equations. Autonomous problems are well understood in the framework of strongly continuous operator semigroups and their generalizations. The Hille-Yosida type theorems settle the question of well-posedness to a great extend, many perturbation and approximation results have been established, and for a large class of problems the asymptotic behaviour can be studied on the basis of spectral theory and transform methods. In these and many other areas semigroup theory has reached a considerable degree of maturity, and its applications thrive in plenty of fields.

For non-autonomous problems, however, we do not yet know of a coherent and general theory. We can rely on several more or less independent, quite sophisticated existence theorems due to P. Acquistapace, H. Amann, G. Da Prato, T. Kato, J.L. Lions, B. Terreni, and others. But these facts cannot be combined into a unified approach. Accordingly, other subjects like perturbation or duality can be treated only in a case by case analysis leaving many questions open. The situation becomes much worse if we look at asymptotic properties because spectral and transform theory cannot be applied directly (with the partial exception of time-periodic problems). The available results are usually restricted to problems 'close' to an equation with known behaviour (in particular an autonomous one). Any attempt to overcome these short-comings must face the difficulties which can be demonstrated by mostly rather simple examples refuting many natural conjectures. We present in Section 3 such examples concerning the asymptotic behaviour, whereas R. Nagel and G. Nickel discuss in [91] many examples concerning the existence theory, see also Section 2.

This state of affairs motivates a different approach to non-autonomous Cauchy problems based on the idea to transform them into autonomous ones. The standard ODE method in this context leads to a nonlinear semigroup which seems difficult to handle, see [44]. G.R. Sell and others introduced the 'hull' of a given equation (compare Section 5) which allows to treat problems possessing certain almost periodicity and compactness properties in the framework of dynamical systems. In the present paper we concentrate on the approach via *evolution semigroups* proposed by J.S. Howland in 1974. Originally devised to study perturbation theory, this concept might help to unify the existence theory of non-autonomous evolution equations, cf. Section 4 and [91]. So far its major merits concern the investigation of asymptotic properties, namely exponential dichotomy, employing the remarkable fact that the spectrum of the evolution semigroup determines the exponential dichotomy of the underlying problem, see Sections 5 and 6.

This survey exposes the above mentioned subjects aiming at a reader with a solid background in evolution equations. We strive for a coherent non-technical description of the results, problems, and methods in the field. The major features of the theorems should become intelligible though we can state only a few of them explicitly.

In Section 2 we give an outline of the existence theory starting with parabolic problems which allow for the most powerful results. Here the efforts of many renowned researchers culminate in the outstanding papers by P. Acquistapace and B. Terreni. On a more general level, T. Kato created his well-known theory which partially extends the Hille-Yosida theorem and is directed to hyperbolic equations. These fundamental contributions are complemented by the operator sum method developed by G. Da Prato, P. Grisvard, and J.L. Lions. We then proceed to discuss exponential dichotomy being one of the most basic asymptotic properties of evolution equations. Evolution semigroups are introduced in Section 4, where their impact on the well-posedness of evolution equations is presented. The latter point is closely related to the operator sum method. Most of the known results on exponential dichotomy rely on certain characterizations of this notion treated in Section 5. Here we emphasize evolution semigroups, but we also explain the connection with the more traditional 'Perron-type' characterizations involving the inhomogeneous problem. Moreover, we present a new proof of Datko's famous stability theorem using evolution semigroups. In the last section results on the asymptotic behaviour of evolution equations are discussed concentrating on exponential dichotomy and qualitative properties of the inhomogeneous problem.

We have compiled a large list of references which nevertheless is far from being complete. In order to keep the bibliography within a reasonable size, we cite no paper dealing exclusively with autonomous problems and very few papers treating the question of (exponential) stability alone. For similar reasons we only give a few hints concerning applications to nonlinear problems, and we usually quote just a sample of an author's publications. The monograph [39] is our standard source for semigroup theory, where one can find unexplained notation, concepts, and results on autonomous problems.

2. Well-posedness of evolution equations

In this section we review existence results for the non-autonomous Cauchy problem

$$(CP)\begin{cases} \frac{d}{dt}u(t) = A(t)u(t) + f(t), & t \geq s,\ t,s \in J, \\ u(s) = x, \end{cases}$$

on a Banach space X, where $A(t)$ are linear operators on X, $x \in X$, $f \in L^1_{\text{loc}}(J, X)$, and $J \subseteq \mathbb{R}$ is a closed interval. The homogeneous problem with $f = 0$ is denoted by $(\mathrm{CP})_0$.

For suitable differential operators $A(t)$ such equations describe, for instance, diffusion processes with time-dependent diffusion coefficients or boundary conditions, quantum mechanical systems with time-varying potential, or non-autonomous hyperbolic systems of first order. Using standard techniques, one can also transform wave equations or retarded differential equations into an evolution equation of the form (CP). Further, partial differential equations on non-cylindrical domains can be formulated as non-autonomous Cauchy problems, [21], [22]. Evolution equations further arise if one solves semi- and quasilinear equations by means of linearization methods; see [63], [64], [136], [137] for hyperbolic problems and [5], [53], [86], [128], [141] for parabolic problems.

In the literature various types of solutions u to (CP) are used depending on the required differentiability properties of u. We always assume that $u : J_s \to X$ is continuous and satisfies $u(s) = x$, where $J_s = J \cap [s, \infty)$. First of all, one can suppose that $u \in C^1(J_s, X)$, $u(t) \in D(A(t))$, and $u'(t) = A(t)u(t) + f(t)$ for $t \in J_s$. Second, if the data $A(t)$ and $f(t)$ are not continuous in t, it is appropriate to look for functions $u \in W^{1,p}_{\text{loc}}(J_s, X)$ satisfying $u(t) \in D(A(t))$ and $u'(t) = A(t)u(t) + f(t)$ for a.e. $t \in J_s$ and some $1 \leq p \leq \infty$. One can also replace the interval J_s by $J'_s = J \cap (s, \infty)$ in both cases. This is suitable for parabolic problems with initial values x not contained in $D(A(s))$. Further notions of solutions are introduced below. In our definition of well-posedness we employ C^1-solutions for simplicity.

Definition 2.1. *The homogeneous problem* $(CP)_0$ *is called* well-posed (on spaces Y_s) *if there are dense subspaces* Y_s, $s \in J$, *of* X *with* $Y_s \subseteq D(A(s))$ *such that for each* $x \in Y_s$ *there is a unique solution* $u = u(\cdot\,; s, x) \in C^1(J_s, X)$ *of* $(CP)_0$ *with* $u(t) \in Y_t$ *for* $t \in J_s$ *and if* $s_n \to s$ *and* $x_n \to x$ *in* X *for* $s_n \in J$, $x_n \in Y_{s_n}$, $x \in Y_s$, *then* $\hat{u}(t\,; s_n, x_n) \to \hat{u}(t\,; s, x)$ *in* X *uniformly for* t *in compact subsets of* J.

Here we set $\hat{u}(t\,; r, y) := u(t\,; r, y)$ for $t \geq r$ and $\hat{u}(t\,; r, y) := y$ for $t \leq r$ and $y \in Y_r$. In other words, we require that there exists a unique solution of the homogeneous problem for sufficiently many initial values x and that the solutions depend continuously on the initial data, cf. [42, § 7.1].

Starting with a well-posed Cauchy problem, we define $U(t,s)x := u(t\,; s, x)$ for $t \geq s$, $t, s \in J$, and $x \in Y_s$. It is not difficult to show that $U(t,s)$ can be extended to a unique bounded linear operator on X (also denoted by $U(t,s)$) such that

(E1) $U(t,s) = U(t,r)U(r,s)$, $U(s,s) = I$, and
(E2) $(t,s) \mapsto U(t,s)$ is strongly continuous

for $t \geq r \geq s$ and $t,s,r \in J$, cf. [91]. This motivates the following definition.

Definition 2.2. *A collection $U(\cdot,\cdot) = (U(t,s))_{t\geq s,t,s\in J} \subseteq \mathcal{L}(X)$ satisfying (E1) and (E2) is called an* evolution family. *If (CP)$_0$ is well posed (on Y_t) with solutions $u = U(\cdot,s)x$, we say that $U(\cdot,\cdot)$* solves *(CP)$_0$ (on Y_t) or that $A(\cdot)$* generates *$U(\cdot,\cdot)$.*

Evolution families are also called evolution systems, evolution operators, evolution processes, propagators, or fundamental solutions in the literature. In contrast to semigroups, it is possible that the mapping $t \mapsto U(t,s)x$ is differentiable only for $x = 0$. This happens if $U(t,s) = \frac{p(t)}{p(s)}$ on $X = \mathbb{C}$ and $p : \mathbb{R} \to [1,2]$ is continuous and nowhere differentiable. Thus an evolution family need not be generated by an operator family $A(\cdot)$. We refer to [91] for further information on these topics.

If (CP)$_0$ is well posed on spaces Y_s, then

$$\begin{aligned}\tfrac{1}{h}(U(t,s+h)x - U(t,s)x) &= U(t,s+h)\,\tfrac{1}{h}(x - U(s+h,s)x)\\ &\to -U(t,s)A(s)x \end{aligned}\tag{2.1}$$

as $h \searrow 0$ for $t > s$ and $x \in Y_s$. Assume that $u \in C^1(J'_s, X)$ solves (CP) for $f \in C(J_s, X)$ and $u(t) \in Y_t$ for $t \in J'_s$. Then (2.1) yields $\frac{\partial^+}{\partial\tau}\,[U(t,\tau)u(\tau)] = U(t,\tau)f(\tau)$ for $\tau \in (s,t)$, and hence

$$u(t) = U(t,s)x + \int_s^t U(t,\tau)f(\tau)\,d\tau, \qquad t \geq s. \tag{2.2}$$

For this reason, the function u defined by (2.2) is called the *mild solution* of (CP) for every $x \in X$ and $f \in L^1_{\mathrm{loc}}(\mathbb{R}_+, X)$.

We now survey existence theorems for (CP) and the methods used to establish them in the case $J = [0,T]$. Most of the results discussed in paragraph (a) and (b) are well presented in the monographs [6], [42], [86], [104], [133], [134].

(a) The parabolic case

Here one assumes that the operators $A(t)$ generate analytic C_0-semigroups of the same type and that $t \mapsto A(t)$ is regular in a sense specified below. Then there exists an evolution family $U(\cdot,\cdot)$ on X solving (CP)$_0$ on $D(A(t))$ such that $U(t,s)X \subseteq D(A(t))$, $\frac{\partial}{\partial t}U(t,s) = A(t)U(t,s)$ in $\mathcal{L}(X)$, and $\|A(t)U(t,s)\| \leq \frac{C}{t-s}$ for $0 \leq s < t \leq T$. The operators $U(t,s)$ can be constructed as solutions to certain integral equations like

$$U(t,s) = e^{(t-s)A(s)} + \int_s^t U(t,\tau)(A(\tau) - A(s))e^{(\tau-s)A(s)}\,d\tau \tag{2.3}$$

(in the case $D(A(t)) \equiv D(A(0))$). This is an abstract version of the method of 'freezing coefficients'; see also the survey given in [2].

If the domains $D(A(t))$ do not depend on t, it suffices to suppose that $A(\cdot) : [0,T] \to \mathcal{L}(Y,X)$ is Hölder continuous, where $Y := D(A(0))$ is endowed with the graph norm. The fundamental results in this direction were established by

P.E. Sobolovskii, [128], and H. Tanabe, [132], around 1960. A. Lunardi treated equations with non-dense $D(A(t))$ in her book [86]. Most of the above mentioned results still hold if $A(\cdot) \in C([0,T], \mathcal{L}(Y,X))$ and the operators $A(t)$ possess some additional space regularity, [18], [19], [33], [43], [109].

The case of time-varying domains was studied by T. Kato and H. Tanabe in [65] imposing (besides another condition) that $A(\cdot)^{-1} \in C^{1+\alpha}([0,T], \mathcal{L}(X))$. Their results were refined by A. Yagi in the seventies, [139]. One can weaken this differentiability assumption to a Hölder condition like (2.4) below if one requires the time-independence of the domain of a fractional power of $A(t)$, [61], [129], or of an interpolation space between $D(A(t))$ and X, [3], [6]. The latter situation is treated in detail in H. Amann's monograph [6]. Finally, P. Acquistapace and B. Terreni introduced in [1] and [4] the assumption

$$\|A(t)R(\lambda, A(t))\,(A(t)^{-1} - A(s)^{-1})\| \leq L\,|t-s|^{\mu}\,|\lambda|^{-\nu} \tag{2.4}$$

for $t, s \in J$, $|\arg \lambda| \leq \phi$, and constants $\phi \in (\frac{\pi}{2}, \pi)$, $L \geq 0$, and $\mu, \nu \in (0,1]$ with $\mu + \nu > 1$ (without assuming the density of $D(A(t))$ in X), see also A. Yagi's papers [140], [141]. The hypothesis (2.4) is logically independent of those used in [65] and [139], see [4, § 7], but in some sense weaker since only a Hölder condition is needed. The approach of P. Acquistapace and B. Terreni is inspired by the work of R. Labbas and B. Terreni on operator sums, see [68], [69], and (c) below, whereas the results of H. Amann and A. Yagi rely on suitable versions of the formula (2.3).

Based on the properties of $U(\cdot,\cdot)$, one can investigate the regularity of the mild solution u to the inhomogeneous problem (CP). One obtains, for instance, $u \in C^1(J_s, X)$ and $u' = A(\cdot)u + f$ if $x \in D(A(s))$ (assumed to be dense) and f is Hölder continuous or bounded with respect to interpolation norms. In [4], [6], [86] it is even shown that u satisfies optimal regularity of Hölder type.

(b) The hyperbolic case

One can partially extend the Hille-Yosida theorem for semigroups to the non-autonomous situation. To that purpose, one assumes that the operators $A(t)$ are densely defined and *stable* in the sense that

$$\|R(\lambda, A(t_n))R(\lambda, A(t_{n-1}))\,\cdots\,R(\lambda, A(t_1))\| \leq M\,(\lambda - w)^{-n} \tag{2.5}$$

for all $0 \leq t_1 \leq \cdots \leq t_n \leq T$, $n \in \mathbb{N}$, $\lambda > w$, and some constants $M \geq 1$ and $w \in \mathbb{R}$. Observe that then $A(t)$ generates a C_0-semigroup by the Hille-Yosida theorem. One further requires the existence of a Banach space Y being continuously and densely embedded in X such that $Y \subseteq D(A(t))$ and the parts of $A(t)$ in Y are also stable for $t \in [0,T]$. If, in addition, $A(\cdot) \in C([0,T], \mathcal{L}(Y,X))$, then T. Kato succeeded to construct an evolution family $U(\cdot,\cdot)$ on X satisfying $\frac{\partial}{\partial t}U(t,s)x_{|t=s} = A(s)x$ for $t \geq s$ and $x \in Y$. He first studied the special case that the operators $A(t)$ have a common domain and generate contraction semigroups in his paper [60] from 1953. The general case was established in 1970, [62]. Kato introduced a time-discretization $A_n(\cdot)$ of $A(\cdot)$, solved the corresponding Cauchy problem by finite products $U_n(t,s)$ of the given operators $e^{\tau A(r)}$, and proved that $U_n(t,s)$ converges strongly to an operator $U(t,s)$ having the asserted properties.

Imposing an additional regularity hypothesis on $t \mapsto A(t)$, see [62, Theorem 6.1], Kato also showed that $U(t,s)Y \subseteq Y$, $U(\cdot,\cdot)$ is strongly continuous on Y, and $\frac{\partial}{\partial t} U(t,s)x = A(t)U(t,s)x$ for $t \geq s$ and $x \in Y$, i.e., $(\mathrm{CP})_0$ is well posed on Y. (This extra assumption holds in particular if $D(A(t)) \equiv Y$ and $A(\cdot)y \in C^1(J,X)$ for $y \in Y$.) One can then verify that the mild solution belongs to $C^1([s,T],X) \cap C([s,T],Y)$ and solves (CP) provided that $f \in C([s,T],Y)$ and $x \in Y$.

These results were extended to strongly continuous $A(\cdot)$ in [66], to strongly measurable $A(\cdot)$ in [58], and to strongly measurable resolvents in [34]. Of course, in the latter two cases (CP) can be solved only for a.e. $t \geq s$. If X is a Hilbert space, results involving conditions on the numerical range of $A(t)$ are proved in [100], [101]. Finally, the case of non-dense domains is investigated in [35], [136], [137], and of so-called C-evolution systems in [135]. Nonlinear versions of Kato's result were shown by M. Crandall and A. Pazy, [29], in 1972 using a resolvent approximation scheme, see also [41].

(c) The operator sum method

In most of the previous two paragraphs, (CP) is treated in two steps. One first constructs the evolution family solving the homogeneous equation and then one establishes the required regularity of the mild solution. There is another approach which directly tackles the inhomogeneous problem. One considers (CP) with $x = 0$ as an equation

$$Gu := -\tfrac{d}{dt}\, u + A(\cdot)u = -f$$

on a function space such as $E = L^p([0,T],X)$ and tries to invert the operator G.* This idea is pursued in the work by G. Da Prato and P. Grisvard both in the hyperbolic and the parabolic case, see in particular [32]. A related approach is developed in [68] and [69].

One says that the problem (CP) has *maximal regularity of type L^p* if the operator G is invertible in $E = L^p([0,T],X)$ on the domain $F = \{f \in W^{1,p}([0,T],X) : f(0) = 0,\ f(t) \in D(A(t))$ for a.e. $t \in [0,T],\ A(\cdot)f(\cdot) \in E\}$. In this case the function $u = -G^{-1}f$ solves (CP) in the $W^{1,p}$-sense. This property holds for many parabolic problems on L^q-spaces (if $1 < p,q < \infty$), see [6], [43], [54], [55], [89], [109], [131], [142], where further references to the autonomous case are given. Following the approach of Da Prato and Grisvard, one can construct the inverse of G by a contour integral involving the resolvents of $\frac{d}{dt}$ and $A(\cdot)$. The main step in the proof is to show that G^{-1} maps into $D(A(\cdot))$.

In the hyperbolic case the operator $G : D(\frac{d}{dt}) \cap D(A(\cdot)) \to E$ is not closed (and thus not invertible) in general. For instance, the closure of $G = -\frac{\partial}{\partial t} + \frac{\partial}{\partial s}$ with $D(G) = W^{1,2}(\mathbb{R}, L^2(\mathbb{R})) \cap L^2(\mathbb{R}, W^{1,2}(\mathbb{R}))$ in $L^2(\mathbb{R}, L^2(\mathbb{R})) \cong L^2(\mathbb{R}^2)$ is the

*It is possible to incorporate nontrivial initial values x, cf. [6, § III.1.5], [34], but for simplicity we concentrate on the case $x = 0$.

derivative in direction $t = s$ with maximal domain. Nevertheless, it was shown in [34] that $G = -\frac{d}{dt} + A(\cdot)$ has an invertible closure if, for instance, the spaces X and Y from paragraph (b) are reflexive, the operators $A(t)$ and the parts of $A(t)$ in Y are stable, $\|A(t)\|_{\mathcal{L}(Y,X)} \leq c$ for $0 \leq t \leq T$, and $R(\lambda, A(\cdot))$ is strongly measurable in X and Y for $\lambda > w$. Moreover, if $f \in L^p([0,T],Y)$, then $u = -\overline{G}^{-1} f$ belongs to $W^{1,p}([0,T],X) \cap L^\infty([0,T],Y)$, $1 < p < \infty$, and u is the $W^{1,p}$-solution of (CP). An inhomogeneity $f \in L^p([0,T],X)$ can be approximated by functions $f_n \in L^p([0,T],Y)$. The $W^{1,p}$-solutions u_n corresponding to f_n then converge uniformly to $u = -\overline{G}^{-1} f$ so that (CP) can be solved approximately in $W^{1,p}$-sense. This line of research was further pursued in [20] and [35]. We will come back to this approach in Section 4.

J.L. Lions developed the operator sum method in the context of weak solutions in Hilbert spaces, see [80, § 3.1, 3.4]. Let $A(t)$ be defined by uniformly bounded and coercive sesquilinear forms a_t on a Hilbert space X with a dense form domain V such that $t \mapsto a_t(x,y)$ is measurable for all $x, y \in V$. Thus, $V \hookrightarrow X \cong X' \hookrightarrow V'$ and $A(t)$ can be extended to a bounded operator from V to V'. Using a duality argument, it is possible to establish the bijectivity of $G : W_0^{1,2}((0,T],V') \cap L^2([0,T],V) \to L^2([0,T],V')$. In other words, (CP) is solved in the larger space V'. Other proofs of this result can be found in [79, § III.1], where a Galerkin scheme is used, and in [133, § 5.5], where the theory of paragraph (a) is applied. It is also possible to obtain weak solutions for wave equations using Galerkin's method, see [80, § 3.8, 3.9]. We finally mention the recent contributions [59] and [130], where form methods are employed for parabolic problems.

We have recalled a multitude of methods, notions of solutions, and existence results for non-autonomous Cauchy problems which are mostly independent of each other. In the autonomous case the Hille-Yosida type theorems provide a powerful characterization of well-posedness which is embedded in a coherent theory. So far, however, no analogues are known for non-autonomous equations. In fact, there are plenty of examples, discussed in [91], which indicate that it is rather difficult to find necessary and sufficient conditions on the operators $A(t)$ for the well-posedness of $(\mathrm{CP})_0$ (except for special situations, see [23], [24]).

Perturbation theory is another important method to solve (CP), see e.g. [38], [53], [62], [65], [113], [126]. Here one considers the Cauchy problem

$$u'(t) = [A(t) + B(t)]\, u(t), \quad t \geq s, \qquad u(s) = x, \tag{2.6}$$

where $A(\cdot)$ generates an evolution family $U(\cdot,\cdot)$ and $B(t)$ is 'small compared with $A(t)$'. If $B(\cdot) \in C([0,T], \mathcal{L}_s(X))$, one obtains a unique evolution family $U_B(\cdot,\cdot)$ such that

$$U_B(t,s)x = U(t,s)x + \int_s^t U(t,\tau)B(\tau)U_B(\tau,s)x\, d\tau, \quad t \geq s, \tag{2.7}$$

by means of the Dyson-Phillips expansion, see [105]. Then $u = U_B(\cdot, s)x$ can be considered as the mild solution of (2.6), cf. (2.2). However, in general (2.6) does not inherit the well-posedness of $(\mathrm{CP})_0$ as is seen by the next example which is a variant of [105, Example 6.4].

Example 2.3. On $X = C_0([0,1))$ we define $A\varphi = \varphi'$ with $D(A) = C_0^1([0,1))$ and

$$(B(t)\varphi)(\xi) = \begin{cases} \varphi(\xi), & 1 > \xi \geq \max\{0, \frac{1}{2} - t\}, \\ 2(\xi + t)\varphi(\xi), & \frac{1}{2} - t > \xi \geq 0, \end{cases}$$

for $\varphi \in X$. Observe that $B(\cdot) \in C_b(\mathbb{R}_+, \mathcal{L}(X))$. The mild solution of (2.6) is

$$u(t)(\xi) := U_B(t,0)\varphi(\xi) = \begin{cases} 0, & \xi + t \geq 1, \\ e^t \varphi(\xi + t), & 1 \leq 2(\xi + t) \leq 2, \\ e^{2(\xi+t)t}\varphi(\xi + t), & 0 \leq 2(\xi + t) \leq 1, \end{cases}$$

for $t \geq 0$ and $0 \leq \xi < 1$. Clearly, $u \in C^1(\mathbb{R}_+, X)$ if and only if $\varphi \in D(A)$ and $\varphi(\frac{1}{2}) = 0$, so that (2.6) is not well-posed.

3. Exponential dichotomy of evolution families

We turn our attention to asymptotic properties of evolution families $U(\cdot,\cdot)$ for $J = [a, \infty)$ or $\mathbb{R}$. The *(uniform exponential) growth bound* of $U(\cdot,\cdot)$ is defined by

$$\omega(U) := \inf\{w \in \mathbb{R} : \exists\, M_w \geq 1 \text{ with } \|U(t,s)\| \leq M_w\, e^{w(t-s)} \text{ for } t \geq s,\ s \in J\}.$$

The evolution family is called *exponentially bounded* if $\omega(U) < \infty$ and *exponentially stable* if $\omega(U) < 0$. For instance, $U(t,s) = \exp(t^2 - s^2)I$ has growth bound $+\infty$ and $V(t,s) = \exp(s^2 - t^2)I$ has growth bound $-\infty$.

A semigroup $T(\cdot)$ is exponentially stable if and only if the spectral radius of one operator $T(t)$, $t > 0$, is less than 1. In the non-autonomous case, however, the following examples show that the location of $\sigma(U(t,s))$ has no influence on the asymptotic behaviour of the evolution family $U(\cdot,\cdot)$, in general.

Example 3.1. Let $S(t)$, $t \geq 0$, be the nilpotent right translation on $X = L^1[0,1]$ and $U(t,s) = e^{t^2 - s^2} S(t-s)$. Then $r(U(t,s)) = 0$ for $t > s$ and $U(s+t, s) = 0$ for $t \geq 1$ and $s \in \mathbb{R}$, but $\|U(s + \frac{1}{2}, s)\| = e^{s+1/4}$ and hence $\omega(U) = +\infty$.

Example 3.2. In Example 3.5 we construct an evolution family such that $r(U(t,s)) = 0$ for $t > s$, but $t \mapsto \|U(t,s)\|$ grows faster than any exponential function as $t \to \infty$.

We add that, due to a standard argument, it is possible to control $\omega(U)$ by means of uniform norm estimates of $U(t,s)$ on strips $0 < t - s \leq t_0$, see [30, § III.4] or [39, p. 479].

One is further interested in exponential decay and increase on invariant subspaces, see [25], [28], [30], [31], [51], [53], [86], [119]. We write $Q = I - P$ for a projection P.

Definition 3.3. *An evolution family $U(\cdot,\cdot)$ on a Banach space X (with $J = \mathbb{R}$ or $[a,\infty)$) has an* exponential dichotomy *(or is called* hyperbolic*) if there are projections $P(t)$, $t \in J$, and constants $N, \delta > 0$ such that $P(\cdot) \in C_b(J, \mathcal{L}_s(X))$ and, for $t \geq s$, $s \in J$,*

a) $U(t,s)P(s) = P(t)U(t,s)$,
b) *the restriction $U_Q(t,s) : Q(s)X \to Q(t)X$ of $U(t,s)$ is invertible (and we set $U_Q(s,t) := U_Q(t,s)^{-1}$),*
c) $\|U(t,s)P(s)\| \leq Ne^{-\delta(t-s)}$ *and* $\|U_Q(s,t)Q(t)\| \leq Ne^{-\delta(t-s)}$.

The existence of an exponential dichotomy gives an important insight into the long-term behaviour of an evolution family. It is also used to study the asymptotic properties of (mild) solutions to the inhomogeneous problem (CP), see Section 6. Further, exponential splittings are preserved under small nonlinear perturbations which leads to principles of linearized (in-)stability and to the construction of stable, unstable, and center manifolds, as exposed in the monographs cited before Definition 3.3.

If the hyperbolic evolution family $U(t,s) = T(t-s)$ is given by a semigroup $T(\cdot)$ for $J = \mathbb{R}$, then $P(t)$ does not depend on t by [121, Corollary 3.3] so that $T(\cdot)$ is hyperbolic in the usual sense. Recall that a semigroup $T(\cdot)$ is hyperbolic if and only if the unit circle $\mathbb{T}$ belongs to $\rho(T(t_0))$ for some/all $t_0 > 0$. Moreover, the dichotomy projection P then coincides with the *spectral projection*

$$P = \frac{1}{2\pi i}\int_{\mathbb{T}} R(\lambda, T(t_0))\, d\lambda. \tag{3.1}$$

Similar results hold for *periodic* evolution families $U(\cdot,\cdot)$, i.e., $U(t+p, s+p) = U(t,s)$ for $t \geq s$, $t, s \in \mathbb{R}$, and some $p > 0$. In that case, one has the equalities

$$\begin{aligned} &\sigma(U(s+p,s)) = \sigma(U(p,0)) \quad \text{and} \\ &U(t,s) = U(t,t-p)^n U(s+\tau,s) = U(t,t-\tau)U(s+p,s)^n, \end{aligned} \tag{3.2}$$

where $t = s + np + \tau$, $n \in \mathbb{N}_0$, and $\tau \in [0,p)$. A periodic evolution family $U(\cdot,\cdot)$ is hyperbolic if and only if $\sigma(U(p,0)) \cap \mathbb{T} = \emptyset$ because of these relations. The dichotomy projections are then given by

$$P(s) = \frac{1}{2\pi i}\int_{\mathbb{T}} R(\lambda, U(s+p,s))\, d\lambda \tag{3.3}$$

for $s \in \mathbb{R}$, cf. [30], [31], [51], [53], [86].

In the autonomous case the exponential dichotomy of a semigroup $T(\cdot)$ generated by A always implies the spectral condition $\sigma(A) \cap i\mathbb{R} = \emptyset$. In particular, the exponential stability of $T(\cdot)$ yields $s(A) := \sup\{\operatorname{Re}\lambda : \lambda \in \sigma(A)\} < 0$. The converse implications fail for general C_0-semigroups, but can be verified if the spectral mapping theorem

$$\sigma(T(t)) \setminus \{0\} = e^{t\sigma(A)}, \qquad t \geq 0, \tag{3.4}$$

holds. Formula (3.4) is satisfied by eventually norm continuous semigroups, and hence by analytic or eventually compact semigroups. Starting from these results a powerful spectral theory for semigroups has been developed, see [39] and the references therein.

In many cases it is thus possible to characterize the exponential dichotomy of a semigroup by the spectrum of its generator. This is an important fact since in applications usually the generator A is the given object. Unfortunately, in the non-autonomous case there is no hope to relate the location of $\sigma(A(t))$ to the asymptotic behaviour of the evolution family $U(\cdot,\cdot)$ generated by $A(\cdot)$ as is shown by the following examples. We point out that the first one, [28, p. 3], [121, Example 3.4], deals with periodic evolution families $U_k(\cdot,\cdot)$ on $X = \mathbb{C}^2$ satisfying

$$s(A_1(t)) = -1 < \omega(U_1) \quad \text{and} \quad s(A_2(t)) = 1 > \omega(U_2) \quad \text{for } t \in \mathbb{R},$$

whereas for semigroups $T(\cdot)$ with generator A one always has $s(A) \le \omega(T)$.

Example 3.4. Let $A_k(t) = D(-t)A_k D(t)$ for $t \in \mathbb{R}$, where

$$D(t) = \begin{pmatrix} \cos t & \sin t \\ -\sin t & \cos t \end{pmatrix}, \qquad A_1 = \begin{pmatrix} -1 & -5 \\ 0 & -1 \end{pmatrix}, \qquad A_2 = \begin{pmatrix} 1 & 0 \\ 0 & -1 \end{pmatrix}.$$

The operators $A_1(t)$ and $A_2(t)$, $t \in \mathbb{R}$, generate the evolution families

$$U_1(t,s) = D(-t)\ \exp\left[(t-s)\begin{pmatrix} -1 & -4 \\ -1 & -1 \end{pmatrix}\right]\ D(s) \quad \text{and}$$
$$U_2(t,s) = D(-t)\ \exp\left[(t-s)\begin{pmatrix} 1 & 1 \\ -1 & -1 \end{pmatrix}\right]\ D(s),$$

respectively. Thus, $\omega(U_1) = 1$ and $\omega(U_2) = 0 = \omega(U_2^{-1})$, but $\sigma(A_1(t)) = \{-1\}$ and $\sigma(A_2(t)) = \{-1, 1\}$ for $t \in \mathbb{R}$. In other words, the exponential stability of $e^{\tau A_1(t)}$ and the exponential dichotomy of $e^{\tau A_2(t)}$ (with constants independent of t) are lost when passing to the non-autonomous problem.

In Example 3.4 we even have $\|R(\lambda, A_k(t))\|_2 = \|R(\lambda, A_k)\|_2$ for $t \in \mathbb{R}$, $\lambda \in \rho(A_k)$, and $k = 1, 2$. This shows that we cannot expect to deduce asymptotic properties of an evolution family from estimates on the resolvent of $A(t)$ along vertical lines as it is possible for a semigroup on a Hilbert space by virtue of the theorem of Gearhart-Howland-Prüss, see e.g. [25, Theorem 2.16]. In the following infinite dimensional example from [121, § 5] we even have $\omega(U) = +\infty$ and $s(A(t)) = -\infty$ for a.e. $t \ge 0$.

Example 3.5. Let $h(s) = 1$ for $s \in [0, \frac{1}{2})$ and $h(s) = 2$ for $s \in [\frac{1}{2}, \frac{3}{4}]$. Set $\mu = h\,ds$, $(\Omega, \nu) = \bigotimes_{n\in\mathbb{N}}([0, \frac{3}{4}], \mu)$, and $X = L^1(\Omega, \nu)$. We define for each $k \in \mathbb{N}$ and $\tau \ge 0$ the right-translation

$$(R_k(\tau)f)(x_1, \cdots) = \begin{cases} f(x_1, \cdots, x_{k-1}, x_k - \tau, x_{k+1}, \cdots), & x_k - \tau \in [0, \frac{3}{4}], \\ 0, & x_k - \tau \notin [0, \frac{3}{4}], \end{cases}$$

for simple functions $f \in X$. In [98, Lemma 2.1] it is shown that these operators can be extended to C_0-semigroups on X with generator A_k such that

$$\begin{aligned} &\|R_k(\tau)\| \le 2 \quad \text{for } \tau \ge 0, \quad R_k(\tau) = 0 \quad \text{for } \tau \ge \tfrac{3}{4}, \\ &\|R_n(\tau_n) \cdots R_m(\tau_m)\| = 2^{n-m+1} \quad \text{for } 0 < \tau_l \le \tfrac{1}{4} \text{ and } m \le l \le n, \\ &Y = \{f \in \bigcap_{k\in\mathbb{N}} D(A_k) : \|f\|_Y := \sup_{k\in\mathbb{N}}\{\|f\|_1\,,\ \|A_k f\|_1\} < \infty\} \quad \text{is dense in } X. \end{aligned} \tag{3.5}$$

Take $(t_k) = (0, 1, 2, \frac{5}{2}, 3, \frac{10}{3}, \frac{11}{3}, 4, \ldots)$ and choose functions $\alpha_k \in C^1(\mathbb{R})$ satisfying $\alpha_k > 0$ on (t_{k-1}, t_k), $\alpha_k = 0$ on $\mathbb{R}_+ \setminus (t_{k-1}, t_k)$, and $\|\alpha_k\|_\infty, \|\alpha_k'\|_\infty \le \frac{1}{4}$ for $k \in \mathbb{N}$. Then $a_k = \int \alpha_k(t)\, dt \le \frac{1}{4}$. We now define

$$A(t) = \alpha_k(t) A_k \quad \text{with } D(A(t)) = D(A_k) \ \text{ for } t_{k-1} < t < t_k \quad \text{and} \quad A(t_{k-1}) = 0$$

for $k \in \mathbb{N}$. Note that the operators $A(t)$ generate uniformly bounded, commuting, positive semigroups which are nilpotent if $t \neq t_k$, and $A(\cdot) \in C_b^1(\mathbb{R}_+, \mathcal{L}(Y, X))$. Moreover, $A(\cdot)$ generates the evolution family

$$U(t,s) = R_{l+1}\Big(\int_{t_l}^t \alpha_{l+1}(r)\,dr\Big)\, R_l(a_l) \,\cdots\, R_{k+1}(a_{k+1})\, R_k\Big(\int_s^{t_k} \alpha_k(r)\,dr\Big),$$

where $t_{k-1} < s \le t_k \le t_l \le t < t_{l+1}$. Finally, $\|U(2n, 0)\| \ge 2^{(n^2)}$ by (3.5), but $s(A(t)) = -\infty$ if $t \neq t_k$, $A(t_k) = 0$, and $r(U(t,s)) = 0$ for $t > s \ge 0$.

4. Evolution semigroups

We now present a semigroup approach to non-autonomous Cauchy problems. Let $U(\cdot,\cdot)$ be an exponentially bounded evolution family on a Banach space X and $J \in \{[a,b], [a,\infty), \mathbb{R}\}$. We define operators $T(t)$, $t \ge 0$, by setting

$$(T(t)f)(s) := \begin{cases} U(s, s-t) f(s-t), & s-t, s \in J, \\ 0, & s \in J, s-t \notin J, \end{cases}$$

on the function spaces $E = L^p(J, X)$, $1 \le p < \infty$, or

$$E = C_{00}(J, X) := \begin{cases} C_0(\mathbb{R}, X), & J = \mathbb{R}, \\ \{f \in C_0([a,\infty), X) : f(a) = 0\}, & J = [a,\infty), \\ \{f \in C([a,b], X) : f(a) = 0\}, & J = [a,b], \end{cases}$$

endowed with the usual p-norm, $1 \le p \le \infty$. It is easily verified that $T(\cdot)$ is a strongly continuous semigroup on E with $\omega(T) = \omega(U)$. We call $T(\cdot)$ the *evolution semigroup* on E associated with $U(\cdot,\cdot)$, and denote its generator by G. Evolution semigroups on the above spaces were introduced in the seventies by J.S. Howland, [56], D.E. Evans, [40], and L. Paquet, [103]. Other early contributions are due to H. Neidhardt, [95], and G. Lumer, [82], [83], [84]. Evolution semigroups can also be defined on other Banach function spaces, [112], [120], and on spaces of (almost) periodic functions, [8], [14], [45], [47], [57], [94]. The recent monograph

[25] by C. Chicone and Y. Latushkin is devoted to the investigation of evolution semigroups with an emphasis on spectral properties and applications to dynamical systems. In the latter field similar concepts have been developed since J. Mather's paper [88] from 1968, see [25, Chapter 6–8].

In this section we indicate how evolution semigroups can be used to solve non-autonomous Cauchy problems. This subject is further treated in [91]. The relationship to the asymptotic behaviour of evolution families is exposed in the next section.

We first give a representation of the generator G in the case that $U(\cdot,\cdot)$ solves a Cauchy problem, see [122, Proposition 3.14] which is a refinement of [25, Theorem 3.12], [71], [83], [103].

Proposition 4.1. *Let $(CP)_0$ be well posed on Y_t with an exponentially bounded evolution family $U(\cdot,\cdot)$, where $J \in \{[a,b],[a,\infty),\mathbb{R}\}$. Then $F = \{f \in C^1(J,X) : f(t) \in Y_t$ for $t \in J,\ f, f', A(\cdot)f \in E\}$ is a core of the generator G of the induced evolution semigroup on $E = C_{00}(J,X)$ and $Gf = -f' + A(\cdot)f(\cdot)$ for $f \in F$.*

We remark that an analogous result can be shown for the evolution semigroup on the space $E = L^p(J,X)$ with $F = \{f \in W^{1,p}(J,X) : f(a) = 0,\ f(t) \in Y_t$ for a.e. $t \in J,\ A(\cdot)f \in E\}$. Here one could also consider Cauchy problems being "well posed in $W^{1,p}$-sense".

We now try to establish a converse of Proposition 4.1: Show first that the sum $-\frac{d}{dt} + A(\cdot)$ on E has a closure which generates a semigroup $T(\cdot)$. (The sum is defined on $F = \{f \in D(A(\cdot)) \cap D(\frac{d}{dt}), f(a) = 0\}$ which is assumed to be dense in E.) Then the implication "(c)⇒(a)" of the following characterization of evolution semigroups provides us with an evolution family $U(\cdot,\cdot)$. Finally, solve (CP) using $U(\cdot,\cdot)$. Observe that the first of these three steps is closely connected to the operator sum method discussed in Section 2.

The next theorem is taken from [110, Theorem 2.4] and [120, Theorem 2.6]. Previous and different versions are contained in [40], [56], [82], [84], [85], [95], [103]. We define for $\varphi : J \to \mathbb{C}$ and $t \geq 0$ the translated function φ_t by $\varphi_t(s) = \varphi(s-t)$ if $s, s-t \in J$ and $\varphi_t(s) = 0$ if $s \in J,\ s-t \notin J$.

Theorem 4.2. *For a C_0-semigroup $T(\cdot)$ on $E = C_{00}(J,X)$ or $L^p(J,X)$, $1 \leq p < \infty$, with generator G, the following assertions are equivalent.*

a) *$T(\cdot)$ is an evolution semigroup given by an evolution family with index set $J \setminus \inf J$.*
b) *$T(t)\,(\varphi f) = \varphi_t T(t) f$ for $f \in E$, $\varphi \in C_b(J)$, $t \geq 0$. $D(G)$ is a dense subset of $C_{00}(J,X)$.*
c) *For all f contained in a core of G and $\varphi \in C^1(J)$ with $\varphi, \varphi' \in C_{00}(J)$, we have $\varphi f \in D(G)$ and $G(\varphi f) = \varphi G f - \varphi' f$. $D(G)$ is a dense subset of $C_{00}(J,X)$.*

Notice that the second condition in (b) and (c) is trivially satisfied if $E = C_{00}(J,X)$. Instead of this condition it was assumed in [110] and [120] that $R(\lambda, G)$

maps E continuously into $C_{00}(J,X)$ with dense range for some $\lambda \in \rho(G)$, but these two assertions are equivalent due to the closed graph theorem.

These characterizations were already used in the earliest papers on evolution semigroups to deduce perturbation results for evolution families from known perturbation theorems for semigroups. We refer to [82], [83], [93], [113] for bounded perturbations, to [110], [127] for relatively bounded perturbations of Miyadera type, and to [40], [56] for perturbation results in the context of scattering theory. We further observe that

$$u \in D(G),\ Gu = -f \iff u = T(r)u + \int_0^r T(\rho)f\,d\rho, \quad r \geq 0,$$

$$\iff u(t) = U(t,s)u(s) + \int_s^t U(t,\tau)f(\tau)\,d\tau, \quad t \geq s, \quad (4.1)$$

for $u, f \in E$, compare [11], [73], [96]. For $J = [a,b]$ or $[a,\infty)$, we arrive at

$$u \in D(G),\ Gu = -f \iff u(t) = \int_a^t U(t,\tau)f(\tau)\,d\tau, \quad t \geq a, \qquad (4.2)$$

since $u(a) = 0$ for $u \in D(G)$. In view of (2.2), (4.1), and (4.2), the elements of $D(G)$ for $J = [s,b]$ coincide with the mild solutions of (CP) with $x = 0$ and $f \in E$, and $U(\cdot,s)x$ with the mild solution of the homogeneous problem. At first glance, this is just a formal correspondence since we have introduced the term 'mild solution' only for well-posed problems. In the present situation, however, we obtain approximative solutions in C^1- or $W^{1,p}$-sense (depending on the choice of E) since G is the closure of $(-\frac{d}{dt} + A(\cdot), F)$. In fact, if $Gu = -f$ on $[s,b]$ (hence $u(s) = 0$), then there are $u_n \in F$ and $f_n \in E$ such that $u_n \to u$ and $f_n \to f$ in E and $u_n' = A(\cdot)u_n + f_n$. Since $u_n = -G^{-1}f_n$, we obtain that $u_n \to u$ uniformly also if $E = L^p$. Further, for $x \in X$, $s \in (a,b)$, and $n \in \mathbb{N}$, we define $u_n = \alpha_n(\cdot)U(\cdot, s - \frac{1}{n})x$, where $\alpha_n \in C^1([a,b])$ is equal to 1 on $[s,b]$ and vanishes on $[a, s - \frac{1}{n}]$ and $U(t,r) := 0$ for $t < r$. It is easy to check that $u_n \in D(G)$ and $Gu_n = -\alpha_n'(\cdot)U(\cdot, s - \frac{1}{n})x$ on $[a,b]$, see [25, p. 64]. By another approximation, we find $v_n \in F$ such that $v_n(t) \to U(t,s)x$ and $v_n' - A(\cdot)v_n \to 0$ on $[s,b]$.

It then remains to determine $D(G)$ or to verify regularity properties of $U(t,s)$ in order to obtain differentiable solutions of (CP). In [85] and [109] we investigated parabolic problems by means of this method. Using maximal regularity of type L^p, we proved that the mild solutions solve (CP) in $W^{1,p}$-sense. We also refer to [90] for a related semigroup approach to parabolic evolution equations.

It is interesting to recapitulate the results by G. Da Prato and M. Iannelli from [34] (see Paragraph (c) of Section 2) from the present point of view. In [34, Corollary 1, Proposition 4] it was shown that $-\frac{d}{dt} + A(\cdot)$ defined on the dense subspace $F = W_0^{1,p}((0,T],X) \cap L^p([0,T],Y)$ has an invertible closure G in $L^p([0,T],X)$ assuming Kato type conditions. Further, $D(G)$ is dense in $C_{00}([0,T],X)$ due to Proposition 3 and Remark 3 of [34]. Let $A_n(t)$ be the Yosida approximation of $A(t)$. Then $G_n = -\frac{d}{dt} + A_n(\cdot)$ converges strongly to G

on F, and the semigroups generated by G_n are uniformly bounded by [34, Proposition 1]. So the Trotter-Kato theorem shows that G generates a C_0-semigroup which is an evolution semigroup by Theorem 4.2. One now obtains approximate solutions as indicated above.

5. Characterizations of exponential dichotomy

Evolution semigroups have attracted a lot of interest in recent years since it was discovered that their spectra characterize the exponential dichotomy of evolution families. This is quite remarkable in view of the discouraging examples in Section 3. We first discuss spectral properties of the evolution semigroup $T(\cdot)$ and its generator G on $E = C_{00}(J,X)$ or $E = L^p(J,X)$. If J is compact, then $T(\cdot)$ is nilpotent so that $\sigma(G) = \emptyset$ and $\sigma(T(t)) = \{0\}$ for $t > 0$. This case is of course irrelevant to the long term behaviour of $U(\cdot,\cdot)$. For unbounded J, we have the following result from, e.g., Section 3.2.2 and 3.3.1 of [25].

Theorem 5.1. *Let $U(\cdot,\cdot)$ be an exponentially bounded evolution family on X and $J = \mathbb{R}$ or $\mathbb{R}_+$. Then the associated evolution semigroup $T(\cdot)$ on $E = C_{00}(J,X)$ or $E = L^p(J,X)$, $1 \le p < \infty$, with generator G has the following properties.*

a) *The spectral mapping theorem $\sigma(T(t)) \setminus \{0\} = e^{t\sigma(G)}$, $t \ge 0$, holds.*
b) *$\sigma(T(t))$, $t > 0$, is rotationally invariant and $\sigma(G)$ is invariant under translations along the imaginary axis. In the case $J = \mathbb{R}_+$, the spectrum of $T(t)$, $t > 0$, is the full disc with center 0 and radius $e^{t\omega(U)}$.*

R. Rau showed part (b) in [116, Proposition 2]. He also established (a) on $L^2(\mathbb{R},X)$ for a Hilbert space X using the Gearhart-Howland-Prüss theorem, see [114, Proposition 2.2] and also [115, Proposition 6]. This idea was previously employed in [74] in a somewhat different context. In the general case, the spectral mapping theorem for $J = \mathbb{R}$ is due to Y. Latushkin and S. Montgomery-Smith who proved it in [70, Theorem 3.1] by means of a reduction to the autonomous case. In [112], we proved (a) first directly on $C_0(\mathbb{R},X)$ and then for a large class of Banach function spaces E on $\mathbb{R}$ (containing $L^p(\mathbb{R},X)$) by showing that the spectra of $T(t)$ and G do not depend on the choice of E. Further direct proofs for the spectral mapping theorem are presented in [71] for the C_0-case and in [25, p. 83] for the L^p-case. All these proofs of (a) (except for Rau's) are based on explicit formulas for approximate eigenfunctions of G using given approximate eigenfunctions of $I - T(t)$ and the definition of $T(t)$. These constructions can be extended to the halfline case, see [120, Theorem 5.3] for $E = C_0$. Theorem 5.1 was also shown independently by A.G. Baskakov in [10] and [11] employing different methods indicated below. The importance of Theorem 5.1 relies on the equivalence (a)$\Leftrightarrow$(b) in the next result, see e.g. [25, § 3.2.3].

Theorem 5.2. *Let $U(\cdot,\cdot)$ be an exponentially bounded evolution family on X and $J = \mathbb{R}$. Let $T(\cdot)$ be the associated evolution semigroup on $E = C_0(\mathbb{R},X)$ or*

$E = L^p(\mathbb{R}, X)$, $1 \le p < \infty$, with generator G. Then the following assertions are equivalent.

a) *$U(\cdot,\cdot)$ has an exponential dichotomy with projections $P(\cdot)$.*
b) *$T(\cdot)$ has an exponential dichotomy with projection $\mathcal{P}$.*
c) *G is invertible.*

If this is the case, then

$$P(\cdot) = \mathcal{P} = \frac{1}{2\pi i}\int_{\mathbb{T}} R(\lambda, T(t))\, d\lambda, \quad t > 0. \tag{5.1}$$

The equivalence of the hyperbolicity of the evolution family and of the induced evolution semigroup on $C_0(\mathbb{R}, X)$ was essentially shown by R. Rau in [116, Theorem 6], see also [111]. The corresponding result for $L^p(\mathbb{R}, X)$ is due to Y. Latushkin, S. Montgomery-Smith, and T. Randolph, see [70, Theorem 3.4] and [72, Theorem 3.6], who used a rather involved discretization technique. We gave an elementary proof in the spirit of Rau's work in [111]. The most important step in each proof is to establish that the spectral projection $\mathcal{P}$ of the hyperbolic evolution semigroup is a multiplication operator which then gives the dichotomy projections for $U(\cdot,\cdot)$. This fact is a consequence of the algebraic structure of $T(t)$. One can also deduce the L^p-case from Rau's result on $C_0(\mathbb{R}, X)$ and the p-independence of the spectra of $T(t)$ and G, see [112]. We further refer to [7], [15], [97] for the case of bounded $A(t)$. Observe that the formula (5.1) gives a substitute for the more direct representations (3.1) and (3.3) valid in the autonomous and periodic case.

In view of (4.1), Theorem 5.2 immediately implies that

$$U(\cdot,\cdot) \text{ is hyperbolic} \iff \forall\, f \in E \;\exists!\, u \in E \;\text{ satisfying (4.1)} \tag{5.2}$$

for $E = C_0(\mathbb{R}, X)$ or $L^p(\mathbb{R}, X)$, [73]. The equivalence (5.2) also holds for $E = C_b(\mathbb{R}, X)$. This can be seen by reducing it to the C_0-case, [25, p. 128]. Moreover, V.V. Zhikov proved it already in 1972 directly by clever manipulations of the equation (4.1), see [76, Chapter 10] and also [28, § 8], [30, § IV.3]. Yet another proof is presented in [73]. Conversely, A.G. Baskakov deduced Theorem 5.1 and 5.2 for $J = \mathbb{R}$ from the equivalence (5.2) on $E = C_b(\mathbb{R}, X)$, [11].

In the same way, Theorem 5.1 for $J = \mathbb{R}_+$ and (4.2) allow to characterize exponential stability of an evolution family, see [25, Theorem 3.26]. We set $(\mathbb{K}f)(t) = \int_0^t U(t,\tau) f(\tau)\, d\tau$ for $t \ge 0$ and $f \in L^1_{\text{loc}}(\mathbb{R}_+, X)$.

Corollary 5.3. *An exponentially bounded evolution family $U(\cdot,\cdot)$ on X with $J = \mathbb{R}_+$ is exponentially stable if and only if $\mathbb{K}$ maps E into E if and only if the generator G of the evolution semigroup on E is invertible, where $E = C_{00}(\mathbb{R}_+, X)$ or $L^p(\mathbb{R}_+, X)$, $1 \le p < \infty$.*

The use of (4.2) in this context goes back to a paper by O. Perron from 1930. For different proofs and similar results (e.g., for $E = C_b(\mathbb{R}_+, X)$) we refer to [10], [17], [30, § III.5], [36, § 5], [96]. A related characterization of exponential stability is due to R. Datko, [36, Theorem 1, Rem.3], see also [25, Corollary 3.24], [30,

Theorem III.6.2], and the references therein. Here we give a new proof of Datko's theorem based on the spectral theory of evolution semigroups.

Theorem 5.4. *Let $U(\cdot,\cdot)$ be an exponentially bounded evolution family on X with $J = \mathbb{R}_+$. Then $U(\cdot,\cdot)$ is exponentially stable if and only if for some $1 \le p < \infty$ and all $x \in X$ and $s \ge 0$ there is a constant M such that*

$$\int_s^\infty \|U(t,s)x\|^p \, dt \le M^p \, \|x\|^p. \tag{5.3}$$

Proof. The necessity of (5.3) is clear. The integral version of Minkowski's inequality, [52, § 202], and (5.3) yield

$$\begin{aligned}\Big(\int_0^\infty \Big\| \int_0^t U(t,s)f(s)\, ds \Big\|^p dt \Big)^{\frac{1}{p}} &\le \int_0^\infty \Big(\int_s^\infty \|U(t,s)f(s)\|^p \, dt \Big)^{\frac{1}{p}} ds \\ &\le M \int_0^\infty \|f(s)\| \, ds = M \, \|f\|_1\end{aligned}$$

for $f \in L^1(\mathbb{R}_+, X)$. Thus, $\mathbb{K}$ maps $L^1(\mathbb{R}_+, X)$ into $L^p(\mathbb{R}_+, X)$. If $p = 1$, then Corollary 5.3 implies the assertion. For arbitrary $p \in [1, \infty)$, the equivalence (4.2) shows that the range of the generator G of the evolution semigroup $T(\cdot)$ on $E = L^p(\mathbb{R}_+, X)$ contains $L^p(\mathbb{R}_+, X) \cap L^1(\mathbb{R}_+, X)$ and is therefore dense in E. Suppose that $0 \in A\sigma(G)$. Then $1 \in A\sigma(T(1))$ by the spectral inclusion theorem, see e.g. [39, Theorem IV.3.6]. From this fact one easily derives the existence of a constant $c > 0$ and functions $f_n \in E$ such that $\|f_n\|_p = 1$ and $\|T(t)f_n\|_p \ge c$ for $t \in [0, n]$ and $n \in \mathbb{N}$. Now (5.3) yields

$$\begin{aligned} n\, c^p &\le \int_0^n \|T(t)f_n\|_p^p \, dt = \int_0^n \int_0^\infty \|U(s+t,s)f_n(s)\|^p \, ds \, dt \\ &\le \int_0^\infty \int_0^\infty \|U(s+t,s)f_n(s)\|^p \, dt \, ds \le M^p \int_0^\infty \|f_n(s)\|^p \, ds = M^p \end{aligned}$$

which is impossible. Hence, $0 \notin A\sigma(G)$ and G is invertible. The theorem is now a consequence of Corollary 5.3. □

Exponential dichotomy on $J = \mathbb{R}_+$ can be characterized in the spirit of (5.2) using the equation

$$u(t) = U(t,0)x + \int_0^t U(t,\tau)f(\tau)\, d\tau, \quad t \ge 0, \tag{5.4}$$

for $x \in X$. In this case, however, for a given $f \in E$ there exists more than one $u \in E$ satisfying (5.4) if $U(\cdot,\cdot)$ has an exponential dichotomy with $P(0) \neq 0$. For matrices $A(t)$ and assuming $\omega(U) < \infty$, it was shown in [28, § 3] or [30, § IV.3] that exponential dichotomy of $U(\cdot,\cdot)$ is equivalent to the existence of bounded solutions u of (5.4) for each $f \in C_b(\mathbb{R}_+, X)$. This leads to Fredholm properties of the operator $\frac{d}{dt} - A(\cdot)$ which are studied in [15], [16], [102]. These results were generalized to certain parabolic equations in [143]. In [96] we proved similar theorems for general evolution families. We remark that these more complicated characterizations can usually be avoided by extending the given problem on $\mathbb{R}_+$ to $\mathbb{R}$, see [121], [124], [125]. Nevertheless, dichotomies on $\mathbb{R}_+$ and $\mathbb{R}_-$ play an important role in dynamical systems, cf. [77], [87], and the references therein.

We mention three more characterizations of hyperbolicity. First, one can replace in the equivalence (a)$\Leftrightarrow$(b) of Theorem 5.2 function spaces on $\mathbb{R}$ by sequence spaces over $\mathbb{Z}$, see [26], [53, § 7.6], [108], and [11], [25], [71], [72], where this approach is used in the context of evolution semigroups. Sequence spaces over $\mathbb{N}$ are considered in [10] and [16].

For $X = \mathbb{C}^n$ exponential dichotomy is equivalent to the existence of a bounded, continuously differentiable Hermitian matrix function H such that

$$\tfrac{d}{dt} H(t) + H(t)A(t) + A^*(t)H(t) \leq -I$$

(where we use the order of symmetric matrices), see [28, § 7]. In Section 4.4 of [25] these results are extended to unbounded operators $A(t)$ on a Hilbert space X which satisfy Kato type conditions, see also [92], [117]. We note that in [25] the characterization is deduced via the evolution semigroup from a more easily accessible result for semigroups.

A third characterization of exponential dichotomy involves the *hull* $\mathcal{A}$ of the operators $A(s)$, i.e., the closure of the translates $\mathbb{R} \ni \tau \mapsto A(t+\tau)$, $t \in \mathbb{R}$, in an appropriate metric space, see [25], [26], [27], [28], [76], [108], [119], and the references therein. This approach requires, besides the existence of $\mathcal{A}$, certain compactness and almost periodicity properties. Then one obtains that the evolution family generated by $A(\cdot)$ is hyperbolic if the equations $u'(t) = B(t)u(t)$, $t \in \mathbb{R}$, do not admit a nontrivial bounded solution for any $B(\cdot) \in \mathcal{A}$, cf. [28], [119]. Due to space limitations we do not treat this approach in detail, see [25, Chapter 6–8], [119], and the bibliography therein for further information.

6. Asymptotic behaviour of evolution equations

In this final section we mainly present sufficient conditions on the operators $A(t)$ implying the existence of an exponential dichotomy for the homogeneous evolution equation. Here we concentrate on parabolic problems possibly having a retarded term. We further mention some results on strong stability of evolution families and discuss the impact of exponential dichotomy on the qualitative behaviour of the inhomogeneous problem.

The following collection is guided by the treatment of exponential dichotomy in the books [28] and [53] by W.A. Coppel and D. Henry, which deal with matrices $A(t)$ and 'lower order' perturbations $A(t) = A + B(t)$ of a sectorial operator A, respectively. The basic idea is to find situations where the spectra of $A(t)$ or of a related operator A_0 allow to describe the asymptotic behaviour of the evolution family $U(\cdot,\cdot)$ generated by $A(\cdot)$. Most of these results are based on the theory discussed in the previous section.

The reader should further recall the results for time-periodic problems which make use of the monodromy operator $U(p,0)$, compare the references given after (3.3). We add that (exponential) stability can also be tackled by direct estimates or by Lyapunov functions; however we do not give specific references concerning this well-known method.

(a) Robustness

We study the perturbed problem (2.6) assuming that $U(\cdot,\cdot)$ is hyperbolic. If the perturbations $B(t)$ are small in suitable norms, one can expect that the evolution family $U_B(\cdot,\cdot)$ given by (2.7) is also hyperbolic. Unfortunately, one has to impose quite restrictive smallness conditions, but for some classes of finite dimensional problems refinements of the standard bounds are known to be optimal, see [99] and the references therein. Among the vast literature on robustness we mention [26], [28], [30], [71], [72] for the case of bounded perturbations, [27], [53], [78], [108], [121] for unbounded perturbations, and [126] for perturbations $B(t) : \underline{X}_t \to \overline{X}_t$ where $\underline{X}_t \hookrightarrow X \hookrightarrow \overline{X}_t$. As a sample we state a theorem (put in a slightly different setting) which follows immediately from an estimate due to C.J.K. Batty and R. Chill, [12, Theorem 4.7], and Theorem 5.2.

Theorem 6.1. *Let $A(t)$ and $B(t)$, $t \in \mathbb{R}$, be operators on X such that $A(t) - w$ and $B(t) - w$ are sectorial of the same type and satisfy* (2.4) *for some $w \geq 0$. Assume that the evolution family $U(\cdot,\cdot)$ generated by $A(\cdot)$ has an exponential dichotomy with projections $P_U(\cdot)$. Then there is a number $\varepsilon > 0$ such that*

$$q := \sup_{t\in\mathbb{R}} \|R(w, A(t)) - R(w, B(t))\| \leq \varepsilon$$

implies that the evolution family $V(\cdot,\cdot)$ generated by $B(\cdot)$ has an exponential dichotomy with projections $P_V(\cdot)$. Moreover, $\dim P_U(t)X = \dim P_V(t)X$ *and* $\dim(I - P_U(t))X = \dim(I - P_V(t))X$.

Proof. Let $T_U(\cdot)$ and $T_V(\cdot)$ be the evolution semigroups on $C_0(\mathbb{R}, X)$ induced by $U(\cdot,\cdot)$ and $V(\cdot,\cdot)$, respectively. In view of Theorem 5.2 one has to show that

$$\|T_U(1) - T_V(1)\| = \sup_{t\in\mathbb{R}} \|U(t+1,t) - V(t+1,t)\| =: \delta$$

is smaller than a certain number $\delta_0 > 0$ depending on the dichotomy constants of $U(\cdot,\cdot)$, see [124, Proposition 2.3]. So the assertion is a consequence of Theorem 4.7 of [12] which says that $\delta \leq cq^\eta$ for some $\eta > 0$ and a constant c independent of q. □

Looking at the results used in the above proof, one sees that ε only depends on the constants in (2.4), w, the type of $A(t) - w$ and $B(t) - w$, and the dichotomy constants of $U(\cdot,\cdot)$. Moreover, $A(t)$ and $B(t)$ need not be densely defined.

Instead of $T_U(1) - T_V(1)$, one can also try to estimate the difference $G_U - G_V$ of the corresponding generators. This approach leads to better numerical values of ε but does not seem to work in the same generality as the above theorem, [121].

Analogous results for functional differential equations are treated in [77] for $X = \mathbb{C}^n$ and in [47], [122], [125] for the infinite dimensional case, where [125] deals with perturbations of the type $B(t)\phi = \phi(-\tau(t))$. In [47] and [125] we employed a generalized 'characteristic equation' which determines the exponential dichotomy of the delay problem. This characteristic equation was derived in [47] from Theorem 5.2. We also refer to [37], [49], [50], and the references therein for further results in the context of (exponential) stability.

Exponential dichotomy is also inherited in some cases if the perturbation is 'integrally small', i.e., a suitable norm of $\int_s^{s+t} B(\tau)\,d\tau$ is small uniformly for $s \geq 0$ and $t \in [0, t_0]$, see [28, § 5], [53, § 7.6].

(b) Asymptotically autonomous equations

Assume that $A(t)$ tends in a suitable sense to an operator A as $t \to \infty$. If A generates a hyperbolic semigroup e^{tA} with projection P, one can expect that $U(\cdot,\cdot)$ has an exponential dichotomy on a halfline $[a, \infty)$. It is not difficult to reduce this problem to case (a), see [76, p. 180] or [124, (3.2)]. Such inheritance properties were established in [76, Chapter 10] for bounded $A(t)$, in [77], [87] for ordinary delay equations (see also [51, § 6.6.3], [106], [107] for the case of a dominant eigenvalue of the autonomous problem), in [12], [124] for parabolic problems (see also [48], [133, § 5.8] for the case of exponential stability), and in [122], [125] for retarded parabolic equations. Further, the operators $U(s+t, s)$ and the dichotomy projections $P(s)$ tend strongly to e^{tA} and P as $s \to \infty$, and $P(s)$ and $I - P(s)$ inherit the rank of P and $I - P$, respectively, due to [124], [125]. In view of (6.1) below, these facts imply the convergence of the mild solution of (CP) if $f(t) \to f_\infty$, see [48], [124], [125], [133, § 5.8]. In [12] also asymptotically periodic problems and the almost periodicity of $U(\cdot,\cdot)$ were studied.

(c) Slowly oscillating coefficients

Example 3.4 shows that the hyperbolicity of the semigroups $e^{\tau A(t)}$ does not imply the exponential dichotomy of $U(\cdot,\cdot)$, in general. Nevertheless, $U(\cdot,\cdot)$ is hyperbolic if in addition the Hölder constant of $t \mapsto A(t)$ is small enough (in an appropriate sense). For bounded $A(t)$ this has been established in [9], [28, Proposition 6.1], and for ordinary delay equations in [81], see also [51, § 12.7]. Extending the approach of [9], we studied parabolic problems in [121] and [123]. There we showed the left- and right-invertibility of the generator G of the induced evolution semigroup making heavy use of parabolic regularity and inter/extrapolation spaces. Thus $U(\cdot,\cdot)$ has an exponential dichotomy, and one can also see that its dichotomy projections have the same rank as those of $e^{\tau A(r)}$.

(d) Rapidly oscillating coefficients

Suppose that $\lim_{T\to\infty} \frac{1}{T} \int_t^{t+T} A(s)\,ds = A$ (in a suitable sense) uniformly for $t \geq 0$ and that $e^{\tau A}$ is hyperbolic. Then it is known in some cases that $A(\omega t)$ generates an hyperbolic evolution family for large ω. This is shown for matrices $A(t)$ in [28, Proposition 5.3], for bounded operators in [76, Chapter 11], and for certain classes of parabolic problems in [53, § 7.6], [75], [76, Chapter 11]; see also [51, § 12.4]. We note that in [28] and [53] the problem is reduced to the robustness of exponential dichotomy under integrally small perturbations.

(e) Strong stability

The Arendt-Batty-Lyubich-Vũ theorem shows that a bounded semigroup $T(\cdot)$ with generator A tends strongly to 0 if $\sigma(A) \cap i\mathbb{R}$ is countable and $P\sigma(A') \cap i\mathbb{R}$ is empty, see e.g. [39, Theorem V.2.21]. This result cannot be applied to evolution semigroups since $\sigma(G)$ consists of vertical strips. However, C.J.K. Batty, R. Chill, and Y. Tomilov could characterize strong stability of a bounded evolution family (i.e., $U(t,s)x \to 0$ as $t \to \infty$ for each $s \geq 0$ and $x \in X$) by the density of the range of G on $L^1(\mathbb{R}_+, X)$, see [13, Theorem 2.2] for this and related facts.

In the time-periodic case the asymptotic behaviour of $U(\cdot, s)x$ is essentially determined by the operator $U(s+p, s)$ due to (3.2). This allows to apply discrete time-versions of the Arendt-Batty-Lyubich-Vũ theorem and its relatives to deduce strong stability or almost periodicity of $U(\cdot, s)x$ and of the solution of the inhomogeneous problem, see [14], [57], [118], [138].

(f) Inhomogeneous problems

Observe that due to (5.2) the exponential dichotomy of $U(\cdot,\cdot)$ implies the strong convergence to 0 of the mild solution u to (CP) with $f \in C_0(\mathbb{R}, X)$. This follows in a more direct way also from the formula

$$u(t) = \int_{-\infty}^{t} U(t,s)P(s)f(s)\,ds - \int_{t}^{\infty} U_Q(t,s)Q(s)f(s)\,ds, \quad t \in \mathbb{R}, \tag{6.1}$$

which is an easy consequence of (4.1) (see e.g. [25, p. 108]). If $U(\cdot,\cdot)$ is periodic, one can further prove the equivalence (5.2) for the space $E = AP(\mathbb{R}, X)$ of almost periodic functions, see [94] and also [57]. As a result, the almost periodicity of the inhomogeneity f is inherited by the mild solution u if $U(\cdot,\cdot)$ is periodic and hyperbolic. Delay equations were treated in an analogous way in [47] by means of evolution semigroups and in [46], [51, Chapter 6], [125] using variants of (6.1).

Related results (also for non hyperbolic $U(\cdot,\cdot)$) are shown by differing methods in [14], [28, Chapter 8], [31, § 6], [53, § 7.6], [57], [67], [76, Chapter 8], [86, § 6.3], [138].

References

[1] P. Acquistapace, *Evolution operators and strong solutions of abstract linear parabolic equations,* Differential Integral Equations **1** (1988), 433–457.

[2] P. Acquistapace, *Abstract linear nonautonomous parabolic equations: A survey,* in: G. Dore, A. Favini, E. Obrecht, A. Venni (Eds.), "Differential Equations in Banach Spaces" (Proc. Bologna 1991), Marcel Dekker, 1993, 1–19.

[3] P. Acquistapace, B. Terreni, *Linear parabolic equations in Banach spaces with variable domain but constant interpolations spaces,* Ann. Scuola Norm. Sup. Pisa Cl. Sci. (4) **13** (1986), 75–107.

[4] P. Acquistapace, B. Terreni, *A unified approach to abstract linear nonautonomous parabolic equations,* Rend. Sem. Mat. Univ. Padova **78** (1987), 47–107.

[5] H. Amann, *Dynamic theory of quasilinear parabolic equations – I. Abstract evolution equations,* Nonlinear Anal. **12** (1988), 895–919.

[6] H. Amann, *Linear and Quasilinear Parabolic Problems. Volume 1: Abstract Linear Theory,* Birkhäuser, 1995.

[7] B. Aulbach, Nguyen Van Minh, P.P. Zabreiko, *Groups of weighted translation operators in L_p and linear nonautonomous differential equations,* J. Math. Anal. Appl. **193** (1995), 622–631.

[8] M.E. Ballotti, J.A. Goldstein, M.E. Parrott, *Almost periodic solutions of evolution equations,* J. Math. Anal. Appl. **138** (1989), 522–536.

[9] A.G. Baskakov, *Some conditions for invertibility of linear differential and difference operators,* Russian Acad. Sci. Dokl. Math. **48** (1994), 498–501.

[10] A.G. Baskakov, *Linear differential operators with unbounded operator coefficients and semigroups of bounded operators,* Math. Notes **59** (1996), 586–593.

[11] A.G. Baskakov, *Semigroups of difference operators in spectral analysis of linear differential operators,* Funct. Anal. Appl. **30** (1996), 149–157.

[12] C.J.K. Batty, R. Chill, *Approximation and asymptotic behaviour of evolution families,* Differential Integral Equations **15** (2002), 477–512.

[13] C.J.K. Batty, R. Chill, Y. Tomilov, *Characterizations of strong stability of uniformly bounded evolution families and semigroups,* as a preprint in: Ulmer Seminare **4** (1999), 122–139.

[14] C.J.K. Batty, W. Hutter, F. Räbiger, *Almost periodicity of mild solutions of inhomogeneous Cauchy problems,* J. Differential Equations **156** (1999), 309–327.

[15] A. Ben-Artzi, I. Gohberg, *Dichotomies of systems and invertibility of linear ordinary differential operators,* Oper. Theory Adv. Appl. **56** (1992), 90–119.

[16] A. Ben-Artzi, I. Gohberg, M.A. Kaashoek, *Invertibility and dichotomy of differential operators on the half-line,* J. Dynam. Differential Equations **5** (1993), 1–36.

[17] C. Buşe, *On the Perron-Bellman theorem for evolutionary processes with exponential growth in Banach spaces,* New Zealand J. Math. **27** (1998), 183–190.

[18] A. Buttu, *On the evolution operator for a class of non-autonomous abstract parabolic equations,* J. Math. Anal. Appl. **170** (1992), 115–137.

[19] A. Buttu, *A construction of the evolution operator for a class of abstract parabolic equations,* Dynam. Systems Appl. **3** (1994), 221–234.

[20] P. Cannarsa, G. Da Prato, *Some results on abstract evolution equations of hyperbolic type,* in: G. Dore, A. Favini, E. Obrecht, A. Venni (Eds.), "Differential Equations in Banach Spaces" (Proceedings Bologna 1991), Marcel Dekker, 1993, 41–50.

[21] P. Cannarsa, G. Da Prato, J.P. Zolésio, *Evolution equations in noncylindrical domains,* Atti Accad. Naz. Lincei Rend. Cl. Sci. Fis. Mat. Natur. (8) **83** (1989), 72–77.

[22] P. Cannarsa, G. Da Prato, J.P. Zolésio, *The damped wave equation in a noncylindrical domain,* J. Differential Equations **85** (1990), 1–14.

[23] O. Caps, *Wellposedness of the time-dependent Cauchy problem in scales of Hilbert spaces,* Preprint Reihe des Fachbereiches Mathematik Universität Mainz **16**, 1998.

[24] O. Caps, *Wellposedness of the time-dependent Cauchy problem in scales of Banach spaces,* Preprint Reihe des Fachbereiches Mathematik Universität Mainz **1**, 1999.

[25] C. Chicone, Y. Latushkin, *Evolution Semigroups in Dynamical Systems and Differential Equations,* Amer. Math. Soc., 1999.

[26] S.-N. Chow, H. Leiva, *Existence and roughness of the exponential dichotomy for skew product semiflow in Banach space,* J. Differential Equations **120** (1995), 429–477.

[27] S.-N. Chow, H. Leiva, *Unbounded perturbations of the exponential dichotomy for evolution equations,* J. Differential Equations **129** (1996), 509–531.

[28] W.A. Coppel, *Dichotomies in Stability Theory,* Springer-Verlag, 1978.

[29] M.G. Crandall, A. Pazy, *Nonlinear evolution equations in Banach spaces,* Israel J. Math. **11** (1972), 57–94.

[30] J.L. Daleckij, M.G. Krein, *Stability of Solutions of Differential Equations in Banach Spaces,* Amer. Math. Soc., 1974.

[31] D. Daners, P. Koch Medina, *Abstract Evolution Equations, Periodic Problems and Applications,* Longman, 1992.

[32] G. Da Prato, P. Grisvard, *Sommes d'opérateurs linéaires et équations différentielles opérationelles,* J. Math. Pures Appl. **54** (1975), 305–387.

[33] G. Da Prato, P. Grisvard, *Equations d'évolution abstraites non linéaires de type parabolique,* Ann. Mat. Pura Appl. **120** (1979), 329–396.

[34] G. Da Prato, M. Iannelli, *On a method for studying abstract evolution equations in the hyperbolic case,* Commun. Partial Differential Equations **1** (1976), 585–608.

[35] G. Da Prato, E. Sinestrari, *Non autonomous evolution operators of hyperbolic type,* Semigroup Forum **45** (1992), 302-321.

[36] R. Datko, *Uniform asymptotic stability of evolutionary processes in a Banach space,* SIAM J. Math. Anal. **3** (1972), 428–445.

[37] W. Desch, I. Győri, G. Gühring, *Stability of nonautonomous delay equations with a positive fundamental solution,* as a preprint in: Tübinger Berichte zur Funktionalanalysis **9** (2000), 125–139.

[38] O. Diekman, M. Gyllenberg, H.R. Thieme, *Perturbing evolutionary systems by step responses and cumulative outputs,* Differential Integral Equations **8** (1995), 1205–1244.

[39] K.J. Engel, R. Nagel, *One-Parameter Semigroups for Linear Evolution Equations,* Springer-Verlag, 2000.

[40] D.E. Evans, *Time dependent perturbations and scattering of strongly continuous groups on Banach spaces,* Math. Ann. **221** (1976), 275–290.

[41] L. C. Evans, *Nonlinear evolution equations in an arbitrary Banach space,* Israel J. Math **26** (1977), 1–42.

[42] H.O. Fattorini, *The Cauchy Problem,* Addison Wesley, 1983.

[43] M. Giga, Y. Giga, H. Sohr, *L^p estimate for abstract linear parabolic equations,* Proc. Japan Acad. Ser. A **67** (1991), 197–202.

[44] J.A. Goldstein, *An example of a nonlinear semigroup,* Nieuw Arch. Wisk. (3) **22** (1974), 170–174.

[45] J.A. Goldstein, *Periodic and pseudo-periodic solutions of evolution equations,* in: H. Brézis, M.G. Crandall, F. Kappel (Eds.), "Semigroups, Theory and Applications I," Longman, 1986, 142–146.

[46] G. Gühring, F. Räbiger, *Asymptotic properties of mild solutions of nonautonomous evolution equations with applications to retarded differential equations,* Abstr. Appl. Anal. **4** (1999), 169–194.

[47] G. Gühring, F. Räbiger, R. Schnaubelt, *A characteristic equation for nonautonomous partial functional differential equations,* to appear in: J. Differential Equations.

[48] D. Guidetti, *On the asymptotic behavior of solutions of linear nonautonomous parabolic equations,* Boll. Un. Mat. Ital. B (7) **1** (1987), 1055–1076.

[49] I. Győri, F. Hartung, *Stability in delay perturbed differential and difference equations,* in: T. Faria, P. Freitas (Eds.): "Topics in Functional Differential and Difference Equations (Proceedings Lisboa 1999)," Amer. Math. Soc., 2001.

[50] I. Győri, F. Hartung, *Preservation of stability in a linear neutral differential equation under delay perturbations,* Dynam. Systems Appl. **10** (2001), 225–242.

[51] J.K. Hale, S.M. Verduyn Lunel, *Introduction to Functional Differential Equations,* Springer-Verlag, 1993.

[52] G.H. Hardy, J.E. Littlewood, G. Pólya, *Inequalities,* Cambridge University Press, Reprint of the 2nd Edition, 1999.

[53] D. Henry, *Geometric Theory of Semilinear Parabolic Equations,* Springer-Verlag, 1981.

[54] M. Hieber, S. Monniaux, *Heat-kernels and maximal L^p-L^q-estimates: the non-autonomous case,* J. Fourier Anal. Appl. **6** (2000), 467–481.

[55] M. Hieber, S. Monniaux, *Pseudo-differential operators and maximal regularity results for non-autonomous parabolic equations,* Proc. Amer. Math. Soc. **128** (2000), 1047–1053.

[56] J.S. Howland, *Stationary scattering theory for time-dependent Hamiltonians,* Math. Ann. **207** (1974), 315–335.

[57] W. Hutter, *Spectral theory and almost periodicity of mild solutions of non-autonomous Cauchy problems,* Ph.D. thesis, Tübingen, 1997.

[58] S. Ishii, *Linear evolution equations $du/dt + A(t)u = 0$: a case where $A(t)$ is strongly uniform-measurable,* J. Math. Soc. Japan **34** (1982), 413-423.

[59] S. Karrmann, *Non-autonomous forms and unique existence of solutions,* as a preprint in: Ulmer Seminare **5** (2000), 202–211.

[60] T. Kato, *Integration of the equation of evolution in a Banach space,* J. Math. Soc. Japan **5** (1953), 208-234.

[61] T. Kato, *Abstract evolution equations of parabolic type in Banach and Hilbert spaces,* Nagoya Math. J. **19** (1961), 93-125.

[62] T. Kato, *Linear evolution equations of "hyperbolic" type,* J. Fac. Sci. Univ. Tokyo **25** (1970), 241-258.

[63] T. Kato, *Quasilinear equations of evolution, with applications to partial differential equations,* in: N.V. Everitt (Ed.), "Spectral Theory and Differential Equations" (Proceedings Dundee 1974), Springer-Verlag, 1975, 25–70.

[64] T. Kato, *Abstract evolution equations, linear and quasi-linear, revisited,* in: H. Komatsu (Ed.), "Functional Analysis and Related Topics" (Proceedings Kyoto 1991), Springer-Verlag, 1993, 103–125.

[65] T. Kato, H. Tanabe, *On the abstract evolution equation,* Osaka Math. J. **14** (1962), 107-133.

[66] K. Kobayasi, *On a theorem for linear evolution equations of hyperbolic type,* J. Math. Soc. Japan **31** (1979), 647-654.

[67] J. Kreulich, *Eberlein-weakly almost-periodic solutions of evolution equations in Banach spaces,* Differential Integral Equations **9** (1996), 1005–1027.

[68] R. Labbas, B. Terreni, *Somme d'opérateurs linéaires de type parabolique.* 1re *Partie.* Boll. Un. Mat. Ital. B(7) **1** (1987), 545–569.

[69] R. Labbas, B. Terreni, *Somme d'opérateurs de type elliptique and parabolique.* 2e *partie: Applications.* Boll. Un. Mat. Ital. B(7) **2** (1988), 141–162.

[70] Y. Latushkin, S. Montgomery-Smith, *Evolutionary semigroups and Lyapunov theorems in Banach spaces,* J. Funct. Anal. **127** (1995), 173–197.

[71] Y. Latushkin, S. Montgomery-Smith, T. Randolph, *Evolutionary semigroups and dichotomy of linear skew-product flows on locally compact spaces with Banach fibers,* J. Differential Equations **125** (1996), 73–116.

[72] Y. Latushkin, T. Randolph, *Dichotomy of differential equations on Banach spaces and an algebra of weighted translation operators,* Integral Equations Operator Theory **23** (1995), 472-500.

[73] Y. Latushkin, T. Randolph, R. Schnaubelt, *Exponential dichotomy and mild solutions of nonautonomous equations in Banach spaces,* J. Dynam. Differential Equations **10** (1998), 489–510.

[74] Y. Latushkin, A. M. Stepin, *Weighted translation operators and linear extension of dynamical systems,* Russian Math. Surveys **46** (1991), 95–165.

[75] V.B. Levenshtam, *Averaging of quasilinear parabolic equations with rapidly oscillating principal part. Exponential dichotomy,* Russian Acad. Sci. Izv. Math. **41** (1993), 95–132.

[76] B.M. Levitan, V.V. Zhikov, *Almost Periodic Functions and Differential Equations,* Cambridge University Press, 1982.

[77] X.B. Lin, *Exponential dichotomies and homoclinic orbits in functional differential equations,* J. Differential Equations **63** (1986), 227–254.

[78] X.B. Lin, *Exponential dichotomy and stability of long periodic solutions in predator-prey models with diffusion,* in: J. Hale, J. Wiener (Ed.), "Partial Differential Equations" (Proceedings Edinburgh (TX), 1991), Pitman, 1992, 101–105.

[79] J.L. Lions, *Optimal Control of Systems Governed by Partial Differential Equations,* Springer-Verlag, 1971.

[80] J.L. Lions, E. Magenes, *Non-homogeneous Boundary Value Problems and Applications I,* Springer-Verlag, 1972.

[81] M. Lizana Peña, *Exponential dichotomy of singularly perturbed linear functional differential equations with small delays,* Appl. Anal. **47** (1992), 213–225.

[82] G. Lumer, *Équations d'évolution, semigroupes en espace-temps et perturbations,* C.R. Acad. Sci. Paris Série I **300** (1985), 169–172.

[83] G. Lumer, *Opérateurs d'évolution, comparaison de solutions, perturbations et approximations,* C.R. Acad. Sci. Paris Série I **301** (1985), 351–354.

[84] G. Lumer, *Equations de diffusion dans les domaines (x,t) non-cylindriques et semigroupes "espace-temps",* Séminaire de Théorie du Potentiel Paris **9** (1989), 161–179.

[85] G. Lumer, R. Schnaubelt, *Local operator methods and time-dependent parabolic equations on non-cylindrical domains,* in: M. Demuth, E. Schrohe, B.-W. Schulze, J. Sjöstrand (Eds.), "Evolution Equations, Feshbach Resonances, Singular Hodge Theory," Wiley-VCH, 1999, 58–130.

[86] A. Lunardi, *Analytic Semigroups and Optimal Regularity in Parabolic Problems,* Birkhäuser, 1995.

[87] J. Mallet-Paret, *The Fredholm alternative for functional differential equations of mixed type,* J. Dynam. Differential Equations **11** (1999), 1–48.

[88] J. Mather, *Characterization of Anosov diffeomorphisms,* Indag. Math. **30** (1968), 479–483.

[89] S. Monniaux, J. Prüss, *A theorem of Dore-Venni type for non-commuting operators,* Trans. Amer. Math. Soc. **349** (1997), 4787-4814.

[90] S. Monniaux, A. Rhandi, *Semigroup methods to solve non-autonomous evolution equations,* Semigroup Forum **60** (2000), 1–13.

[91] R. Nagel, G. Nickel, *Wellposedness for nonautonomous abstract Cauchy problems,* appears in this volume.

[92] R. Nagel, A. Rhandi, *Positivity and Lyapunov stability conditions for linear systems,* Adv. Math. Sci. Appl. **3** (1993), 33–41.

[93] R. Nagel, A. Rhandi, *A characterization of Lipschitz continuous evolution families on Banach spaces,* Oper. Theory Adv. Appl. **75** (1995), 275–286.

[94] T. Naito, Nguyen Van Minh, *Evolution semigroups and spectral criteria for almost periodic solutions of periodic evolution equations,* J. Differential Equations 152 (1999), 338–376.

[95] H. Neidhardt, *On abstract linear evolution equations I,* Math. Nachr. **103** (1981), 283–293.

[96] Nguyen Van Minh, F. Räbiger, R. Schnaubelt, *Exponential stability, exponential expansiveness, and exponential dichotomy of evolution equations on the half-line,* Integral Equations Operator Theory **32** (1998), 332–353.

[97] Nguyen Van Minh, *Semigroups and stability of nonautonomous differential equations in Banach spaces,* Trans. Amer. Math. Soc. **345** (1994), 223–241.

[98] G. Nickel, R. Schnaubelt, *An extension of Kato's stability condition for nonautonomous Cauchy problems,* Taiwanese J. Math. **2** (1998), 483–496.

[99] D.G. Obert, *Examples and bounds in the general case of exponential dichotomy roughness,* J. Math. Anal. Appl. **174** (1993), 231–241.

[100] N. Okazawa, *Remarks on linear evolution equations of hyperbolic type in Hilbert space,* Adv. Math. Sci. Appl. **8** (1998), 399–423.

[101] N. Okazawa, A. Unai, *Linear evolution equations of hyperbolic type in Hilbert spaces,* SUT J. Math. **29** (1993), 51–70.

[102] K.J. Palmer, *Exponential dichotomy and Fredholm operators*, Proc. Amer. Math. Soc. **104** (1988), 149–156.

[103] L. Paquet, *Semigroupes géneralisés et équations d'évolution,* Séminaire de Théorie du Potentiel Paris **4** (1979), 243–263.

[104] A. Pazy, *Semigroups of Linear Operators and Applications to Partial Differential Equations,* Springer-Verlag, 1983.

[105] R.S. Phillips, *Perturbation theory for semi-groups of linear operators,* Trans. Amer. Math. Soc. **74** (1953), 199–221.

[106] M. Pituk, *Asymptotic behaviour of solutions of a differential equation with asymptotically constant delay,* Nonlinar Anal. **30** (1997), 1111–1118.

[107] M. Pituk, *The Hartman-Wintner theorem for functional differential equations,* J. Differential Equations **155** (1999), 1–16.

[108] V.A. Pliss, G.R. Sell, *Robustness of exponential dichotomies in infinite-dimensional dynamical systems,* J. Dynam. Differential Equations **11** (1999), 471–514.

[109] J. Prüss, R. Schnaubelt, *Solvability and maximal regularity of parabolic evolution equations with coefficients continuous in time,* J. Math. Anal. Appl. **256** (2001), 405–430.

[110] F. Räbiger, A. Rhandi, R. Schnaubelt, J. Voigt, *Non-autonomous Miyadera perturbations,* Differential Integral Equations **13** (1999), 341–368.

[111] F. Räbiger, R. Schnaubelt, *A spectral characterization of exponentially dichotomic and hyperbolic evolution families,* Tübinger Berichte zur Funktionalanalysis **3** (1994), 222–234.

[112] F. Räbiger, R. Schnaubelt, *The spectral mapping theorem for evolution semigroups on spaces of vector-valued functions,* Semigroup Forum **52** (1996), 225–239.

[113] F. Räbiger, R. Schnaubelt, *Absorption evolution families and exponential stability of non-autonomous diffusion equations,* Differential Integral Equations **12** (1999), 41–65.

[114] R. Rau, *Hyperbolic evolution semigroups,* Ph.D. thesis, Tübingen, 1992.

[115] R. Rau, *Hyperbolic evolution groups and dichotomic evolution families,* J. Dynam. Differential Equations **6** (1994), 335–350.

[116] R. Rau, *Hyperbolic evolution semigroups on vector valued function spaces,* Semigroup Forum **48** (1994), 107–118.

[117] A. Rhandi, *Positivity methods for dichotomy of linear skew-product flows on Hilbert spaces,* Quast. Math. **17** (1994), 499–512.

[118] W.M. Ruess, W.H. Summers, *Weak almost periodicity and the strong ergodic limit theorem for periodic evolution systems,* J. Funct. Anal. **94** (1990), 177–195.

[119] R.J. Sacker, G.R. Sell, *Dichotomies for linear evolutionary equations in Banach spaces,* J. Differential Equations **113** (1994), 17–67.

[120] R. Schnaubelt, *Exponential bounds and hyperbolicity of evolution families,* Ph.D. thesis, Tübingen, 1996.

[121] R. Schnaubelt, *Sufficient conditions for exponential stability and dichotomy of evolution equations,* Forum Math. **11** (1999), 543–566.

[122] R. Schnaubelt, *Exponential dichotomy of non-autonomous evolution equations,* Habilitation thesis, Tübingen, 1999.

[123] R. Schnaubelt, *A sufficient condition for exponential dichotomy of parabolic evolution equations,* in: G. Lumer, L. Weis (Eds.), "Evolution Equations and their Applications in Physical and Life Sciences (Proceedings Bad Herrenalb, 1998)," Marcel Dekker, 2000, 149–158.

[124] R. Schnaubelt, *Asymptotically autonomous parabolic evolution equations,* J. Evol. Equ. **1** (2001), 19–37.

[125] R. Schnaubelt, *Parabolic evolution equations with asymptotically autonomous delay,* Report No. 02 (2001), Fachbereich Mathematik und Informatik, Universität Halle (preprint).

[126] R. Schnaubelt, *Feedbacks for non-autonomous regular linear systems,* Report No. 14 (2001), Fachbereich Mathematik und Informatik, Universität Halle (preprint).

[127] R. Schnaubelt, J. Voigt, *The non-autonomous Kato class,* Arch. Math. **72** (1999), 454–460.

[128] P.E. Sobolevskii, *Equations of parabolic type in a Banach space,* Amer. Math. Soc. Transl. **49** (1966), 1–62. (First published 1961 in Russian.)

[129] P.E. Sobolevskii, *Parabolic equations in a Banach space with an unbounded variable operator, a fractional power of which has a constant domain of definition,* Soviet Math. Dokl. **2** (1961), 545–548.

[130] W. Stannat, *The theory of generalized Dirichlet forms and its applications in analysis and stochastics,* Memoirs Amer. Math. Soc. **142** (1999), no. 678.

[131] Ž. Štrkalj, *$\mathcal{R}$-Beschränkheit, Summensätze abgeschlossener Operatoren und operatorwertige Pseudodifferentialoperatoren,* Ph.D. thesis, Karlsruhe, 2000.

[132] H. Tanabe, *Remarks on the equations of evolution in a Banach space,* Osaka Math. J. **112** (1960), 145–166.

[133] H. Tanabe, *Equations of Evolution,* Pitman, 1979.

[134] H. Tanabe, *Functional Analytic Methods for Partial Differential Equations,* Marcel Dekker, 1997.

[135] N. Tanaka, *Linear evolution equations in Banach spaces,* Proc. Lond. Math. Soc. **63** (1991), 657–672.

[136] N. Tanaka, *Semilinear equations in the "hyperbolic" case,* Nonlinear Anal. **24** (1995), 773–788.

[137] N. Tanaka, *Quasilinear evolution equations with non-densely defined operators,* Differential Integral Equations **9** (1996), 1067–1108.

[138] Vũ Quôc Phóng, *Stability and almost periodicity of trajectories of periodic processes,* J. Differential Equations **115** (1995), 402–415.

[139] A. Yagi, *On the abstract evolution equations of parabolic type,* Osaka J. Math. **14** (1977), 557-568.

[140] A. Yagi, *Parabolic equations in which the coefficients are generators of infinitely differentiable semigroups II,* Funkcial. Ekvac. **33** (1990), 139–150.

[141] A. Yagi, *Abstract quasilinear evolution equations of parabolic type in Banach spaces,* Boll. Un. Mat. Ital. B (7) **5** (1991), 351–368.

[142] Y. Yamamoto, *Solutions in L^p of abstract parabolic equations in Hilbert spaces,* J. Math. Kyoto Univ. **33** (1993), 299–314.

[143] W. Zhang, *The Fredholm alternative and exponential dichotomies for parabolic equations,* J. Math. Anal. Appl. **191** (1995), 180–201.

Roland Schnaubelt
FB Mathematik und Informatik
Martin-Luther-Universität
Theodor-Lieser-Str. 5
D-06120 Halle, Germany
e-mail: roland@hilbert.mathematik.uni-halle.de

Progress in Nonlinear Differential Equations
and Their Applications, Vol. 50, 339–351

Abstract Hyperbolic Equations in Non-reflexive Spaces

Eugenio Sinestrari

0. Introduction

Let X be a Banach space and $A : D(A) \subset X \to X$ a linear operator and consider homogeneous linear abstract Cauchy problem:

$$(P_0) \qquad \begin{cases} u'(t) = Au(t) \;, & t \geq 0 \\ u(0) = u_0. \end{cases}$$

Various results show that this problem is "well posed" (according to a suitable definition) if and only if A is the infinitesimal generator of a semigroup e^{At} of linear operators: see e.g. Theorem 1.3 of [35] or cor. 6.9 of [11].

Now the theorem of Hille-Yosida gives necessary and sufficient conditions for a linear operator A to be the generator of a semigroup: among them there is the density of $D(A)$ in X: but this condition is violated when (P_0) is the abstract version of some initial-boundary value problem for a partial differential equation studied in classical spaces such as continuous or essentially bounded functions spaces.

From this consideration originated the study of problem (P_0) when A has a non-dense domain: if A verifies all the other conditions to be the generator of an analytic semigroup, the theory is well established (for the research up to 1995 we refer to the monograph [23]). In this case it is possible to define a semigroup of operators in the whole X although not strongly continuous up to $t = 0$ and for the non-homogeneous Cauchy problem

$$(P) \qquad \begin{cases} u'(t) = Au(t) + f(t) \;, & t \in [0,T] \\ u(0) = u_0 \end{cases}$$

it is possible to obtain maximal regularity results i.e. to find some spaces $\mathcal{F}$ of functions from $[0,T]$ to X such that if $f \in \mathcal{F}$ then there exists a solution such that u', $Au \in \mathcal{F}$.

The situation is completely different when A is a Hille-Yosida operator i.e. satisfies the conditions of Hille-Yosida theorem except the density of its domain. In this paper we want to summarize the main results about problem (P) when A has these properties and X is a non-reflexive space. We also add the results of the time-dependent case and some applications. We take this occasion to present a new proof of the temporal regularity result, essentially due to T. Kato ([17]) and contained in a personal letter to the author of September 1986.

In this way I wish to join the tribute to the memory of Brunello Terreni to that of Tosio Kato.

In this paper $(X, \|\cdot\|)$ will denote a Banach space and $\mathcal{L}(X)$ the Banach space of linear bounded operators of X into itself. If A is a linear operator in X, $D(A)$ denotes its domain and $\rho(A)$ its resolvent set.

We will need the usual spaces of functions $u : [0,T] \to X$ such as $C^n(0,T;X)$, $n = 0,1,2$ or the spaces of (classes) of functions $L^p(0,T;X)$ and the Sobolev space $W^{1,p}(0,T;X)$. We set $\mathbf{N}_* = \{1,2,3,\dots\}$.

1. Hille-Yosida operators and approximants

Definition 1. *Let X be a Banach space. A linear operator $A : D(A) \subset X \to X$ is called a Hille-Yosida (H.Y.) operator if there exist ω, $M \in \mathbf{R}$ such that if $\lambda \succ \omega$ then $\lambda \in \rho(A)$ and for each $n \in \mathbf{N}_*$*

$$\|(\lambda - A)^{-n}\|_{\mathcal{L}(X)} \leq \frac{M}{(\lambda-\omega)^n}\,. \tag{1.1}$$

It is known that A is the generator of semigroup if and only if we have also $\overline{D(A)} = X$: now this property is a consequence of (1.1) when X is reflexive (see [16]).

In any case it is possible to associate to the H.Y.-operator A a semigroup of operators in the space

$$X_0 = \overline{D(A)}$$

by considering the part of A in X_0 i.e.

$$\begin{cases} A_0 : D(A_0) \subset X_0 \to X_0 \\ D(A_0) = \{x \in D(A)\ ; \quad Ax \in \overline{D(A)}\} \\ A_0 x = Ax\ , \quad x \in D(A_0)\,. \end{cases}$$

It is known (see [15]) that $D(A_0)$ is dense in X_0 and A_0 verifies condition (1.1) with the same constants ω, M: hence it generates a semigroup $e^{A_0 t}$ in X_0.

For a H.Y.-operator its Yosida approximants A_n can be defined for each $n \in \mathbf{N}_*$, $n > \omega$

$$A_n = nA(n-A)^{-1} = n^2(n-A)^{-1} - n\,.$$

They have the following properties (see [35]):

$$\lim_{n\to\infty} A_n x = A_0 x\ , \qquad x \in D(A_0)\,. \tag{1.2}$$

Given $\omega' > \omega$ there exists $n' \in \mathbf{N}_*$ such that for each $n > n'$ and $t > 0$

$$\|e^{A_n t}\|_{\mathcal{L}(X)}\| \leq M e^{\omega' t}\,. \tag{1.3}$$

For each $T > 0$ and compact set $K_0 \subset X_0$ we have also

$$\lim_{n\to\infty} e^{A_n t} x_0 = e^{A_0 t} x_0 \tag{1.4}$$

uniformly for $t \in [0,T]$ and $x_0 \in K_0$.

We will use also the following properties of A_n which are direct consequences of its definition:

Proposition 1. *Let A verify (1.1). If there exists $A^{-1} \in \mathcal{L}(X)$ then:*

$$\text{there exists } A_n^{-1} = A^{-1} - n^{-1} \quad \text{for} \quad n > \omega \tag{1.5}$$

$$\sup_{n>\omega} \|A_n^{-1}\|_{\mathcal{L}(X)} < \infty \tag{1.6}$$

$$\lim_{n\to\infty} \|A_n^{-1} - A^{-1}\|_{\mathcal{L}(X)} = 0 \tag{1.7}$$

$$\begin{cases} \text{for each } T > 0 \text{ and compact set } K \subset X \text{ we have} \\ \lim_{h\to\infty} e^{A_n t} A_n^{-1} x = e^{A_0 t} A^{-1} x \\ \text{uniformly for } t \in [0,T] \text{ and } x \in K\,. \end{cases} \tag{1.8}$$

$$\begin{cases} \text{If } v, v_n \in C(0,T;X) \text{ and } \lim_{n\to\infty} \|v_n - v\|_\infty = 0 \text{ then} \\ \lim_{n\to\infty} \|A_n^{-1} v_n - A^{-1} v\|_\infty = 0 \end{cases} \tag{1.9}$$

$$\begin{cases} \text{If } v, u_n \in C(0,T;X) \text{ and } \lim_{n\to\infty} \|A_n u_n - v\|_\infty = 0 \text{ then} \\ \lim_{n\to\infty} \|u_n - A^{-1} v\|_\infty = 0\,. \end{cases} \tag{1.10}$$

2. Temporal regularity

The first result about the non-homogeneous Cauchy problem (P) goes back to 1953 and since then is referred as a classic in the text-books on semigroup theory (see e.g. [35])

Theorem 1 (Phillips 1953). *Let $A_0 : D(A_0) \subset X_0$ be the generator of a semigroup $e^{A_0 t}$ in the Banach space X_0: if $x_0 \in D(A_0)$ and $f \in C^1(0,T;X_0)$ then problem (P) has a unique solution $u \in C^1(0,T;X)$.*

For a proof see [36].

Now in 1985 this theorem has been considerably extended by proving that the same conclusion can be reached by eliminating the density of the domain and by allowing f' to exist only in distributional sense. More precisely

Theorem 2 (Da Prato-Sinestrari 1985). *Let $A : D(A) \subset X \to X$ be a Hille-Yosida operator verifying* (1.1). *Given $u_0 \in D(A)$ and $f \in W^{1,1}(0,T;X)$ such that*

$$u_1 := Au_0 + f(0) \in \overline{D(A)} \tag{2.1}$$

there exists a unique solution $u \in C^1(0,T;X)$ *of problem* (P) *and we have for each* $t \in [0,T]$

$$\|u(t)\| \le Me^{\omega t}\left(\|u_0\| + \int_0^t \|e^{-\omega s}f(s)\|ds\right) \tag{2.2}$$

$$\|u'(t)\| \le Me^{\omega t}\left(\|u_1\| + \int_0^t \|e^{-\omega s}f'(s)\|ds\right) . \tag{2.3}$$

Remark 1. It must be noted that to have a C^1-solution of problem (P), one must suppose $u_0 \in D(A)$ and $u_1 = u'(0) \in \overline{D(A)}$; so that only the condition on f is not necessary: in fact it can be changed (see Section 4) but not reduced to $f \in C(0,T;X)$ (see [35] p. 106).

The proof of this theorem given in [8] is based on a method which consists in writing (P) as an equation in a space of functions of $t \in [0,T]$ and by approximating the time derivative (considered as an operator in this space) by its Yosida approximants.

After the communication of this result in the Tübingen conference of June 1985 attended also by T. Kato, I received a new proof by him [17]: the main idea is to substitute A with its Yosida approximants in the non-homogeneous equation (P) and not in (P_0) as in the usual proof of the Hille-Yosida theorem.

In this way not all the results of paper [8] have been recovered but in any case this approach seems to be the more direct because it is based only on the elementary properties of the Yosida approximants. So it deserves to be known. We will give a version which contains few variants from the original one.

Proof of Theorem 2.2. Let us first show the uniqueness of a solution $u \in C^1(0,T;X)$ of (P): it will be sufficient to assume that such a solution exists when $u_0 = 0$ and $f \equiv 0$ and to prove that it vanishes on $[0,T]$. If $0 \le s \le t$ and $n > \omega$ we deduce from the equation in (P):

$$\frac{d}{ds}[e^{A_n(t-s)}u(s)] = -A_n e^{A_n(t-s)}u(s) + e^{A_n(t-s)}u'(s) = e^{A_n(t-s)}(A - A_n)u(s) .$$

By integrating we get

$$u(t) = \int_0^t e^{A_n(t-s)}(A - A_n)u(s)ds .$$

Fixing $\lambda \in \rho(A)$ we deduce

$$(\lambda - A)^{-1}u(t) = \int_0^t e^{A_n(t-s)}[\lambda(\lambda - A)^{-1} - I](A - n)^{-1}Au(s)ds .$$

By virtue of (1.1) there exists $c > 0$ independent on $n > \omega$ such that

$$\|(\lambda - A)^{-1}u(t)\| \le \frac{c}{n - \omega} .$$

For $n \to +\infty$ we deduce $u(t) = 0$ and the uniqueness is proved.

To prove the existence of the solution we can suppose that $\omega = 0$ and

$$A^{-1} \in \mathcal{L}(X)$$

by changing (if necessary) the unknown u with ue^{ct} with suitable $c \in \mathbf{R}$. Let us first suppose that $f \in C^2(0,T;X)$. Setting for $n \in \mathbf{N}_*$:

$$u_{0n} := u_0 - n^{-1} A u_0 \tag{2.4}$$

the approximating problem

$$u_n'(t) = A_n u_n(t) + f(t)\,, \qquad t \in [0,T] \tag{2.5}$$

$$u_n(0) = u_{0n} \tag{2.6}$$

has a solution given by

$$u_n(t) = e^{A_n t} u_{0n} + \int_0^t e^{A_n s} f(t-s)ds\,. \tag{2.7}$$

As (1.1) holds with $\omega = 0$ we have

$$\|e^{A_n t}\| \leq M$$

hence

$$\|u_n(t)\| \leq M \left(\|u_{0n}\| + \int_0^t \|f(s)\|ds \right)\,. \tag{2.8}$$

We will prove now that u_n converges in $C^1(0,T;X)$ to a solution of (P). By using

$$u_{0n} = A_n^{-1} A u_0 \tag{2.9}$$

and integrating by parts we obtain

$$\begin{aligned} u_n(t) &= e^{A_n t} A_n^{-1} A u_0 + \int_0^t \frac{d}{ds}(e^{A_n s}) A_n^{-1} f(t-s)ds \\ &= e^{A_n t} A_n^{-1} u_1 - A_n^{-1} f(t) + \int_0^t e^{A_n s} A_n^{-1} f'(t-s)ds\,. \end{aligned} \tag{2.10}$$

From (1.7)–(1.8) we deduce that there exists uniformly for $t \in [0,T]$

$$u(t) = \lim_{n\to\infty} u_n(t) = e^{A_0 t} A^{-1} u_1 - A^{-1} f(t) + \int_0^t e^{A_0 s} A^{-1} f'(t-s)ds \tag{2.11}$$

and from this

$$u(0) = u_0 \tag{2.12}$$

From (2.9)–(2.10) and integrating by parts twice we get

$$\begin{aligned} A_n u_n(t) &= e^{A_n t} A u_0 + \int_0^t \frac{d}{ds}(e^{A_n s}) f(t-s) ds \\ &= e^{A_n t} u_1 - f(t) + \int_0^t e^{A_n s} f'(t-s) ds \\ &= e^{A_n t} u_1 - f(t) + \int_0^t \frac{d}{ds}(e^{A_n s}) A_n^{-1} f'(t-s) ds \\ &= e^{A_n t} u_1 - f(t) + e^{A_n t} A_n^{-1} f'(0) - A_n^{-1} f'(t) \\ &\quad + \int_0^t e^{A_n s} A_n^{-1} f''(t-s) ds . \end{aligned} \tag{2.13}$$

From (1.4), (1.7) and (1.8) we deduce that there exists uniformly for $t \in [0, T]$.

$$\lim_{n\to\infty} A_n u_n(t) = e^{A_0 t} u_1 - f(t) + e^{A_0 t} A^{-1} f'(0) - A^{-1} f'(t) + \int_0^t e^{A_0 s} A^{-1} f''(t-s) ds. \tag{2.14}$$

From equation (2.5) we have that $\{u_n'\}$ converges uniformly in $[0, T]$ hence $u \in C^1(0, T; X)$ and

$$u'(t) = \lim_{n\to\infty} u_n'(t) = \lim_{n\to\infty} A_n u_n(t) + f(t) . \tag{2.15}$$

As

$$\|u_n - A^{-1}(u' - f)\|_\infty \le \|A_n^{-1}(A_n u_n - u' + f)\|_\infty + \|(A_n^{-1} - A^{-1})(u' - f)\|_\infty$$

we deduce from (1.6), (1.7) and (2.15) for $n \to \infty$

$$u(t) = A^{-1}(u'(t) - f(t)) .$$

In conclusion u is a solution of (P).

By using (2.13) we get

$$u_n'(t) = A_n u_n(t) + f(t) = e^{A_n t} u_1 + \int_0^t e^{A_n s} f'(t-s) ds \tag{2.16}$$

and so

$$\|u_n'(t)\| \le M \left(\|u_1\| + \int_0^t \|f'(s)\| ds \right) . \tag{2.17}$$

Now (2.8) and (2.17) imply estimates (2.2)–(2.3) with $\omega = 0$.

Let us suppose now that $f \in W^{1,1}(0, T; X)$ i.e. there exists $g \in L^1(0, T; X)$ such that

$$f(t) = f(0) + \int_0^t g(s) ds , \qquad t \in [0, T] .$$

Let $\{g_n\} \in C^1(0,T;X)$ be such that $\lim_{n\to\infty} \|g_n - g\|_{L^1(0,T;X)} = 0$. Setting

$$f_n(t) = f(0) + \int_0^t g_n(s)ds$$

we have $f_n \in C^2(0,T;X)$, $f_n(0) = f(0)$ and

$$\lim_{n\to\infty} \|f_n - f\|_\infty = 0\,, \qquad \lim_{n\to\infty} \|f_n' - f'\|_{L^1} = 0\,.$$

As $Au_0 + f_n(0) = Au_0 + f(0) \in \overline{D(A)}$ we can use the first part of the proof to deduce the existence of a unique solution $v_n \in C^1(0,T;X)$ of

$$\begin{cases} v_n'(t) = Av_n(t) + f_n(t) \\ v_n(0) = u_0 \end{cases} \tag{2.18}$$

and

$$\begin{cases} \|v_n(t)\| \le M\left(\|u_0\| + \int_0^t \|f_n(s)\|ds\right) \\ \|v_n'(t)\| \le M\left(\|u_1\| + \int_0^t \|f_n'(s)\|ds\right)\,. \end{cases} \tag{2.19}$$

By virtue of these estimates one proves that v_n converges in $C^1(0,T;X)$ to a function $v \in C^1(0,T;X)$. Using the equation (2.18) and the closedness of A we deduce that v is a solution of (P); estimates (2.2)–(2.3) are a consequence of (2.19).

3. Other approaches to temporal regularity

The result of Theorem 2 was proved later by three different methods. They are less direct than Kato's one because they are based on the non-linear semigroups', the integrated semigroups' and the extrapolation spaces' theory respectively.

With non-linear semigroups' method a partial result was first found by Da Prato ([4], Prop. III.3.17); in [3] Bénilan and Egberts use analogous methods to consider (under the assumptions of the theorem) problem (P) with f and u_0 substituted by f' and u_1: by virtue of Crandall-Liggett's theorem of non-linear semigroup theory, this problem has a mild solution. By using an equivalence result stated in [2] they deduce the existence and uniqueness of a strict solution of (P). The result is generalized to the case of time-dependent operator $A(t)$ by Tanaka in [38]: see Section 7 later.

Another approach is based on the theory of integrated semigroups. After some partial results (see Prop. (5.5) of [1]), H. Kellermann and M. Hieber ([18]) proved by using some theorems of W. Arendt ([1]) that the Hille-Yosida operators are the generators of locally Lipschitz continuous integrated semigroups: then a regularity theorem for the convolution of the integrated semigroups with f can be used to prove the temporal regularity result of Theorem 2.

Finally the extrapolation theory's methods enable to consider problem (P) in a larger space X_{-1} (called the extrapolation space associated to A) and to extend A into a generator A_{-1} of a semigroup in X_{-1}: after this, in [24] a product space $\mathcal{X}$ is introduced to reduce the non-homogeneous problem (P) to a homogeneous one:

in this way the result of Theorem 2 is a consequence of the semigroup properties of the invariant subspaces of $\mathcal{X}$ (see Corollary (2.8) of [24]).

It must be noted that the extrapolation method gives also an extension of Theorem 2. In fact denoting by F and F_{-1} the Favard classes generated by A and A_{-1} respectively we have the following

Theorem 1 (R. Nagel, E. Sinestrari 1991). *Let $A : D(A) \subset X \to X$ be a Hille-Yosida operator. For each $u_0 \in F$ and $f \in W^{1,1}(0,T;F_{-1})$ such that*

$$A_{-1}u_0 + f(0) \in \overline{D(A)}$$

the Cauchy problem

$$u'(t) = A_{-1}u(t) + f(t) \ , \quad t \in [0,T]$$

$$u(0) = u_0$$

has a unique solution $u \in C^1(0,T;X) \cap C(0,T;F)$.

See Corollary (3.5) of [24].

4. Spatial regularity

In [8] it is proved the existence of a strict solution when f is regular with respect to the space in the sense that it is required to have values in a subspace of X.

Theorem 1 (Da Prato-Sinestrari 1985). *Let $f \in L^p(0,T;D(A)$, $u_0 \in D(A)$ and $Au_0 \in \overline{D(A)}$. Then there exists a unique $u \in W^{1,p}(0,T;X) \cap C(0,T;D(A))$ solution of (P) a.e. in $[0,T]$. If in addition $f \in C(0,T;X)$ then we have also $u \in C^1(0,T;X)$ and (P) holds for each $t \in [0,T]$. Moreover we have*

$$\|u(t)\| \leq Me^{\omega t}\left(\|u_0\| + \int_0^t e^{-\omega s}\|f(s)\|ds\right) \tag{4.1}$$

$$\|Au(t)\| \leq Me^{\omega t}\left(\|Au_0\| + \int_0^t e^{-\omega s}\|Af(s)\|ds\right) . \tag{4.2}$$

It can be seen that even if f and u_0 have values in X_0 (where A generates a semigroup), the result does not derive from the Phillips theorem applied to problem (P) and considered in X_0.

With the methods of nonlinear semigroups theory the first part of this theorem was proved in Prop. III.3.14 of [4] and the second one (except the estimates) in [3].

5. Generalized solutions

In [8] were considered two types of solutions less regular than those obtained in the previous sections. They replace the mild solutions defined when $D(A)$ is dense (see [35]) through the variation of constants' formula.

Definition 1. *Let $f \in L^1(0,T,X)$ and $u_0 \in E$. A function $u \in C(0,T;X)$ is called an integral solution of problem (P) if for each $t \in [0,T]$ we have*

$$\int_0^t u(s)ds \in D(A) \tag{5.1}$$

$$u(t) = u_0 + A\int_0^t u(s)ds + \int_0^t f(s)ds\,. \tag{5.2}$$

Definition 2. *Let $f \in L^p(0,T;X)$ and $u_0 \in X$. A function $u \in L^p(0,T;X)$ is called a F-solution in L^p of (P) if there exist $u_k \in W^{1,p}(0,T;X) \cap L^p(0,T;D(A))$ such that setting $f_k = u_k' - Au_k$ we have*

$$\lim_{k\to\infty} u_k = u \quad \textit{and} \quad \lim_{k\to\infty} f_k = f \ \textit{in}\ L^p(0,T;X) \tag{5.3}$$

$$\lim_{k\to\infty} u_k(0) = u_0 \quad \textit{in} \quad X\,. \tag{5.4}$$

By replacing $L^p(0,T;X)$, $L^p(0,T;D(A))$ and $W^{1,p}(0,T;X)$ with $C(0,T;X)$, $C(0,T;D(A))$ and $C^1(0,T;X)$ respectively we obtain the definition of F-solution in C.

The definition of integral solution was given for the first time in [8] and later used by other authors (see [2], [11], [40]).

The F-solutions were introduced by G. Da Prato in [5] under the name of "solutions faibles" and called "solutions fortes" in [6] where their existence in some particular case was proved.

In [8] several equivalence, existence, uniqueness results and a priori estimates for the generalized solutions are proved. They can be summarized in the following theorems proved in Sections 4–9 of [8].

Theorem 1. *For each $u_0 \in \overline{D(A)}$ and $f \in L^p(0,T;E)$ there exists a unique u, F-solution in L^p of problem (P). Moreover*

(i) $u \in C(0,T;\overline{D(A)})$
(ii) *u verifies estimate (4.1)*
(iii) *if in addition $f \in C(0,T;E)$ then u is an F-solution in C of problem (P)*

Theorem 2. *Let $u_0 \in \overline{D(A)}$ and $f \in L^p(0,T;E)$. An integral solution is an F-solution in L^p and viceversa. If an integral solution belongs to $W^{1,p}(0,T;E)$ or to $L^p(0,T;D(A))$ then it is strict (i.e. satisfies the equation in (P) a.e. in $[0,T]$).*

Let $u_0 \in \overline{D(A)}$ and $f \in C(0,T;E)$. An integral solution is an F-solution in C and viceversa. If an integral solution belongs to $C^1(0,T;E)$ or to $C(0,T;D(A))$ then it is strict (i.e. satisfies the equation in (P)).

Let us recall an interesting variation of constants formula for the integral solutions proved by H. Thieme (Corollary 1.6 of [40]).

Theorem 3 (H. Thieme 1990). *Let u be an integral solution of (P) and $e^{A_0 t}$ the semigroup generated by the part of A in $\overline{D(A)}$ (see Section 1). Then for $t \in [0,T]$ we have*

$$u(t) = e^{A_0 t}u_0 + \lim_{\lambda \to +\infty} \int_0^t e^{A_0(t-s)}\lambda(\lambda - A)^{-1} f(s)ds\,. \tag{5.5}$$

6. Applications

In [8] there are many examples of Hille-Yosida operators defined by means of differential operators in non-reflexive spaces as those of continuous or L^∞-functions. The most important applications are: the ultraparabolic partial differential equations as e.g.

$$u_t + u_\tau = \Delta_x u + f$$

where u is a function of t, $\tau \in [0,T]$ and $x \in \Omega \subseteq \mathbf{R}$ and Δ_x is the Laplace operator in Ω and the generalized Laplacian in infinite-dimensional spaces with applications to the stochastic control.

The abstract theory can be successfully applied to the study of the transverse vibration of an elastic string with fixed ends and subject to an external force:

$$\begin{cases} w_{tt}(t,x) = w_{xx}(t,x) + f(t,x)\,, & w_t(0,x) = w_1(x) \\ w(0,x) = w_0(x)\,, & w(t,0) = w(t,1) = 0\,. \end{cases}$$

It is possible to obtain a C^2-solution by reducing this problem to an abstract one as (P) in a suitable product space of continuous functions: see [14].

By using the extrapolation theory's approach to the Hille-Yosida operators one can study integro-differential equation such as

$$u'(t) = Au(t) + \int_0^t k(t-s)Au(s)ds + f(t)\,, \qquad t \geq 0$$

where K is a scalar function ([24]) as well as the nonlinear one ([25])

$$u'(t) = Au(t) + \int_0^t H(t,s,Au(s))ds + f(t), t \geq 0\,.$$

By different methods they have been studied by R. Grimmer, J. Liu, H. Oka and N. Tanaka ([12], [13], [28]–[34]).

Applications have been made to delay equations by Y. Lei and H. Thieme ([19], [40]) and in [37] to hyperbolic partial differential equations with delay in the highest order derivatives as the wave equation with memory effects:

$$\begin{cases} w_{tt}(t,x) = w_{xx}(t,x) + \int_{-r}^0 k(s)w_{xx}(t+s,x)ds + f(t,x)\,, & t \geq 0,\ x \in [0,1] \\ w(t,x) = z(t,x) & -r \leq t \leq 0,\ x \in [0,1] \\ w_t(t,x) = w_1(x) & x \in [0,1] \\ w(t,x') = 0 & t \geq 0,\ x' = 0,1\,. \end{cases}$$

Other types of equations of second order in time have been investigated by Y. Lin and H. Oka ([20], [27], [30], [31]).

7. The non-autonomous case

The first attempt to study problem (P) when the Hille-Yosida operator depends on time was made in [10] under the Kato's conditions (except for the density of the domains) which can be described as follows:

(i) For all $t \in [0,T]$, $A(t) : D \to X$ is a linear operator from the Banach space $(D, \|\cdot\|_D)$ to the Banach space $(X, \|\cdot\|)$

(ii) $D \subset X$ and there is $c > 0$ such that for all $t \in [0,T]$ and $x \in D$

$$c^{-1}\|x\|_D \leq \|x\| + \|A(t)x\| \leq c\|x\|_D$$

(iii) there exist ω, $M \in \mathbf{R}$ such that $\rho(A(t)) \supset]\omega, +\infty[, t \in [0,T]$ and for each $n \in \mathbf{N}$ we have

$$\|(\lambda - A(t_1))^{-1} \dots (\lambda - A(t_n))^{-1}\|_{\mathcal{L}(X)} \leq M(\lambda - \omega)^{-n}$$

when $0 \leq t_n \leq \dots t_1 \leq T$ and $\lambda > \omega$.

But in [10] the existence of generalized and strict solution of problem

$$(P') \qquad \begin{cases} u'(t) = A(t)u(t) + f(t) \ , & t \in [0,T] \\ u(0) = u_0 \end{cases}$$

was assured under additional assumptions on $A(t)$, which although easily checked in all the applications given, nevertheless are redundant as it was shown in Theorem 4.2 of [38].

Theorem 1 (N. Tanaka, 1995). *Let (i)–(iii) hold and for each $x \in D$, let $A(\cdot)x \in C^1(0,T;X)$. Then for each $x \in D$ and $f \in W^{1,1}(0,T;X)$ such that $A(0)x + f(0) \in \overline{D}$, there exist a unique $u \in C(0,T;D) \cap C^1(0,T;X)$ solution of (P').*

The proof given in [38] is based on the construction of a generalized variation of constants formula by the methods of nonlinear semigroup theory: this enables also the study of the semilinear version (P')

$$u'(t) = A(t)u(t) + B(t, u(t)) \ , \qquad t \in [0,T]$$

where B is a non-linear operator on $[0,T] \times X$.

By the same methods or those of the integrated semigroup theory, many types of equations have been studied: quasilinear (see [32], [39]) integro-differential (see [21], [33], [39]) and of second order in time (see [27]).

References

[1] W. Arendt: "Vector-valued Laplace transforms and Cauchy problems", *Israel J. Math.* **59** (1987) 327–352.

[2] P. Benilan, M. Crandall, A. Pazy: *"Bonnes solutions" d'un problème d'évolution semi-linéaire*, C.R.A.S. Paris 306 (1988) 527–530.

[3] P. Benilan, P. Egberts: "Mild solutions", ("Workshop on operator semigroups and evolution equations", Blaubeuren, Nov. 1989), Semesterbericht Funktionanalysis, Tübingen, 1989.

[4] G. Da Prato: *Applications croissantes et équations d'évolutions dans les espaces de Banach*, Academic Press, 1976.

[5] G. Da Prato: "Equations opérationnelles dans les espaces de Banach et applications", C.R.A.S. Paris 266 (1968) 60–63.

[6] G. Da Prato, P. Grisvard: "Sommes d'opérateurs linéaires et équations différentielles opérationelles", *J. Math. Pures Appl.* **54** (1975) 305–387.

[7] G. Da Prato, E. Sinestrari: *On the Phillips and Tanabe regularity theorems*, ("Aspects of positivity in functional analysis", Tübingen, June 1985), Semesterbericht Funktionalanalysis Tübingen 8 (1985) 117–124.

[8] G. Da Prato, E. Sinestrari: *Differential operators with non-dense domain*, Ann. Sc. Norm. Sup. Pisa 14 (1987) 285–344.

[9] G. Da Prato, E. Sinestrari: *Time-dependent differential equations in non-reflexive Banach spaces*, ("Differential equations with applications in Biology, Physics, and Engineering", Retzhof, June 1989), Dekker Lecture Notes in Pure and Applied Mathematics n. 133 (1991) 79–89.

[10] G. Da Prato, E. Sinestrari: *Non-autonomous evolution operators of hyperbolic type*, Semigroup Forum 45 (1992) 302–321.

[11] K.-J. Engel, R. Nagel: *One-parameter semigroups for linear evolution equations*, Springer-Verlag 2000.

[12] R. Grimmer, J. Liu: *Integrodifferential equations with nondensely defined operators*, ("Differential equations with applications in Biology, Physics, and Engineering", Retzhof, June 1989), Dekker Lecture Notes in Pure and Applied Mathematics n. 133 (1991) 185–199.

[13] R. Grimmer, J. Liu: "Integrated semigroups and integro-differential equations", *Semigroup Forum* **48** (1994) 79–95.

[14] R. Grimmer, E. Sinestrari: "Maximum norm in one-dimensional hyperbolic problems", *Diff. Int. Eq.* **5** (1992) 421–432.

[15] E. Hille, R.S. Phillips: *Functional analysis and semi-groups*, A.M.S., 1957.

[16] T. Kato: "Remarks on pseudo-resolvents and infinitesimal generators of semigroups", *Proc. Japan Ac.* **35** (1959) 467–468.

[17] T. Kato: *Private Letter* (Sept. 22, 1986).

[18] H. Kellermann, M. Hieber: "Integrated semigroups", *J. Funct. Anal.* **84** (1989) 160–180.

[19] Y. Lei: "Semilinear functional differential equations in Banach spaces with Hille-Yosida operators", *Non-linear Anal.* **41** (2000) 989–1004.

[20] Y. Lin: "Time-dependent perturbation theory for abstract evolution equations of second order", *Studia Math.* **130** (1998) 263–274.

[21] Y. Lin, N. Tanaka: "Non-linear abstract wave equations with strong damping", *J. Math. Anal. Appl.* **225** (1998) 46–61.

[22] Y. Lin, N. Tanaka: "Abstract hyperbolic Volterra integro-differential equations", *J. Int. Eq. Appl.* **10** (1998) 195–218.

[23] A. LUNARDI: *Analytic semigroups and optimal regularity in parabolic problems*, Birkhauser 1995.

[24] R. NAGEL, E. SINESTRARI: *Inhomogeneous Volterra integro-differential equations for Hille-Yosida operators* ("Symposium zur Funktionalanalysis", Essen, November 1991), Lecture Notes in Pure and Applied Mathematics n. 150 (1994) 51–70.

[25] R. NAGEL, E. SINESTRARI: "Non-linear hyperbolic Volterra integro-differential operators", *Non-linear Anal.* **27** (1996) 167–186.

[26] H. OKA: "Integrated resolvent operators", *J. Int. Eq. Appl.* **7** (1995) 193–232.

[27] H. OKA: "A class of complete second order linear differential equations", *Proc. Amer. Math. Soc.* **124** (1996) 3143–3150.

[28] H. OKA: "Non-autonomous integro-differential equations of hyperbolic type", *Diff. Int. Eq.* **8** (1995) 1823–1831.

[29] H. OKA: "Second order linear Volterra integro-differential equations", *Semigroup Forum* **53** (1996) 25–43.

[30] H. OKA: "Linear Volterra equations and integrated solution families", *Semigroup Forum* **53** (1996) 278–297.

[31] H. OKA: "Second order linear Volterra equations governed by a sine family", *J. Int. Eq. Appl.* **8** (1996) 447–456.

[32] H. OKA: "Abstract quasilinear Volterra integro-differential equations", *Non-linear Anal.* **28** (1997) 1019–1045.

[33] H. OKA, N. TANAKA: "Non-autonomous integro-differential equations of hyperbolic type", *Diff. Int. Eq.* **8** (1995) 1823–1831.

[34] H. OKA, N. TANAKA: "Abstract quasilinear integro-differential equations of hyperbolic type", *Non-linear Anal.* **29** (1997) 903–925.

[35] B. PAZY: *Semigroups of linear operators and applications to partial differential equations*, Springer-Verlag 1993.

[36] R.S. PHILLIPS: "Perturbation theory of semigroups of linear operators", *Trans. Am. Math. Soc.* **74** (1953) 199–221.

[37] E. SINESTRARI: "Wave equation with memory", *Discr. Cont. Dyn. Syst.* **5** (1999) 881–896.

[38] N. TANAKA: "Semilinear equations in the "hyperbolic" case", *Non-linear Anal.* **24** (1995) 773–788.

[39] N. TANAKA: "Quasilinear evolution equations with non-densely defined operators", *Diff. Int. Eq.* **9** (1996) 1067–1106.

[40] H. THIEME: "Semiflows generated by Lipschitz perturbations of non-densely defined operators", *Diff. Int. Eq.* **3** (1990) 1035–1066.

Eugenio Sinestrari
Dipartimento di Matematica
Università di Roma "La Sapienza"
P.le Aldo Moro 7
I-00185 Roma
e-mail: sinestrari@axcasp.caspur.it

Progress in Nonlinear Differential Equations
and Their Applications, Vol. 50, 353–379

Min-Max Game Theory and Optimal Control with Indefinite Cost under a Singular Estimate for $e^{At}B$ in the Absence of Analyticity*

Roberto Triggiani

0. Introduction

In this paper, we continue the study of the abstract dynamics of [4, 5, 7, 8, 10] – characterized by a singular estimate for $e^{At}B$, in the absence of analyticity for e^{At} – in two natural directions, both over an infinite horizon. They are: (i) the min-max game theory problem, where in part we extend and in part we complement the theory of [9, Chapter 6, Part II, Sections 6.19 through 6.26, pp. 608–630], in the stable case; and (ii) the optimal control problem, with indefinite quadratic cost, where likewise in part we extend and in part we complement the theory of [9, Chapter 6, Appendix 6A, pp. 630–638], in the stable case. In turn, each topic may be seen as an extension of [4, 7, 10]. Throughout this paper, and for both topics (i) and (ii), the following strategy is applied. The first part of our analysis – Sections 1 through 3 for the min-max problem; and Proposition 6.1 through Proposition 6.5 for the indefinite cost problem – relies *only* on the regularity properties of the operators L, L^*, W, W^*, etc., given in (1.11)–(1.14) in Section 1 below, which continue to hold true under the assumed singular estimate for $e^{At}B$, as in the abstract parabolic case of [9, Volume 1], in spite of the lack of analyticity for e^{At}. Therefore, with this key observation in mind, the corresponding abstract *parabolic* treatment of [9, Chapter 6], which hinges only on said regularity properties of L, L^*, W, W^*, goes through verbatim. By contrast, the final stage of our analysis – which in [9, Chapter 6], made explicit use of the analyticity of the s.c. semigroup e^{At}, a property presently not available – needs now appropriate modifications, in both the proofs and the final statements, of the type already performed in the final stage of the analysis of [10]. Accordingly, the treatment of this paper is brisk, as it heavily hinges either on [9, Chapter 6] or on [10]. For clarity, the corresponding building blocks of the respective theories are explicitly singled out and displayed in formal statements, for both problems. This paper aims at serving at least the following purpose: in giving an affirmative answer to open questions raised over the

* Research partially supported by the National Science Foundation under Grant DMS-9804056.

past few years in control-theoretic circles about a possible 'extension' of the min-max game theory problem and of the indefinite cost problem to systems of PDE's consisting of a parabolic-like component strongly coupled with a hyperbolic-like component, such as they arise in the "structural acoustic problem." The general solution provided here is based on the abstract assumption that a singular estimate for $e^{At}B$ holds true, as in [4, 7, 8, 10, 11, 13]. As seen in [5, 4, 7, 8, 11, 13], such an assumption is, in fact, a dynamical *property* naturally satisfied in acoustic problems, with either a structurally elastic or a thermoelastic or a composite (sandwich) flexible wall. Once this key feature is realized and extracted at the abstract level, then a solution of both problems treated here readily follows from the work of [9, Volume 1], modulo the additional analysis of [10]. As the "acoustic models" of the type that motivate [10, 11, 13] and the present paper are naturally stable, we limit our study to the case where the free dynamics semigroup e^{At} is indeed uniformly (exponentially) stable. A corresponding study with no stability assumption is surely available by invoking instead [9, Chapter 6, Sections 6.1 through 6.18] for the min-max theory, rather than [9, Sections 6.19 through 6.26].

The topic treated here has been deliberately chosen because it is in the mainstream of that area of research where Brunello Terreni devoted his last years and energy [1, 2], before his hasty and sorely missed departure. Thus, this paper is intended as a tribute to his memory, befitting the occasion of the present volume.

Part I: Min-Max Game Problem

1. Mathematical setting; formulation of the min-max game problem

Dynamics. Let U (control), V (disturbance), Y (state) be separable Hilbert spaces. In this Part I, we consider the following abstract state equation

$$\dot{y}(t) = Ay(t) + Bu(t) + Gw(t) \text{ in } [\mathcal{D}(A^*)]'; \quad y(0) = y_0 \in Y. \tag{1.1}$$

Here, the function $u \in L_2(0, \infty; U)$ is the control and $w \in L_2(0, \infty; V)$ is a deterministic disturbance. The dynamics (1.1) is subject to the following assumptions, to be maintained throughout Part I:

(H.1) $A : Y \supset \mathcal{D}(A) \to Y$ is the infinitesimal generator of a s.c. semigroup e^{At} on Y. Moreover, e^{At} is (exponentially) uniformly stable: that is, there exist constants $M \geq 1$, $\omega > 0$, such that

$$\|e^{At}\|_{\mathcal{L}(Y)} \leq Me^{-\omega t}, \quad t \geq 0. \tag{1.2}$$

(H.2) B is a linear operator $U \equiv \mathcal{D}(B) \to [\mathcal{D}(A^*)]'$, the dual space of the domain $\mathcal{D}(A^*)$ with respect to the pivot space Y. Here, A^* is the adjoint of A in Y. Thus, e^{At} can be extended as a s.c. semigroup on $[\mathcal{D}(A^*)]'$, as well.

(H.3) G is a linear operator $V \equiv \mathcal{D}(G) \to [\mathcal{D}(A^*)]'$.

(H.4) There exist constants $0 \le r < 1$ and $T > 0$, such that the following singular estimates hold true:

$$\begin{cases} e^{At}Bu \in {}_rC([0,T];Y), \quad \forall\, u \in U; & (1.3a)\\ \|e^{At}B\|_{\mathcal{L}(U;Y)} = \|B^*e^{A^*t}\|_{\mathcal{L}(Y;U)} \le \frac{c_T}{t^r}, \quad 0 < t \le T, & (1.3b)\end{cases}$$

$$\begin{cases} e^{At}Gv \in {}_rC([0,T];Y), \quad \forall\, v \in V; & (1.4a)\\ \|e^{At}G\|_{\mathcal{L}(V;Y)} \equiv \|G^*e^{A^*t}\|_{\mathcal{L}(Y;V)} \le \frac{c_T}{t^r}, \quad 0 < t \le T, & (1.4b)\end{cases}$$

where $(Bu, y)_Y = (u, B^*y)_U$, $u \in U$, $y \in \mathcal{D}(B^*) \supset \mathcal{D}(A^*)$; and $(Gv, y)_Y = (v, G^*y)_V$, $v \in V$, $y \in \mathcal{D}(G^*) \supset \mathcal{D}(A^*)$. In this Part I, we shall study the min-max game-theoretic problem with indefinite cost described below for the above abstract dynamics, where the observation operator R in (1.6) below is subject to the assumption

(H.5)

$$R \in \mathcal{L}(Y;Z), \tag{1.5}$$

where Z is another Hilbert space.

Min-max game theory problem. For a fixed $\gamma > 0$, we associate with (1.1) the cost function

$$J(u,w) = J(u,w,y(u,w)) = \int_0^\infty [\|Ry(t)\|_Z^2 + \|u(t)\|_U^2 - \gamma^2\|w(t)\|_V^2]dt, \tag{1.6}$$

where $y(t) = y(t; y_0)$ is the solution to (1.1) due to $u(t)$ and $w(t)$, see below in (1.10). The aim of this Part I is to study the following game-theory problem

$$\sup_w \inf_u J(u,w), \tag{1.7}$$

where the infimum is taken over all $u \in L_2(0,\infty;U)$ for w fixed, and the supremum is taken over all $w \in L_2(0,\infty;V)$.

Remark 1.1. We emphasize, e^{At} is definitely *not* assumed to be analytic. Moreover, a more general cost functional such as the one in [9, Chapter 6, Eqn. (6.1.2.4), p. 559] may be used in place of (1.6). □

We readily obtain the following consequence of assumption (H.4) on B, see (1.3), and on G, see (1.4): for any $0 < \omega_0 < \omega$ (the latter defined in (1.2)), there exists a corresponding constant $k > 0$ (depending on M, C_T, ω, ω_0, T, r) such that

$$\left\|e^{At}B\right\|_{\mathcal{L}(U;Y)} = \left\|B^*e^{A^*t}\right\|_{\mathcal{L}(Y;U)} \le \frac{ke^{-\omega_0 t}}{t^r}, \quad \forall\, t > 0;\ 0 < \omega_0 < \omega, \tag{1.8}$$

$$\left\|e^{At}G\right\|_{\mathcal{L}(V;Y)} \equiv \left\|G^*e^{A^*t}\right\|_{\mathcal{L}(Y;V)} \le \frac{ke^{-\omega_0 t}}{t^r}, \quad \forall\, t > 0;\ 0 < \omega_0 < \omega. \tag{1.9}$$

Preliminaries. The solution to problem (1.1) is given by

$$y(t) = e^{At}y_0 + (Lu)(t) + (Ww)(t); \tag{1.10}$$

$$\begin{aligned}(Lu)(t) &= \int_0^t e^{A(t-\tau)}Bu(\tau)d\tau && (1.11a)\\ &: \text{continuous } L_2(0,\infty;U) \to L_2(0,\infty;Y) && (1.11b)\\ &: \text{continuous } C([0,\infty];U) \to C([0,\infty];Y); && (1.11c)\end{aligned}$$

$$\begin{aligned}(Ww)(t) &= \int_0^t e^{A(t-\tau)}Gw(\tau)d\tau && (1.12a)\\ &: \text{continuous } L_2(0,\infty;V) \to L_2(0,\infty;Y) && (1.12b)\\ &: \text{continuous } C([0,\infty];V) \to C([0,\infty];Y). && (1.12c)\end{aligned}$$

The regularity properties noted in (1.11b–c) and (1.12b–c), due to the Young's inequality [12], as well as assumption (H.3) on B and G, were formalized in [10, Proposition 3.1.2], as well as in [9, Volume 1, Theorem 6.23.1, p. 620]. The L_2-adjoints of L and W are

$$\begin{aligned}(L^*f)(t) &= \int_t^\infty B^*e^{A^*(\tau-t)}f(\tau)d\tau && (1.13a)\\ &: \text{continuous } L_2(0,\infty;Y) \to L_2(0,\infty;U) && (1.13b)\\ &: \text{continuous } C([0,\infty];Y) \to C([0,\infty];U); && (1.13c)\end{aligned}$$

$$\begin{aligned}(W^*v)(t) &= \int_t^\infty G^*e^{A^*(\tau-t)}v(\tau)d\tau && (1.14a)\\ &: \text{continuous } L_2(0,\infty;Y) \to L_2(0,\infty;V) && (1.14b)\\ &: \text{continuous } C([0,\infty];Y) \to C([0,\infty];V). && (1.14c)\end{aligned}$$

Remark 1.2. Regarding the operators A and B, the setting is precisely the same as the one in [7, 10]. The additional disturbance operator G behaves like B. □

2. Statement of main results

The main result of the present Part I is the following theorem.

Theorem 2.1. Assume (H.1)–(H.5). Then, there exists a critical value $\gamma_c \geq 0$ defined explicitly in terms of the problem data by Eqn. (3.17) below, such that:

(a) If $\gamma_c > 0$ and $0 < \gamma < \gamma_c$, then taking the supremum in w as in (1.7) leads to $+\infty$ for all initial conditions $y_0 \in Y$; that is, there is no finite solution of the game theory problem (1.7) (see Theorem 3.3.3(iii)).

(b) If $\gamma > \gamma_c$, then:

(i) there exists a unique solution $\{u^*(\,\cdot\,;y_0); w^*(\,\cdot\,;y_0); y^*(\,\cdot\,;y_0)\}$ of the game-theory problem (1.7) (see Theorem 3.3(ii), and Eqn. (3.23)). These quantities, along with the corresponding cost $J^*(y_0)$, are given by explicit formulas directly in terms of the problem's data by Theorem 3.5.

(ii) there exists a bounded, non-negative, self-adjoint operator $P = P^* \in \mathcal{L}(Y)$, that satisfies the following algebraic Riccati equation ARE$_\gamma$, for all $x \in Y$, $z \in \mathcal{D}(A)$, or else all $x \in \mathcal{D}(A_F)$, $z \in Y$ (see Proposition 4.6 below):

$$\begin{aligned}(PAx, z)_Y + (Px, Az)_Y + (Rx, Rz)_Y \\ = (B^*Px, B^*Pz)_U - \gamma^{-2}(G^*Px, G^*Pz)_U \end{aligned} \tag{2.1}$$

(see Theorem 4.7 below). Moreover, see Lemma 4.1(ii) and (4.13) below,

$$B^*P \in \mathcal{L}(Y;U); \ G^*P \in \mathcal{L}(Y;V). \tag{2.2}$$

(iii) The following pointwise feedback relations hold:

$$u^*(t;y_0) = -B^*Py^*(t;y_0) \in L_2(0,\infty;U) \cap C([0,\infty];U) \tag{2.3}$$

$$\gamma^2 w^*(t;y_0) = G^*Py^*(t;y_0) \in L_2(0,\infty;V) \cap C([0,\infty];V), \tag{2.4}$$

see Lemma 4.1(i), (iii) below.

(iv) The operator (the subindex F stands for "feedback") with maximal domain

$$A_F = A - BB^*P + \gamma^{-2}GG^*P, \tag{2.5a}$$

$$\mathcal{D}(A_F) = \{x \in Y : [I - A^{-1}BB^*P + \gamma^{-2}A^{-1}GG^*P] \in \mathcal{D}(A)\}, \tag{2.5b}$$

is the generator of a s.c. semigroup $e^{A_F t}$ on Y, which is, moreover, uniformly stable on Y (see (4.13), Proposition 4.2, Theorem 3.8 below).
In fact, for $y_0 \in Y$, we have (see Theorem 3.8)

$$y^*(t;y_0) = e^{A_F t}y_0 = e^{(A-BB^*P+\gamma^{-2}GG^*P)t}y_0 \tag{2.6a}$$

$$\in L_2(0,\infty;Y) \cap C([0,\infty];Y). \tag{2.6b}$$

(v) The operators $(A - BB^*P)$ and $(A + \gamma^{-2}GG^*P)$, with maximal domains, generate s.c. semigroups, the first of which, $e^{(A-BB^*P)t}$, is, moreover, stable (as in [9, Chapter 6, Prop. 6.16.1, p. 604]). □

Theorem 2.2. Conversely, suppose $P = P^* \geq 0$ is an operator in $\mathcal{L}(Y)$ such that:

(a) the operator $A_F = A - BB^*P + \gamma^{-2}GG^*P$ is the generator of a s.c. uniformly stable semigroup on Y for some $\gamma > 0$;

(b) P is a solution of the corresponding ARE$_\gamma$ in (2.1) for all $x, z \in \mathcal{D}(A)$, with the properties that $B^*P \in \mathcal{L}(Y;U)$, and $G^*P \in \mathcal{L}(Y;V)$.

Then, the operators $(A - BB^*P)$ and $(A - \gamma^{-2}GG^*P)$ are likewise the generators of s.c. semigroups on Y, the first of which is uniformly stable and, moreover, the game problem (1.7) has a finite cost functional for all $y_0 \in Y$, so that then $\gamma \geq \gamma_c$.

Additional results are given in the treatment below.

3. Proof of Theorem 2.1: Results based on regularity properties for L, L^*, W, W^*

Orientation. Under assumption (H.4) on B, see (1.3) for $0 < t \leq T$ and its implication in (1.9) for all $t > 0$, the operators L and L^* in (1.11a) and (1.13a) have the regularity properties noted in (1.11b–c) and (1.13b–c). This was formalized in [10, Proposition 3.1.2]. Similarly, for the operators W and W^* in (1.12a) and (1.14a), whose regularity is noted in (1.12b–c) and (1.14b–c), under the counterpart assumption (H.4) on G for $0 < t \leq T$, see (1.4), and its implication in (1.9) for all $t > 0$. Thus, all results in the treatment of the min-max game theory problem under stability, see (1.2), given in [9, Chapter 6, Part II], which rely on the regularity results for L, L^*, W, W^* such as those given in (1.11b–c), (1.13b–c), (1.12b–c), (1.14b–c) – and more generally in Proposition 3.1.2 of [10] for L and its counterpart for W – continue to hold true verbatim in the present case. For ease in following the present development, we shall retrace the necessary steps from [9, Chapter 6, Part II], until we hit the point where the property of analyticity of e^{At} – now not available – was made use of there. At this point, we shall then switch into the treatment presented in [10], which does not need analyticity of e^{At}. In short: the present analysis results by combining the relevant contribution of [9, Chapter 6, Part II], with the relevant contributions of [10]. □

We begin with the counterpart of [9, Theorem 6.20.1.1, Chapter 6, p. 613].

Theorem 3.1. (Minimiation of J over u for w fixed) Assume (H.1)–(H.5). With reference to the minimization problem

$$\inf_{u \in L_2(0,\infty;U)} J(u, w; y_0) \text{ holding } w \in L_2(0, \infty; V) \text{ fixed,} \tag{3.1}$$

for the dynamics (1.1), or (1.10), the following results hold true:

(i) For each $y_0 \in Y$, and $w \in L_2(0, \infty; V)$, there exists a unique optimal pair denoted by $\{u_w^0(\cdot\,; y_0), y_w^0(\cdot\,; y_0)\}$, with corresponding optimal cost denoted by

$$\begin{aligned} J_w^0(y_0) &= J(u_w^0(\cdot\,; y_0), y_w^0(\cdot\,; y_0)) \\ &= \int_0^\infty [\|Ry_w^0(t; y_0)\|_Z^2 + \|u_w^0(t; y_0)\|_U^2 - \gamma^2 \|w(t)\|_V^2]\, dt. \end{aligned} \tag{3.2}$$

(ii) The optimal pair is related by

$$u_w^0(\cdot\,; y_0) = -L^* R^* R y_w^0(\cdot\,; y_0), \tag{3.3}$$

and is explicitly given in terms of the problem data by the following formulas:

$$-u_w^0(\cdot\,; y_0) = [I + L^*R^*RL]^{-1} L^*R^*R[e^{A\cdot} y_0 + Ww] \in L_2(0, \infty; U) \tag{3.4a}$$

$$= -u_{w=0}^0(\cdot\,; y_0) - u_w^0(\cdot\,; y_0 = 0); \tag{3.4b}$$

$$y_w^0(\cdot\,; y_0) = [I + LL^*R^*R]^{-1}[e^{A\cdot} y_0 + Ww] \in L_2(0, \infty; Y) \tag{3.5a}$$

$$= y_{w=0}^0(\cdot\,; y_0) + y_w^0(\cdot\,; y_0 = 0); \tag{3.5b}$$

$$Ry_w^0(\cdot\,; y_0) = [I + RLL^*R^*]^{-1}[Re^{A\cdot} y_0 + RWw] \in L_2(0, \infty; Z), \tag{3.6}$$

where L, L^*, W, W^* are given by (1.11) through (1.14), and where the inverse operators in (3.4a), (3.5a), (3.6) are well defined as bounded operators on all of $L_2(0,\infty;U)$, $L_2(0,\infty;Y)$, and $L_2(0,\infty;Z)$, respectively (for $[I + LL^*R^*R]^{-1}$, see [9, Chapter 2, Appendix 2A of Volume 1, p. 167]). Moreover, the optimal dynamics is

$$y_w^0(t;y_0) = e^{At}y_0 + \{Lu_w^0(\cdot\,;y_0)\}(t) + \{Ww(\cdot)\}(t) \in L_2(0,\infty;Y). \tag{3.7}$$

(iii) The optimal cost $J_w^0(y_0)$ in (3.2) is given explicitly in terms of the data by the following formulas:

$$J_w^0(y_0) = \left(Re^{A\cdot}y_0 + RWw, [I + RLL^*R^*]^{-1}[Re^{A\cdot}y_0 + RWw]\right)_{L_2(0,\infty;Z)} \tag{3.8a}$$
$$-\gamma^2(w,w)_{L_2(0,\infty;V)}$$
$$= J_{w=0}^0(y_0) + J_w^0(y_0 = 0) + \mathcal{X}_{y_0,w}; \tag{3.8b}$$

$$J_{w=0}^0(y_0) = (P_{w=0}y_0, y_0)_Y = \left(Re^{A\cdot}y_0, [I + RLL^*R^*]^{-1}Re^{A\cdot}y_0\right)_{L_2(0,\infty;Z)}, \tag{3.9}$$

where $P_{w=0}$ is the Riccati operator in the case $w = 0$, established in [10],

$$J_w^0(y_0 = 0) = (RWw, [I + RLL^*R^*]^{-1}RWw)_{L_2(0,\infty;Z)}$$
$$- \gamma^2(w,w)_{L_2(0,\infty;V)} \tag{3.10}$$
$$= -(w, E_\gamma w)_{L_2(0,\infty;V)} = \text{quadratic in } w, \tag{3.11}$$

$$E_\gamma = \gamma^2 I - W^*R^*[I + RLL^*R^*]^{-1}RW = \gamma^2 I - S \in \mathcal{L}(L_2(0,\infty;V)); \tag{3.12}$$
$$S \equiv W^*R^*[I + RLL^*R^*]^{-1}RW \in \mathcal{L}(L_2(0,\infty;V) \tag{3.13a}$$
$$= \text{non-negative, self-adjoint operator in } \mathcal{L}(L_2(0,\infty;V)). \tag{3.13b}$$

The cross term $\mathcal{X}_{y_0,w}$ in (3.8b) is linear in w:

$$\mathcal{X}_{y_0,w} = 2(Re^{A\cdot}y_0, [I + RLL^*R^*]^{-1}RWw)_{L_2(0,\infty;Z)}$$
$$= \text{linear in } w. \tag{3.14}$$

Finally, the following identity holds true:

$$[I + RLL^*R^*]^{-1}R = R[I + LL^*R^*R]^{-1} \in \mathcal{L}(L_2(0,\infty;Y); L_2(0,\infty;Z)), \tag{3.15}$$

so that we may rewrite S in (3.13a) as

$$S = W^*R^*R[I + LL^*R^*R]^{-1}W \in \mathcal{L}(L_2(0,\infty;V)). \tag{3.16}$$

Proof. See [9, Chapter 6, Section 6.20, Proof of Theorem 6.20.1.1, p. 613 and Remark 6.20.1.2]. This proof relies on the regularity properties of L, L^*, W, W^* in (1.11b)–(1.14b), as pointed out in the Orientation. □

As in [9, Chapter 6, Section 6.20.2, Eqns. (6.20.2.1–2), of Volume 1, p. 615], we define the critical value γ_c of γ by

$$\gamma_c^2 \equiv \|S\|_{\mathcal{L}(L_2(0,\infty;V))} = \|W^*R^*[I + RLL^*R^*]^{-1}RW\|^2_{\mathcal{L}(L_2(0,\infty;V))} \tag{3.17}$$
$$= \sup_{\|w\|=1} (Sw, w)_{L_2(0,\infty;V)}, \tag{3.18}$$

where the norm of w is that of $L_2(0,\infty;V)$, since S is a non-negative, self-adjoint bounded operator. We then obtain the counterpart of [9, Corollary 6.20.2.1, Chapter 6, p. 615].

Corollary 3.2. (Strict positive-definiteness of the operator E_γ for $\gamma > \gamma_c$) Assume (H.1)–(H.5). Then, the self-adjoint operator $E_\gamma \in \mathcal{L}(L_2(0,\infty;V))$ in (3.12) is strictly positive self-adjoint if and only if $\gamma > \gamma_c$ (defined by (3.17)), in which case:

$$(E_\gamma w, w)_{L_2(0,\infty;V)} \geq (\gamma^2 - \gamma_c^2)\|w\|^2_{L_2(0,\infty;V)}, \tag{3.19a}$$

in which case

$$E_\gamma^{-1} \in \mathcal{L}(L_2(0,\infty;V)). \tag{3.19b}$$

□

Next, we return to the optimal $J_w^0(y_0)$ in (3.8) for $w \in L_2(0,\infty;V)$ and consider the problem:

$$\text{maximize } J_w^0, \text{ equivalently, minimize } -J_w^0(y_0), \text{ over all } w \in L_2(0,\infty;V). \tag{3.20}$$

Theorem 3.3. (Maximization of $J_w^0(y_0)$ over w) Assume (H.1)–(H.5).

(i) For $\gamma > \gamma_c$ (defined in (3.17)), and with reference to (3.8), the following estimate holds true for any $\epsilon > 0$ and every $w \in L_2(0,\infty;V)$:

$$-J_w^0(y_0) \geq [\gamma^2 - (\gamma_c^2 + \epsilon)]\|w\|^2_{L_2(0,\infty;V)} - J_{w=0}^0(y_0) - C_\epsilon\|y_0\|_Y^2. \tag{3.21}$$

(ii) For $\gamma > \gamma_c$ (defined in (3.17)), there exists a unique optimal solution $w^*(\,\cdot\,;y_0) \in L_2(0,\infty;V)$ for the optimal problem (3.20),

$$\max_{w\in L_2(0,\infty;V)} J_w^0(y_0) = J_{w=w^*}^0(y_0) \equiv J^*(y_0). \tag{3.22}$$

(iii) Let $\gamma_c > 0$. If $0 < \gamma < \gamma_c$, then $\sup_w J_w^0(y_0) = +\infty$ for all initial conditions $y_0 \in Y$.

Proof. See [9, Chapter 6, Section 6.21, Proof of Theorem 6.21.1, p. 616] and Orientation. □

With the optimal w^* provided by Theorem 3.3(ii), we return to the optimal pair $\{u_w^0, y_w^0\}$ over u of Theorem 3.1, and set, along with (3.22):

$$\begin{aligned} u^*(\,\cdot\,;y_0) &= u_{w=w^*}^0(\,\cdot\,;y_0) \in L_2(0,\infty;U); \\ y^*(\,\cdot\,;y_0) &= y_{w=w^*}^0(\,\cdot\,;y_0) \in L_2(0,\infty;Y). \end{aligned} \tag{3.23}$$

Theorem 3.4. (Explicit formulas of u^*, y^* in terms of w^*) Assume (H.1)–(H.5). Let $y_0 \in Y$.

(i) The unique optimal $w^*(\,\cdot\,;y_0)$ provided by Theorem 3.3 is given explicitly in terms of the problem data by (see (1.14) (3.23), (3.7))

$$\gamma^2 w^*(\,\cdot\,;y_0) = W^*R^*Ry^*(\,\cdot\,;y_0) \in L_2(0,\infty;V), \quad \gamma > \gamma_c. \tag{3.24}$$

(ii) Thus, for $\gamma > \gamma_c$ (defined by (3.17)), the original min-max problem (1.7) has a unique solution $\{u^*(\cdot\,; y_0), y^*(\cdot\,; y_0), w^*(\cdot\,; y_0)\}$ satisfying (3.24) and given by

$$-u^*(\cdot\,; y_0) = [I + L^*R^*RL]^{-1}L^*R^*R[e^{A\cdot}y_0 + Ww^*(\cdot\,; y_0)] \in L_2(0,\infty; U); \quad (3.25)$$

$$y^*(\cdot\,; y_0) = [I + LL^*R^*R]^{-1}[e^{A\cdot}y_0 + Ww^*(\cdot\,; y_0)] \in L_2(0,\infty; Y); \quad (3.26)$$

$$u^*(\cdot\,; y_0) = -L^*R^*Ry^*(\cdot\,; y_0) \in L_2(0,\infty; U); \quad (3.27)$$

$$Ry^*(\cdot\,; y_0) = [I + RLL^*R^*R]^{-1}[Re^{A\cdot}y_0 + RWw^*(\cdot\,; y_0)] \in L_2(0,\infty; Z), \quad (3.28)$$

with optimal dynamics

$$y^*(t; y_0) = e^{At}y_0 + \{Lu^*(\cdot\,; y_0)\}(t) + \{Ww^*(\cdot\,; y_0)\}(t), \quad (3.29)$$

which therefore satisfies

$$\{[I + LL^*R^*R - \gamma^{-2}WW^*R^*R]y^*(\cdot\,; y_0)\}(t) = e^{At}y_0; \quad (3.30)$$

$$\{[I + RLL^*R^* - \gamma^{-2}RWW^*R^*]Ry^*(\cdot\,; y_0)\}(t) = Re^{At}y_0. \quad (3.31)$$

Proof. See [9, Chapter 6, Section 6.21, proof of Theorem 6.21.2, p. 617]. □

The counterpart of [9, Proposition 6.22.1 and of Corollary 6.22.2, Chapter 6, Section 6.22] follows next.

Theorem 3.5. (Explicit expressions of $\{u^*, y^*, w^*\}$ and P for $\gamma > \gamma_c$ via E_γ^{-1}) Assume (H.1)–(H.5). Let $\gamma > \gamma_c$ (defined by (3.17). Then, the following formulas explicitly in terms of the data hold true:

$$\begin{aligned} w^*(\cdot\,; y_0) &= E_\gamma^{-1}W^*R^*R[I + LL^*R^*R]^{-1}(e^{A\cdot}y_0) \in L_2(0,\infty; V) &(3.32a)\\ &= E_\gamma^{-1}W^*R^*[I + RLL^*R^*]^{-1}Re^{A\cdot}y_0; &(3.32b)\end{aligned}$$

$$\begin{aligned} &-u^*(\cdot\,; y_0)\\ &= [I + L^*R^*RL]^{-1}L^*R^*\{I + RWE_\gamma^{-1}W^*R^*[I + RLL^*R^*]^{-1}\}Re^{A\cdot}y_0; \end{aligned} \quad (3.33)$$

$$Ry^*(\cdot\,; y_0) = [I+RLL^*R^*]^{-1}\{I+RWE_\gamma^{-1}W^*R^*[I+RLL^*R^*]^{-1}\}Re^{A\cdot}y_0; \quad (3.34)$$

$$y^*(\cdot\,; y_0) = [I + LL^*R^*R]^{-1}\{I + WE_\gamma^{-1}W^*R^*R[I + LL^*R^*R]^{-1}\}e^{A\cdot}y_0; \quad (3.35)$$

$$\begin{aligned} J^*(y_0) &= J^0_{w=w^*}(y_0) = (Py_0, y_0)_Y &(3.36)\\ &= (Re^{A\cdot}y_0 + RWw^*, [I + RLL^*R^*]^{-1}[Re^{A\cdot}y_0 + RWw^*])_{L_2(0,\infty;Z)} &(3.37)\\ &= (Re^{A\cdot}y_0, [I + RLL^*R^*]^{-1}Re^{A\cdot}y_0)_{L_2(0,\infty;Z)}\\ &+ (RWE_\gamma^{-1}W^*R^*[I + RLL^*R^*]^{-1}Re^{A\cdot}y_0,\\ &\quad [I + RLL^*R^*]^{-1}RWE_\gamma^{-1}W^*R^*[I + RLL^*R^*]^{-1}Re^{A\cdot}y_0)_{L_2(0,\infty;Z)}\\ &+ 2(Re^{A\cdot}y_0, [I + RLL^*R^*]^{-1}RWE_\gamma^{-1}W^*R^*\\ &\quad [I + RLL^*R^*]^{-1}Re^{A\cdot}y_0)_{L_2(0,\infty;Z)}. &(3.38)\end{aligned}$$

Thus, (3.9) and (3.36)–(3.38) yield the following relationship, expected from the definition of the min-max problem (1.7),

$$\begin{aligned} J^*(y_0) &= (Py_0, y_0)_Y \geq J^0_{w=0}(y_0) = (P_{w=0}y_0, y_0)_Y \\ &= (Re^{A\,\cdot}\, y_0, [I + RLL^*R^*]^{-1} Re^{A\,\cdot}\, y_0)_{L_2(0,\infty;Z)}, \end{aligned} \tag{3.39}$$

where the non-negative, self-adjoint operator P is defined in (3.36) and the non-negative, self-adjoint operator $P_{w=0}$ is defined in (3.9).

Proof. For (3.32) on w^*, see [9, Chapter 6, Section 6.22, proof of Proposition 6.22.1, p. 618]. Inserting such w^* into the right sides of (3.25), (3.26), (3.28), (3.37) produces the noted expressions for u^*, y^*, Ry^*, J^*, and P, see [9, Chapter 6, Corollary 6.22.2]. □

We now give the counterpart of [9, Corollary 6.23.2, Chapter 6, p. 621].

Theorem 3.6. (Regularity of $\{u^*, y^*, w^*\}$) Assume (H.1)–(H.5). With reference to the unique solution $\{u^*, y^*, w^*\}$, we have for any $y_0 \in Y$:

$$u^*(\,\cdot\,; y_0) \in C([0,\infty]; U);\;\; y^*(\,\cdot\,; y_0) \in C([0,\infty]; Y);\;\; w^*(\,\cdot\,; y_0) \in C([0,\infty]; V). \tag{3.40}$$

Proof. Both operators L and W, as well as L^* and W^*, satisfy the regularity properties of [10, Proposition 3.1.2], under the present assumptions (H.4) in (1.3) and (1.4) for B and G. This is the counterpart of [9, Theorem 6.23.1 of Chapter 6, p. 620]. Thus, the desired conclusions in (3.40) follow by the usual boot-strap argument, exactly as in the proof of [9, Corollary 6.23.2 (or Corollary 6.9.3) in Chapter 6]. □

Theorem 3.7. (Transition property for w^* for $\gamma > \gamma_c$) Assume (H.1)–(H.5). Let $\gamma > \gamma_c$ (defined in (3.17)). Then, with reference to w^* in (3.32), we have:

(i)

$$w^*(t+\sigma; y_0) = w^*(\sigma; y^*(t; y_0)), \quad \forall\, t, \sigma > 0, \tag{3.41}$$

for t fixed, the equality being intended in $C([0,\infty]; Y)$ in σ.

(ii)

$$\{Ww^*(\,\cdot\,; y_0)\}(t+\sigma) = \{Ww^*(\,\cdot\,; y_0)\}(\sigma) - e^{A\sigma}\{Ww^*(\,\cdot\,; y_0)\}(t) \equiv 0. \tag{3.42}$$

Proof. See [9, Chapter 6, Section 6.24, proof of Theorem 6.24.1 for (i), and proof of Corollary 6.24.4 for (ii)]. □

We next define (as in [9, Eqn. (6.25.1) of Chapter 6, p. 626]) the operator $\Phi(t)$ (which depends on γ) by

$$y^*(t; x) \equiv \Phi(t)x \in C([0,\infty]; Y) \cap L_2(0,\infty; Y), \quad \forall\, x \in Y, \tag{3.43}$$

and obtain that $\Phi(t)$ defines a s.c. semigroup on Y which, moreover, is uniformly stable.

Next, the counterpart of [9, Theorem 6.25.1 and of Corollary 6.25.3 of Chapter 6].

Theorem 3.8. (Semigroup property for y^* and stability, for $\gamma > \gamma_c$) Assume (H.1)–(H.5). Let $\gamma > \gamma_c$ (defined in (3.17)), and $y_0 \in Y$. Then:

(i) for $t, \sigma > 0$ we have

$$y^*(t+\sigma; y_0) = y^*(\sigma; y^*(t; y_0)) \in C([0,\infty]; Y), \tag{3.44}$$

so that $\Phi(t)$ in (3.43) defines a s.c. semigroup on Y;

(ii) the semigroup $\Phi(t)$ is uniformly (exponentially) stable: there exist constants $M_F \geq 1$, $\omega_F > 0$, such that

$$\|\Phi(t)\|_{\mathcal{L}(Y)} \leq M_F e^{-\omega_F t}, \quad t \geq 0. \tag{3.45}$$

Proof. See the proof of [9, Theorem 6.25.1 and of Corollary 6.25.3 of Chapter 6]. In particular, part (ii) follows from part (i) since $y^*(t;x) = \Phi(t)x \in L_2(0,\infty;Y)$ for all $x \in Y$ by optimality (3.26), so that a well-known result [6] yields (3.45). □

We now abandon the technical treatment as given in [9, Chapter 6, Section 6.26] – which explicitly uses the analyticity property of e^{At}, presently non-available – and replace it with the corresponding treatment of [10].

4. Conclusion of proof of Theorem 2.1, $\gamma > \gamma_c$

First, for $\gamma > \gamma_c$, we define the operator $P \in \mathcal{L}(Y)$ (which will turn up to be the same as the operator P in (3.36), see (4.12) below) by setting

$$Px = \int_0^\infty e^{A^*t} R^* R y^*(t;x)dt = \int_0^\infty e^{A^*t} R^* R\Phi(t)x\,dt \tag{4.1a}$$

$$= \int_t^\infty e^{A^*(\tau-t)} R^* R\Phi(\tau-t)x\,d\tau, \quad x \in Y, \tag{4.1b}$$

where $\Phi(\cdot)$ is defined in (3.43). Boundedness of P in $\mathcal{L}(Y)$ follows from (4.1a) via (1.2) and (3.43). We begin with some preliminary critical properties involving P.

Lemma 4.1. Assume (H.1)–(H.5). Let $\gamma > \gamma_c$. Then:

(i) With reference to (3.27), we have for $y_0 \in Y$,

$$\begin{aligned} -u^*(t;y_0) &= \{L^*R^*Ry^*(\cdot\,;y_0)\}(t) \\ &= \int_t^\infty B^* e^{A^*(\tau-t)} R^* R y^*(\tau; y_0)d\tau && (4.2) \\ &= \int_t^\infty B^* e^{A^*(\tau-t)} R^* R\Phi(\tau)y_0 d\tau \\ &= \int_t^\infty B^* e^{A^*(\tau-t)} R^* R\Phi(\tau-t)\Phi(t)y_0 d\tau && (4.3) \\ &= B^*P\Phi(t)y_0 \in L_2(0,\infty;U) \cap C([0,\infty];U) && (4.4) \end{aligned}$$

$$B^*P\Phi(t) : \text{ continuous } Y \to L_2(0,\infty;U) \cap C([0,\infty];U). \tag{4.5}$$

(ii) With reference to (4.1) we have: $P \in \mathcal{L}(Y)$ and moreover B^*P: continuous $Y \to U$; G^*P: continuous $Y \to V$:

$$\begin{cases} B^*P \in \mathcal{L}(Y;U); \; G^*P \in \mathcal{L}(Y;V) & (4.6a) \\ B^*Px = \int_0^\infty B^* e^{A^*t} R^* R\Phi(t)x\, dt \in U, \quad x \in Y; & (4.6b) \\ G^*Px = \int_0^\infty G^* e^{A^*t} R^* R\Phi(t)x\, dt \in V, \quad x \in Y. & (4.6c) \end{cases}$$

(iii) With reference to (3.24) we likewise have for $y_0 \in Y$,

$$\begin{aligned} \gamma^2 w^*(t;y_0) &= \{W^*R^*Ry^*(\,\cdot\,;y_0)\}(t) \\ &= \int_t^\infty G^* e^{A^*(\tau-t)} R^* R y^*(\tau;y_0) d\tau & (4.7) \\ &= \int_t^\infty G^* e^{A^*(\tau-t)} R^* R\Phi(\tau) y_0 d\tau & (4.8) \\ &= \int_t^\infty G^* e^{A^*(\tau-t)} R^* R\Phi(\tau - t)\Phi(t) y_0 d\tau = G^*P\Phi(t)y_0 \\ &\in L_2(0,\infty;V) \cap C([0,\infty];V); & (4.9) \end{aligned}$$

$$G^*P\Phi(t): \text{ continuous } Y \to L_2(0,\infty;V) \cap C([0,\infty];V). \tag{4.10}$$

(iv) The operator $P \in \mathcal{L}(Y)$ satisfies the symmetric relation for $x_1, x_2 \in Y$:

$$(Px_1, x_2)_Y = \int_0^\infty [(Ry^*(t;x_1), Ry^*(t;x_2))_Y + (u^*(t;x_1), u^*(t;x_2))_U - \gamma^2 (w^*(t;x_1), w^*(t;x_2))_V] dt, \tag{4.11}$$

from which it follows that P is a self-adjoint operator: $P = P^*$ on Y, and that the optimal cost of problem (1.7) is

$$\begin{aligned} (Py_0, y_0) &= J^*(y_0) \text{ [optimal cost in (3.22)]} & (4.12a) \\ &= J(u^*(\,\cdot\,;y_0), y^*(\,\cdot\,;y_0), w^*(\,\cdot\,;y_0)), & (4.12b) \end{aligned}$$

in line with (3.36), which, by (3.37) or (3.39), identified P as non-negative, self-adjoint in Y, $P \geq 0$.

Proof. (i), (ii), (iii) The steps from (4.2) to (4.4) follow from (3.27) via the definition (1.13a) of L^*, identity (3.43), the semigroup property of $\Phi(\,\cdot\,)$ noted in Theorem 3.8, and finally the definition (4.1a) of P.

All the integral terms from (4.2) to (4.6c) are well defined by virtue of assumptions (H.1) = (1.2) and (H.4) = (1.3) on B and (H.4) = (1.4) on G, leading to properties (1.8) and (1.9), respectively; precisely as in the proof of [10, Lemma 3.2.2]. These same facts are likewise responsible for obtaining (4.6a-b-c).

(iv) The proof of identity (4.11) follows precisely as in the proof of [9, Chapter 6, Corollary 6.26.12(iii), p. 628].

(v) Then, (iv) yields (v). □

We now turn to the 'feedback' generator A_F, that is to the infinitesimal generator of the s.c. uniformly stable semigroup $\Phi(t)$, guaranteed by Theorem 3.8, so that

$$\Phi(t)x \equiv e^{A_F t}x,\ x \in Y;\ \frac{d\Phi(t)x}{dt} = A_F\Phi(t)x = \Phi(t)A_F x,\ x \in \mathcal{D}(A_F). \tag{4.13}$$

We next provide information about A_F essentially as a consequence of inserting $u^*(t;y_0) = -B^*P\Phi(t)y_0$ and $\gamma^2 w^*(t;y_0) = G^*P\Phi(t)y_0$, see (4.4) and (4.9), in Eqn. (1.1).

Proposition 4.2. (Identification of A_F) Assume (H.1)–(H.5). Let $\gamma > \gamma_c$. For $x \in Y$ and all $t > 0$:

(i)

$$\frac{d\Phi(t)x}{dt} = [A - BB^*P + \gamma^{-2}GG^*P]\Phi(t)x \in [\mathcal{D}(A^*)]'; \tag{4.14}$$

(ii)

$$\begin{aligned}[A - BB^*P + \gamma^{-2}GG^*P]\Phi(t)x &= A_F\Phi(t)x \\ &= \Phi(t)A_F x \in Y,\ x \in \mathcal{D}(A_F),\ t \geq 0;\end{aligned} \tag{4.15a}$$

$$[A - BB^*P + \gamma^{-2}GG^*P]x = A_F x \in Y,\ x \in \mathcal{D}(A_F); \tag{4.15b}$$

$$\Phi(t)x \equiv e^{A_F t}x \equiv e^{(A-BB^*P+\gamma^{-2}GG^*P)t}x. \tag{4.15c}$$

Proof. See the proof of [9, Chapter 6, Theorem 6.14.1, p. 601, which also works as a proof of Theorem 6.26.2.1 (in the stable case)]. We return to the min-max solution dynamics

$$y^*(t;y_0) = \Phi(t)y_0 = e^{At}y_0 + \int_0^t e^{A(t-\tau)}Bu^*(\tau;y_0)d\tau + \int_0^t e^{A(t-\tau)}Gw^*(\tau;y_0)d\tau, \tag{4.16}$$

take the inner product of (4.16) with $y \in \mathcal{D}(A^*)$ and differentiate in t with $y_0 \in \mathcal{D}(A_F)$, thus obtaining

$$\begin{aligned}\left(\frac{dy^*(t;y_0)}{dt}, y\right)_Y &= \left(\frac{d\Phi(t)y_0}{dt}, y\right)_Y = (\Phi(t)A_F y_0, y)_Y \\ &= (\Phi(t)y_0, A^*y)_Y + (Bu^*(t;y_0), y)_Y + (Gw^*(t;y_0), y)_Y \\ \text{(by (4.4), (4.9))} \quad &= ([A - BB^*P + \gamma^{-2}GG^*P]\Phi(t)y_0, y)_Y, \\ &\qquad y_0 \in \mathcal{D}(A_F),\ y \in \mathcal{D}(A^*),\end{aligned} \tag{4.17}$$

after using, in the last step,

$$u^*(t;y_0) = -B^*P\Phi(t)y_0, \text{ and } \gamma^2 w^*(t;y_0) = G^*P\Phi(t)y_0,$$

from (4.4) and (4.9), respectively. Then, (4.17) readily shows (4.14) and (4.15). □

As in [10, Section 3.3], our next task is to complete the description of the dynamics $e^{A_F t}$ by transferring the singular estimates from $e^{At}B$ and $e^{At}G$ to $e^{A_F t}B$ and $e^{A_F t}G$, for $t > 0$. However, before doing this, we need the following critical result.

Proposition 4.3. Assume (H.1)–(H.5). Then, the regularity results for L and L^* in [10, Proposition 3.3.1, Eqns. (3.3.1)–(3.3.5)] hold true; and, in addition, the following counterpart results for W, W^* hold true:

(i) for $0 < s < 1$, r as in (1.4), and any $0 < T < \infty$:

$$L, W : \text{ continuous } {}_sC([0,T];V) \to {}_{(s+r-1)}C([0,T];Y); \tag{4.18}$$

(ii) for $s > 0$ and $\epsilon > 0$ arbitrary:

$$L^*, W^* : \text{ continuous } {}_sC([0,T];Y) \to {}_{(s+r-1+\epsilon)}C([0,T];V). \tag{4.19}$$

(iii) Finally, with reference to the strictly positive, self-adjoint operator E_γ in (3.12), (3.19) for $\gamma > \gamma_c$, we have that, for any $0 \le s < 1$:

$$E_\gamma^{-1} \in \mathcal{L}({}_sC([0,T];V)), \qquad \gamma > \gamma_c, \ 0 \le s < 1. \tag{4.20}$$

Proof. Eqns. (4.18), (4.19) for W, W^* are the same results as those for L, L^* in [10, Proposition 3.3.1, Eqns. (3.3.1), (3.3.2)] due to the assumption (H.3) = (1.4) for G. Thus, it remains to prove (iii).

(iii) The proof of (iii) follows the same pattern as the proof of [10, Proposition 3.3.1, Eqn. (3.3.4)], in asserting that

$$[I + L^*R^*RL]^{-1}, [I + LL^*R^*R]^{-1} \in \mathcal{L}({}_sC([0,T];Y)), \ 0 \le s < 1, \tag{4.21}$$

which is the original pattern of the proof of [9, Chapter 1, Theorem 1.4.4.4, p. 40, *mutatis mutandis*]. Recalling $E_\gamma = \gamma^2 I - S$ from (3.12), (3.13), with $S = W^*R^*[I + RLL^*R^*]^{-1}RW$, we have preliminarily that W and W^* are smoothing and reduce the singularity, see (4.18), (4.19) with $r < 1$, while $[I + RLL^*R^*]^{-1}$ preserves the order of singularity, see (4.21). Thus S in (3.16) is smoothing. *A-fortiori*, $S \in \mathcal{L}(C[0,T];V)$; see also (1.12c), (1.14c), and [10, Eqn. (3.1.14)]. More precisely, the following two preliminary results hold true:

(a) Given any $h \in {}_sC([0,T];V)$, $0 \le s < 1$, there exists a positive integer $n_0(s)$, depending on s, such that

$$S^{n_0}h \in C([0,T];V); \tag{4.22}$$

(b) given any $z \in L_2(0,T;V)$, there exists a positive integer $n_1(s)$, depending on s, such that

$$S^{n_1}z \in C([0,T];V). \tag{4.23}$$

These results are obtained by using (4.18), (4.19), and (4.21) for part (a), to obtain (4.22); and by [10, Proposition 3.1.2, Eqn. (3.1.10)–(3.1.20)], for part (b), to obtain (4.23).

After these preliminaries, we may begin the proof of (4.20). To this end, we follow the proof of [9, Chapter 1, Theorem 1.4.4.4, p. 41]. Let $h \in {}_sC([0,T];V)$. We seek a unique $g \in {}_sC([0,T];V)$ such that

$$E_\gamma g = h \ \text{ or } \ \gamma^2 g - Sg = h, \tag{4.24}$$

with S as in (3.13).

Step 1. We apply part (a) above, so that by (4.22), we have

$$S^{n_0}h \in C([0,T];V) \subset L_2(0,T;V).$$

Hence, there exists a unique $v \in L_2(0,T;V)$ such that

$$E_\gamma v \equiv \gamma^2 v - Sv = S^{n_0}h \in C([0,T];V) \subset L_2(0,T;V), \tag{4.25}$$

since E_γ is boundedly invertible on $L_2(0,T;V)$ by (3.19b).

Step 2. We shall show that, in fact,

$$v \in C([0,T];V) \text{ or that } E_\gamma^{-1} : \text{ continuous } C([0,T];V) \to \text{itself.} \tag{4.26}$$

In fact, we apply S^j to (4.25), with $j = 0, 1, \dots, n_1 - 1$, thus obtaining

$$\gamma^2 S^j v - S^{j+1} v = S^{n_0+j}h \in C([0,T];V), \qquad j = 0, 1, \dots, n_1 - 1, \tag{4.27}$$

where the regularity on the right of (4.27) is a consequence of (4.25) and of the regularity $S \in \mathcal{L}(C[0,T];V)$, as noted below (4.21).

Starting from $j = n_1 - 1$ in (4.27) and $v \in L_2(0,T;V)$, we first obtain $S^{j+1}v = S^{n_1}v \in C([0,T];V)$ by (4.23); hence $S^{n_1-1}v \in C([0,T];V)$ by (4.27). Next, using this latter information in (4.27), this time with $r = n_1 - 2$, leads to $S^{n_1-2}v \in C([0,T];V)$. By repeating this procedure a finite number of times, we arrive at (4.26) that, in fact, $v \in C([0,T];V)$.

Step 3. Starting from the given h and v obtained in (4.25) satisfying (4.26), we shall finally define a finite sequence of functions called g_{n_0-1}, g_{n_0-2}, g_{n_0-3}, $\dots$ g_1, g, whose last element g will be precisely the sought-after unique solution of (4.24). We define recursively

$$\gamma^2 g_{n_0-1} = S^{n_0-1}h + v \in {}_sC([0,T];V), \tag{4.28_{n_0-1}}$$
$$\gamma^2 g_{n_0-2} = S^{n_0-2}h + g_{n_0-1} \in {}_sC([0,T]V); \tag{4.28_{n_0-2}}$$
$$\dots\dots\dots\dots\dots$$
$$\gamma^2 g_1 = Sh + g_2 \in {}_sC([0,T];V); \tag{4.28_1}$$
$$\gamma^2 g = h + g_1 \in {}_sC([0,T];V). \tag{4.28_0}$$

The (conservative) regularity noted on the right of (4.28_{n_0-1}) is a consequence of [10, Proposition 3.3.1 (i)] applied to $h \in {}_sC([0,T];Y)$ and of (4.26). In particular, $g \in {}_sC([0,T];V)$. It is now an easy matter to see that g is the unique, sought-after solution of (4.24). Part (iii) is proved. □

Theorem 4.4. Assume (H.1)–(H.5). Let $\gamma > \gamma_c$. With reference to the semigroup $\Phi(t) = e^{A_F t}$, guaranteed by Theorem 3.8, we have:

(i)

$$\begin{cases} e^{A_F t}Bu \in {}_rC([0,T];Y), \quad \forall\, u \in U; & (4.29a) \\ \|e^{A_F t}B\|_{\mathcal{L}(U;Y)} \equiv \left\|B^* e^{A_F^* t}\right\|_{\mathcal{L}(Y;U)} \le c_T \frac{1}{t^r}, \quad 0 < t \le T; & (4.29b) \end{cases}$$

$$\begin{cases} e^{A_F t}Gv \in \; {}_rC([0,T];Y), \quad \forall\, v \in V; & (4.30a) \\ \|e^{A_F t}G\|_{\mathcal{L}(V;Y)} \equiv \left\|G^* e^{A_F^* t}\right\|_{\mathcal{L}(Y;V)} \leq c_T \frac{1}{t^r}, \quad 0 < t \leq T, & (4.30b) \end{cases}$$

where $0 < r < 1$ is the constant in assumption (H.4), Eqns. (1.3), (1.4).

(ii) For any $0 < \omega_1 < \omega_F$ (defined in (3.45)), there exists a constant k_1 (depending on C_F, ω_F, ω_1, T, r) such that

$$\begin{aligned} \left\|e^{A_F t}B\right\|_{\mathcal{L}(U;Y)} &= \left\|B^* e^{A_F^* t}\right\|_{\mathcal{L}(Y;U)} \\ &\leq k_1 \frac{e^{-\omega_1 t}}{t^r}, \; \forall\, t > 0, \; 0 < \omega_1 < \omega_F; \end{aligned} \tag{4.31}$$

$$\begin{aligned} \left\|e^{A_F t}G\right\|_{\mathcal{L}(V;Y)} &= \left\|G^* e^{A_F^* t}\right\|_{\mathcal{L}(Y;V)} \\ &\leq k_1 \frac{e^{-\omega_1 t}}{t^r}, \; \forall\, t > 0, \; 0 < \omega_1 < \omega_F. \end{aligned} \tag{4.32}$$

Proof. (i) First, we return to the explicit formula (3.35) giving $y^*(\,\cdot\,; y_0) = \Phi(\,\cdot\,)y_0$, $y_0 \in Y$ in terms of the data, which we apply with Bu, $u \in U$ in place of y_0, thus obtaining

$$\begin{aligned} \Phi(t)Bu &= e^{A_F t}Bu = \{[I + LL^*R^*R]^{-1}(e^{A\cdot}Bu)\}(t) \\ &+ \{[I + LL^*R^*R]^{-1}WE_\gamma^{-1}W^*R^*R[I + LL^*R^*R]^{-1}(e^{A\cdot}Bu)\}(t). \end{aligned} \tag{4.33}$$

Next, we invoke (4.21); that is, [10, Proposition 3.3.1(v), Eqn. (3.3.5)] and obtain under present assumptions

$$[I + LL^*R^*R]^{-1} : \text{ continuous } \; {}_rC([0,T];Y) \to \text{itself}, \tag{4.34}$$

where $0 < r < 1$ is the constant in (1.3), and hence, by (1.3),

$$[I + LL^*R^*R]^{-1}(e^{A\cdot}Bu) \in \; {}_rC([0,T];Y), \tag{4.35}$$

continuously in $u \in U$. Plainly, the second term in (4.33) is smoother than the first, as we now verify. Indeed, E_γ^{-1} and $[I + LL^*R^*R]^{-1}$ preserve the order s of singularity, see (4.20) and (4.21), while W^* and W are smoothing, see (4.19) and (4.18). By assumption (1.4) on G, the operators W and W^* in (1.12), (1.14) satisfy the regularity properties (4.18), (4.19). Thus, combining (4.35) with (4.19) with $s = r$, we obtain

$$W^*R^*R[I + LL^*R^*R]^{-1}(e^{A\cdot}Bu) \in \; {}_{(2r-1+\epsilon)}C([0,T];V), \tag{4.36}$$

continuously in $u \in U$. Next, we invoke the regularity property for E_γ^{-1} in (4.20) with $s = 2r - 1 + \epsilon$ and conclude that

$$E_\gamma^{-1}W^*R^*R[I + LL^*R^*R]^{-1}(e^{A\cdot}Bu) \in \; {}_{(2r-1+\epsilon)}C([0,T];V), \tag{4.37}$$

continuously in $u \in U$. Next, by (4.18) and (4.37), we obtain

$$WE_\gamma^{-1}W^*R^*R[I + LL^*R^*R]^{-1}(e^{A\cdot}Bu) \in \; {}_{(3r-2+\epsilon)}C([0,T];Y), \tag{4.38}$$

continuously in $u \in U$. Finally, by (4.21) once more, we conclude that

$$[I + LL^*R^*R]^{-1}WE_\gamma^{-1}W^*R^*R[I + LL^*R^*R]^{-1}(e^{A\cdot}Bu) \in {}_{(3r-2+\epsilon)}C([0,T];Y), \tag{4.39}$$

continuously in $u \in U$, where $(3r - 2 + \epsilon) < r$, since $r < 1$, and $\epsilon > 0$ is arbitrarily small. We conclude that, as predicted, the second term in (4.33) is smoother than the first. Ultimately we obtain via (4.35) and (4.39) in (4.33):

$$\Phi(t)Bu = e^{A_F t}Bu \in {}_rC([0,T];Y), \tag{4.40}$$

and part (i) is proved for B. Part (i) for G is identical.

(ii) Part (i) readily implies part (ii), as in going from (1.3), (1.4), over $0 < t \leq T$ to (1.8), (1.9) over $t > 0$: see [10, Lemma 1.1]. The proof of Theorem 4.4 is complete. □

As a consequence of Theorem 4.4, we obtain

Proposition 4.5. Assume (H.1)–(H.5). Let $\gamma > \gamma_c$. Then, the s.c. semigroup $\Phi(t) = e^{A_F t}$ in (4.13) is strongly differentiable on $\mathcal{D}(A)$ for $t > 0$: that is, more precisely, if $x \in \mathcal{D}(A)$ and $t > 0$, then:

$$\frac{de^{A_F t}x}{dt} = e^{A_F t}A_F x = e^{A_F t}(A - BB^*P + \gamma^{-2}GG^*P)x \tag{4.41}$$

$$= e^{A_F t}Ax - e^{A_F t}B(B^*Px) + \gamma^{-2}e^{A_F t}G(G^*Px), \quad x \in \mathcal{D}(A), \ t > 0, \tag{4.42}$$

and, in fact,

$$\left\| \frac{d}{dt}e^{A_F t}x \right\|_Y \leq M_F e^{-\omega_F t}\|Ax\|_Y + \frac{k_1 e^{-\omega_1 t}}{t^r}\left[\|B^*P\|_{\mathcal{L}(Y;U)} + \|G^*P\|_{\mathcal{L}(Y;V)}\right]\|x\|_Y, \quad t > 0, \tag{4.43}$$

where $0 \leq r < 1$ is the constant in (H.4), so that (4.43) is integrable over $[0, T]$.

Proof. Similar to the proof of [10, Proposition 3.3.3]. The steps in (4.41), (4.42) are self-explanatory, after recalling Proposition 4.2, as well as the boundedness of B^*P and G^*P in (4.6). Of course, (4.41) makes sense, at least, in $[\mathcal{D}(A_F^*)]'$. The point is that, under present assumptions, (4.41) is well defined actually in Y by Theorem 4.4. Estimate (4.43) makes use of the exponential decay in (3.45) for $\Phi(t) = e^{A_F t}$, as well as of (4.31) and (4.32) for $e^{A_F t}B$ and $e^{A_F t}G$; and (4.6) for the boundedness of B^*P and G^*P. □

Remark 4.1. This is the counterpart of Remark 3.3.1 in [10]. Proposition 4.5 may be used to derive the Algebraic Riccati Equation (ARE) for P on $\mathcal{D}(A)$. However, in the infinite horizon case, $T = \infty$, as in Section 3 of [10], a simpler approach is available to the derivation of the ARE: this will be given below in Theorem 4.7, and will be based on the regularity properties of P in Proposition 4.7, a far simpler result to prove than Proposition 4.5.

We now derive the ARE for P defined in (4.1a) as a consequence of Proposition 4.5, as in [11, Remark 3.3.1].

Let, at first, $x, z \in Y$. From the definition of P in (4.1a) we have

$$(Px,z)_Y = \int_0^\infty (Re^{A_Ft}x, Re^{At}z)_Z dt = \int_t^\infty (Re^{A_F(\tau-t)}x, Re^{A(\tau-t)}z)_Z d\tau,\ x,y \in Y. \tag{4.44}$$

We next specialize to $x, z \in \mathcal{D}(A)$ and differentiate (4.44) in t. We obtain, recalling Proposition 4.5:

$$\begin{aligned} 0 = & -\int_t^\infty \left(Re^{A_F(\tau-t)}A_Fx, Re^{A(\tau-t)}z\right)_Z d\tau \\ & -\int_t^\infty \left(Re^{A_F(\tau-t)}x, Re^{A(\tau-t)}Az\right)_Z d\tau - (Rx, Rz)_Z,\ x,z \in \mathcal{D}(A). \end{aligned} \tag{4.45}$$

We recall that the singularity in (4.43) is integrable. Invoking (4.44), we rewrite (4.45) first as

$$(PA_Fx, z)_Y + (Px, Az)_Y + (Rx, Rz)_Z = 0, \quad x, z \in \mathcal{D}(A), \tag{4.46}$$

and, then, since $A_F = A - BB^*P + \gamma^{-2}GG^*P$ from (4.15b), as

$$\begin{aligned} & (P(A - BB^*P + \gamma^{-2}GG^*P)x, z)_Y \\ & \quad + \quad (Px, Az)_Y + (Rx, Rz)_Z = 0, \quad x, z \in \mathcal{D}(A). \end{aligned} \tag{4.47}$$

Finally, recalling the boundedness of B^*P and G^*P in (4.6a), we obtain from (4.47),

$$\begin{aligned} & (PAx, z)_Y + (A^*Px, z)_Y + (Rx, Rz)_Z \\ & \quad = \quad (B^*Px, B^*Pz)_U - \gamma^{-2}(G^*Px, G^*Pz)_V, \quad x, z \in \mathcal{D}(A), \end{aligned} \tag{4.48}$$

where each term is well defined. This is precisely a first proof of Theorem 2.1(ii) for $x, z \in \mathcal{D}(A)$. □

A simpler approach to the derivation of the ARE for P is given next.

Proposition 4.6. Assume (H.1)–(H.5). Let $\gamma > \gamma_c$. Then, the following identities hold true: for P defined by (4.1a) (with $\Phi(t) \equiv e^{A_Ft}$, see (4.13)); that is

$$Px = \int_0^\infty e^{A^*t}R^*Re^{A_Ft}x\, dt, \quad x \in Y. \tag{4.49}$$

(i)

$$A^*Px = -R^*R - PA_Fx \in Y, \quad \forall\, x \in \mathcal{D}(A_F), \tag{4.50a}$$

and so

$$A^*P: \text{ continuous } \mathcal{D}(A_F) \to Y; \tag{4.50b}$$

(ii)

$$A_F^*Pz = -R^*Rz - PAz \in Y, \quad \forall\, z \in \mathcal{D}(A), \tag{4.51a}$$

(compare with (4.46)), and so

$$A_F^* P : \text{ continuous } \mathcal{D}(A) \to Y. \tag{4.51b}$$

(iii)

$$\mathcal{D}(A) \subset \mathcal{D}(A^* P); \quad \mathcal{D}(A_F) \subset \mathcal{D}(PA). \tag{4.52}$$

Proof. Same proof as that of [10, Proposition 3.4.1]. □

A first proof that P satisfies the ARE on $\mathcal{D}(A)$ was given in Remark 4.1. A simpler proof, based on Proposition 4.6, follows next.

Theorem 4.7. Assume (H.1)–(H.5). Then the operator P defined by (4.49) = (4.1a) – which was noted in Lemma 4.1(iv) to be non-negative, self-adjoint: $P = P^* \geq 0$ – satisfies the following Algebraic Riccati Equation on $\mathcal{D}(A)$, or on $\mathcal{D}(A_F)$: that is

$$\begin{aligned}(Px, Az)_Y + (Ax, Pz)_Y + (Rx, Rz)_Y = (B^*Px, B^*Pz)_U - \gamma^{-2}(G^*Px, G^*Pz)_V \\ \forall\, x \in Y,\ z \in \mathcal{D}(A);\ \text{or else } \forall\, x \in \mathcal{D}(A_F),\ \forall\, z \in Y.\end{aligned} \tag{4.53}$$

Proof. (Similar to the proof of [10, Theorem 3.5.1].)

Let $z \in \mathcal{D}(A)$ and $x \in Y$. Taking the inner product of (4.51a) with x yields

$$(PA_Fx, z)_Y + (Px, Az)_Y + (Rx, Rz)_Z = 0, \quad x \in Y,\ z \in \mathcal{D}(A) \tag{4.54}$$

(compare with (4.46)). Recalling $A_F = A - BB^*P + \gamma^{-2}GG^*P$ from (4.15b), we rewrite (4.54) as

$$\begin{aligned}(P(A - BB^*P + \gamma^{-2}GG^*P)x, z)_Y + (Px, Az)_Y + (Rx, Rz)_Z \\ = 0, \quad x \in Y,\ z \in \mathcal{D}(A),\end{aligned} \tag{4.55}$$

or

$$\begin{aligned}&(PAx, z)_Y + (Px, Az)_Y + (Rx, Rz)_Z \\ &\quad = (B^*Px, B^*Pz)_U - \gamma^{-2}(G^*Px, G^*Pz)_U, \quad x \in Y,\ z \in \mathcal{D}(A),\end{aligned} \tag{4.56}$$

where each term is well defined: the first term by (4.52), so that $A^*Pz \in Y$ for $z \in \mathcal{D}(A)$; the last two terms by $B^*P,\ G^*P \in \mathcal{L}(Y; U)$ in (4.6a).

Similarly, let $x \in \mathcal{D}(A_F)$ and $z \in Y$. Taking the inner product of (4.50a) with z yields

$$(PA_Fx, z)_Y + (A^*Px, z)_Y + (Rx, Rz)_Y = 0, \quad x \in \mathcal{D}(A_F),\ z \in Y; \tag{4.57}$$

$$\begin{aligned}(P[A - BB^*P + \gamma^{-2}GG^*P]x, z)_Y + (A^*Px, z)_Y + (Rx, Rz)_Z \\ = 0, \quad x \in \mathcal{D}(A_F),\ z \in Y;\end{aligned} \tag{4.58}$$

$$\begin{aligned}&(PAx, z)_Y + (A^*Px, z)_Y + (Rx, Rz)_Z \\ &\quad = (B^*Px, B^*Pz)_U - \gamma^{-2}(G^*Px, G^*Pz)_U, \quad x \in \mathcal{D}(A_F),\ z \in Y,\end{aligned} \tag{4.59}$$

where each term is well defined; the first one by (4.52), so that $PAx \in Y$ for $x \in \mathcal{D}(A_F)$; the second by (4.50b). Theorem 4.7 is proved. □

Part II: Optimal Control Problem with Nondefinite Quadratic Cost

Orientation. In this conclusive Part II, we return to the optimal control problem with nondefinite quadratic cost, which was studied in [9, Appendix A to Chapter 6] *under analyticity* of the free dynamics semigroup. The goal is to show that a combination of the treatment of [9, Appendix A to Chapter 6] and of parts of the analysis of [10] permits us to provide a full theory also in the present case, where analyticity is no longer available, but assumption (H.3) = (1.3) holds true in its place. Indeed, in that Appendix 6A, analyticity of the s.c. semigroup e^{At} was used only in the following few spots:

(i) in Proposition 6A.E.1, Eqn. (6A.E.2), which is no longer claimed in the present setting;

(ii) in Corollary 6A.F.3 in both providing the form of the domain of the feedback generator A_P in (6A.F.100), as well as in asserting that the s.c. semigroup $\Phi(t) = e^{A_P t}$ is, moreover, analytic on Y, for $t > 0$. Both these conclusions are no longer valid in the present setting;

(iii) in Theorem 6A.6.1 in deriving the validity of the ARE (6A.G.1) for all $x, y \in \mathcal{D}((-A)^\epsilon)$, as well as in asserting the inclusion $\mathcal{D}(A_P) \subset \mathcal{D}((-A)^{1-\gamma})$. In the present setting, the latter claim is no longer valid; moreover, the derivation of the ARE has to be done by following [10] instead.

Accordingly, it will suffice to give a brisk exposition of the new theory in the present setting.

5. Mathematical Setting. Assumptions

Dynamics. We consider the same dynamics of Section 1 of the present chapter with no disturbance

$$\dot{y}(t) = Ay(t) + Bu(t) \in [\mathcal{D}(A^*)]', \; y(0) = y_0 \in Y; \text{ or } y(t) = e^{At}y_0 + (Lu)(t), \quad (5.1)$$

under the same hypotheses (H.1), (H.2), (H.4) of Section 1 ((H.3) is no longer applicable); in particular, the stability assumption (1.2) of the s.c. semigroup e^{At}, and the singular estimate assumption (1.3) for $e^{At}B$. The operator L has the usual meaning defined by (1.11), with its adjoint L^* defined by (1.13).

Cost functional. In line with [9, Appendix A to Chapter 6], we presently consider a more general quadratic cost

$$F(x, u) = (F_1x, x) + 2 \operatorname{Re}(F_2x, u) + (u, u), \quad (5.2)$$

which is a continuous Hermitian form on $Y \times U$, under the same assumptions

$$F_1 = F_1^* \in \mathcal{L}(Y); \quad F_2 \in \mathcal{L}(Y; U), \quad (5.3)$$

as in [9, (6A.A.2), (6A.A.3)]. Thus, the quadratic cost is now

$$J(y_0; u) = \int_0^\infty [(F_1y, y) + 2 \operatorname{Re}(F_2y, u) + \|u\|^2]dt, \quad (5.4)$$

as in [9, (6A.A.4)], with y the solution to the dynamics (5.1) due to $y_0 \in Y$ and $u \in L_2(0,\infty;U) \equiv \mathcal{U}$. We likewise set $L_2(0,\infty;Y) \equiv \mathcal{Y}$.

Optimal Control Problem. The corresponding optimal control problem (OCP) is as follows: given $y_0 \in Y$, seek $u^0 = u^0(\,\cdot\,; y_0) \in \mathcal{U}$ such that

$$J^0(y_0) \equiv J(y_0; u^0) = \inf_{u \in \mathcal{U}} J(y_0; u) > -\infty. \tag{5.5}$$

Any such u^0 is then called 'an optimal control,' and its corresponding trajectory y^0 obtained via (5.1) 'an optimal solution.' Assumptions (H.1), (H.2), (H.4) of Section 1, as well as assumption (5.3) above *are in force throughout the present Part II, and will not be repeated.*

6. Results directly based on the regularity properties of L, L^*, as given in (1.11), (1.13)

The results given in this section are all proved exactly as their original and corresponding counterparts in [9, Chapter 6, Appendix A]. Indeed, they are all based directly on the very same regularity properties of the basic operators L and L^*, given in (1.11) and (1.13), which continue to hold true under the singular estimate assumption (H.4) = (1.3), in place of analyticity, as noted in the Orientation statement of Part II. This remark will not be repeated.

Proposition 6.1. (Existence, uniqueness)

(i) For $y_0 \in Y$ and $u \in \mathcal{U}$, the cost $J(y_0; u)$ in(5.4) may be rewritten as a continuous form on $Y \times \mathcal{U}$ as

$$J(y_0; u) = (\Lambda u, u)_{\mathcal{U}} + 2\,\mathrm{Re}(K y_0, u)_{\mathcal{U}} + (S y_0, y_0)_Y, \tag{6.1}$$

where the bounded operators Λ (self-adjoint), K, and S (self-adjoint) are given by

$$\Lambda = I + L^* F_1 L + F_2 L + L^* F_2^* \in \mathcal{L}(\mathcal{U}); \tag{6.2}$$

$$K = [L^* F_1 + F_2] e^{A\,\cdot} \in \mathcal{L}(Y; \mathcal{U}); \tag{6.3}$$

$$S = \int_0^\infty e^{A^* t} F_1 e^{At} dt \in \mathcal{L}(Y). \tag{6.4}$$

(ii) Let $y_0 \in Y$. The OCP has a solution $u^0 = u^0(\,\cdot\,; y_0)$ such that (5.5) holds true if and only if the following two conditions are satisfied:

$$(\Lambda u, u)_{\mathcal{U}} \geq 0, \quad \forall\, u \in \mathcal{U}; \text{ and } \Lambda u^0 + K y_0 = 0, \tag{6.5}$$

in which case

$$J(y_0, u) = \left\| \Lambda^{\frac{1}{2}} (u - u^0) \right\|^2 + (S y_0, y_0) - (\Lambda u^0, u^0); \tag{6.6}$$

$$\min J(y_0; u) = J(y_0; u^0) = (S y_0, y_0) - (\Lambda u^0, u^0). \tag{6.7}$$

(iii) Let $y_0 \in Y$. Assume that the corresponding OCP in (5.5) has a unique solution $u^0 = u^0(\,\cdot\,; y_0) \in \mathcal{U}$. Then, the self-adjoint operator Λ in (6.2) is positive definite:

$$(\Lambda u, u)_{\mathcal{U}} > 0, \quad \forall u \text{ s.t.}, \quad 0 \neq u \in \mathcal{U}. \tag{6.8}$$

Thus, Λ is injective and its inverse Λ^{-1} exists as an operator from the range $\mathcal{R}(\Lambda)$ of Λ onto $\mathcal{U}$, where $\overline{\mathcal{R}(\Lambda)} = Y$.

Proof. See proof of [9, Proposition 6A.B.1 and Proposition 6A.B.2, Appendix A of Chapter 6, p. 632]. □

Proposition 6.2. (Optimality condition, explicit formulas) Assume that the OCP in (5.5) admits a unique optimal control $u^0 = u^0(\,\cdot\,; y_0) \in \mathcal{U}$, for all $y_0 \in Y$, with corresponding optimal trajectory $y^0 = y^0(\,\cdot\,; y_0) \in \mathcal{Y}$ obtained via (5.1). Then

(i)

$$[I + L^* F_2^*] u^0 = -[L^* F_1 + F_2] y^0; \tag{6.9}$$

(ii)

$$\Lambda u^0(\,\cdot\,; y_0) = -[L^* F_1 + F_2] e^{A\,\cdot} y_0 = -K y_0 \in \mathcal{Y}; \tag{6.10}$$

(iii)

$$\Lambda^{-1} K = \Lambda^{-1} [L^* F_1 + F_2] e^{A\,\cdot} \in \mathcal{L}(Y; \mathcal{U}); \tag{6.11}$$

(iv)

$$u^0(\,\cdot\,; y_0) = -\Lambda^{-1} K y_0 = -\Lambda^{-1} [L^* F_1 + F_2] e^{A\,\cdot} y_0 \in \mathcal{U}; \tag{6.12}$$

$$y^0(\,\cdot\,; y_0) = e^{A\,\cdot} y_0 - L \Lambda^{-1} K y_0 = \{I - L\Lambda^{-1}[L^* F_1 + F_2]\} e^{A\,\cdot} y_0 \in \mathcal{Y}; \tag{6.13}$$

$$J^0(y_0) = (P y_0, y_0)_Y; \quad P = S - K^* \Lambda^{-1} K = P^* \in \mathcal{L}(Y); \tag{6.14}$$

$$J(y_0; u) = (\Lambda u, u) - 2\,\mathrm{Re}(\Lambda u, u^0) + (S y_0, y_0); \tag{6.15}$$

$$J(y_0; u) - J(y_0; u^0) = (\Lambda [u - u^0], [u - u^0])_{\mathcal{U}}; \tag{6.16}$$

$$J^0(y_0) = -(\Lambda u^0, u^0) + (S y_0, y_0). \tag{6.17}$$

Proof. See proof of [9, Proposition 6A.B.4 in Appendix A of Chapter 6, p. 634]. □

Henceforth we *assume that* the OCP has a *unique optimal control* $u^0 = u^0(\,\cdot\,; y_0) \in \mathcal{U} = L_2(0, \infty; U)$ for all $y_0 \in Y$ *with corresponding optimal trajectory* $y^0 = y^0(\,\cdot\,; y_0)$.

Proposition 6.3. (Continuity of optimal pair) The following regularity properties hold true for the optimal pair, with $y_0 \in Y$:

$$u^0(\,\cdot\,; y_0) \in C([0, \infty]; U); \quad y^0(\,\cdot\,; y_0) \in C([0, \infty]; Y), \tag{6.18}$$

in addition to $u^0 \in L_2(0, \infty; U) = \mathcal{U}$ and $y^0 \in L_2(0, \infty; Y) = \mathcal{Y}$ as in (6.12), (6.13).

Proof. See proof of [9, Proposition 6A.C.1 in Appendix A of Chapter 6, p. 635]. This is based on the regularity properties of L and L^* as in (1.11b) and (1.13b), which continue to hold true under assumption (H.4) = (1.3) of Section 1. □

Proposition 6.4. (Transitivity properties) The following transitivity properties hold true for $u^0(\,\cdot\,; y_0)$, $y^0(\,\cdot\,; y_0)$, $y_0 \in Y$,

$$u^0(t + \tau; y_0) = u^0(\tau; y^0(t; y_0)); \; y^0(t + \tau; y_0) = y^0(\tau; y^0(t; y_0)). \tag{6.19}$$

Thus, setting

$$y^0(t;y_0) = \Phi(t)y_0 \in L_2(0,\infty;Y) \cap C([0,\infty];Y), \tag{6.20}$$

we have that $\Phi(t)$ is a s.c. semigroup which, moreover, is exponentially stable: there exist constants $C \geq 1$, $\omega_P > 0$, such that

$$\|\Phi(t)\|_{\mathcal{L}(Y)} \leq Ce^{-\omega_P t}, \; t \geq 0. \tag{6.21}$$

Proof. See the proof of [9, Proposition 6A.D.1 in Appendix A of Chapter 6, p. 635]. □

Regarding the operator $P = P^* \in \mathcal{L}(Y)$ defined in (6.14), we have

Proposition 6.5. For any $x \in Y$, and recalling $y^0(t;x) = \Phi(t)x$ by (6.20), we have

$$Px = \int_0^\infty e^{A^*t}[F_1\Phi(t)x + F_2^*u^0(t;x)]dt, \tag{6.22}$$

where $u^0(t;x)$ is given explicitly in terms of the problem's data by (6.12).

Proof. See the proof of [9, Proposition 6A.E.1, part (i), Appendix A of Chapter 6, p. 635]. □

Proposition 6.6. (Boundedness of B^*P) With reference to (6.15) we have

$$\begin{cases} B^*P \in \mathcal{L}(Y;U); & (6.23)\\ B^*Px = \displaystyle\int_0^\infty B^*e^{A^*t}[F_1\Phi(t)x + F_2^*u^0(t;x)]dt, \; x \in Y. & (6.24)\end{cases}$$

Proof. The representation (6.22) for P yields (6.24), from which we deduce the boundedness of B^*P in (6.23), by invoking estimate (1.8) (a consequence of assumptions (H.1) and (H.4)) as well as the regularities of the optimal pair in (6.18) [or see (6.21)] and (5.3):

$$\begin{aligned} \|B^*Px\|_Y &\leq \int_0^\infty \left\|B^*e^{A^*t}\right\| [\|F_1\| \; \|\Phi(t)x\| + \|F_2^*\| \; \|u^0(t;x)\|] \, dt & (6.25)\\ &\leq \left(\int_0^\infty \frac{ke^{-\omega_0 t}}{t^\gamma} \, dt\right) [\|F_1\| \; \|\Phi(\,\cdot\,)x\|_{C([0,\infty];Y)} \\ &\quad + \|F_2^*\| \; \|u^0(\,\cdot\,;x)\|_{C([0,\infty];U)} & (6.26)\\ &\leq \text{const}\|x\|_Y, \forall \; x \in Y. & (6.27)\end{aligned}$$

Then (6.27) establishes (6.23). □

Proposition 6.7. (Feedback synthesis of optimal pair) For $y_0 \in Y$ we have

$$u^0(t;y_0) = -[F_2 + B^*P]\Phi(t)y_0 \in L_2(0,\infty;U) \cap C([0,\infty];U). \tag{6.28}$$

Proof. Same as the proof of [9, Proposition 6A.F.1, Appendix A to Chapter 6]. □

According to (6.28) we can rewrite formula (6.24) for P.

Corollary 6.8. For $x \in Y$,

$$Px = \int_0^\infty e^{A^*t}[F_1 - F_2^*(F_2 + B^*P)]\Phi(t)x\,dt, \tag{6.29}$$

where $\Phi(t)x$ is expressed explicitly from the data via (6.13).

Proof. We insert (6.28) into (6.22). □

Corollary 6.9. (Feedback generator) For $x \in Y$ and all $t > 0$:

(i)

$$\frac{d\Phi(t)x}{dt} = [A - B(F_2 + B^*P)]\Phi(t)x \in [\mathcal{D}(A^*)]'; \tag{6.30}$$

(ii)

$$[A - B(F_2 + B^*P)]\Phi(t)x = A_P\Phi(t)x = \Phi(t)A_Px \in Y,\ x \in \mathcal{D}(A_P); \tag{6.31}$$

(iii)

$$[A - B(F_2 + B^*P)]x = A_Px \in Y,\ x \in \mathcal{D}(A_P); \tag{6.32}$$

$$\mathcal{D}(A_P) = \{x \in Y :\ [I - A^{-1}B(F_2 + B^*P)]x \in \mathcal{D}(A)\}; \tag{6.33}$$

$$\Phi(t)x = e^{A_Pt}x = e^{[A-B(F_2+B^*P)]t}. \tag{6.34}$$

Proof. Same as the proof of [9, Corollary 6A.F.3 in Appendix A of Chapter 6, p. 635]. □

Our next task in the derivation of the Algebraic Riccati Equation for P. In preparation for this, we have the following result.

Proposition 6.10. The following identities hold true:

(i)

$$A^*Px = F_2^*(F_2 + B^*P)x - F_1x - PA_Px, \quad \forall\, x \in \mathcal{D}(A_P), \tag{6.35a}$$

and so

$$A^*P:\ \text{continuous } \mathcal{D}(A_P) \to Y; \tag{6.35b}$$

(ii)

$$A_P^*Pz = [F_2 + B^*P]^*F_2z - F_1z - PAz, \quad \forall\, z \in \mathcal{D}(A), \tag{6.36a}$$

and so

$$A_P^*P:\ \text{continuous } \mathcal{D}(A) \to Y. \tag{6.36b}$$

Proof. The proof is similar to that of [10, Proposition 3.4.1].

(i) Let $x \in \mathcal{D}(A_P)$. Then, recalling the definition of P in (6.29), with $\Phi(t) = e^{A_P t}$ by (6.34), we integrate by parts and obtain

$$\begin{aligned} A^*Px &= \int_0^\infty A^* e^{A^*t}[F_1 - F_2^*(F_2 + B^*P)]e^{A_P t}x\,dt &(6.37)\\ &= \left[e^{A^*t}[F_1 - F_2^*(F_2+B^*P)]e^{A_P t}x\right]_{t=0}^{t=\infty} \\ &\quad - \int_0^\infty e^{A^*t}[F_1 - F_2^*(F_2 + B^*P)]e^{A_P t}A_P x\,dt &(6.38)\\ \text{(by (6.29))} \quad &= -[F_1 - F_2^*(F_2 + B^*P)]x - PA_P x \in Y, &(6.39) \end{aligned}$$

recalling (H.1) = (1.2) and (6.21) at $t = \infty$. Then (6.39) establishes (6.35a), from which (6.35b) follows by the closed graph theorem.

(ii) Taking the Y-adjoint of (6.29) we obtain since $F_1 = F_1^*$ by (5.3):

$$Px = P^*x = \int_0^\infty e^{A_P^*t}[F_1 - (F_2 + B^*P)^*F_2]e^{At}x, \; x \in Y, \tag{6.40}$$

since $P = P^*$ as already noted in (6.14). Thus (6.40) provides an alternative expression for P. Next, we apply the counterpart argument to that employed in (i). For $x \in \mathcal{D}(A)$, we integrate by parts and obtain

$$\begin{aligned} A_P^*Px &= \int_0^\infty A_P^* e^{A_P^*t}[F_1 - (F_2 + B^*P)^*F_2]e^{At}x &(6.41)\\ &= \left[e^{A_P^*t}[F_1 - (F_2+B^*P)^*F_2]e^{At}x\right]_{t=0}^{t=\infty} \\ &\quad - \int_0^\infty e^{A_P^*t}[F_1 - (F_2 + B^*P)^*F_2]e^{At}Ax &(6.42)\\ \text{(by (6.41))} \quad &= -[F_1 - (F_2 + B^*P)^*F_2]x - PAx \in Y. &(6.43) \end{aligned}$$

Thus, (6.43) establishes (6.35a) from which (6.35b) follows by the closed graph theorem. □

We can now establish that the operator P in (6.14) or (6.29) satisfies the ARE.

Theorem 6.11. The operator P in (6.14) or (6.29) satisfies the following Algebraic Riccati Equation for all $x, z \in \mathcal{D}(A)$ [or else for all $x, z \in \mathcal{D}(A_P)$]:

$$(A^*Px, z)_Y + (PAx, z)_Y + (F_1 x, z)_Y = ([F_2 + B^*P]x, [F_2 + B^*P]z)_U. \tag{6.44}$$

Proof. We return to (6.36a) with $z \in \mathcal{D}(A)$ and take the inner product with $x \in Y$, thus obtaining

$$(PA_P x, z)_Y + (x, F_1 z)_Y + (x, PAz)_Y = ([F_2 + B^*P]x, F_2 z)_U, \ x \in Y, \ z \in \mathcal{D}(A). \tag{6.45}$$

We now specialize to $x \in \mathcal{D}(A)$, recall $A_P = A - B(F_2 + B^*P)$ from (6.32), and finally that $B^*P \in \mathcal{L}(Y; U)$, by (6.23), thus obtaining since $F_1^* = F_1$:

$$\begin{aligned} (P[A - B(F_2 + B^*P)]x, z)_Y + (F_1 x, z)_Y + (x, PAz)_Y & \\ = ([F_2 + B^*P]x, F_2 z)_U, \quad x, z \in \mathcal{D}(A); & \quad (6.46) \\ (PAx, z)_Y + (Px, Az)_Y + (F_1 x, z)_Y & \\ = ([F_2 + B^*P]x, [F_2 + B^*P]z)_U, \quad x, z \in \mathcal{D}(A), & \quad (6.47) \end{aligned}$$

and (6.44) is proved for $x, z \in \mathcal{D}(A)$.

Similarly, we use (6.35a) to obtain (6.44) for $x, z \in \mathcal{D}(A_P)$. □

References

[1] P. Acquistapace and B. Terreni, Classical solutions of non-autonomous Riccati equations, arising in parabolic boundary control problems, Part I, *Appl. Math. and Optimiz.* 39 (1999), 361–410.

[2] P. Acquistapace and B. Terreni, Classical solutions of non-autonomous Riccati equations, arising in parabolic boundary control problems, Part II, *Appl. Math. and Optimiz.* 41 (2000), 199–226.

[3] G. Avalos, The exponential stability of a coupled hyperbolic-parabolic system arising in structural acoustics, *Abstr. Appl. Anal.* 1(2) (1996), 203–217.

[4] G. Avalos and I. Lasiecka, Differential Riccati equation for the active control of a problem in structural acoustics, *J. Optim. Theory Appl.* 91(3) (1996), 695–728.

[5] F. Bucci, I. Lasiecka, and R. Triggiani, Singular estimate and uniform stability for coupled systems of hyperbolic/hyperbolic PDEs, Abstract & Applied Analysis, to appear.

[6] R. Datko, Extending a theorem of Liapunov to Hilbert spaces, *J. Math. Anal. and Appl.* 32 (1970), 610–616.

[7] I. Lasiecka, *Mathematical Control Theory of Coupled PDE Systems: NSF-CMBS Lecture Notes*, SIAM, 2001.

[8] I. Lasiecka, Optimization problems for structural acoustic models with thermoelasticity and smart materials, *Discussiones Mathematicae* 20 (2000), 113–140.

[9] I. Lasiecka and R. Triggiani, *Control Theory for Partial Differential Equations, Volume 1*, Cambridge University Press, 2000.

[10] I. Lasiecka and R. Triggiani, Optimal control and Algebraic Riccati Equations under singular estimates for $e^{At}B$ in the absence of analyticity, *Differential Equations and Optimal Control*, edited by S. Aizicovici and N. Pavel; Marcel Dekker, Lectures Notes in Pure and Applied Mathematics, vol. 225, pp. 193–220, International Workshop held at Ohio University, Athens, Ohio, May 2000.

[11] I. Lasiecka and R. Triggiani, Optimal control and differential Riccati equations under singular estimates for $e^{At}B$ in the absence of analycity, *Nonlinear Systems in Aviation, Aerospace, Aeronautics adn Astronautics*, Gordon and Breach Science Publishers, volume in honour of A.V. Balakrishnan.

[12] C. Sadosky, *Interpolation of Operators and Singular Integrals*, Marcel Dekker Monographs in Pure and Applied Mathematics, 1979.

[13] R. Triggiani, The coupled PDE system of a composite (sandwich) beam revisited: reduction to a thermoelastic system. Known and new results, 2001, to appear.

Roberto Triggiani
Department of Mathematics
University of Virginia
Charlottesville, VA 22903, USA
e-mail: `rt7u@virginia.edu`

Progress in Nonlinear Differential Equations
and Their Applications, Vol. 50, 381–397

Quasilinear Abstract Parabolic Evolution Equations with Applications

Atsushi Yagi

1. Introduction

We are concerned with the Cauchy problem of a quasilinear parabolic evolution equation

$$\begin{cases} \dfrac{dU}{dt} + A(U)U = F(U), \quad 0 < t \leq T, \\ U(0) = U_0 \end{cases} \tag{QE}$$

in a Banach space X. Here, $A(U)$ is the negative generator of an analytic semigroup on X with the domain $\mathcal{D}(A(U))$, $A(U)$ is defined for each $U \in K = \{U \in Z; \|U\|_Z < R\}$, $R > 0$, where Z is a second Banach space such that $\mathcal{D}(A(U)) \subset Z \subset X$. $F\colon K \to X$ is a non-linear mapping. U_0 is an initial value which belongs at least to K.

The author has already studied the problem in the paper [10]. Existence and uniqueness results were obtained there in a rather general framework. The Schauder fixed point theorem was utilized to show the existence of X-valued $\mathcal{C}^1$ local solutions to (QE) and the Gronwall inequality to show the uniqueness of such solutions under a number of assumptions, respectively. But in the point of view of applications these results are not necessarily convenient because in applications we have to show the existence and uniqueness simultaneously. In these notes we will present a theorem which assures the existence of a unique local solution by means of an analogous method. To avoid complexities, we will keep the framework as simple as possible. For example, we will content ourselves with treating the coefficient operators $A(U)$ which do not depend on t.

Among others the following Lipschitz condition of the form

$$\|A(U)(\lambda - A(U))^{-1}\{A(U)^{-1} - A(V)^{-1}\}\|_{\mathcal{L}(X)} \leq \frac{N\|U-V\|_Y}{(|\lambda|+1)^\nu}, \quad U, V \in K$$

is the crucial assumption of our method, where Y is a third Banach space such that $Z \subset Y \subset X$ and $\nu \in (0,1]$ is some exponent. There is however a very convenient way to verify this condition. We shall explain this also in these notes.

Since the paper [10] was published, our method has been applied to some non-linear diffusion systems in biology. Mathematical models of the population dynamics of competitive species and the aggregation process induced by chemotaxis were handled in [11] and [12], respectively. Not only these, our results are applicable to other various diffusion systems with interactions and reactions describing non-linear phenomena, such as the drift diffusion in semiconductors [13], the adsorbate-induced phase transition [6], and the pattern formation induced by chemotaxis and growth [7].

Notation. The norm of a Banach space X is denoted by $\|\cdot\|_X$. For two Banach spaces X and Y, $\mathcal{L}(X,Y)$ is the space of all bounded linear operators from X to Y with the uniform operator norm; when $X=Y$, $\mathcal{L}(X,X)$ is abbreviated as $\mathcal{L}(X)$. For a positive closed linear operator A, A^θ, $-\infty<\theta<\infty$, denotes the fractional power of A, see [14, 15]. Let I be an interval, $\mathcal{C}(I;\, X)$, $\mathcal{C}^\sigma(I;\, X)$, $0<\sigma<1$, and $\mathcal{C}^1(I;\, X)$ denote, respectively, the Banach spaces of all continuous functions, all Hölder continuous functions with exponent σ, and all continuously differentiable functions on I with values in a Banach space X. $\mathcal{B}(I;\, X)$ is the Banach space of all bounded (not necessarily measurable) functions on I. Let $\Omega\subset\mathbb{R}^n$ be a domain. $L_p(\Omega)$, $1\le p\le\infty$, is the usual L_p space. $W_p^s(\Omega)$, $0\le s<\infty$, is the Sobolev space with the fractional exponent s, see [16]. $W_2^s(\Omega)$ is denoted by $H^s(\Omega)$. $\mathcal{C}(\overline{\Omega})$ is the space of all continuous functions over $\overline{\Omega}$. By C we denote a universal constant which is determined by other initial constants, so C may change from occurrence to occurrence. When C depends on some parameter, say θ, it is denoted by C_θ.

2. Linear evolution operators

We consider a family of densely defined, closed linear operators $A(t)$, $0\le t\le T$, acting in a Banach space X. We assume that the spectral set $\sigma(A(t))$ of $A(t)$ is contained in $\Sigma=\{\lambda\in\mathbb{C};\ |\arg\lambda|\le\phi\}$, $0\le\phi<\frac{\pi}{2}$, and

$$\|(\lambda-A(t))^{-1}\|_{\mathcal{L}(X)}\le\frac{M}{|\lambda|+1},\quad \lambda\notin\Sigma,\ 0\le t\le T \tag{L.A.i}$$

with some constant $M>0$. In addition, $A(t)$ is assumed to satisfy a Lipschitz condition of the form

$$\begin{aligned}&\|A(t)(\lambda-A(t))^{-1}\{A(t)^{-1}-A(s)^{-1}\}\|_{\mathcal{L}(X)}\\ &\qquad\le\frac{N|t-s|^\mu}{(|\lambda|+1)^\nu},\quad \lambda\notin\Sigma,\ 0\le s,\, t\le T\end{aligned} \tag{L.A.ii}$$

with some exponents μ, $\nu\in(0,1]$ and constant $N>0$. These exponents satisfy the relations

$$0<\mu\le1,\ 0<\nu\le1,\ \text{ and }\ 1<\mu+\nu. \tag{L.Ex}$$

Under (L.A.i,ii) and (L.Ex), an evolution operator $U(t,s)$, $0\le s\le t\le T$, is constructed for the family $A(t)$. In this section we shall review various properties of the evolution operator. For the proofs we refer the reader to [8, 9]. In what follows C denotes a universal constant determined by the initial constants.

(L.Aii) first implies that $\mathcal{D}(A(s)) \subset \mathcal{D}(A(t)^\theta)$ for any $\theta \in [0,\nu)$ with the estimate

$$\|A(t)^\theta\{A(t)^{-1} - A(s)^{-1}\}\|_{\mathcal{L}(X)} \leq C_\theta|t-s|^\mu, \quad 0 \leq \theta < \nu, 0 \leq s, t \leq T. \tag{2.1}$$

More generally, this implies that, for $0 \leq \theta < \nu$ and $1-\nu+\theta < \varphi$,

$$\|A(t)^\theta\{A(t)^{-\varphi} - A(s)^{-\varphi}\}\|_{\mathcal{L}(X)} \leq C_{\theta,\varphi}|t-s|^\mu, \quad 0 \leq s, t \leq T. \tag{2.2}$$

(L.A.i) yields that each $A(t)$ is the negative generator of an analytic semigroup and the semigroup satisfies

$$\|A(t)^\theta e^{-\tau A(t)}\|_{\mathcal{L}(X)} \leq C_\theta \tau^{-\theta}, \quad \theta > 0, \tau > 0. \tag{2.3}$$

From this the following estimate

$$\|\{e^{-\tau A(t)} - 1\}A(t)^{-\theta}\|_{\mathcal{L}(X)} \leq C\tau^\theta, \quad 0 \leq \theta \leq 1, \tau > 0 \tag{2.4}$$

is also verified immediately.

As shown in [8, 9], we can construct under (L.A.i,ii) and (L.Ex) an evolution operator $U(t,s)$, $0 \leq s \leq t \leq T$, as a solution to the Volterra integral equation. $U(t,s)$ enjoys the following fundamental properties: a) $U(t,s)U(s,r) = U(t,r)$ for $0 \leq r \leq s \leq t \leq T$, $U(s,s) = 1$ for $0 \leq s \leq T$; b) $U(t,s)$ (resp. $A(t)U(t,s)$) is strongly continuous for $0 \leq s \leq t \leq T$ (resp. $0 \leq s < t \leq T$) with the estimate $\|U(t,s)\|_{\mathcal{L}(X)} \leq C$ (resp. $\|A(t)U(t,s)\|_{\mathcal{L}(X)} \leq C(t-s)^{-1}$); $U(t,s)$ is strongly differentiable in t for $s < t$ with $\partial U(t,s)/\partial t = -A(t)U(t,s)$; and d) $U(t,s)$ is strongly differentiable in s for $s < t$ on the domain $\mathcal{D}(A(s))$ with $\partial U(t,s)/\partial s = U(t,s)A(s)$.

Let $0 \leq s < T$ and consider the Cauchy problem of a linear evolution equation

$$\begin{cases} \dfrac{dU}{dt} + A(t)U = F(t), & s < t \leq T, \\ U(s) = U_s \end{cases} \tag{LE$_s$}$$

in X, where $U_s \in X$ is an initial value and $F\colon [s,T] \to X$ is a continuous function. If U is an X-valued $\mathcal{C}^1$ solution such that

$$U \in \mathcal{C}([s,T];\, X) \cap \mathcal{C}^1((s,T];\, X), \quad AU \in \mathcal{C}((s,T];\, X), \tag{2.5}$$

then U must be written in the form

$$U(t) = U(t,s)U_s + \int_s^t U(t,\tau)F(\tau)d\tau, \quad s \leq t \leq T. \tag{2.6}$$

Conversely, if $F \in \mathcal{C}^\sigma([s,T];\, X)$, $\sigma > 0$, then the function given by (2.6) is a unique solution to (LE$_s$) in the space (2.5). In particular,

$$\left\| A(t)\int_s^t U(t,\tau)F(\tau)d\tau \right\|_X \leq C_\sigma \|F\|_{\mathcal{C}^\sigma([s,T];X)}, \quad s \leq t \leq T. \tag{2.7}$$

Remark 2.1. *We have assumed the condition* (L.Aii) *which was first introduced by Acquistapace and Terreni* [1, 2, 3]. *In order to establish more elaborate properties of* $U(t,s)$ *than those in* [3], *we used the classical method of employing the Volterra integral equations. For another different method we refer the reader to* [5].

In the study of quasilinear evolution equations, we need the following estimates of $U(t,s)$:

$$\|A(t)^\theta U(t,s)\|_{\mathcal{L}(X)} \le C(t-s)^{-\theta}, \quad 0 \le \theta \le 1,\ 0 \le s < t \le T, \tag{2.8}$$

$$\|U(t,s)A(s)^\theta\|_{\mathcal{L}(X)} \le C_\theta(t-s)^{-\theta}, \quad 0 \le \theta < \mu,\ 0 \le s < t \le T, \tag{2.9}$$

$$\|A(t)U(t,s)A(s)^{-\theta}\|_{\mathcal{L}(X)} \le C(t-s)^{\theta-1}, \quad 0 \le \theta \le 1,\ 0 \le s < t \le T, \tag{2.10}$$

$$\|A(t)^\theta U(t,s)A(s)^{-\theta}\|_{\mathcal{L}(X)} \le C, \quad 0 \le \theta \le 1,\ 0 \le s < t \le T. \tag{2.11}$$

For the proofs, see [10,Sec. 2]. As for the difference of $U(t,s)$ and $e^{-(t-s)A(t)}$ or $e^{-(t-s)A(s)}$, we verify that

$$\|A(t)^\theta U(t,s)A(s)^{-\theta} - e^{-(t-s)A(s)}\|_{\mathcal{L}(X)} \le C(t-s)^{\mu+\nu-1}, \qquad 0 \le \theta \le 1,\ 0 \le s < t \le T, \tag{2.12}$$

$$\|A(t)^\theta\{U(t,s) - e^{-(t-s)A(t)}\}A(s)^{-\varphi}\|_{\mathcal{L}(X)} \le C(t-s)^{\varphi-\theta+\mu+\nu-1}, \quad 0 \le \theta, \varphi \le 1,\ 0 \le s < t \le T, \tag{2.13}$$

see (2.8) and (2.12) in [10].

3. Quasilinear evolution equations

Let X denote a Banach space. In X we consider the Cauchy problem of a quasilinear parabolic evolution equation

$$\begin{cases} \dfrac{dU}{dt} + A(U)U = F(U), & 0 < t \le T, \\ U(0) = U_0. \end{cases} \tag{QE}$$

Here, $A(U)$ are densely defined, closed linear operators in X, $A(U)$ are defined for

$$U \in K = \{U \in Z;\ \|U\|_Z < R\},$$

where Z is another Banach space and $R > 0$ is some constant. $F\colon K \to X$ is a non-linear mapping. U_0 is an initial value in Z.

We make the following assumptions. The spectral set $\sigma(A(U))$ of $A(U)$ is contained in $\Sigma = \{\lambda \in \mathbb{C};\ |\arg\lambda| \le \phi\}$, $0 \le \phi < \frac{\pi}{2}$, and

$$\|(\lambda - A(U))^{-1}\|_{\mathcal{L}(X)} \le \frac{M}{|\lambda|+1}, \quad \lambda \notin \Sigma,\ U \in K \tag{A.i}$$

with some constant $M > 0$. In addition, $A(U)$ satisfies the Lipschitz condition of the form

$$\|A(U)(\lambda - A(U))^{-1}\{A(U)^{-1} - A(V)^{-1}\}\|_{\mathcal{L}(X)} \le \frac{N\|U-V\|_Y}{(|\lambda|+1)^\nu}, \quad \lambda \notin \Sigma,\ U, V \in K \tag{A.ii}$$

with some exponent $\nu \in (0, 1]$ and constant $N > 0$, Y being a third Banach space. $F(U)$ also satisfies the Lipschitz condition

$$\|F(U) - F(V)\|_X \leq L\|U - V\|_Y \quad U, V \in K \tag{F}$$

with some constant $L > 0$. The spaces Z, Y and X satisfy the following conditions:

$$Z \subset Y \subset X \tag{Sp.i}$$

with continuous embeddings. Z is a reflexive Banach space:

$$Z^* = Z. \tag{Sp.ii}$$

For $U \in K$, $\mathcal{D}(A(U)^\alpha) \subset Y$ and $\mathcal{D}(A(U)^\beta) \subset Z$ with the estimates

$$\|\cdot\|_Y \leq D_1\|A(U)^\alpha \cdot\|_X \text{ and } \|\cdot\|_Z \leq D_2\|A(U)^\beta \cdot\|_X, \tag{Sp.iii}$$

where $0 \leq \alpha < \beta < 1$ and $D_i > 0\,(i = 1, 2)$. The exponents satisfy the relations

$$0 < \nu \leq 1, \ \ 0 \leq \alpha < \beta < 1, \ \text{ and } \ \alpha + 1 < \beta + \nu. \tag{Ex}$$

Finally, the initial value U_0 satisfies that

$$U_0 \in \mathcal{D}(A(U_0)^\beta) \text{ with } \|U_0\|_\beta \equiv \|A(U_0)^\beta U_0\|_X < \tfrac{R}{D_2}. \tag{In}$$

Then, the following result is proved.

Theorem 3.1. *Under* (A.i,ii), (F), (Sp.i,ii,iii), (Ex), *and* (In), *there exists a unique local solution to* (QE) *in the function space*

$$\begin{cases} U \in \mathcal{C}([0, T_{U_0}]; X) \cap \mathcal{C}^1((0, T_{U_0}]; X) \cap \mathcal{B}([0, T_{U_0}]; Z), \\ A(U)U \in \mathcal{C}((0, T_{U_0}]; X), \ \ t^{1-\beta}A(U)U \in \mathcal{B}((0, T_{U_0}]; X), \end{cases} \tag{3.1}$$

where $T_{U_0} > 0$ *is some positive number depending on* U_0.

Proof. The proof consists of several steps. C denotes a universal constant which is determined by the initial constants.

Step 1. For S such that $0 < S \leq T$, we set a Banach space $\mathcal{Y}(S) = \mathcal{C}([0, S]; Y)$ and a subset

$$\mathcal{K}(S) = \{U \in \mathcal{C}^\mu([0, S]; Y) \cap \mathcal{B}([0, S]; Z); \, U(0) = U_0, \\ \sup_{0 \leq s < t \leq S} \tfrac{\|U(t) - U(s)\|_Y}{|t-s|^\mu} \leq 1 \text{ and } \sup_{0 \leq t \leq S} \|U(t)\|_Z \leq D\}$$

of $\mathcal{Y}(S)$, where μ is some fixed exponent so that $1 - \nu < \mu < \beta - \alpha$, D is a fixed constant so that

$$D_2\|U_0\|_\beta < D < R. \tag{3.2}$$

Then, $\mathcal{K}(S)$ is non-empty closed subset of $\mathcal{Y}(S)$. Indeed, let $\{U_n\}_{n=1,2,\dots}$ be a sequence in $\mathcal{K}(S)$ which is convergent to $\overline{U}$ in $\mathcal{Y}(S)$. For each $0 \leq t \leq S$, $\|U_n(t)\|_Z \leq D$; (Sp.ii) then implies that some subsequence $\{U_{n_k}(t)\}$ of $\{U_n(t)\}$ converges to $\overline{U}_t$ weakly in Z with $\|\overline{U}_t\|_Z \leq D$; as a consequence it follows from (Sp.i) that $U_{n_k}(t) \to \overline{U}_t$ weakly in Y also; therefore, $\overline{U}(t) = \overline{U}_t$ and $\|\overline{U}(t)\|_Z \leq D$. On the other hand, it is clear that $\|\overline{U}(t) - \overline{U}(s)\|_Y \leq |t - s|^\mu$. Hence, $\overline{U} \in \mathcal{K}(S)$.

Step 2. For each $V \in \mathcal{K}(S)$, let us consider a linear problem

$$\begin{cases} \dfrac{dU}{dt} + A_V(t)U = F_V(t), & 0 < t \le S, \\ U(0) = U_0, \end{cases}$$

where $A_V(t) = A(V(t))$ and $F_V(t) = F(V(t))$ for $0 \le t \le S$. It is easy to observe that $A_V(t)$ satisfies (L.A.i,ii) and (L.Ex) in Section 2 and that $F_V \in \mathcal{C}^\mu([0,S]; X)$. Therefore, there exists a unique solution in the space

$$U \in \mathcal{C}([0,S]; X) \cap \mathcal{C}^1((0,S]; X), \quad A_V U \in \mathcal{C}((0,S]; X).$$

U is given by the formula

$$U(t) = U_V(t,0)U_0 + \int_0^t U_V(t,s)F_V(s)ds, \quad 0 \le t \le S,$$

here $U_V(t,s)$ is the evolution operator for $A_V(t)$.

We then arrive at defining a correspondence $\Phi(V)$ from $\mathcal{K}(S)$ to $\mathcal{Y}(S)$ by setting $\Phi(V)(t) = U(t)$, $0 \le t \le S$, for each $V \in \mathcal{K}(S)$.

Step 3. If $S > 0$ is sufficiently small, then Φ maps the set $\mathcal{K}(S)$ into itself. Indeed, for $U = \Phi(V)$, we can write as

$$A_V(t)^\beta U(t) = A(U_0)^\beta U_0 + \{e^{-tA(U_0)} - 1\}A(U_0)^\beta U_0$$
$$+ \{A_V(t)^\beta U_V(t,0)A_V(0)^{-\beta} - e^{-tA_V(0)}\}A_V(0)^\beta U_0 + \int_0^t A_V(t)^\beta U_V(t,s)F_V(s)ds.$$

Then, from (2.8), (2.12) and (In), it follows that

$$\|A_V(t)^\beta U(t)\|_X \le \|U_0\|_\beta + \|\{e^{-tA(U_0)} - 1\}A(U_0)^\beta U_0\|_X$$
$$+ Ct^{\mu+\nu-1}\|U_0\|_\beta + C\int_0^t (t-s)^{-\beta}ds.$$

Therefore, in view of (Sp.iii) and (3.2),

$$\begin{aligned} \|U(t)\|_Z &\le D_2\|A_V(t)^\beta U(t)\|_X \\ &\le D_2\|U_0\|_\beta + D_2 \sup_{0\le t\le S} \|\{e^{-tA(U_0)} - 1\}A(U_0)^\beta U_0\|_X \\ &\quad + C(S^{\mu+\nu-1} + S^{1-\beta})(\|U_0\|_\beta + 1) \le D, \quad 0 \le t \le S, \end{aligned}$$

provided $S > 0$ is sufficiently small.

Utilizing (2.6), we next write as

$$U(t) - U(s) = \{U_V(t,s) - 1\}U(s) + \int_s^t U_V(t,\tau)F_V(\tau)d\tau$$
$$= \big[\{U_V(t,s) - e^{-(t-s)A_V(t)}\}A_V(s)^{-\beta} + \{e^{-(t-s)A_V(t)} - 1\}$$
$$\times \{A_V(s)^{-\beta} - A_V(t)^{-\beta}\} + \{e^{-(t-s)A_V(t)} - 1\}A_V(t)^{-\beta}\big]A_V(s)^{\beta}U(s)$$
$$+ \int_s^t U_V(t,\tau)F_V(\tau)d\tau, \quad 0 \le s < t \le S.$$

Then, from (2.13), it is seen that

$$\|A_V(t)^{\alpha}\{U_V(t,s) - e^{-(t-s)A_V(t)}\}A_V(s)^{-\beta}\|_{\mathcal{L}(X)} \le C(t-s)^{\beta-\alpha+\mu+\nu-1}.$$

Similarly, from (2.2) and (2.4),

$$\|A_V(t)^{\alpha}\{e^{-(t-s)A_V(t)} - 1\}\{A_V(s)^{-\beta} - A_V(t)^{-\beta}\}\|_{\mathcal{L}(X)}$$
$$\le \|\{e^{-(t-s)A_V(t)} - 1\}A_V(t)^{\alpha-\alpha'}\|_{\mathcal{L}(X)}\|A_V(t)^{\alpha'}\{A_V(s)^{-\beta} - A_V(t)^{-\beta}\}\|_{\mathcal{L}(X)}$$
$$\le C(t-s)^{\alpha'-\alpha+\mu}$$

with some α' such that $\alpha < \alpha' < \beta + \nu - 1$. Finally,

$$\|A_V(t)^{\alpha}\{e^{-(t-s)A_V(t)} - 1\}A_V(t)^{-\beta}\|_{\mathcal{L}(X)} \le C(t-s)^{\beta-\alpha},$$
$$\left\|A_V(t)^{\alpha}\int_s^t U_V(t,\tau)F_V(\tau)d\tau\right\|_X \le C(t-s)^{1-\alpha}.$$

Therefore, in view of (Sp.iii), we obtain that

$$\|U(t) - U(s)\|_Y \le D_1\|A_V(t)^{\alpha}\{U(t) - U(s)\}\|_X$$
$$\le C(S^{\beta-\alpha-\mu} + S^{\alpha'-\alpha})(D+1)(t-s)^{\mu} \le (t-s)^{\mu}, \quad 0 \le s < t \le S,$$

provided $S > 0$ is sufficiently small.

Step 4. If $S > 0$ is sufficiently small, then the mapping $\Phi\colon \mathcal{K}(S) \to \mathcal{K}(S)$ is a contraction with respect to the norm $\|\cdot\|_{\mathcal{Y}(S)}$. Indeed, for $U_i = \Phi(V_i)$, $V_i \in \mathcal{K}(S)$, $i = 1, 2$, we have

$$U_1(t) - U_2(t) = \{U_{V_1}(t,0) - U_{V_2}(t,0)\}U_0$$
$$+ \int_0^t \{U_{V_1}(t,s) - U_{V_2}(t,s)\}F_{V_1}(s)ds + \int_0^t U_{V_2}(t,s)\{F_{V_1}(s) - F_{V_2}(s)\}ds.$$

Here we notice the following lemma.

Lemma 3.1. *Fix* ρ *so that* $\max\{\alpha - \beta + 1, 1 - \mu\} < \rho < \nu$. *Then,*

$$\|A_{V_1}(t)^{\alpha}\{U_{V_1}(t,0) - U_{V_2}(t,0)\}U_0\|_X$$
$$\le Ct^{\rho+\beta-\alpha-1}\|U_0\|_{\beta}\|V_1 - V_2\|_{\mathcal{Y}(S)}, \quad 0 \le t \le S,$$

and

$$\left\| A_{V_1}(t)^\alpha \int_0^t \{U_{V_1}(t,s) - U_{V_2}(t,s)\} F_{V_1}(s) ds \right\|_X \\ \leq C t^{\rho-\alpha} \|F_{V_1}\|_{\mathcal{C}^\mu([0,t];\, X)} \|V_1 - V_2\|_{\mathcal{Y}(S)}, \quad 0 \leq t \leq S.$$

Proof. Essentially this lemma has been verified in [10,Lemma 3.2]. We note that

$$A_{V_1}(t)^\alpha \{U_{V_1}(t,0) - U_{V_2}(t,0)\} A_{V_2}(0)^{-\beta} = \int_0^t A_{V_1}(t)^\alpha U_{V_1}(t,s) A_{V_1}(s)^{1-\rho} \\ \times A_{V_1}(s)^\rho \{A_{V_1}(s)^{-1} - A_{V_2}(s)^{-1}\} A_{V_2}(s) U_{V_2}(s,0) A_{V_2}(0)^{-\beta} ds.$$

Here we use (2.8), (2.9), (2.10) and the estimate

$$\|A_{V_1}(s)^\rho \{A_{V_1}(s)^{-1} - A_{V_2}(s)^{-1}\}\|_{\mathcal{L}(X)} \leq C \|V_1(s) - V_2(s)\|_Y, \tag{3.3}$$

then it is verified that

$$\|A_{V_1}(t)^\alpha \{U_{V_1}(t,0) - U_{V_2}(t,0)\} A_{V_2}(0)^{-\beta} A_{V_2}(0)^\beta U_0\|_X \\ \leq C \int_0^t (t-s)^{\rho-\alpha-1} s^{\beta-1} ds \, \|U_0\|_\beta \, \|V_1 - V_2\|_{\mathcal{Y}(S)}.$$

On the other hand, we can write as

$$A_{V_1}(t)^\alpha \int_0^t \{U_{V_1}(t,s) - U_{V_2}(t,s)\} F_{V_1}(s) ds = \int_0^t \int_s^t A_{V_1}(t)^\alpha U_{V_1}(t,\tau) A_{V_1}(\tau)^{1-\rho} \\ \times A_{V_1}(\tau)^\rho \{A_{V_1}(\tau)^{-1} - A_{V_2}(\tau)^{-1}\} A_{V_2}(\tau) U_{V_2}(\tau,s) F_{V_1}(s) d\tau ds \\ = \int_0^t A_{V_1}(t)^\alpha U_{V_1}(t,\tau) A_{V_1}(\tau)^{1-\rho} A_{V_1}(\tau)^\rho \{A_{V_1}(\tau)^{-1} - A_{V_2}(\tau)^{-1}\} \\ \times A_{V_2}(\tau) \int_0^\tau U_{V_2}(\tau,s) F_{V_1}(s) ds d\tau.$$

From (2.7) we observe that

$$\left\| A_{V_2}(\tau) \int_0^\tau U_{V_2}(\tau,s) F_{V_1}(s) ds \right\|_X \leq C \|F_{V_1}\|_{\mathcal{C}^\mu([0,t];\, X)},$$

therefore it follows from (3.3) that

$$\left\| A_{V_1}(t)^\alpha \int_0^t \{U_{V_1}(t,s) - U_{V_2}(t,s)\} F_{V_1}(s) ds \right\|_X \\ \leq C \int_0^t (t-\tau)^{\rho-\alpha-1} d\tau \|F_{V_1}\|_{\mathcal{C}^\mu([0,t];\, X)} \|V_1 - V_2\|_{\mathcal{Y}(S)}.$$

□

Let us now complete the proof of this step. It is easy to see that

$$\left\| A_{V_2}(t)^\alpha \int_0^t U_{V_2}(t,s)\{F_{V_1}(s) - F_{V_2}(s)\}ds \right\|_X$$
$$\leq C\int_0^t (t-s)^{-\alpha}\|V_1(s) - V_2(s)\|_Y ds \leq Ct^{1-\alpha}\|V_1 - V_2\|_{\mathcal{Y}(S)}.$$

In view of (Sp.iii), this together with the lemma above yields that

$$\|U_1(t) - U_2(t)\|_Y \leq CS^{\rho+\beta-\alpha-1}\|V_1 - V_2\|_{\mathcal{Y}(S)}, \quad 0 \leq t \leq S.$$

Hence, we have verified that Φ is a contraction if $S > 0$ is sufficiently small.

Step 5. Let $T_{U_0} = S > 0$ so that the results of Steps 3 and 4 are valid. Then, there exist a unique fixed point $U \in \mathcal{K}(S)$ of Φ. Since U satisfies the formula

$$U(t) = U_U(t,0)U_0 + \int_0^t U_U(t,s)F_U(s)ds, \quad 0 \leq t \leq S, \tag{3.4}$$

U is shown to be a solution to (QE) which belongs to the space (3.1).

Step 6. Finally we verify the uniqueness of solution. Let $\widetilde{U}$ be another solution to (QE) in the space (3.1). Then, for $0 < t < S\,(\leq T_{U_0})$,

$$\tfrac{\partial}{\partial s}U_U(t,s)\widetilde{U}(s) = U_U(t,s)\{A_U(s) - A(\widetilde{U}(s))\}\widetilde{U}(s) + U_U(t,s)F(\widetilde{U}(s)), \quad 0 < s < t.$$

Integrating this identity in $s \in (0,t)$, we have

$$\widetilde{U}(t) - U(t) = \int_0^t U_U(t,s)A_U(s)^{1-\rho}A_U(s)^\rho\{A_U(s)^{-1} - A_{\widetilde{U}}(s)^{-1}\}A_{\widetilde{U}}(s)\widetilde{U}(s)ds$$
$$+ \int_0^t U_U(t,s)\{F_{\widetilde{U}}(s) - F_U(s)\}ds,$$

where ρ is the same exponent used in the proof of Lemma 3.1.

We can then repeat the same argument as in Step 4 to obtain that

$$\|A_U(t)^\alpha\{\widetilde{U}(t) - U(t)\}\|_X \leq C_{\widetilde{U}}S^{\rho+\beta-\alpha-1}\|\widetilde{U} - U\|_{\mathcal{Y}(S)}, \quad 0 \leq t \leq S.$$

This in turn shows that $\widetilde{U}(t) = U(t)$ for all $t \in [0,S]$ if $S > 0$ is sufficiently small. As a matter of fact, we have shown by this argument that the set $\{S \in (0,T_{U_0}];\ \widetilde{U}(t) = U(t)$ for all $t \in [0,S]\}$ is non-empty and open in $(0,T_{U_0}]$. On the other hand, it is clear that the set is closed; therefore, $\widetilde{U}(t) = U(t)$ for all $t \in [0,T_{U_0}]$. □

Remark 3.1. *In the proof,* T_{U_0} *was determined by the norm* $\|U_0\|_\beta$ $(\leq \frac{R}{D_2})$ *and the modulus of continuity*

$$\omega(S) = \sup_{0\leq t\leq S}\|\{e^{-tA(U_0)} - 1\}A(U_0)^\beta U_0\|_X.$$

If $U_0 \in \mathcal{D}(A(U_0)^{\beta'})$ with some $\beta' > \beta$, then T_{U_0} will be determined by the norm $\|A(U_0)^{\beta'} U_0\|_X$ only, since

$$\omega(S) \leq \sup_{0\leq t\leq S} \|\{e^{-tA(U_0)} - 1\}A(U_0)^{\beta-\beta'}\|_{\mathcal{L}(X)}\|A(U_0)^{\beta'}U_0\|_X \leq CS^{\beta'-\beta}\|U_0\|_{\beta'}.$$

This then means that the global existence of the solution will be established if we can verify the a priori estimate $\|U(t)\|_{\beta'} \leq C$ for every local solution to (QE).

Remark 3.2. *Let U be the solution to* (QE). *Since Z is reflexive,* (3.1) *implies that, as $t \to 0$, $U(t)$ converges to U_0 in the weak topology of Z. On the other hand, it was shown in the proof that $A(U(t))^\beta U(t)$ converges to $A(U_0)^\beta U_0$ in X. In some favorable case it can be verified that $U(t)$ is continuous at $t = 0$ in the topology of Z.*

Remark 3.3. *Let $\mathcal{D}(A(U)^\beta) \equiv \mathcal{D}(A(0)^\beta)$ be independent of $U \in K$ with the estimate $\|A(U)^\beta A(V)^{-\beta}\|_{\mathcal{L}(X)} \leq C$ for $U, V \in K$, and let β, ν satisfy the relation $\beta < \nu$. Then, it is shown that the solution $U(t)$ of* (QE) *is continuous at $t = 0$ with respect to the graph norm of $\mathcal{D}(A(0)^\beta)$. In fact, from* (3.4) *it is written as*

$$\begin{aligned} A_U(t)^\beta\{U(t) - U_0\} &= A_U(t)^\beta\{U_U(t,0) - e^{-tA_U(t)}\}A_U(0)^{-\beta}A_U(0)^\beta U_0 \\ &\quad + A_U(t)^\beta\{e^{-tA_U(t)} - 1\}A_U(0)^{-\beta}A_U(0)^\beta U_0 \\ &\quad + \int_0^t A_U(t)^\beta U_U(t,s)F_U(s)ds = J_1(t) + J_2(t) + J_3(t). \end{aligned}$$

From (2.13), $\lim_{t\to 0} J_1(t) = 0$ *in X. From* (2.1) *and* (2.4) *we have*

$$\begin{aligned} &\|A_U(t)^\beta\{e^{-tA_U(t)} - 1\}A_U(0)^{-1}\|_{\mathcal{L}(X)} \\ &\qquad \leq \|\{e^{-tA_U(t)} - 1\}A_U(t)^{\beta-\beta'}\|_{\mathcal{L}(X)}\|A_U(t)^{\beta'}A_U(0)^{-1}\|_{\mathcal{L}(X)} \leq Ct^{\beta'-\beta} \end{aligned}$$

with some $\beta' \in (\beta, \nu)$. Then, the uniform boundedness of $A_U(t)^\beta\{e^{-tA_U(t)} - 1\}A_U(0)^{-\beta}$ together with the density of $\mathcal{D}(A_U(0))$ yields that $\lim_{t\to 0} J_2(t) = 0$ in X. It is clear that $\lim_{t\to 0} J_3(t) = 0$. Therefore, $\lim_{t\to 0}\|A(0)^\beta\{U(t) - U_0\}\|_X = 0$.

We now prove continuous dependence of the solutions on the initial values.

Theorem 3.2. *Let U_0, V_0 be two initial values satisfying* (In), *and let U, V be the solutions obtained in Theorem* 3.1 *corresponding to the initial values U_0, V_0 respectively. Then,*

$$\sup_{0\leq t\leq T_0} \|A(U(t))^\alpha\{U(t) - V(t)\}\|_X \leq C\|A(U_0)^\alpha\{U_0 - V_0\}\|_X,$$

where $T_0 = \min\{T_{U_0}, T_{V_0}\}$.

Proof. We can write as

$$\begin{aligned} U(t) - V(t) &= U_U(t,0)(U_0 - V_0) + \{U_U(t,0) - U_V(t,0)\}V_0 \\ &\quad + \int_0^t \{U_U(t,s) - U_V(t,s)\}F_V(s)ds + \int_0^t U_V(t,s)\{F_U(s) - F_V(s)\}ds. \end{aligned}$$

As in the proof of Lemma 3.1, we observe that

$$\|A_U(t)^\alpha\{U_U(t,0) - U_V(t,0)\}V_0\|_X \leq C\int_0^t (t-s)^{\rho-\alpha-1}s^{\beta-1}\|A_U(s)^\alpha\{U(s) - V(s)\}\|_X ds,$$

and

$$\left\|A_U(t)^\alpha \int_0^t \{U_U(t,s) - U_V(t,s)\}F_U(s)ds\right\|_X \leq C\int_0^t (t-s)^{\rho-\alpha-1}\|A_U(s)^\alpha\{U(s) - V(s)\}\|_X ds$$

with the same ρ as before. On the other hand, by (2.11),

$$\|A_U(t)^\alpha U_U(t,0)(U_0 - V_0)\|_X \leq C\|A(U_0)^\alpha(U_0 - V_0)\|_X$$

and

$$\left\|\int_0^t A_U(s)^\alpha U_V(t,s)\{F_U(s) - F_V(s)\}ds\right\|_X \leq C\int_0^t (t-s)^{-\alpha}\|A_U(s)^\alpha\{U(s) - V(s)\}\|_X ds.$$

Therefore, we obtain that

$$t^{\beta-1}\|A_U(t)^\alpha\{U(t) - V(t)\}\|_X \leq Ct^{\beta-1}\|A(U_0)^\alpha\{U_0 - V_0\}\|_X + C\int_0^t (t-s)^{\rho+\beta-\alpha-2}s^{\beta-1}\|A_U(s)^\alpha\{U(s) - V(s)\}\|_X ds.$$

As $1+\alpha < \rho + \beta$, the desired estimate follows immediately. □

4. Verification of the condition (A.ii)

In applications it is necessary to fix the Banach spaces X, Y and Z suitably for which all the conditions (A.i,ii), (F), (Sp.i,ii,iii) and (Ex) are fulfilled. Especially the condition (A.ii) is crucial.

In order to verify (A.ii) we can use a trick. Consider a case where X is a Hilbert space and $A(U)$ are determined by bilinear forms on another Hilbert space V such that $V \subset X$ densely and continuously. For each $U \in K \subset Z$, let $a(U;\cdot,\cdot)$ be a bilinear form on V satisfying

$$\begin{cases} \Re\mathrm{e}\, a(U;\widetilde{U},\widetilde{U}) \geq \delta\|\widetilde{U}\|_V^2, & \widetilde{U} \in V, \\ |a(U;\widetilde{U},\widetilde{V})| \leq C\|\widetilde{U}\|_V\|\widetilde{V}\|_V, & \widetilde{U},\widetilde{V} \in V \end{cases}$$

with some constants $\delta > 0$ and $C \geq 0$. By the relation

$$A(U;\widetilde{U},\widetilde{V}) = (A(U)\widetilde{U},\widetilde{V})_X \quad \text{for all } \widetilde{V} \in V,$$

$A(U)$ is defined as a closed linear operator in X with a dense domain. $A(U)$ then satisfies (A.i).

Let us assume that $a(U;\cdot,\cdot)$ satisfies the following Lipschitz condition

$$|a(U_1;\widetilde{U},\widetilde{V}) - a(U_2;\widetilde{U},\widetilde{V})| \le N\|U_1 - U_2\|_Y \|A(U_2)\widetilde{U}\|_X \|\widetilde{V}\|_V,$$
$$U_1,\, U_2 \in K;\ \widetilde{U} \in \mathcal{D}(A(U_2));\ \widetilde{V} \in V \quad (4.1)$$

with the norm $\|\cdot\|_Y$ of some Banach space Y such that $Z \subset Y \subset X$. Taking $\widetilde{U} = A(U_2)^{-1}\widetilde{F}$ and $\widetilde{V} = A(U_1)^{*-1}\widetilde{G}$ in

$$a(U_1;\widetilde{U},\widetilde{V}) = (\widetilde{U}, A(U_1)^*\widetilde{V})_X \quad \text{and} \quad a(U_2;\widetilde{U},\widetilde{V}) = (A(U_2)\widetilde{U},\widetilde{V})_X,$$

respectively, we observe that

$$a(U_1; A(U_2)^{-1}\widetilde{F}, A(U_1)^{*-1}\widetilde{G}) = (A(U_2)^{-1}\widetilde{F},\widetilde{G})_X, \quad \widetilde{F},\widetilde{G} \in X,$$
$$a(U_2; A(U_2)^{-1}\widetilde{F}, A(U_1)^{*-1}\widetilde{G}) = (A(U_1)^{-1}\widetilde{F},\widetilde{G})_X, \quad \widetilde{F},\widetilde{G} \in X.$$

Therefore,

$$(\{A(U_1)^{-1} - A(U_2)^{-1}\}\widetilde{F},\widetilde{G})_X$$
$$= a(U_2; A(U_2)^{-1}\widetilde{F}, A(U_1)^{*-1}\widetilde{G}) - a(U_1; A(U_2)^{-1}\widetilde{F}, A(U_1)^{*-1}\widetilde{G}).$$

As a consequence, we obtain that

$$(A(U_1)(\lambda - A(U_1))^{-1}\{A(U_1)^{-1} - A(U_2)^{-1}\}\widetilde{F},\widetilde{G})_X$$
$$= a(U_2; A(U_2)^{-1}\widetilde{F}, (\overline{\lambda} - A(U_1)^*)^{-1}\widetilde{G})$$
$$- a(U_1; A(U_2)^{-1}\widetilde{F}, (\overline{\lambda} - A(U_1)^*)^{-1}\widetilde{G}), \quad \widetilde{F},\widetilde{G} \in X.$$

Hence, using the fact that

$$\|(\lambda - A(U_1)^*)^{-1}\|_{\mathcal{L}(X,V)} \le \frac{C}{(|\lambda| + 1)^{\frac{1}{2}}}, \quad \lambda \notin \Sigma,$$

see [14,Chap. 6,Lemma 6.1], we show that (4.1) implies (A.ii) with $\nu = \frac{1}{2}$.

Such an argument is available even in Banach space cases. Consider an elliptic differential operator

$$A(x, U(x); D)\widetilde{U} = -\sum_{i,j=1}^{n} \frac{\partial}{\partial x_i}\left\{a_{ij}(x, U(x))\frac{\partial \widetilde{U}}{\partial x_j}\right\} + c\widetilde{U},$$

in a bounded region $\Omega \subset \mathbb{R}^n$ with a $\mathcal{C}^2$ boundary $\partial\Omega$, and a boundary operator

$$B(x, U(x); D)\widetilde{U} = \sum_{i,j=1}^{n} a_{ij}(x, U(x))\gamma_i(x)\frac{\partial \widetilde{U}}{\partial x_j}$$

on $\partial\Omega$, where $\gamma(x) = (\gamma_1(x), \dots, \gamma_n(x))$ denotes the outer normal vector at a point $x \in \partial\Omega$. $a_{ij}(x, U)\,(1 \le i, j \le n)$ are real-valued smooth functions defined for $(x, U) \in \overline{\Omega} \times (\mathbb{R} + i\mathbb{R})$ such that $a_{ij} = a_{ji}$ and that

$$\sum_{i,j=1}^{n} a_{ij}(x, U)\xi_i\xi_j \ge \delta|\xi|^2, \quad \xi = (\xi_1, \dots, \xi_n) \in \mathbb{R}^n \tag{4.2}$$

with some constant $\delta > 0$. $c > 0$ is a positive constant.

We set $X = L_p(\Omega)$ with $n < p < \infty$ and set $Z = W_p^h(\Omega)$ with an arbitrarily fixed exponent $h \in (1 + \frac{n}{p}, 2)$ so that $Z \subset \mathcal{C}^1(\overline{\Omega})$. Then, for $U \in K = \{U \in Z;\ \|U\|_Z < R\}$, we can define a linear operator

$$\begin{cases} \mathcal{D}(A(U)) = \{\widetilde{U} \in W_p^2(\Omega);\ B(x, U(x); D)\widetilde{U} = 0 \text{ on } \partial\Omega\}, \\ \qquad A(U)\widetilde{U} = A(x, U(x); D)\widetilde{U} \end{cases} \tag{4.3}$$

in X. $A(U)$ then satisfies (A.i) provided c is sufficiently large. In the similar way as above we obtain that

$$\begin{aligned} &\langle A(U_1)(\lambda - A(U_1))^{-1}\{A(U_1)^{-1} - A(U_2)^{-1}\}\widetilde{F}, \widetilde{G}\rangle_{L_p \times L_q} \\ &\quad = \sum_{i,j=1}^{n} \int_\Omega \{a_{ij}(x, U_2(x)) - a_{ij}(x, U_1(x))\}\frac{\partial}{\partial x_j}A(U_2)^{-1}\widetilde{F} \\ &\qquad\qquad \times \frac{\partial}{\partial x_i}(\lambda - A(U_1)^*)^{-1}\widetilde{G}\,dx, \quad \widetilde{F} \in X,\, \widetilde{G} \in X', \end{aligned}$$

where $X' = L_q(\Omega)$, $\frac{1}{p} + \frac{1}{q} = 1$, and $A(U_1)^*$ is the adjoint operator of $A(U_1)$ acting in X'. Then all the integrals are estimated by

$$\begin{aligned} &C\|a_{ij}(x, U_1) - a_{ij}(x, U_2)\|_{L_p}\|\tfrac{\partial}{\partial x_j}A(U_2)^{-1}\widetilde{F}\|_{\mathcal{C}}\|\tfrac{\partial}{\partial x_i}(\lambda - A(U_1)^*)^{-1}\widetilde{G}\|_{L_q} \\ &\qquad \le C\|U_1 - U_2\|_{L_p}\|A(U_2)^{-1}\widetilde{F}\|_{W_p^2}\|(\lambda - A(U_1)^*)^{-1}\widetilde{G}\|_{W_q^1}. \end{aligned}$$

Since

$$\begin{aligned} \|(\lambda - A(U_1)^*)^{-1}\widetilde{G}\|_{W_q^1} &\le C\|(\lambda - A(U_1)^*)^{-1}\widetilde{G}\|_{W_q^2}^{\frac{1}{2}}\|(\lambda - A(U_1)^*)^{-1}\widetilde{G}\|_{L_q}^{\frac{1}{2}} \\ &\le \frac{C}{(|\lambda| + 1)^{\frac{1}{2}}}\|\widetilde{G}\|_{L_q}, \end{aligned}$$

this shows that (A.ii) is fulfilled with $Y = X$ and $\nu = \frac{1}{2}$.

According to [10,Appendix] it is known in the present case that

$$W_{p,B(U)}^{2\theta_3}(\Omega) \subset \mathcal{D}(A(U)^{\theta_2}) \subset W_{p,B(U)}^{2\theta_1}(\Omega) \tag{4.4}$$

for any $0 \le \theta_1 < \theta_2 < \theta_3 \le 1$, where

$$W^{2\theta}_{p,B(U)}(\Omega) = \begin{cases} W^{2\theta}_p(\Omega) & \text{if } 0 \le \theta < \frac{p+1}{2p}, \\ \{\widetilde{U} \in W^{2\theta}_p(\Omega);\ \sum_{ij=1}^n a_{ij}(x, U(x))\gamma_i(x)\frac{\partial \widetilde{U}}{\partial x_j} = 0 \text{ on } \partial\Omega\} & \text{if } \frac{p+1}{2p} < \theta \le 1. \end{cases}$$

Then (Sp.iii) as well as (Ex) is also fulfilled with $\alpha = 0$ and $\beta \in (\frac{h}{2}, 1)$.

Remark 4.1. *The exponent ν is in some sense a measure of dependence of the domain $\mathcal{D}(A(U))$ on U. The more temperately the domain varies with U, the larger ν can be taken. If the domain is constant, then ν may be 1. The relations (4.4) show that $\mathcal{D}(A(U)) \subset \mathcal{D}(A(V)^{\theta_2})$, $U, V \in K$ for any $\theta_2 < \frac{p+1}{2p}$. In other words we can expect that ν may be taken arbitrarily close to $\frac{p+1}{2p}$. In fact, this is true if Y is set as $Y = W^1_p(\Omega)$, see* [10, Proposition 5.1].

In verifying the condition (A.ii), the most favorable case is that the domain $\mathcal{D}(A(U))$ is independent of $U \in K$. In this case it is clear that the Lipschitz condition

$$\|A(U)\{A(U)^{-1} - A(V)^{-1}\}\|_{\mathcal{L}(X)} = \|\{A(U) - A(V)\}A(V)^{-1}\|_{\mathcal{L}(X)} \le N\|U - V\|_Y, \quad U, V \in K \qquad \text{(A.ii}')$$

implies (A.ii) with $\nu = 1$ and that (Ex) reduces to $0 \le \alpha < \beta < 1$.

Consider a differential operator

$$A(U(x); D) = \begin{pmatrix} -a\Delta + c & b\nabla\cdot\{u(x)\nabla\cdot\} \\ 0 & -d\Delta + g \end{pmatrix}$$

in a bounded region $\Omega \subset \mathbb{R}^3$ with a $\mathcal{C}^2$ boundary $\partial\Omega$, where a, b, c, d, and g are positive constants. Let X be a product Hilbert space $X = L_2(\Omega) \times L_2(\Omega)$ and $Z = H^{\frac{3}{2}+\varepsilon}(\Omega) \times H^{\frac{3}{2}+\varepsilon}(\Omega)$ with an arbitrarily fixed $\varepsilon \in (0, \frac{1}{2})$. For $U = \begin{pmatrix} u \\ \rho \end{pmatrix} \in K = \{U \in Z;\ \|U\|_Z < R\}$, we define the linear operator $A(U)$ in X by

$$\begin{cases} \mathcal{D}(A(U)) = \left\{\widetilde{U} = \begin{pmatrix} \widetilde{u} \\ \widetilde{\rho} \end{pmatrix} \in H^2(\Omega) \times H^2(\Omega);\ \dfrac{\partial \widetilde{u}}{\partial n} = \dfrac{\partial \widetilde{\rho}}{\partial n} = 0 \text{ on } \partial\Omega\right\} \\ A(U)\widetilde{U} = A(U(x); D)\widetilde{U}. \end{cases} \tag{4.5}$$

Actually, $A(U)$ does not depend on ρ of $U = \begin{pmatrix} u \\ \rho \end{pmatrix}$, and the domain $\mathcal{D}(A(U))$ is independent of U. As will be mentioned below, such a differential operator arises in chemotactic models in mathematical biology.

We clearly observe that

$$\{A(U_1) - A(U_2)\}\widetilde{U} = \begin{pmatrix} b\nabla\cdot\{(u_1 - u_2)\nabla\widetilde{\rho}\} \\ 0 \end{pmatrix},$$

$$\nabla\cdot\{(u_1 - u_2)\nabla\widetilde{\rho}\} = (u_1 - u_2)\Delta\widetilde{\rho} + \nabla(u_1 - u_2)\cdot\nabla\widetilde{\rho}.$$

Furthermore,

$$\|(u_1-u_2)\Delta\widetilde{\rho}\|_{L_2} \le C\|u_1-u_2\|_{\mathcal{C}}\|\widetilde{\rho}\|_{H^2},$$
$$\|\nabla(u_1-u_2)\cdot\nabla\widetilde{\rho}\|_{L_2} \le \|\nabla(u_1-u_2)\|_{L_3}\|\nabla\widetilde{\rho}\|_{L_6} \le C\|u_1-u_2\|_{H^{\frac{3}{2}}}\|\widetilde{\rho}\|_{H^2}.$$

Therefore, since $H^{\frac{3}{2}+\varepsilon'}(\Omega) \subset \mathcal{C}(\overline{\Omega})$ for $\varepsilon' > 0$, $A(U)$ satisfies (A.ii′) with $Y = H^{\frac{3}{2}+\varepsilon'}(\Omega) \times H^{\frac{3}{2}+\varepsilon'}(\Omega)$. In addition, according to [12, Proposition 3.3 and Theorem A.1], it is known that

$$\mathcal{D}(A(U)^\theta) = H_N^{2\theta}(\Omega) \times H_N^{2\theta}(\Omega) \tag{4.6}$$

where

$$H_N^{2\theta}(\Omega) = \begin{cases} H^{2\theta}(\Omega) & \text{if } 0 \le 2\theta < \frac{3}{2}, \\ \{u \in H^{2\theta}(\Omega);\ \frac{\partial u}{\partial n} = 0 \text{ on } \partial\Omega\} & \text{if } \frac{3}{2} < 2\theta \le 2. \end{cases}$$

This shows that (Sp.iii) and (Ex) are also fulfilled with $\alpha = \frac{1}{2}(\frac{3}{2}+\varepsilon')$ and $\beta = \frac{1}{2}(\frac{3}{2}+\varepsilon)$ if $0 < \varepsilon' < \varepsilon < \frac{1}{2}$.

5. Applications

We shall handle two concrete problems to sketch the way of applications of our abstract results.

Consider the Cauchy problem of a parabolic equation

$$\begin{cases} \dfrac{\partial U}{\partial t} = \displaystyle\sum_{ij=1}^{n} \frac{\partial}{\partial x_i}\left\{a_{ij}(x,U)\frac{\partial U}{\partial x_j}\right\} - cU + f(x,U) & \text{in } (0,T]\times\Omega, \\ \displaystyle\sum_{ij=1}^{n} a_{ij}(x,U)\gamma_i(x)\frac{\partial U}{\partial x_j} = 0 & \text{on } (0,T]\times\partial\Omega, \\ U(x,0) = U_0(x) & \text{in } \Omega \end{cases}$$

in a bounded region $\Omega \subset \mathbb{R}^n$ of $\mathcal{C}^2$ class. Here, $\gamma(x) = (\gamma_1(x), \ldots, \gamma_n(x))$ is the outer normal vector at the point $x \in \partial\Omega$. $a_{ij}(x,U)$ $(1 \le i,j \le n)$ are real-valued smooth functions defined for $(x,U) \in \overline{\Omega} \times (\mathbb{R}+i\mathbb{R})$ such that $a_{ij} = a_{ji}$ and satisfy (4.2). $f(x,U)$ is also smooth function of $(x,U) \in \overline{\Omega} \times (\mathbb{R}+i\mathbb{R})$. $c > 0$ is a positive constant.

We formulate the problem in $X = L_p(\Omega)$, $n < p < \infty$. Set $Z = W_p^h(\Omega)$ with an arbitrarily fixed $h \in (1+\frac{n}{p}, 2)$ and $K = \{U \in Z;\ \|U\|_Z < R\}$, $R > 0$. For $U \in K$, a linear operator $A(U)$ is defined by (4.3). If $c > 0$ is sufficiently large, then $A(U)$ fulfils (A.i). In addition, as verified in the preceding section, (A.ii) is also fulfilled with $Y = X = L_p(\Omega)$ and $\nu = \frac{1}{2}$. It is easy to see that

$$\|f(x,U_1) - f(x,U_2)\|_{L_p} \le C\|U_1 - U_2\|_{L_p}, \quad U_1, U_2 \in K,$$

therefore (F) is valid. We thus conclude that, for any $U_0 \in W^b_{p,B(U_0)}(\Omega)$, $h < b < 2$, there exists a unique solution in the function space

$$\begin{cases} U \in \mathcal{C}([0, T_{U_0}]; L_p(\Omega)) \cap \mathcal{C}^1((0, T_{U_0}]; L_p(\Omega)) \cap \mathcal{B}([0, T_{U_0}]; W^h_p(\Omega)), \\ A(U)U \in \mathcal{C}((0, T_{U_0}]; L_p(\Omega)), \ t^{1-\beta}A(U)U \in \mathcal{B}((0, T_{U_0}]; L_p(\Omega)). \end{cases}$$

In view of (4.4) it suffices to take β so that $\frac{h}{2} < \beta < \frac{b}{2}$.

For a different approach to the present problem we refer the reader to [4].

Consider next the Cauchy problem of a chemotaxis-growth diffusion system

$$\begin{cases} \dfrac{\partial u}{\partial t} = a\Delta u - b\nabla \cdot \{u\nabla\rho\} + \varphi(u), & \text{in } (0, T] \times \Omega, \\ \dfrac{\partial \rho}{\partial t} = d\Delta\rho + fu - g\rho, & \text{in } (0, T] \times \Omega, \\ \dfrac{\partial u}{\partial n} = \dfrac{\partial \rho}{\partial n} = 0 & \text{on } (0, T] \times \partial\Omega, \\ u(x, 0) = u_0(x), \quad \rho(x, 0) = \rho_0(x) & \text{in } \Omega \end{cases}$$

in a bounded region $\Omega \subset \mathbb{R}^3$ of $\mathcal{C}^2$ class. This system was presented by Mimura et al. [7] to describe the pattern formation of biological individuals induced by chemotaxis and growth. $u = u(x, t)$ and $\rho = \rho(x, t)$ denote the population density and the concentration of chemical substance respectively. The interaction term $b\nabla \cdot \{u\nabla\rho\}$ describes the chemotaxis, ρ is produced with the rate fu, and u grows with the rate $\varphi(u)$. a, b, d, f, and g are positive constants, and $\varphi(u)$ is a smooth function of $u \in \mathbb{R} + i\mathbb{R}$ with $\varphi(0) = 0$.

Let us formulate the problem in the product space $X = L_2(\Omega) \times L_2(\Omega)$. Set $Z = H^{\frac{3}{2}+\varepsilon}(\Omega) \times H^{\frac{3}{2}+\varepsilon}(\Omega)$ with an arbitrarily fixed $\varepsilon \in (0, \frac{1}{2})$ so that $Z \subset \mathcal{C}(\overline{\Omega}) \times \mathcal{C}(\overline{\Omega})$ and $K = \{U \in Z; \|U\|_Z < R\}$, $R > 0$. For $U = \begin{pmatrix} u \\ \rho \end{pmatrix} \in K$, the linear operator $A(U)$ is defined by (4.5). $F(U)$ is defined by

$$F(U) = \begin{pmatrix} cu + \varphi(u) \\ fu \end{pmatrix}, \quad U = \begin{pmatrix} u \\ \rho \end{pmatrix} \in K.$$

As verified in the preceding section, $A(U)$ satisfies (A.i) and (A.ii) with $Y = H^{\frac{3}{2}+\varepsilon'}(\Omega) \times H^{\frac{3}{2}+\varepsilon'}(\Omega)$, $0 < \varepsilon' < \frac{1}{2}$ and $\nu = 1$. It is also easy to show that (F) is fulfilled with the same space Y. In view of (4.6), if we set $\alpha = \frac{1}{2}(\frac{3}{2} + \varepsilon')$ and $\beta = \frac{1}{2}(\frac{3}{2} + \varepsilon)$ with $0 < \varepsilon' < \varepsilon < \frac{1}{2}$, then (Sp.iii) and (Ex) are fulfilled.

In this way we conclude that, for any u_0, $\rho_0 \in H^{\frac{3}{2}+\varepsilon}(\Omega)$, $0 < \varepsilon < \frac{1}{2}$, there exists a unique local solution in the function space

$$\begin{cases} u, \rho \in \mathcal{C}([0, T]; H^{\frac{3}{2}+\varepsilon}(\Omega)) \cap \mathcal{C}^1((0, T]; L_2(\Omega)) \cap \mathcal{C}((0, T]; H^2_N(\Omega)), \\ \qquad t^{\frac{1}{4}(1-2\varepsilon)}u, \ t^{\frac{1}{4}(1-2\varepsilon)}\rho \in \mathcal{B}((0, T]; H^2_N(\Omega)). \end{cases}$$

Since $\beta < \nu = 1$, we can use the results noticed in Remark 3.3.

References

[1] P. Acquistapace and B. Terreni, *On the fundamental solutions for abstract parabolic equations*, Differential Equations in Banach Spaces, Bologna, 1985, Lecture Notes in Math. **1233** (ed. A. Favini, E. Obrecht), Springer-Verlag, 1986, 1–11.

[2] P. Acquistapace and B. Terreni, *A unified approach to abstract linear non-autonomous parabolic equations*, Rend. Sem. Univ. Padova **78**(1987), 47–107.

[3] P. Acquistapace and B. Terreni, *Evolution operators and strong solutions of abstract linear parabolic equations*, Differential Integral Equations **1**(1988), 433–457.

[4] H. Amann, *Quasilinear parabolic systems under non-linear boundary conditions*, Arch. Rational Mech. Anal. **92**(1986), 153–192.

[5] H. Amann, *On abstract parabolic fundamental solutions*, J. Math. Soc. Japan **39**(1987), 93–116.

[6] M. Hildebrand, M. Kuperman, H. Wio, A. S. Mikhailov and G. Ertl, *Self-organized chemical nanoscal microreactors*, Physical Review Letters **83**(1999), 1475–1478.

[7] M. Mimura and T. Tujikawa, *Aggregating pattern dynamics in a chemotaxis model including growth*, Physica A **230**(1996), 499–543.

[8] A. Yagi, *Fractional powers of operators and evolution equations of parabolic type*, Proc. Japan Acad. Ser. A **64**(1988), 227–230.

[9] A. Yagi, *Parabolic evolution equations in which the coefficients are the generators of infinitely differential semigroups, II*, Funkcial. Ekvac. **33**(1990), 139–150.

[10] A. Yagi, *Abstract quasilinear evolution equations of parabolic type in Banach spaces*, Boll. Un. Mat. Ital. **5-B**(1991), 341–368.

[11] A. Yagi, *Global solutions to some quasilinear parabolic system in population dynamics*, Non-linear Anal. **21**(1993), 603–630.

[12] A. Yagi, *Norm behavior of solutions to a parabolic system of chemotaxis*, Math. Japonica **45**(1997), 241–265.

[13] S. M. Sze, *Physics of Semiconductor Devices*, Wiley, New York, 1981.

[14] H. Tanabe, *Hatten Hoteishiki* (in Japanese), Iwanami, Tokyo, 1975; *Equation of Evolution* (English translation), Pitman, London, 1979.

[15] H. Tanabe, *Functional Analytic Methods for Partial Differential Equations*, Marcel Dekker, New York, 1997.

[16] H. Triebel, *Interpolation Theory, Function Spaces, Differential Operators*, North-Holland, Amsterdam, 1978.

Atsushi Yagi
Department of Applied Physics
Osaka University
Suita, Osaka 565-0871, Japan
e-mail: yagi@ap.eng.osaka-u.ac.jp

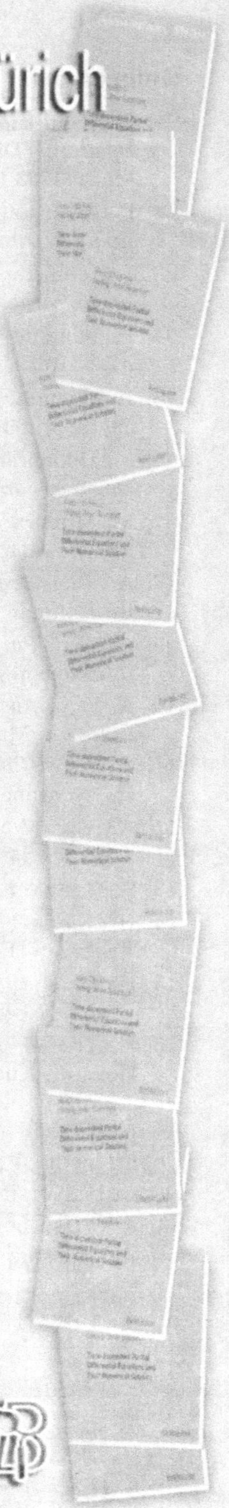

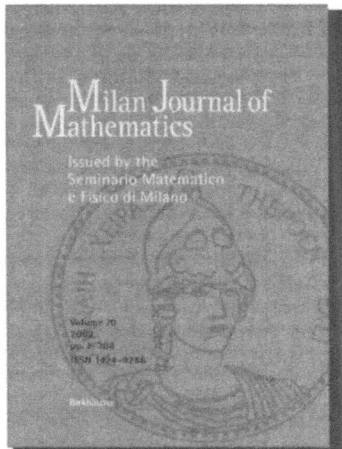